# Science for the Protection of Indonesian Coastal Ecosystems (SPICE)

# Science for the Protection of Indonesian Coastal Ecosystems (SPICE)


Edited by

Tim C. Jennerjahn

Leibniz Centre for Tropical Marine Research (ZMT), Bremen, Germany;
Faculty of Geoscience, University of Bremen, Bremen, Germany

Tim Rixen

Leibniz Centre for Tropical Marine Research (ZMT), Bremen, Germany;
Institute of Geology, Universität Hamburg, Germany

Hari Eko Irianto

Research and Development Center for Marine and Fisheries Product Processing and Biotechnology (Balai Besar Riset Pengolahan Produk dan Bioteknologi Kelautan dan Perikanan), Jakarta, Indonesia

Joko Samiaji

University of Riau, Pekanbaru, Indonesia



ELSEVIER

Elsevier
Radarweg 29, PO Box 211, 1000 AE Amsterdam, Netherlands
The Boulevard, Langford Lane, Kidlington, Oxford OX5 1GB, United Kingdom
50 Hampshire Street, 5th Floor, Cambridge, MA 02139, United States



**Library of Congress Cataloging-in-Publication Data**
A catalog record for this book is available from the Library of Congress

**British Library Cataloguing-in-Publication Data**
A catalogue record for this book is available from the British Library

ISBN: 978-0-12-815050-4



*Publisher:* Candice Janco
*Acquisitions Editor:* Louisa Munro
*Editorial Project Manager:* Kathy Padilla
*Production Project Manager:* Bharatwaj Varatharajan
*Cover Designer:* Greg Harris

*Cover image courtesy:* Sebastian Ferse, Marion Glaser, Tim Jennerjahn, Inga Nordhaus and Tim Rixen

Typeset by TNQ Technologies

# Contents

Contributors xiii
Reviewers xxi
Foreword xxiii

1. Introduction—Science for the Protection of Indonesian Coastal Ecosystems (SPICE) 1
   *Tim C. Jennerjahn, Tim Rixen, Hari Eko Irianto and Joko Samiaji*
   1.1 Rationale 2
   1.2 Development and implementation of the research and education program SPICE 4
   1.3 Research, education, and outreach activities 5
   1.4 Summary and synthesis of SPICE results 8
   Acknowledgments 9
   References 10

2. Physical environment of the Indonesian Seas with focus on the western region 13
   *Bernhard Mayer, Herbert Siegel, Monika Gerth, Thomas Pohlmann, Iris Stottmeister, Mutiara Putri and Agus Setiawan*
   2.1 Introduction 14
   2.2 The marine circulation 17
   2.3 Seasonal variability and long-term changes 22
   2.4 Water residence times 26
   2.5 Sources and sinks of freshwater 28
   2.6 Remote sensing methods applied in coastal process studies 31

Acknowledgments 40

References 40

**3. Human interventions in rivers and estuaries of Java and Sumatra 45**

*Tim C. Jennerjahn, Antje Baum, Ario Damar, Michael Flitner, Jill Heyde, Ingo Jänen, Martin C. Lukas, Muhammad Lukman, Mochamad Saleh Nugrahadi, Tim Rixen, Joko Samiaji and Friedhelm Schroeder*

3.1 Introduction 46

3.2 Drivers of environmental change affecting river fluxes 48

3.3 Natural factors, human interventions, and extreme events controlling river fluxes 49

3.4 Governance and management programs 71

Acknowledgments 74

References 74

**4. Carbon cycle in tropical peatlands and coastal seas 83**

*Tim Rixen, Francisca Wit, Andreas A. Hutahaean, Achim Schlüter, Antje Baum, Alexandra Klemme, Moritz Müller, Widodo Setiyo Pranowo, Joko Samiaji and Thorsten Warneke*

4.1 Introduction 85

4.2 Background information 86

4.3 Indonesian peatlands 89

4.4 Peat carbon losses 97

4.5 Land–ocean continuum 99

4.6 Estuaries and the ocean 112

4.7 Ecosystem $CO_2$ emissions 120

4.8 Evaluation of $CO_2$ emissions 123

4.9 Socioeconomic implications 127
4.10 Outlook 131
References 132

5. Coral reef social–ecological systems under pressure in Southern Sulawesi 143

*Hauke Reuter, Annette Breckwoldt, Tina Dohna, Sebastian Ferse, Astrid Gärdes, Marion Glaser, Filip Huyghe, Hauke Kegler, Leyla Knittweis, Marc Kochzius, Wiebke Elsbeth Kraemer, Johannes Leins, Muhammad Lukman, Hawis Madduppa, Agus Nuryanto, Min Hui, Sara Miñarro, Gabriela Navarrete Forero, Sainab Husain Paragay, Jeremiah Plass-Johnson, Hajaniaina Andrianavalonarivo Ratsimbazafy, Claudio Richter, Yvonne Sawall, Kathleen Schwerdtner Máñez, Mirta Teichberg, Janne Timm, Rosa van der Ven and Jamaluddin Jompa*

5.1 Introduction—coral reefs in Indonesia and the Spermonde Archipelago 145
5.2 Functioning of coral reefs 148
5.3 Genetic connectivity of reefs in the Coral Triangle region 154
5.4 Social systems associated with the use of coral-based resources and reef-specific challenges 165
5.5 Modeling to support the management of reef systems 171
5.6 Summary and outlook 178
Acknowledgments 180
References 180
Appendix A5 189

6. Ecology of seagrass beds in Sulawesi—Multifunctional key habitats at the risk of destruction 201

*Harald Asmus, Dominik Kneer, Claudia Pogoreutz, Sven Blankenhorn, Jamaluddin Jompa, Nadiarti Nurdin and Dody Priosambodo*

6.1 General introduction to tropical Southeast Asian seagrass meadows 203
6.2 The current distribution of seagrasses in the Spermonde Archipelago 206
6.3 Seagrass ecology 211
6.4 Tropical seagrass beds as key habitat for fish species 225
6.5 Human–seagrass interactions 230
6.6 Conclusions and outlook 236
Acknowledgments 239
References 239

7. Mangrove ecosystems under threat in Indonesia: the Segara Anakan Lagoon, Java, and other examples 251

*Tim C. Jennerjahn, Erwin Riyanto Ardli, Jens Boy, Jill Heyde, Martin C. Lukas, Inga Nordhaus, Moh Husein Sastranegara, Kathleen Schwerdtner Máñez and Edy Yuwono*

7.1 Introduction 252
7.2 The study areas 254
7.3 Environmental setting and natural resource use 257
7.4 Environmental change in the Segara Anakan Lagoon region: causes, drivers, and impacts 269
7.5 Threats to mangrove forests and their ecosystem services in Indonesia 274
7.6 Management programs 276
Acknowledgments 279
References 279

8. Impact of megacities on the pollution of coastal areas—the case example Jakarta Bay 285

*Andreas Kunzmann, Jan Schwarzbauer, Harry W. Palm, Made Damriyasa, Irfan Yulianto, Sonja Kleinertz, Vincensius S.P. Oetam, Muslihudeen A. Abdul-Aziz, Grit Mrotzek, Haryanti Haryanti, Hans Peter Saluz, Zainal Arifin, Gunilla Baum, Larissa Dsikowitzky, Dwiyitno, Hari Eko Irianto, Simon van der Wulp, Karl J. Hesse, Norbert Ladwig and Ario Damar*

8.1 Introduction 287

8.2 Hydrological system and nutrient dispersion 289

8.3 Organic and inorganic pollution in Jakarta Bay 296

8.4 Water quality and biological responses 303

8.5 Microbial diversity in Indonesian fish and shrimp: a comparative study on different ecological conditions 310

8.6 Fish parasites in Indonesian waters: new species findings, biodiversity patterns, and modern applications 317

8.7 Seafood consumption and potential risk 321

8.8 Implications 330

Acknowledgments 333

References 333

9. Late quaternary environmental history of Indonesia 347

*Mahyar Mohtadi, Andreas Lückge, Stephan Steinke, Haryadi Permana, Susilohadi Susilohadi, Rina Zuraida and Tim C. Jennerjahn*

9.1 Introduction 348

9.2 Tools for reconstructing the environmental history of Indonesia 349

9.3 Environmental history of western and southern Indonesia 352

Acknowledgments 364

References 365

10. Decision tool for assessing marine finfish aquaculture sites in Southeast Asia 371

*Roberto Mayerle, Ketut Sugama, Simon van der Wulp, Poerbandono and Karl-Heinz Runte*

10.1 Introduction 372
10.2 Methods for sustainable management of marine finfish aquaculture 374
10.3 Decision support system for sustainable management of marine finfish aquaculture 375
10.4 Marine finfish aquaculture site in Bali, Indonesia 376
10.5 Surveys and monitoring at the aquaculture site in Bali 378
10.6 Dynamic model of the aquaculture site in Bali 379
10.7 Assessment of operation of the marine finfish aquaculture site in Bali 380
10.8 Conclusion 384
Acknowledgments 385
References 385

11. Decision tool for estimating energy potential from tidal resources 389

*Roberto Mayerle, Kadir Orhan, Wahyu W. Pandoe, Poerbandono and Peter Krost*

11.1 Introduction 390
11.2 Investigation sites in Indonesia 392
11.3 Guidelines for setting up circulation models in data-poor sites 393
11.4 Procedure for assessing the energy potential from tidal currents 397
11.5 Decision support system for assessing the energy potential from tidal currents 401

11.6 Assessing the energy potential from tidal currents within the target sites in Indonesia 401
11.7 Conclusion 402
Acknowledgments 404
References 404

12. The governance of coastal and marine social–ecological systems: Indonesia and beyond 407

*Marion Glaser, Luky Adrianto, Annette Breckwoldt, Nurliah Buhari, Rio Deswandi, Sebastian Ferse, Philipp Gorris, Sainab Husain Paragay, Bernhard Glaeser, Neil Mohammad, Kathleen Schwerdtner Máñez and Dewi Yanuarita*

12.1 Introduction 409
12.2 Marine and coastal governance: sectors, issues, and options for intervention 411
12.3 The need for integrated coastal and marine planning 425
12.4 Conclusions and outlook 430
12.5 Final remarks 434
Acknowledgments 435
References 436

Index 445

# Contributors

**Muslihudeen A. Abdul-Aziz** Leibniz Institute for Natural Product Research and Infection Biology, Jena, Germany; Australian Centre for Ancient DNA, University of Adelaide, Adelaide, SA, Australia

**Luky Adrianto** Faculty of Fisheries and Marine Sciences, IPB University, Bogor, Indonesia

**Erwin Riyanto Ardli** Leibniz Centre for Tropical Marine Research (ZMT), Bremen, Germany; Fakultas Biologi, Universitas Jenderal Soedirman, Purwokerto, Java, Indonesia

**Zainal Arifin** Research Centre for Oceanography-LIPI, Ancol Timur, Jakarta, Indonesia

**Harald Asmus** Alfred Wegener Institute, Helmholtz Centre for Polar and Marine Research, List/Sylt, Germany

**Gunilla Baum** Institute of Geology and Geochemistry of Petroleum and Coal, RWTH Aachen University, Germany

**Antje Baum** Leibniz Centre for Tropical Marine Research (ZMT), Bremen, Germany

**Sven Blankenhorn** Alfred Wegener Institute, Helmholtz Centre for Polar and Marine Research, List/Sylt, Germany

**Jens Boy** Institute of Soil Science, Leibniz Universität, Hannover, Germany

**Annette Breckwoldt** Leibniz Centre for Tropical Marine Research (ZMT), Bremen, Germany

**Nurliah Buhari** Mataram University, West Nusa Tenggara, Indonesia

**Ario Damar** Center for Coastal and Marine Resources Studies, Bogor Agricultural University (IPB University), Indonesia

**Made Damriyasa** Udayana University, Faculty of Veterinary Sciences, Badung, Bali, Indonesia

**Rio Deswandi** Leibniz Centre for Tropical Marine Research (ZMT), Bremen, Germany; Center for Regulation Policy and Governance, Bogor, Indonesia

**Tina Dohna** Marum - Center for Marine Environmental Sciences, University of Bremen, Germany

**Larissa Dsikowitzky** Leibniz Centre for Tropical Marine Research (ZMT), Bremen, Germany

**Dwiyitno** Research and Development Center for Marine and Fisheries Product Processing and Biotechnology (Balai Besar Riset Pengolahan Produk dan Bioteknologi Kelautan dan Perikanan), Jakarta, Indonesia

**Sebastian Ferse** Leibniz Centre for Tropical Marine Research (ZMT), Bremen, Germany

**Michael Flitner** Sustainability Research Center (ARTEC), University of Bremen, Germany

**Gabriela Navarrete Forero** Leibniz Centre for Tropical Marine Research (ZMT), Bremen, Germany; Avenida Río Toachi y Calle Bambúes, Santo Domingo de los Colorados, Ecuador

**Astrid Gärdes** Leibniz Centre for Tropical Marine Research (ZMT), Bremen, Germany

**Monika Gerth** Leibniz Institute for Baltic Sea Research, Rostock-Warnemünde, Germany

**Bernhard Glaeser** Society for Human Ecology and Free University of Berlin, Germany

**Marion Glaser** Leibniz Centre for Tropical Marine Research (ZMT), Bremen, Germany; Institute of Geography, University of Bremen, Bremen, Germany

**Philipp Gorris** University of Osnabrück, Germany

**Haryanti Haryanti** Gondol Research Institute for Mariculture GRIM, Singaraja, Bali, Indonesia

Karl J. Hesse ECOLAB - Group Coastal Ecosystems, Research and Technology Centre Westcoast (FTZ), University of Kiel, Büsum, Germany

Jill Heyde Sustainability Research Center (ARTEC), University of Bremen, Germany

Min Hui Institute of Oceanology, Chinese Academy of Sciences, Qingdao, China

Andreas A. Hutahaean Ministry of Maritime Affairs, Jakarta, Indonesia

Filip Huyghe Ecology & Biodiversity - Marine Biology, Vrije Universiteit Brussel (VUB), Brussels, Belgium

Hari Eko Irianto Research and Development Center for Marine and Fisheries Product Processing and Biotechnology (Balai Besar Riset Pengolahan Produk dan Bioteknologi Kelautan dan Perikanan), Jakarta, Indonesia

Ingo Jänen Leibniz Centre for Tropical Marine Research (ZMT), Bremen, Germany

Tim C. Jennerjahn Leibniz Centre for Tropical Marine Research (ZMT), Bremen, Germany; Faculty of Geoscience, University of Bremen, Bremen, Germany

Jamaluddin Jompa Department of Marine Science, Hasanuddin University (UNHAS), Makassar, Indonesia; Faculty of Marine Science and Fisheries, Hasanuddin University, Makassar, South Sulawesi, Indonesia

Hauke Kegler Leibniz Centre for Tropical Marine Research (ZMT), Bremen, Germany

Sonja Kleinertz Aquaculture and Sea-Ranching, Faculty of Agricultural and Environmental Sciences, University of Rostock, Germany

Alexandra Klemme Institute for Environmental Physics, University of Bremen, Germany

Dominik Kneer Alfred Wegener Institute, Helmholtz Centre for Polar and Marine Research, List/Sylt, Germany

Leyla Knittweis Department of Biology, University of Malta, Msida, Malta

Marc Kochzius Ecology & Biodiversity - Marine Biology, Vrije Universiteit Brussel (VUB), Brussels, Belgium

Wiebke Elsbeth Kraemer Ecology & Biodiversity - Marine Biology, Vrije Universiteit Brussel (VUB), Brussels, Belgium

Peter Krost Coastal Research & Management (CRM), Kiel, Germany

Andreas Kunzmann Leibniz Centre for Tropical Marine Research (ZMT), Bremen, Germany

Norbert Ladwig ECOLAB - Group Coastal Ecosystems, Research and Technology Centre Westcoast (FTZ), University of Kiel, Büsum, Germany

Johannes Leins Leibniz Centre for Tropical Marine Research (ZMT), Bremen, Germany; Helmholtz Centre for Environmental Research (UFZ), Leipzig, Germany

Andreas Lückge Federal Institute for Geosciences and Natural Resources, Hannover, Germany

Martin C. Lukas Leibniz Centre for Tropical Marine Research (ZMT), Bremen, Germany; Sustainability Research Center (ARTEC), University of Bremen, Germany

Muhammad Lukman Department of Marine Science, Hasanuddin University (UNHAS), Makassar, Indonesia

Hawis Madduppa Agricultural University Bogor (IPB), Bogor, Indonesia

Kathleen Schwerdtner Máñez Leibniz Centre for Tropical Marine Research (ZMT), Bremen, Germany; Place Nature Consultancy, Ashausen, Germany

Bernhard Mayer Institute of Oceanography, University of Hamburg, Germany

Roberto Mayerle Research and Technology Centre Westcoast (FTZ), University of Kiel, Büsum, Germany

Sara Miñarro Leibniz Centre for Tropical Marine Research (ZMT), Bremen, Germany; Institute of Environmental Science and Technology (ICTA), Universitat Autònoma de Barcelona, Bellaterra, Spain

**Neil Mohammad** Hasanuddin University, Makassar, Indonesia

**Mahyar Mohtadi** MARUM-Center for Marine Environmental Sciences, University of Bremen, Germany

**Grit Mrotzek** Leibniz Institute for Natural Product Research and Infection Biology, Jena, Germany

**Moritz Müller** Swinburne University of Technology, Sarawak Campus, Kuching, Sarawak, Malaysia

**Inga Nordhaus** Leibniz Centre for Tropical Marine Research (ZMT), Bremen, Germany

**Mochamad Saleh Nugrahadi** Agency for the Assessment and Application of Technology (BPPT), Jakarta, Indonesia

**Nadiarti Nurdin** Faculty of Marine Science and Fisheries, Hasanuddin University, Makassar, South Sulawesi, Indonesia

**Agus Nuryanto** Jenderal Soedirman University (UNSOED), Purwokerto, Indonesia

**Vincensius S.P. Oetam** Leibniz Institute for Natural Product Research and Infection Biology, Jena, Germany; Max Planck Institute for Chemical Ecology, Jena, Germany; Friedrich Schiller University of Jena, Germany

**Kadir Orhan** Research and Technology Centre Westcoast (FTZ), University of Kiel, Büsum, Germany

**Harry W. Palm** Aquaculture and Sea-Ranching, Faculty of Agricultural and Environmental Sciences, University of Rostock, Germany

**Wahyu W. Pandoe** Agency for the Assessment and Application of Technology (BPPT), Jakarta, Indonesia

**Sainab Husain Paragay** Enlightening Indonesia, Makassar, South Sulawesi, Indonesia

**Haryadi Permana** Research Center for Geotechnology, Indonesia Institute of Sciences (LIPI), Bandung, Indonesia

**Jeremiah Plass-Johnson** Leibniz Centre for Tropical Marine Research (ZMT), Bremen, Germany

**Poerbandono** Institute of Technology Bandung (ITB), Bandung, Jawa Barat, Indonesia

**Claudia Pogoreutz** Red Sea Research Center (RSRC), Biological, Environmental Science and Engineering Division (BESE), King Abdullah University of Science and Technology (KAUST), Thuwal, Saudi Arabia

**Thomas Pohlmann** Institute of Oceanography, University of Hamburg, Germany

**Widodo Setiyo Pranowo** Research & Development Center for Marine & Coastal Resources (P3SDLP), Jakarta, Indonesia

**Dody Priosambodo** Department of Biology, Faculty of Mathematics and Natural Sciences, Hasanuddin University, Makassar, South Sulawesi, Indonesia

**Mutiara Putri** Institut Teknologi Bandung (ITB), Bandung, Jawa Barat, Indonesia

**Hajaniaina Andrianavalonarivo Ratsimbazafy** Systems Ecology & Resource Management Unit, Univérsité Libre de Bruxelles (ULB), Brussels, Belgium

**Hauke Reuter** Leibniz Centre for Tropical Marine Research (ZMT), Bremen, Germany; Faculty for Biology and Chemistry, University of Bremen, Bremen, Germany

**Claudio Richter** Division Biosciences/Bentho-Pelagic Processes, Alfred Wegener Institute, Bremerhaven, Germany

**Tim Rixen** Leibniz Centre for Tropical Marine Research (ZMT), Bremen, Germany; Institute of Geology, Universität Hamburg, Germany

**Karl-Heinz Runte** Research and Technology Centre Westcoast (FTZ), University of Kiel, Büsum, Germany

**Hans Peter Saluz** Leibniz Institute for Natural Product Research and Infection Biology, Jena, Germany; Friedrich Schiller University of Jena, Germany

**Joko Samiaji** University of Riau, Pekanbaru, Indonesia

**Moh Husein Sastranegara** Fakultas Biologi, Universitas Jenderal Soedirman, Purwokerto, Java, Indonesia

**Yvonne Sawall** Bermuda Institute of Ocean Sciences (BIOS), St. George's, Bermuda

**Achim Schlüter** Leibniz Centre for Tropical Marine Research (ZMT), Bremen, Germany

**Friedhelm Schroeder** Institute for Coastal Research, Helmholtz Centre Geesthacht, Germany

**Jan Schwarzbauer** Institute of Geology and Geochemistry of Petroleum and Coal, RWTH Aachen University, Germany

**Agus Setiawan** Marine Research Centre, Agency for Marine and Fisheries Research and Human Resources, Ministry of Marine Affairs and Fisheries, Jakarta, Indonesia

**Herbert Siegel** Leibniz Institute for Baltic Sea Research, Rostock-Warnemünde, Germany

**Stephan Steinke** Department of Geological Oceanography and State Key Laboratory of Marine Environmental Science (MEL), College of Ocean and Earth Sciences, Xiamen University, China

**Iris Stottmeister** Leibniz Institute for Baltic Sea Research, Rostock-Warnemünde, Germany

**Ketut Sugama** Centre for Aquaculture Research and Development (CARD), Ministry of Marine Affairs and Fisheries, Jakarta, Indonesia

**Susilohadi Susilohadi** Marine Geological Institute (MGI), Bandung, Indonesia

**Mirta Teichberg** Leibniz Centre for Tropical Marine Research (ZMT), Bremen, Germany

**Janne Timm** Faculty for Biology and Chemistry, University of Bremen, Bremen, Germany

**Rosa van der Ven** Ecology & Biodiversity - Marine Biology, Vrije Universiteit Brussel (VUB), Brussels, Belgium

**Simon van der Wulp** ECOLAB - Group Coastal Ecosystems, Research and Technology Centre Westcoast (FTZ), University of Kiel, Büsum, Germany

**Thorsten Warneke** Institute for Environmental Physics, University of Bremen, Germany

**Francisca Wit** Leibniz Centre for Tropical Marine Research (ZMT), Bremen, Germany

**Dewi Yanuarita** Hasanuddin University, Makassar, Indonesia

**Irfan Yulianto** Faculty of Fisheries and Marine Sciences, Bogor Agricultural University, Bogor, Indonesia

**Edy Yuwono** Fakultas Biologi, Universitas Jenderal Soedirman, Purwokerto, Java, Indonesia; Universitas Wanita Internasional, Bandung, Jawa Barat, Indonesia

**Rina Zuraida** Center for Geological Survey, Bandung, Indonesia

# Reviewers

Dedi Adhuri

Derek Armitage

Helge Arz

David Burdige

Arnold Gordon

Victor de Jonge

Shuh-Ji Kao

Ella Kari-Muhl

Patrick Martin

David Pearton

Jeremy Pittman

Husnah Samhudi

Charles Sheppard

Frida Sidik

Dirk Steenbergen

R. Dwi Susanto

Richard Unsworth

Arie Vonk

Xianfeng Wang

Eric Wolanski

# Foreword

Venugopalan Ittekkot, University of Bremen, Germany
ittekkot@uni-bremen.de
Rubiyanto Misman, University Jenderal Soedirman, Purwokerto, Indonesia
Rubiyanto.misman@yahoo.com

Coastal ecosystems, particularly in the tropics, are under enormous pressure from both climate change and human activities. Ocean warming and acidification, shifting ocean currents, and rising sea levels lead to loss or geographic shifts of habitats including those supporting fisheries. Changing pattern of land and water use shrink wetlands and pollute estuaries and coastal seas. Frequent extreme events further exacerbate these pressures undermining ocean's regulatory, social, and economic functions. The chapters in this book address these themes using results from the "Science for the Protection of Indonesian Coastal Marine Ecosystems" (SPICE) program implemented under the framework of German–Indonesian science and technology cooperation.

The agreement between Indonesia and Germany in the field of Scientific Research and Technological Development dates to 1979. Until the late 1990s, ocean research cooperation between the two countries had been mostly in the field of marine geosciences conducted on the cruises of the German Research Vessel SONNE in Indonesian Seas. Individual contact and research opportunities in other marine fields were available through the DAAD (German Academic Exchange Service). SPICE program has its origins in the new contacts established between Indonesian and German scientists and science administrators in the late 1990s. The period saw an added awareness of the seas in the Southeast Asia region and a better appreciation of their social and economic dimension as livelihood providers. Particularly in Indonesia, there was a renewed recognition of ocean's potential to support national development and of the need for governance measures to sustainably manage ocean's resources. One of the pressing needs was the expansion of the country's ocean research and technology base and the availability of trained personnel to run it. Creation of the Indonesian Ministry of Maritime Exploration in 1999 (now the Ministry of Maritime Affairs and Fisheries) provided additional impetus to fulfilling this need.

The newly established Indonesian–German contacts led in 1998 to a joint workshop in Hamburg supported by RISTEK's BPPT (Agency for the Assessment and Application of Technology, Indonesia) and the BMBF (Federal German Ministry of Education and Research) to explore and exchange ideas and themes for future cooperation in the field of earth and ocean sciences. An outcome of the workshop, which brought together scientists and science administrators from both countries, was the recognition of the need for expanding the scope of bilateral cooperation and for revising its mode. The ensuing bilateral discussion led to the creation of a Joint Steering Committee in the field of Earth and Ocean Sciences in 2000, which recommended the preparation of an action plan for cooperation with inputs from broader Indonesian and German earth and ocean science communities. From our respective positions in Hamburg and Bremen in Germany, and from Purwokerto in Indonesia, we had the opportunity, with support from policy makers and science managers in both countries to coordinate the formulation of the plan and to facilitate the development of bilateral projects for implementation.

As a first step, a bilateral workshop was held in 2001 in close coordination with a DAAD Alumni Conference at UNSOED in Purwokerto to discuss and prioritize research and development topics that could go into future cooperative efforts. The workshop brought together about 300 participants including senior academic and research personnel from all major ocean research and academic Institutions in the two countries as well as the alumni of DAAD who had completed their graduate and postgraduate studies at various German universities. Among the selected topics were management of living and nonliving resources, geothermal energy, and early warning systems. SPICE provided the umbrella for projects related to the first topic.

Capacity building has been a key component of the SPICE program. The program could build on and support a BMBF special scholarship program for marine science studies implemented by the DAAD. The program also created several new opportunities for scientists and students from Indonesia, Germany and their international network partners to participate in research and to pursue graduate and postgraduate education. More than a hundred talents from Indonesia and Germany have completed their graduate and postgraduate education under the program with the SPICE—scientists taking supervisory roles. Some of them are among the contributors to this book.

Though conceived and implemented as a bilateral program, SPICE maintained contacts to regional networks and international organizations and their programs that contributed to building capacity for managing coastal marine resources in developing countries in the tropics. Of special mention are the links to the NAM S&T Centre, IGBP-LOICZ (the predecessor of Future Coasts) as well as to a network of scientists and institutions focusing on the Aquatic Ecosystems of the Monsoon Asia (ACEMON).

The editors and several of the contributors of the book have been with the SPICE program since the very beginning. The book's makeup shows how the program has evolved over the years. As it progressed, SPICE has attracted new partners from institutions in Germany and Indonesia. We received invaluable support and advice from Professor Indroyono Soesilo in his various roles in Indonesia (Deputy Chairman of BBPT, Chairman of BRKP, and the Maritime Coordinating Minister) and at the FAO in developing the program and expanding the SPICE network. Similar advice and support came from the Project Management of Office of Research Centre Julich (PT-Julich) and from the German experts stationed at BPPT and RISTEK. For their contribution to the coordination efforts at various phases of the project, we like to thank our colleagues Farid Ma'ruf and Ardito Kodijat from Indonesia, and Petra Westhaus-Ekau, Karin Gaertner, Eberhard Krain, and Claudia Schultz from Germany.

We congratulate the editors for bringing out this book at an important moment in the study of the Ocean. It is a timely contribution to the UN Decade of the Ocean coordinated by UNESCO-IOC. SPICE research themes and its capacity building efforts as well as its cooperative implementation strategy are among the critical targets of SDG 14 of the UN Development Agenda 2030. This book will also serve as an important background document for Indonesia, as the country prepares to implement international cooperation projects with support from the Indonesian Science Fund. We hope Ocean will occupy its rightful place among the funded projects.

1

# Introduction—Science for the Protection of Indonesian Coastal Ecosystems (SPICE)

Tim C. Jennerjahn[1,2], Tim Rixen[1], Hari Eko Irianto[4], Joko Samiaji[3]

[1]LEIBNIZ CENTRE FOR TROPICAL MARINE RESEARCH (ZMT), BREMEN, GERMANY; [2]FACULTY OF GEOSCIENCE, UNIVERSITY OF BREMEN, BREMEN, GERMANY; [3]UNIVERSITY OF RIAU, PEKANBARU, INDONESIA; [4]RESEARCH AND DEVELOPMENT CENTER FOR MARINE AND FISHERIES PRODUCT PROCESSING AND BIOTECHNOLOGY (BALAI BESAR RISET PENGOLAHAN PRODUK DAN BIOTEKNOLOGI KELAUTAN DAN PERIKANAN), JAKARTA, INDONESIA

## Abstract

*Indonesia, with its more than 17,000 islands, is an extraordinary place on this planet. It is among the countries with the highest river fluxes of dissolved and particulate substances into the ocean, the most abundant mangrove forests, seagrass meadows and coral reefs, and the highest marine biodiversity. However, Indonesia is also vulnerable to man-made environmental change and the outcomes of climate change. In order to face these challenges, the Indonesian – German inter- and transdisciplinary research and capacity building program "Science for the Protection of Indonesian Coastal Ecosystems" (SPICE) was set up. It addressed the scientific, social and economic issues related to the management of the Indonesian coastal ecosystems and their resources in three phases over a period of 12 years. This chapter outlines the major goals and themes of the program, its organizational structure, and it briefly synthesizes its major findings.*

## Chapter outline

**1.1 Rationale** ........ 2
**1.2 Development and implementation of the research and education program SPICE** ........ 4
**1.3 Research, education, and outreach activities** ........ 5
**1.4 Summary and synthesis of SPICE results** ........ 8
**Acknowledgments** ........ 9
**References** ........ 10

Science for the Protection of Indonesian Coastal Ecosystems (SPICE). https://doi.org/10.1016/B978-0-12-815050-4.00013-4

## 1.1 Rationale

Indonesia, with its more than 17,000 islands also called the "Maritime Continent," is an extraordinary place on this planet in many respects. People from all over the world come to visit it, because they want to enjoy its culture and the beauty of the landscape and the coastal zone. Because of its unique location in a major tropical biogeographic transition zone and the fact that it has the second longest coastline by country on Earth, Indonesia's coastal zone is of particular importance on a national and global scale with regard to its natural wealth, the natural processes shaping the environment, as well as the natural resource potential and resource uses. Twirling cities and calm villages as well as dense forests set the impression of its landscape. Tropical rain forests cover the flanks of soaring volcanoes, while peat swamp forests and mangroves thrive in coastal lowlands. Due to the high rainfall, the water discharge of the numerous small- and medium-sized Indonesian rivers is almost as high as that of the Amazon in South America, the world's largest river. Indonesia is among the countries with the highest river fluxes of dissolved and particulate substances into the ocean (Milliman and Farnsworth, 2011), the most abundant mangrove forests, seagrass meadows and coral reefs, and the highest marine biodiversity (Fig. 1.1). The country can build on a wealth of coastal ecosystem services supplied to society and sustaining the livelihoods of its people.

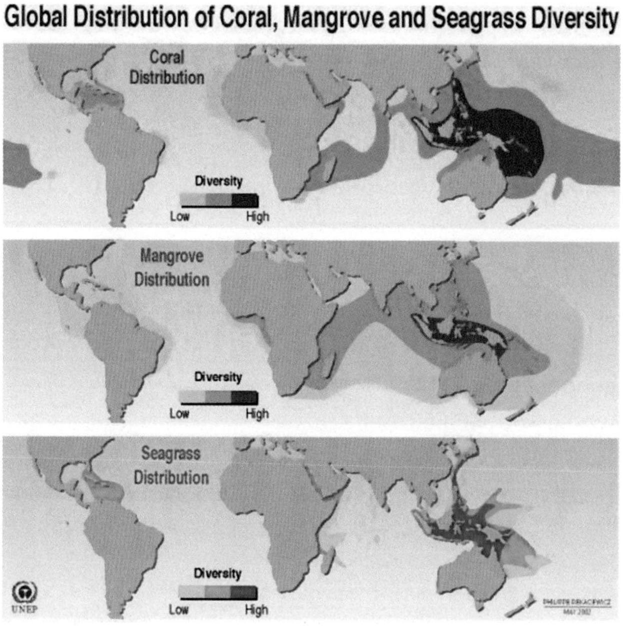


**FIGURE 1.1** Distribution and biodiversity of coral, mangrove, and seagrass. *Image created by Philippe Rekacewicz in May 2002, from data compiled by UNEP-WCMC, 2001.*

The Indonesian archipelago is also relevant for global climate and ocean circulation, in particular through the Indonesian Throughflow, which links the Pacific Ocean and the Indian Ocean through a network of small gateways (Gordon, 2005). They form a bottleneck of the ocean's so-called global thermohaline circulation (THC), which is a key element of our climate system since it controls the heat transport in the ocean and its heat exchange with the atmosphere. In the past decades, the ocean accounted for 93% of the human-induced increase of the Earth energy inventory, and the poorly predictable behavior of the THC decreases the reliability of climate predictions (Rhein et al., 2013; Mauritsen and Pincus, 2017). Moreover, the world famous El Niño–Southern Oscillation (ENSO) causes pronounced interannual variations of coupled ocean–atmosphere circulation and rainfall, which are associated with floods and droughts in Indonesia.

However, global climate change, the growing economic demands of a large and rapidly increasing population, competitive and destructive resource uses, and export-oriented commercialization also make the Indonesian coastal zone vulnerable to ecosystem degradation and the loss of ecosystem functions and services (Fig. 1.2). Economic growth and a sustainable use of natural resources is a global challenge, which has to be solved regionally and requires a solid scientific basis.

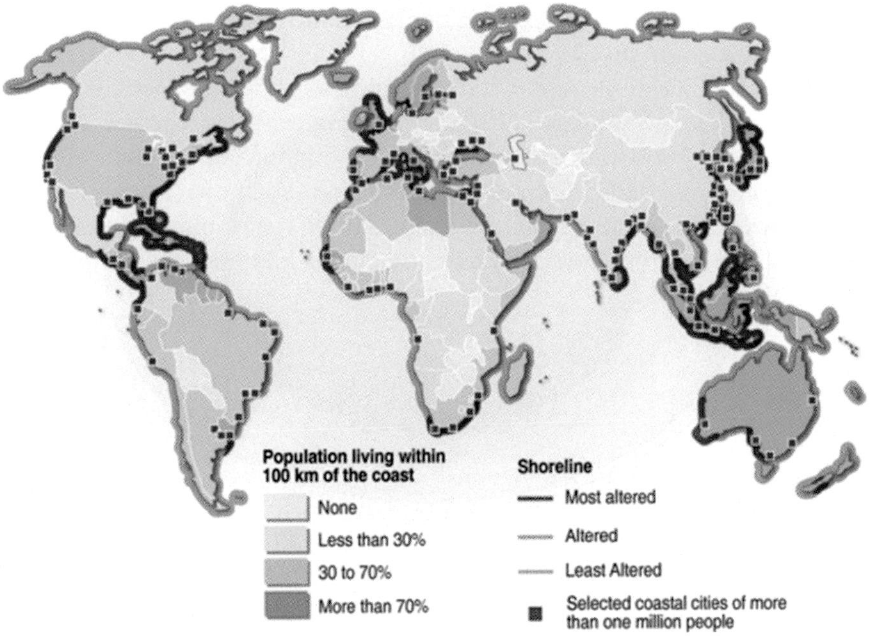


**FIGURE 1.2** Global distribution of population living within 100 km of the coast and degree of shoreline degradation. *Sources: From Burke et al. (2001) and Harrison and Pearce (2001).*

## 1.2 Development and implementation of the research and education program SPICE

Indonesia's vulnerability was underscored by the 1998 EL Niño, the largest ever since then, which led to countless forest and peat fires in Indonesia. It was also the year of a joint Indonesian–German workshop during which scientists and science administrators met to exchange ideas for future cooperation in the field of earth and ocean sciences to expand and diversify a bilateral collaboration that started in the 1970s. It became evident that developing measures toward a sustainable use of Indonesia's rich natural resources of the coastal zone requires a solid scientific basis. The multi- and interdisciplinary "Science for the Protection of Indonesian Coastal Ecosystems" program, in short called "SPICE," was born. It was supported and funded by the German Ministry of Education and Research (BMBF, Bundesministerium für Bildung und Forschung), the Indonesian Ministry of Research and Technology (RISTEK, Kementerian Riset dan Teknologi), and the Indonesian Ministry for Marine Affairs and Fisheries (KKP, Kementerian Kelautan dan Perikanan). A Steering Committee jointly led by the supporting ministries and including representatives of the participating research and education organizations provided and evaluated the thematic frame of the program during annual meetings. The overarching goal of the SPICE program was to address the scientific, social, and economic issues related to the management of the Indonesian coastal ecosystems and their resources. In addition to strengthening the existing scientific database on coastal ecosystems, the program should promote capacity and infrastructure building in the maritime sector in Indonesia and Germany, and it was supposed to contribute to education and public awareness raising. The program was carried out jointly by numerous partners from universities, research organizations, and companies in Indonesia and Germany. SPICE became an internationally known research and capacity building program that was officially conducted in three major phases between 2003 and 2015 (Fig. 1.3), but its spirit and resulting individual bilateral collaboration activities remain until today.

**FIGURE 1.3** Logos of the three phases of the SPICE program. *SPICE*, Science for the Protection of Indonesian Coastal Ecosystems.

For the strategic planning of SPICE, researchers from both countries identified thematic and regional focus areas in a so-called "Action Plan." As a result, the first phase of SPICE was conceived and implemented between 2003 and 2007 in five thematically organized and regionally spread collaborative research clusters. It provided baseline information on the status and ecosystem health and the challenges the coastal zone is facing from climate change and human activities. The information on the structure and functioning of coastal ecosystems and their alterations due to human interventions obtained during SPICE I formed the basis of SPICE II (2007–10). The focus on natural sciences in the first phase was complemented by social science in the second phase with the aim to better understand the social dimension of coastal zone changes and to improve the links between research and decision-making. As a consequence, an additional sixth cluster was added during the second SPICE phase, in which relevant social science work was conducted by economists, sociologists, anthropologists, geographers, and political scientists. This second phase identified major thematic issues in several focal geographical areas and generated a deeper and integrated understanding of the complex social–ecological dynamics in the coastal zone of these regions. In consequence, the third phase of SPICE (2012–15) adopted a more solution-oriented and holistic approach explicitly aimed at supporting the application of the newly obtained knowledge in fostering more sustainable modes of natural resource management. A rigorous evaluation in April 2010 resulted in the recommendation of the Joint Indonesian–German Steering Committee in the Field of Earth and Ocean Sciences to continue the program and implement a synthesis phase. The third phase of SPICE was accordingly planned and implemented in a fully inter- and transdisciplinary way, integrating the natural and the social sciences, as well as partly involving stakeholder groups (Table 1.1).

## 1.3 Research, education, and outreach activities

Basic research on ecosystem structure, functions, and services and on the complexity of social–ecological systems and their dynamics and governance were the primary focus of the program, but the activities were manifold and included stakeholder involvement and in particular capacity building. More than 130 students from over 10 countries obtained a PhD, MSc, or BSc degree during the SPICE program, and dozens of students conducted internships. Academic education activities were to some extent coordinated with and supported by the German Academic Exchange Service (DAAD, Deutscher Akademischer Austauschdienst) and the Indonesian Ministry for Education (DIKTI, Direktorat Jenderal Pendidikan Tinggi Kementerian Pendidikan dan Kebudayaan). Numerous special courses were held by individual scientists or bilateral teams in Indonesian universities and research organizations. Moreover, several international summer schools and training courses were organized. An annual Coastal Management Master Program at the Hasanuddin University in Makassar, Sulawesi, was regularly

**Table 1.1** Thematic foci and geographical locations of the research clusters during the three SPICE phases.

| Phase-cluster | Theme | Geographical area |
|---|---|---|
| I-1 | Coral reef–based ecosystems and resources | Sulawesi, Spermonde Archipelago |
| I-2 | Mangrove ecology—strategies for sustainable use of living resources and mariculture | Java |
| I-3 | Coastal ecosystem health | Sumatra, Indonesian Seas |
| I-4 | Marine natural products | Bali |
| I-5 | Marine geology and biogeochemistry | Java, Indian Ocean off Java |
| II-1 | Coral reef–based ecosystems and resources | Sulawesi, Spermonde Archipelago |
| II-2 | Mangrove ecology—strategies for sustainable use of living resources and mariculture | Java |
| II-3 | Coastal ecosystem health | Sumatra, Indonesian Seas |
| II-4 | Ecology and aquaculture | Bali |
| II-5 | Biogeochemical fluxes in Indonesian Seas | Indian Ocean off Java |
| II-6 | Governance and management in coastal social–ecological systems | Sumatra, Java |
| III-1 | Impacts of marine pollution on biodiversity and coastal livelihoods | Java, Jakarta Bay |
| III-2 | Carbon sequestration in the Indonesian Seas and its global significance | Sumatra, Java Sea |
| III-3 | Understanding and managing the resilience of coral reefs and associated social systems | Sulawesi, Spermonde Archipelago |
| III-4 | Terrestrial influences on mangrove ecology and sustainability of their resources | Java, Sulawesi, Kalimantan, Sumatra |
| III-5 | Climate versus anthropogenic forcing of Late Holocene environmental change affecting Indonesian marine, coastal, and terrestrial ecosystems | Java, Java Sea, Indian Ocean off Java and Sumatra |
| III-6 | Potentials of ocean renewable energy in the Indonesian Seas | Java Sea and Indian Ocean off Sumatra, Java and the Lesser Sunda Islands |

supported with a social science course with field expedition and methods training. In addition, a conference and summer school on “Coastal and Disaster Risk Management for Extreme Events Impact Mitigation” and a workshop reporting research results to local and regional decision-makers were held at the Hasanuddin University. An international workshop and training course on “Coastal Ecosystems: Hazards Management and Rehabilitation” with participants from 14 countries was organized in collaboration with the Science and Technology Center of the Non-Aligned Nations movement (NAM) at Jenderal Soedirman University in Purwokerto, Java.

The research results were presented in numerous international conferences and workshops, some of which were held in Indonesia. At the end of phase I, SPICE organized the “Southeast Asia Coastal Governance and Management Forum: Science Meets Policy for Coastal Management and Capacity Building” in collaboration with LOICZ

(Land–Ocean Interactions in the Coastal Zone, a core project of IGBP, the International Geosphere-Biosphere program), ATSEF (Arafura and Timor Seas Expert Forum), and SEACORM (Southeast Asia Center for Ocean Research and Monitoring). During this conference, which was held in Bali in fall 2006, results of SPICE phase I were presented to and discussed with international experts from the region. In May 2009, Indonesia invited for the "World Ocean Conference" in Manado, Sulawesi, where the whole SPICE consortium presented results of phase II to the international audience. And finally, a SPICE synthesis conference was held in Bali in January 2016.

The research results were also jointly published in numerous papers in peer-reviewed international as well as in regional journals. In accordance with and to highlight their interdisciplinary nature and wide thematic coverage, individual SPICE clusters published special issues of regional and international journals. Cluster 2 published a special issue entitled "Segara Anakan, Java, Indonesia, a mangrove-fringed coastal lagoon affected by human activities" with phase I results in the journal *Regional Environmental Change* (Jennerjahn and Yuwono, 2009). A special issue of the *Asian Journal for Water, Environment and Pollution* reported phase I and II results of Cluster 5 on "Environmental change affecting the Brantas River and Madura Strait coastal zone, Java, Indonesia" (Jennerjahn et al., 2013). In the same journal, Cluster 3 published phase I and II results on "The Siak, a Blackwater River in the Central Sumatran Province Riau, Indonesia" (Rixen and Baum, 2014). The phase III results of then Cluster 1 on "Impacts of Megacities on Tropical Coastal Ecosystems—The Case of Jakarta, Indonesia" were published in a special issue of the Marine Pollution Bulletin (Dsikowitzky et al., 2016). Phase III results of Cluster 3 were published in the Research Topic "Small Islands and Big Impacts—Studying Global and Local, Ecological and Social Effects on Coral Reef Ecosystems" in the journal *Frontiers in Marine Science* in 2018 (https://www.frontiersin.org/research-topics/4899/small-islands-and-big-impacts—studying-global-and-local-ecological-and-social-effects-on-coral-ree#overview). A new framework for the assessment of small island sustainability was presented with reference to the SPICE work in the Spermonde Archipelago off South Sulawesi (Glaser et al., 2018).

As mentioned in the title, an overall goal of the program was to make the wealth of extraordinary new research findings available for the protection of Indonesian coastal ecosystems and their sustainable use in the future. Therefore, numerous dissemination and outreach activities were conducted to share and discuss the generated knowledge with diverse stakeholder groups including, representatives of governmental and nongovernmental organizations and corporations, farmers, fishers, and the general public. Many of the clusters conducted annual conferences to which some of them also invited various stakeholders for an exchange of information on the thematic issues dealt with in the respective cluster. For example, a delegation of scientists of phase I and II Cluster 3 on "Coastal Ecosystem Health" was invited by the provincial Government of Riau, Sumatra, for an exchange of information. Stakeholder workshops were organized by Cluster 6 in phase II and during phase III by Clusters 1–4. A policy paper on watershed and coastal management and the potential role of Payments for

Environmental Services was published in both English and Indonesian (Heyde et al., 2017a,b). Along with Indonesian translations of research papers on the causes of watershed degradation and coastal sedimentation (Lukas, 2017a,b), they were disseminated to more than 30 governmental and nongovernmental organizations across all levels from subdistrict administrations to the national government and discussed with a diverse range of key actors in the frame of a policy workshop. A Policy Brief on "Water quality and bacterial community management in shrimp ponds in Rembang, Indonesia: Toward sustainable shrimp aquaculture" highlights the environmental concerns related to operation of aquaculture facilities and provides recommendations to decision-makers how aquaculture management can be improved (Alfiansah and Gärdes, 2019).

## 1.4 Summary and synthesis of SPICE results

This book summarizes and synthesizes the research findings in 11 thematic chapters, which cover the spectrum of themes addressed in 12 years of SPICE. Chapter 2 summarizes the major features of the physical oceanography and remote sensing methods applied in coastal process studies in the western region of the Indonesian Seas (Mayer et al., 2021). It highlights the relevance of the Indonesian Seas in global circulation as well as the distribution of phytoplankton biomass related to circulation and climate in the region. Chapter 3 focuses on the effects of land use change and regulations of hydrology on the water quality and biogeochemistry of rivers and adjacent coastal seas in Java and Sumatra, featuring the Brantas River on Java and the Siak River on Sumatra. It also addresses the underlying socioeconomic dynamics and possible governance and management options and mechanisms (Jennerjahn et al., 2021a). The fourth chapter puts a focus on the peatlands and their relevance for the global carbon cycle as well as on their role for the implementation of the Paris agreement and associated socioeconomic conflicts (Rixen et al., 2021). Chapter 5 features the complex interactions in coral reef social–ecological systems at the example of the Spermonde Archipelago in south Sulawesi (Reuter et al., 2021). It delineates how intensive exploitation of marine resources results in the degradation of coral reef systems, thereby addressing ecological processes as well as the underlying socioeconomic dynamics relating to resource use, social networks, and governance structures. In a similar manner, Chapter 6 addresses the status of and the threats to seagrass beds, again at the example of the Spermonde Archipelago (Asmus et al., 2021). The generally high diversity and abundance of seagrasses and the related ecosystem services are endangered mainly by sedimentation, dredging, and aquaculture operation. Mangrove forests are also ecologically and economically important coastal ecosystems of the tropics, which often occur in association with seagrass beds and coral reefs. Using examples from Java, but also from Kalimantan and north Sulawesi, Chapter 7 addresses the functions of and current threats to mangrove forests and examines underlying socioeconomic and political dynamics as well as possible governance and management options and mechanisms (Jennerjahn et al., 2021b).

Chapter 8 reports on the impact of megacities on the pollution of coastal waters as illustrated by the city of Jakarta and adjacent Jakarta Bay (Kunzmann et al., 2021). It spans a wide range of pollutants and their dispersion modes, including nutrients, organic pollutants and microorganisms, and the response of organisms and ecosystems to the exposure of one or more anthropogenic stressors. Chapter 9 delineates the history of climate and environment development during the late Quaternary period, ranging from recent centuries and millennia to the past 300,000 years (Mohtadi et al., 2021). It demonstrates how large-scale climatic phenomena and oscillations like the monsoons and the ENSO triggered and governed environmental change. By reconstructing past climate variability, paleoclimate research contributes to improving the simulation of future climate in Indonesia. Chapter 10 summarizes the development of a decision support system (DSS) for spatial planning of marine finfish aquaculture sites in Southeast Asia at the example of a site in Bali (Mayerle et al., 2021a). Aquaculture is an important economic sector in Indonesia, which may also negatively affect coastal ecosystems and therefore requires careful planning toward a sustainable management. Another important requirement for the large Indonesian population is the supply of energy, most importantly the supply of renewable energy. Chapter 11 focuses on the ocean renewable energy potential, specifically explores the use of tidal energy, and develops a DSS (Mayerle et al., 2021b). Chapter 12 closes the book with an overarching social science view on changes, issues, and sectors of coastal governance in Indonesia (Glaser et al., 2021). The final chapter also provides a set of generic suggestions for integrated inclusive governance approaches, as well as specific policy recommendations regarding the major issues and sectors identified in this book.

This synthesis documents a large-scale research and education effort of numerous scientists and students from Indonesia, Germany, and other countries, and it takes an important step forward in understanding the complex dynamics of Indonesia's coastal social–ecological systems. It provides a largely improved knowledge base for developing a sustainable use of coastal resources and opens avenues for a better societal use of the newly obtained research results. In this context, besides providing the best available knowledge, all chapters identify gaps in knowledge and future directions of research and depict implications and recommendations for policy. In this respect, the book serves science and society.

## Acknowledgments

We thank all the researchers and the countless enthusiastic students who participated in the SPICE program and contributed to the production of all the knowledge summarized in this book. It would have been impossible without the financial and logistic support of the Indonesian Ministry of Research and Technology (RISTEK, Kementerian Riset dan Teknologi), the Indonesian Ministry for Marine Affairs and Fisheries (KKP, Kementerian Kelautan dan Perikanan), and the German Federal Ministry of Education and Research (BMBF, Bundesministerium für Bildung und Forschung; Grant Nos. 03F0396A, 03F0470A, 03F0628A). We are grateful for the support of the Indonesian Embassy in Berlin and the German Embassy in Jakarta. We appreciate the thorough reviews of the manuscripts provided by numerous colleagues around the globe, which greatly improved the quality of the content of this book. We also thank the "fathers" of SPICE, Venugopalan Ittekkot and Indroyono Soesilo, for making this large-scale collaborative research and education program possible. Finally, we thank the coordinators on both sides, who organized this large-scale program: Farid Maruf, Ardito Kodijat, Nada Marsudi, Petra Westhaus-Ekau, Karin Gärtner, Georg Heiss, Eberhard Krain, and Claudia Schultz. We also thank Elsevier Acquisitions Editor Louisa Munro and several Editorial Project Managers, in particular Mona Zahir, who guided us through the production process of this book.

# References

Alfiansah, J.R., Gärdes, A., 2019. Water Quality and Bacterial Community Management in Shrimp Ponds in Rembang, Indonesia: Towards Sustainable Shrimp Aquaculture. Policy Brief. Leibniz Centre for Tropical Marine Research, Bremen. https://doi.org/10.21244/zmt.2019.002.

Asmus, H., Kneer, D., Pogoreutz, C., Blankenhorn, S., Jompa, J., Nurdin, N., et al., 2021. Ecology of seagrass beds in Sulawesi – multifunctional key habitats at the risk of destruction. In: Jennerjahn, T.C., Rixen, T., Irianto, H.E., Samiaji, J. (Eds.), Science for the Protection of Indonesian Coastal Ecosystems (SPICE). Elsevier, Amsterdam (pp. xxx-xxx).

Burke, L., Kura, Y., Kassem, K., Ravenga, C., Spalding, M., McAllister, D., 2001. Pilot Assessment of Global Ecosystems: Coastal Ecosystems. World Resources Institute, Washington, D.C., p. 77

Dsikowitzky, L., Ferse, S., Schwarzbauer, J., Vogt, T.S., Irianto, H.E., 2016. Editorial – impacts of megacities on tropical coastal ecosystems — the case of Jakarta, Indonesia. Marine Pollution Bulletin 110, 621–623.

Glaser, M., Breckwoldt, A., Carruthers, T., Forbes, D., Costanzo, S., Kelsey, H., et al., 2018. Towards a framework to support coastal change governance in small islands. Environmental Conservation 45, 227–237. https://doi.org/10.1017/S0376892918000164.

Glaser, M., Adrianto, L., Breckwoldt, A., Buhari, N., Deswandi, R., Ferse, S., et al., 2021. The governance of coastal and marine social-ecological systems: Indonesia and beyond. In: Jennerjahn, T.C., Rixen, T., Irianto, H.E., Samiaji, J. (Eds.), Science for the Protection of Indonesian Coastal Ecosystems (SPICE). Elsevier, Amsterdam (pp. xxx-xxx).

Gordon, A.L., 2005. Oceanography of the Indonesian seas and their throughflow. Oceanography 18, 14–27.

Harrison, P., Pearce, F., 2001. AAAS Atlas of Population and Environment. American Association for the Advancement of Science, University of California Press, Berkeley, p. 215.

Heyde, J., Lukas, M.C., Flitner, M., 2017a. Payments for Environmental Services: A New Instrument to Address Long-Standing Problems? Policy Paper. Artec-Paper 213, Sustainability Research Center (Artec). University of Bremen.

Heyde, J., Lukas, M.C., Flitner, M., 2017b. Pembayaran Jasa Lingkungan: Instrumen baru untuk mengatasi masalah lingkungan berkepanjangan di daerah aliran sungai dan pesisir di Indonesia? Naskah Kebijakan. Artec-paper 214, Sustainability Research Center (artec). University of Bremen.

Jennerjahn, T.C., Adi, S., Schroeder, F., 2013. Editorial – human activities and natural disasters affecting water quality and ecology of the Brantas River and Madura Strait coastal waters, Java, Indonesia. Asian Journal of Water, Environment and Pollution 10, 1–3.

Jennerjahn, T.C., Ardli, E.R., Boy, J., Heyde, J., Lukas, M.C., Nordhaus, I., et al., 2021a. Mangrove ecosystems under threat in Indonesia: the Segara Anakan Lagoon, Java, and other examples. In: Jennerjahn, T.C., Rixen, T., Irianto, H.E., Samiaji, J. (Eds.), Science for the Protection of Indonesian Coastal Ecosystems (SPICE). Elsevier, Amsterdam (pp. xxx-xxx).

Jennerjahn, T.C., Baum, A., Damar, A., Flitner, M., Heyde, J., Jänen, I., et al., 2021b. Human interventions in rivers and estuaries of Java and Sumatra. In: Jennerjahn, T.C., Rixen, T., Irianto, H.E., Samiaji, J. (Eds.), Science for the Protection of Indonesian Coastal Ecosystems (SPICE). Elsevier, Amsterdam (pp. xxx-xxx).

Jennerjahn, T.C., Yuwono, E., 2009. Editorial – Segara Anakan, Java, Indonesia, a mangrove-fringed coastal lagoon affected by human activities. Regional Environmental Change 9, 231–233.

Kunzmann, A., Schwarzbauer, J., Palm, H.W., Damriyasa, M., Yulianto, I., Kleinertz, S., et al., 2021. Impact of megacities on the pollution of coastal areas – the case example Jakarta Bay. In: Jennerjahn, T.C., Rixen, T., Irianto, H.E., Samiaji, J. (Eds.), Science for the Protection of Indonesian Coastal Ecosystems (SPICE). Elsevier, Amsterdam (pp. xxx-xxx).

Lukas, M.C., 2017a. Konservasi daerah aliran sungai di Pulau Jawa, Indonesia: Terjebak dalam konflik sumberdaya hutan. artec-Paper 212, Sustainability Research Center (artec). University of Bremen.

Lukas, M.C., 2017b. Widening the scope: linking coastal sedimentation with watershed dynamics in Java, Indonesia. Regional Environmental Change 17 (3), 901–914. https://doi.org/10.1007/s10113-016-1058-4. Online Resource 1: Memperluas jangkauan: Menghubungkan sedimentasi pesisir dengan dinamika daerah aliran sungai di Jawa.

Mauritsen, T., Pincus, R., 2017. Committed warming inferred from observations. Nature Climate Change 7, 652–655.

Mayer, B., Siegel, H., Gerth, M., Pohlmann, T., Stottmeister, I., Putri, M., et al., 2021. Physical environment of the Indonesian seas with focus on the western region. In: Jennerjahn, T.C., Rixen, T., Irianto, H.E., Samiaji, J. (Eds.), Science for the Protection of Indonesian Coastal Ecosystems (SPICE). Elsevier, Amsterdam (pp. xxx-xxx).

Mayerle, R., Orhan, K., Pandoe, W.W., Poerbandono, K.P., 2021a. Decision tool for estimating energy potential from tidal resources. In: Jennerjahn, T.C., Rixen, T., Irianto, H.E., Samiaji, J. (Eds.), Science for the Protection of Indonesian Coastal Ecosystems (SPICE). Elsevier, Amsterdam (pp. xxx-xxx).

Mayerle, R., Sugama, K., van der Wulp, S., Poerbandono, R.K.-H., 2021b. Decision tool for assessing marine finfish aquaculture sites in Southeast Asia. In: Jennerjahn, T.C., Rixen, T., Irianto, H.E., Samiaji, J. (Eds.), Science for the Protection of Indonesian Coastal Ecosystems (SPICE). Elsevier, Amsterdam (pp. xxx-xxx).

Milliman, J.D., Farnsworth, K.L., 2011. River Discharge to the Coastal Ocean – a Global Synthesis. Cambridge University Press, Cambridge.

Mohtadi, M., Lückge, A., Steinke, S., Permana, H., Susilohadi, S., Zuraida, R., et al., 2021. Late quaternary environmental history of Indonesia. In: Jennerjahn, T.C., Rixen, T., Irianto, H.E., Samiaji, J. (Eds.), Science for the Protection of Indonesian Coastal Ecosystems (SPICE). Elsevier, Amsterdam (pp. xxx-xxx).

Reuter, H., Breckwoldt, A., Dohna, T., Ferse, S., Gärdes, A., Glaser, M., et al., 2021. Coral reef social-ecological systems under pressure in southern Sulawesi. In: Jennerjahn, T.C., Rixen, T., Irianto, H. E., Samiaji, J. (Eds.), Science for the Protection of Indonesian Coastal Ecosystems (SPICE). Elsevier, Amsterdam (pp. xxx-xxx).

Rhein, M., Rintoul, S.R., Aoki, S., Campos, E., Chambers, D., Feely, R.A., et al., 2013. Observations: ocean. In: Stocker, T.F., Qin, D., Plattner, G.-K., Tignor, M., Allen, S.K., Boschung, J., et al. (Eds.), Climate Change 2013: The Physical Science Basis. Contribution of Working Group I to the Fifth Assessment Report of the Intergovernmental Panel on Climate Change. Cambridge University Press, Cambridge, New York, pp. 255–316.

Rixen, T., Baum, A., 2014. Editorial – the Siak, a Blackwater River in the Central Sumatran Province Riau, Indonesia. Asian Journal of Water, Environment and Pollution 11, 1–2.

Rixen, T., Wit, F., Hutahaean, A.A., Schlüter, A., Baum, A., Klemme, A., et al., 2021. Carbon cycle in tropical peatlands and coastal seas. In: Jennerjahn, T.C., Rixen, T., Irianto, H.E., Samiaji, J. (Eds.), Science for the protection of Indonesian Coastal Ecosystems (SPICE). Elsevier, Amsterdam (pp. xxx-xxx).

# 2

# Physical environment of the Indonesian Seas with focus on the western region

Bernhard Mayer[1], Herbert Siegel[2], Monika Gerth[2], Thomas Pohlmann[1], Iris Stottmeister[2], Mutiara Putri[3], Agus Setiawan[4]

[1]*INSTITUTE OF OCEANOGRAPHY, UNIVERSITY OF HAMBURG, GERMANY;* [2]*LEIBNIZ INSTITUTE FOR BALTIC SEA RESEARCH, ROSTOCK-WARNEMÜNDE, GERMANY;* [3]*INSTITUT TEKNOLOGI BANDUNG (ITB), BANDUNG, JAWA BARAT, INDONESIA;* [4]*MARINE RESEARCH CENTRE, AGENCY FOR MARINE AND FISHERIES RESEARCH AND HUMAN RESOURCES, MINISTRY OF MARINE AFFAIRS AND FISHERIES, JAKARTA, INDONESIA*

## Abstract

*The Indonesian Seas play an important role in the marine global environment, from the oceanographic point of view as well as from the bioenvironmental perspective, since a large population of nearly 400 million people is living directly or indirectly in contact with these waters affecting the marine environment. In this chapter, we describe the history of oceanographic research in this region, the role of the Indonesian Seas in the global circulation, the most important features of the regional circulation, the distribution of tides, water exchange rates, and freshwater sources and sinks for certain subregions, and the effect of global warming as it is detectable nowadays already. Furthermore, we explain how satellite-derived data in connection with biooptical in situ measurements can support the investigation of the marine environment in a vast region, especially in coastal waters with respect to river discharge, transport of suspended matter, and primary production.*

## Abstrak

*Perairan Indonesia memegang peranan penting dalam lingkungan laut global, mulai dari sisi oseanografis hingga ke perspektif lingkungan hidupnya, karena hampir 400 juta orang hidup berinteraksi baik secara langsung maupun tidak langsung dengan perairan yang berdampak pada lingkungan laut. Dalam bab ini dijelaskan sejarah penelitian oseanografi di wilayah ini, peran perairan Indonesia dalam sirkulasi global, fitur paling penting dari sirkulasi regional, distribusi pasang surut, percampuran air dan sumber air tawar dan pengurangannya untuk sub-daerah tertentu, serta efek dari pemanasan global yang sudah dapat dideteksi saat ini. Selanjutnya, kami menjelaskan bagaimana data satelit yang berhubungan dengan pengukuran in situ bio-optik dapat mendukung penelitian lingkungan laut di wilayah yang luas, terutama di perairan pesisir dengan adanya debit sungai, pengangkutan bahan-bahan tersuspensi dan produksi primer.*

**Science for the Protection of Indonesian Coastal Ecosystems (SPICE). https://doi.org/10.1016/B978-0-12-815050-4.00007-9**

**Chapter outline**

**2.1 Introduction** ..... **14**
**2.2 The marine circulation** ..... **17**
2.2.1 The global context ..... 17
2.2.2 The regional circulation ..... 18
2.2.3 Tides ..... 19
**2.3 Seasonal variability and long-term changes** ..... **22**
2.3.1 Seasonality of circulation ..... 22
2.3.2 Seasonality of temperature and salinity ..... 25
2.3.3 Long-term development of sea surface temperature and sea surface salinity ..... 25
**2.4 Water residence times** ..... **26**
**2.5 Sources and sinks of freshwater** ..... **28**
**2.6 Remote sensing methods applied in coastal process studies** ..... **31**
2.6.1 Available satellite data ..... 31
2.6.2 Ocean color and its variation in Indonesian coastal waters ..... 32
2.6.3 Satellite-based studies of phytoplankton and coastal processes ..... 35
*2.6.3.1 Distribution of phytoplankton* ..... 35
*2.6.3.2 Coastal discharge and influence of tidal and monsoon phases* ..... 36
*2.6.3.3 Climatological aspects* ..... 37
**Acknowledgments** ..... **40**
**References** ..... **40**

## 2.1 Introduction

The Indonesian waters have received a growing interest in recent years, mainly because the economic value of the marine resources of these waters is increasingly recognized. Moreover, the exploitation of these resources in combination with the impact of global climate change causes an increasing threat, which is accounted for by a growing number of research projects on the national but also the international levels.

The history of oceanography of the Indonesian waters can be subdivided into three major periods (Pariwono et al., 2005), namely precolonial (until end of 16th century), colonial (from beginning of 17th century to Indonesian independence), and post-independence (starting from 1945). In the precolonial period, scientific investigations were mostly limited to those, which were essential to operate the merchant and military fleets. They relied totally on wind power. Accordingly, the gathered oceanographic information was limited to tides, monsoon-driven currents and the ocean circulation.

During the colonial period, about 38 expeditions were carried out in the Indonesian Archipelago by 10 different countries, since the colonial power, the Netherlands, had not enough resources to perform all necessary research of this vast territory on their own. The expedition of the colonial period, which had the strongest impact on the following research work, was the Snellius Expedition in 1929–30. It covered the entire central and eastern part of the Indonesian Seas with 374 stations in total, performing more than 33,000 echo soundings (van Aken, 2005).

In the post–independence period after World War II, the number of expeditions increased drastically to 75 until the year 1990. Forty of them were led by Indonesia (Pariwono et al., 2005). The most important publication of this period was a monograph written by Wyrtki (1961). In this report, the so-called Naga Report, all information on physical oceanography of the colonial period was thoroughly collected and combined with latest results partly from own investigations. Interestingly, the Naga Report remained the most accepted and cited general reference for the physical properties of the Indonesian archipelago until the beginning of the 21st century.

Between the 1960 and 1980s studies in the Indonesian waters were mostly of localized nature and in most cases related to fisheries. Starting from the 1980s, the interest in the Indonesian waters grew drastically, to a large degree because the importance of these waters with respect to their influence on the global ocean circulation was increasingly acknowledged. It turned out that the Indonesian Throughflow (ITF) is one of the few bottlenecks of the large-scale conveyor belt circulation, which connects all major parts of the global ocean. The latest campaign, the international INSTANT program (Sprintall et al., 2004), was entirely dedicated to the investigation of this throughflow and its temporal and spatial variability.

In recent years, the response of the Indonesian Seas to the climate change attracts increasing interest of climate scientists, because the global ocean warm pool is located at the entrance to these waters, and they play an important role in the El Niño–Southern Oscillation system. The response of these specific features to the global warming is presently not fully understood, which is also reflected by the fact that some climate projections expect an increase of El Niño events in strength and frequency, whereas other estimates show the opposite (Van Oldenborgh et al., 2013).

In this chapter, we will also describe some interesting findings related to the following subdomains of the Indonesian Seas located on the shallow Sunda Shelf:

- Inner Sunda Shelf with an average depth of 49 m consisting of the southern South China Sea (SCS) with depths less than 200 m, the Karimata Strait and regions listed in the following:
- Gulf of Thailand with an average depth of 42 m
- Malacca Strait with an average depth of 51 m
- Java Sea with an average depth of 40 m

Geometrical measures, average values, and trends for certain parameters in these subdomains are listed in Table 2.1.

This book utilizes results of our jointly performed investigations within the scientific 10 years lasting German–Indonesian cooperation SPICE ("Science for the Protection of

**Table 2.1** Some geometric measures, average values, and trends for subregions of the Indonesian Seas.

| Parameter | Unit | I. Sunda Shelf[a] | Gulf of Thailand | Malacca Strait | Java Sea |
|---|---|---|---|---|---|
| Depth ∅ | m | 48.8 | 41.7 | 50.6 | 40.3 |
| Area | $km^2$ | 1,893,307 | 315,437 | 139,330 | 436,274 |
| Volume | $km^3$ | 92,374 | 13,161 | 7052 | 17,576 |
| **Regionally averaged sea surface temperature (SST) and salinity (SSS)** | | | | | |
| SST ∅[b] | °C | 29.0 | 29.2 | 28.8 | 29.2 |
| SST trend[c] | $°C\ c.^{-1}$ | +1.05 | +1.47 | +1.08 | +0.96 |
| SSS ∅[b] | psu | 31.8 | 30.9 | 30.0 | 32.0 |
| SSS trend[c] | $psu\ c.^{-1}$ | −1.34 | −0.96 | −2.82 | −1.79 |
| **Regionally averaged freshwater input/output** | | | | | |
| Precipitation ∅[b] | $m^3\ s^{-1}$ | 119,044 | 19,650 | 9888 | 23,305 |
| Precipitation trend[c] | $m^3\ s^{-1}\ c.^{-1}$ | +70,900 | +6200 | +6000 | +27,500 |
| Evaporation ∅[b] | $m^3\ s^{-1}$ | 43,300 | 8350 | 2531 | 11,939 |
| Evaporation trend[c] | $m^3\ s^{-1}\ c.^{-1}$ | −14 600 | −900 | −600 | −5500 |
| River discharge ∅[b] | $m^3\ s^{-1}$ | 64,188 | 4762 | 6528 | 10,433 |
| River discharge trend[c] | $m^3\ s^{-1} c.^{-1}$ | +8900 | +300 | +2600 | +1600 |
| **Regionally averaged freshwater balance** | | | | | |
| Freshwater ∅ sum | $L\ day^{-1}\ km^{-2}$ | 6.4 | 4.4 | 8.6 | 4.3 |
| Freshwater trends sum | $L\ day^{-1}\ km^{-2}\ c.^{-1}$ | 3.0 | 1.5 | 5.0 | 4.7 |

[a]Inner Sunda Shelf, containing all subregions listed here plus SCS with depth <200 m plus the Karimata Strait.
[b]Average values (∅), calculated from model results (Mayer et al., 2018) for the period 1965–2014.
[c]Linear trends, calculated from model results (Mayer et al., 2018) for the period 1965–2014, given in units per century ($1c.^{-1}$). Average and trend values for SST and SSS from Mayer et al. (2018).

the Indonesian Coastal Ecosystems"). The research projects and their measurements were focused mainly on the peatland areas draining to the eastern coast of Sumatra located in the western Indonesian Seas. Therefore, also our investigation of the physical environment partly focuses on the western region of the Indonesian Seas.

For the description of certain phenomena and parameter distributions or time series, we will use—among other data—also simulation results of the numerical hydrodynamical model HAMSOM (Hamburg Shelf Ocean Model), which has been applied to the Indonesian Seas within several research projects to simulate the long-term circulation for the past 50–60 years. References to our publications are given, where model results are displayed in this chapter. HAMSOM is a free surface baroclinic and prognostic regional numerical model working with finite differences on an Arakawa C grid. The model domain included more than the entire Indonesian Seas covering the area of about 90–135°E and 15°S to 17°N with a resolution of 0.1 degree (about 11 km). The depth was resolved in z-coordinates with 39 layers of increasing thickness from 5 m at the surface to 550 m for the lowest layers. The domain corresponds approximately to the map shown in Fig. 2.1, lower panel.

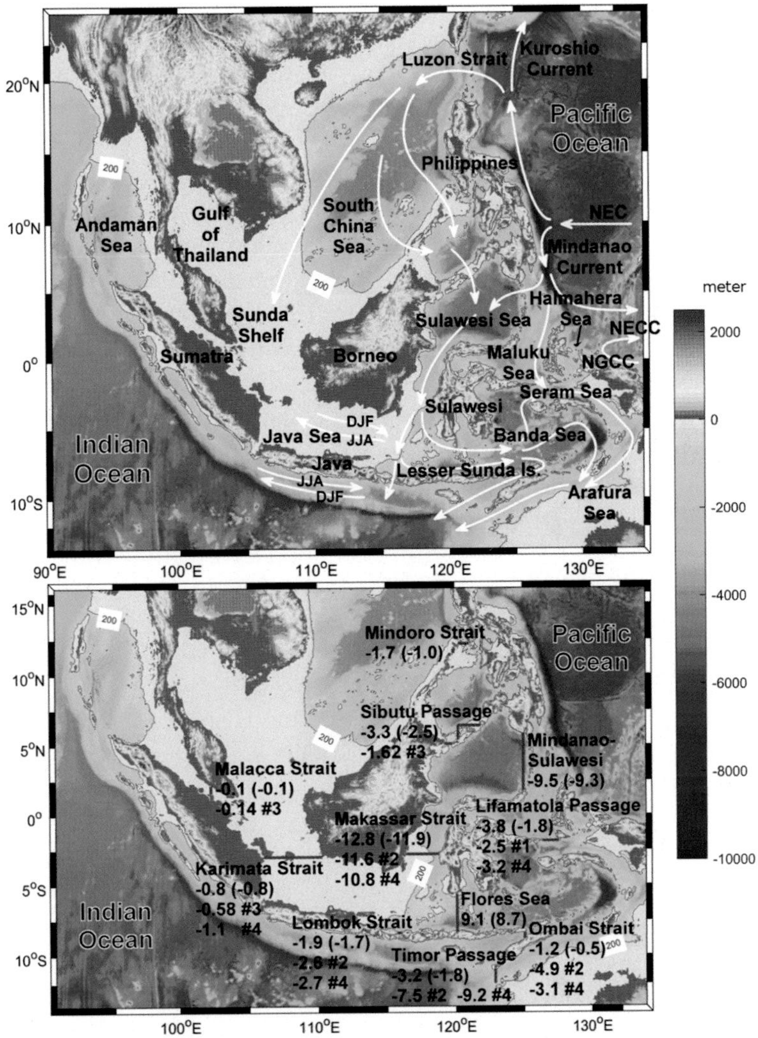


**FIGURE 2.1** The Indonesian Seas and their vicinity with main currents and water transports: The upper panel shows the region of the Maritime Continent with most important ocean currents. They partly change their directions with the monsoon season. Lower panel gives the most important annual mean water volume transports in the Indonesian Seas, positive north- or eastward, negative south- or westward, in Sverdrup, averaged for 2004–06 (1970–2006). #1 to #4 refer to literature: #1 to van Aken et al. (2009); #2 to Sprintall et al. (2009); #3 to Gordon et al. (2012); #4 to Feng et al. (2013). *After Mayer et al. (2018).*

# 2.2 The marine circulation

## 2.2.1 The global context

The Indonesian Seas cover the larger part of the Maritime Continent. They consist of the marine environment enclosed by Sumatra in the west, Philippines to New Guinea to

Torres Strait in the east, approximately the 10°N latitude in the north, and Java and the Lesser Sunda Islands in the south (about 10°S)—officially, most of this area belongs to the Pacific Ocean. Some coastal parts of the enclosing islands belong to the Indian Ocean. The region shows a very complex bathymetry with approximately 17,805 small and large islands, more than 84,000 km coastline, ocean basins up to 7000 m deep with sometimes very steep topographic gradients, with shelf seas, sills, and many wide or narrow straits connecting all the different elements.

The region provides the only natural low-latitude connection between two oceans in the world and plays an important role in the global thermohaline circulation (also called "global conveyor belt"), which is an ocean water transport band connecting the Pacific, Indian, Atlantic, and Southern Oceans. On its way from the Pacific to the Indian Oceans, it passes through the Indonesian Seas. Therefore, the general flow through the Indonesian Seas (Fig. 2.1) is directed from the Pacific to the Indian Ocean. This part of the global conveyor belt is the ITF, an average transport of water masses from the northern and southern tropical Pacific Ocean to the tropical Indian Ocean with a climatological rate of approximately 15 Sv (Gordon et al., 2010; Oppo and Rosenthal, 2010; Sprintall et al., 2014). 1 Sv (Sverdrup) is defined as a volume transport of 1 million $m^3\ s^{-1}$.

## 2.2.2 The regional circulation

The general pattern of the ITF in the Indonesian Seas is shown in Fig. 2.1 including the annual mean volume transports through particular key sections to give an idea about transport rates. The ITF consists essentially of four routes. The western route (main branch) is an extension of the Mindanao Current, transporting North Pacific water through the Sulawesi Sea, Makassar Strait, and then directly through Lombok Strait into the Indian Ocean or along the Lesser Sunda Islands through the Flores Sea and either through Ombai Strait or through Banda Sea and Timor Passage into the Indian Ocean. Another smaller amount of volume transport adds coming from the Luzon Strait via the SCS leaving through the Malacca Strait into the Andaman Sea or into the Karimata Strait leaving directly through the Sunda Strait between Sumatra and Java into the Indian Ocean or adding to the Makassar Strait throughflow. The Karimata Strait throughflow—and with this, the direction of water flow in the Java Sea—shows strong seasonality with a general westward flow during SE monsoon season (June, July, August) and eastward flow during NW monsoon season (December, January, February). Its annual average volume transport is directed to the east (Fang et al., 2009, 2010; Susanto et al., 2013, 2016).

The two eastern branches transport North Pacific water via the Maluku Sea into the Seram Sea and Banda Sea and/or Arafura Sea and South Pacific water from the New Guinea Coastal Current via the Halmahera Sea into the Seram and Arafura Seas both joining the exit path through Timor Passage. Additionally, small amounts of Deep South Pacific water overflow the Lifamatola Sill and fill the deep parts of the Seram Sea and

Banda Sea basins. Oppo and Rosenthal (2010) nicely describe the further pathway of the ITF water in the Indian and to the Atlantic Ocean.

In summary, the inflow passages from the Pacific Ocean consist from west to east of the regions Karimata Strait, Makassar Strait, Lifamatola Sill, and Halmahera Sea, while the outflow passages into the Indian Ocean are the Malacca Strait, Sunda Strait, Ombai Strait, and Timor Passage. The small gaps between the Lesser Sunda Islands are further leakages into the Indian Ocean.

Coastal upwelling regions play an important role for local ecosystems. They depend also on the seasons. During southern monsoon season, major local upwelling regions are detected in the Indian Ocean south of Sumatra, Java, and Lesser Sunda Islands, connected with the westward directed South Java Current and the alongshore winds during this season. A distinct upwelling center moves from the central southern Javanese coast in June toward the Sunda Strait in August and further to Sumatran western coast in September/October (Susanto et al., 2001). Further upwelling is visible from satellite-derived ocean color in the eastern Banda Sea and along the western coast of Irian Jaya (Papua New Guinea) in the Arafura Sea, occurring also during the SE monsoon season (Susanto et al., 2006; Susanto and Marra, 2005). Both upwelling processes are interrupted and partly even reversed to downwelling during the northwest monsoon season.

In the Indonesian Seas, ocean water changes its properties along its path. Due to high tidal mixing also from internal tides and subsequent horizontal and vertical turbulent diffusion (Nagai et al., 2017; Nagai and Hibiya, 2015; Ray and Susanto, 2016) as well as wind-induced mixing and upwelling, a downward heat flux within the water column and a strong atmospheric heat flux into the ocean surface water is observed. Furthermore, because of high tropical precipitation rates and freshwater runoff from land, a freshening of the upper ocean water occurs. A collection of TS diagrams for different areas of the Indonesian Seas, which are part of the ITF pathway, shows clearly this transformation, particularly for salinity (Sprintall et al., 2014). Therefore, the water exiting the Indonesian Seas into the Indian Ocean is supposed to be fresher and cooler than the water entering from the Pacific Ocean (Godfrey, 1996), which can be detected even in maps of ocean temperature and salinity within the upper thermocline (Gordon, 2005). These maps show the trace of relatively cool and low-saline ITF water after it has left the Indonesian Seas toward the Indian Ocean.

### 2.2.3 Tides

Tides play an important role in the Indonesian Seas, because they cause high-energy tidal currents as well as internal waves and strong vertical turbulent mixing (Nugroho et al., 2017; Ray et al., 2005; Ray and Susanto, 2016). The great importance of this tidally induced change of water properties even for the climate in the SE Asian–Australian region has been shown using coupled numerical ocean–atmosphere circulation models (Koch-Larrouy et al., 2010).

To get an impression on the relative importance of semidiurnal or diurnal tidal forms, we compare the amplitudes of the four major partial tides, which are as follows:

- $M_2$: semidiurnal, principal lunar, period of 12.421 = 12:25 h
- $S_2$: semidiurnal, principal solar, period of 12.0 = 12:00 h
- $K_1$: diurnal, declination lunisolar, period of 23.934 = 23:56 h
- $O_1$: diurnal, principal lunar, period of 25.819 = 25:49 h

From the single partial tidal amplitudes, a tidal form factor can be derived telling us, which oscillation type from pure semidiurnal to pure diurnal tide will be the dominant one:

$$F = (K_1 + O_1)/(M_2 + S_2). \tag{2.1}$$

Depending on the value of F, we can distinguish certain tidal forms:

- $0.0 \leq F < 0.25$: semidiurnal tides, the daily two high water levels (or low water levels) have similar amplitudes with a maximum, when moon passes the local meridian
- $0.25 \leq F < 1.5$: mixed but predominantly semidiurnal tides, daily two high and two low water levels with quite different magnitudes and a maximum after extremes of moon declination
- $1.5 \leq F > 3.0$: mixed but predominantly diurnal tides, daily usually two very different high waters and two very different low waters—there might be days with only one high and one low water (after extremes of moon declination)
- $F \geq 3.0$: diurnal tidal form, one daily high water and one low water—when moon passes the equatorial declination, it is neap tide and two high waters might occur

The amplitudes of the aforementioned partial tides were taken from the Ocean State University (OSU) website. They provide numerical results of a global barotropic tide model with fine resolution completions for certain regions and assimilation of TOPEX/POSEIDON satellite altimeter data (OSU Tidal Inversion Software (OTIS) (Egbert and Erofeeva, 2002)). Barotropic tidal data as calculated by this software can defer from baroclinic tidal data, which would include the effect of the ocean's stratification, particularly in shallow areas, where bottom friction and nonlinear interaction may influence the tidal currents significantly. Fig. 2.2 presents the horizontal distribution of the amplitudes of the aforementioned four tidal constituents as well as the percentage of $M_2 + S_2$ in all four partial tides and the tidal form factor F for the Indonesian Seas.

It is clear from the figure that the semidiurnal tidal periods are more or less predominant in the Pacific and Indian Oceans as well as in the eastern Indonesian Seas, while the diurnal tidal signal predominates in the SCS, Sulu Sea, and most parts of the Gulf of Thailand and the southwestern Indonesian Seas. There are some regions like the Karimata Strait with only diurnal tidal signal.

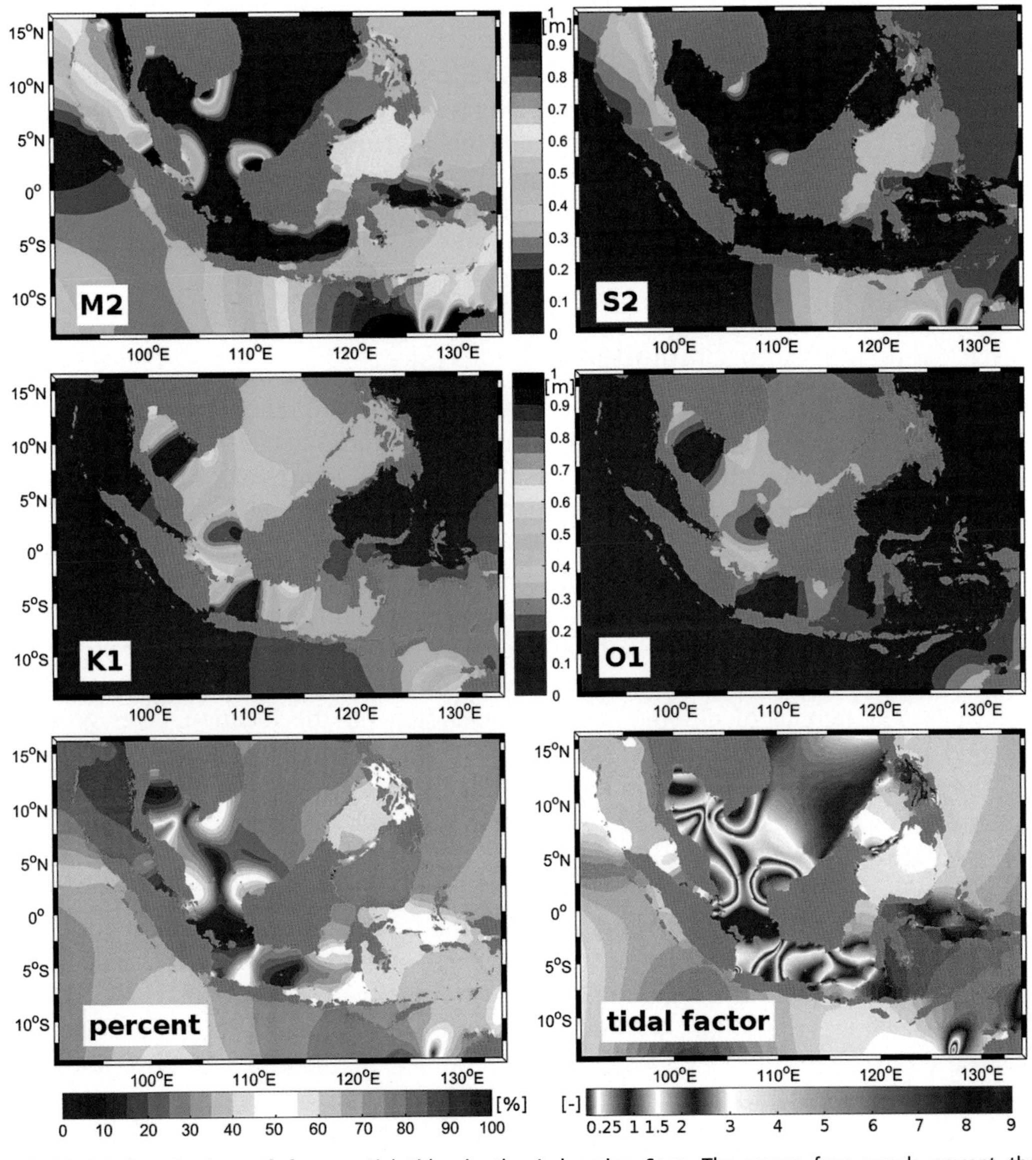


**FIGURE 2.2** Contributions of four partial tides in the Indonesian Seas: The upper four panels present the distribution of amplitudes due to the partial tides $M_2$, $S_2$, $K_1$, and $O_1$. Lower panel shows the percentage of $M_2 + S_2$ within all four tidal amplitudes and the distribution of the tidal form factor $F = (K_1 + O_1)/(M_2 + S_2)$ with 0–0.25: semidiurnal; 0.25–1.5: mixed with predominantly semidiurnal; 1.5–3.0: mixed with predominantly diurnal; larger 3.0: diurnal tides (for calculation, see text). *Amplitudes of partial tides calculated with OTPS from Ocean State University (Egbert and Erofeeva, 2002).*

The distribution of all the tidal constituents and signals and the resulting forms and periods of the local tidal movement are influenced by the local geometry and bathymetry of ocean basins and straits. There is no general rule for this except the fact that tidal waves often generate Kelvin waves moving alongshore with the coast on the right in the northern and on the left in the southern hemisphere. However, due to the topographic complexity, tides and their amphidromic points (points in the ocean without any tidal movement) are unpredictable without applying numerical models.

## 2.3 Seasonal variability and long-term changes

### 2.3.1 Seasonality of circulation

There is a clear seasonal signal in the major flow patterns within the Indonesian Seas and the volume transport of the ITF. The Indonesian Seas are effected by three meteorological wind systems: the northeasterly trade winds (northern Hadley cell), the intertropical convergence zone (ITCZ), and the southeasterly trade winds (southern Hadley cell), all of which are moving northward and southward following the sun's zenith point. These general wind systems are superimposed by the seasonally altering southeast asian monsoon system, leading to prevailing monsoon winds from SE to SW during May/June to September and NW to NE during November to March on the climatological average (meaning an average over at least 30 years). This is accompanied by winds and rainfall of different excess, depending on the location. Average wind directions are shown by arrows in Fig. 2.3.

The seasonality of the monsoon winds results in a seasonality of the surface current system in the Indonesian Seas and subsequently in the ITF volume transport. As a representative, Fig. 2.4 shows a 15-year average of the Makassar Strait throughflow from numerical model results (Mayer et al., 2018). The figure displays monthly averages over the period 2001–2015 as blue vertical bars for the total depth and for different depth ranges. Additionally, red vertical lines display the standard deviations from the corresponding mean values expressing the immense variability of the current system. During the boreal winter, prevailing northwesterly winds push surface water from the SCS eastward into the Java Sea, Makassar Strait, Flores Sea, and Banda Sea. Since the water cannot leave the Indonesian Seas fast enough via the narrow straits into the Indian Ocean, it is piling up in the southern Indonesian Seas during this time of the year. This leads to a barotropic pressure gradient slowing down the southward ITF flow, particularly its major branch though the Makassar Strait. As soon as this gradient is balanced by the decelerated inflow from the north (November), the major Makassar Strait throughflow at depths of 50–300 m arrives at a lower level. Below this level at 300–700 m, even a return flow occurs. In April, the monsoon winds change, and the blocking situation ends, leading to a full release of these backed-up water masses (largest flow at depths of 50–300 m, where the main part of the Makassar Strait throughflow occurs). Three months later (July to August), when the southward wind direction supports the

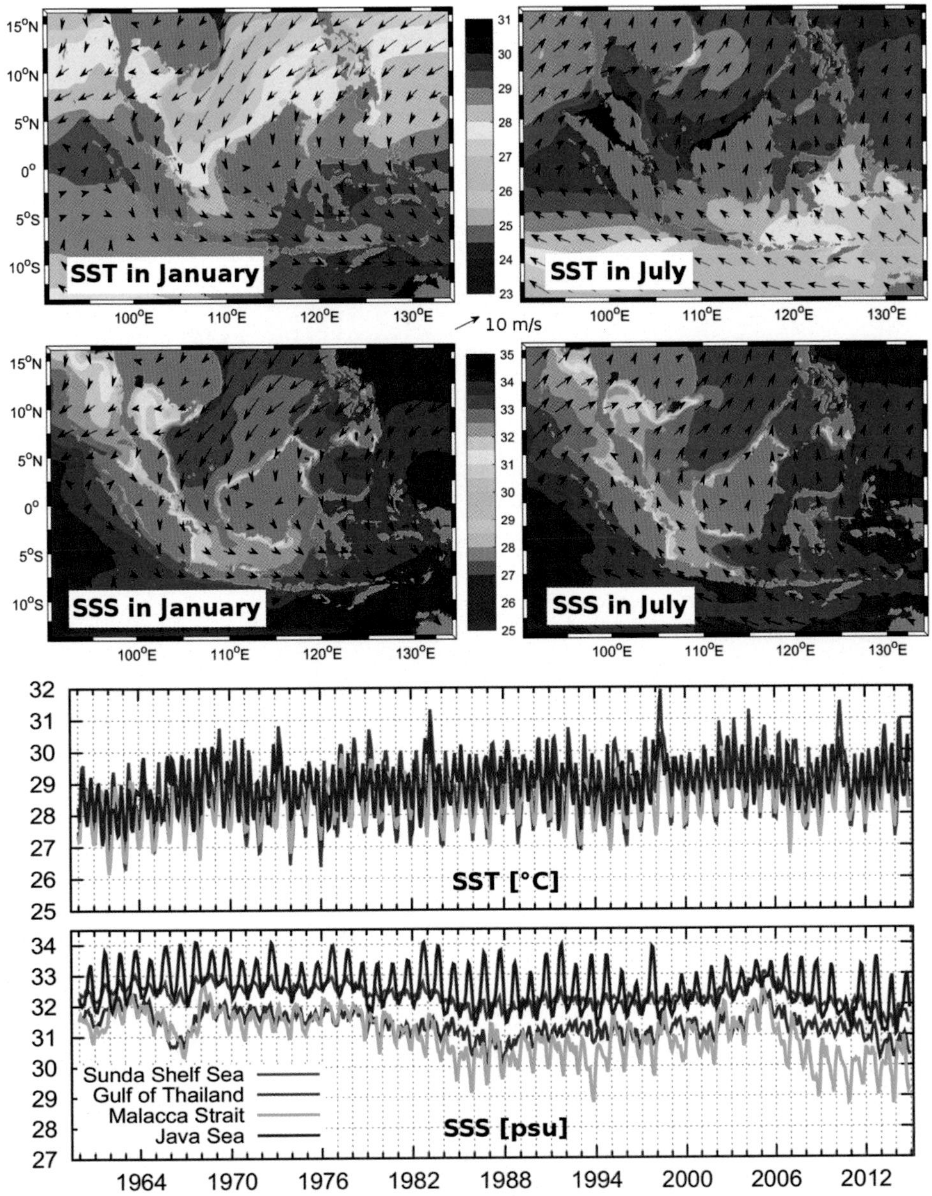


**FIGURE 2.3** Seasonality of wind, sea surface temperature (SST), and sea surface salinity (SSS) in the Indonesian Seas: Horizontal distributions of SST and SSS as monthly average (1990–2016) for January and July, superimposed with surface wind arrows from monthly average (1981–2010) for January (NW monsoon season) and July (SE monsoon season). Time series show monthly and spatially averaged SST and SSS for different regions (numerical model results) for the period from 1960 to 2014. *Horizontal SST/SSS data from Storto and Masina (2016a,b), wind data from NCEP/NCAR (Kalnay et al., 1996).*

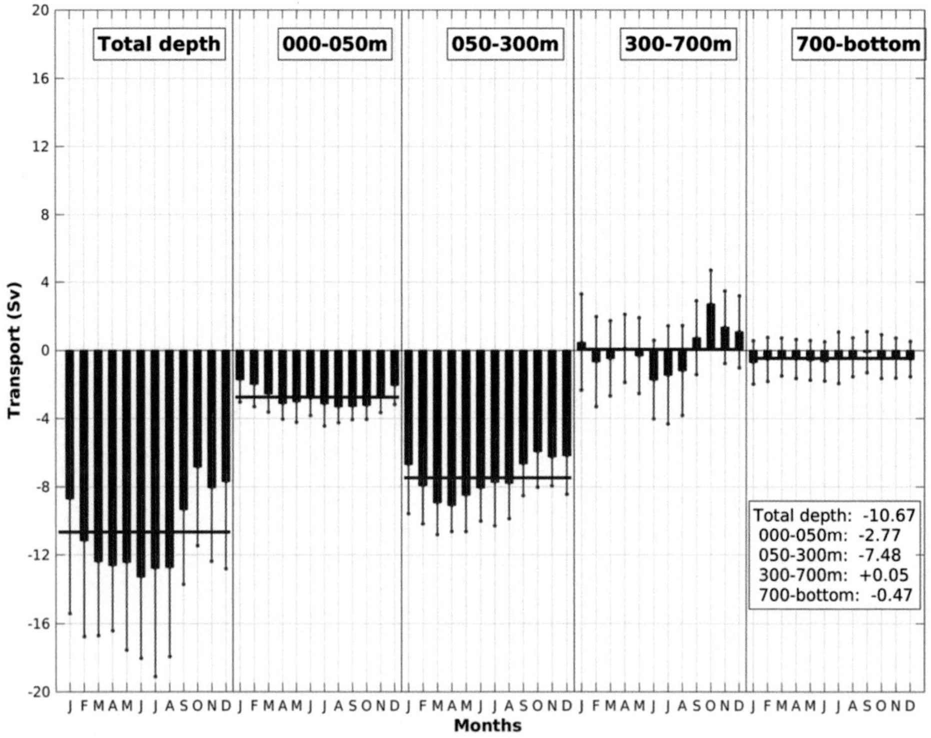


**FIGURE 2.4** Climatology of the water volume transport of the western ITF branch through the Makassar Strait: Monthly mean volume transports through the Labani Channel section of Makassar Strait as depicted in Fig. 2.1 in Sv for the entire depth and different depth ranges. Blue bars are monthly mean values for the time range 2001–15, and red lines are standard deviations from mean. The blue horizontal lines show the overall average transports (negative = southward). *Data from own simulations according to Mayer et al. (2018).*

throughflow, it reaches its peak values of a year. During this season, the general flow in the Java Sea is directed to the west. In September/October, the blocking situation starts to develop again leading to a minimum transport in October, before the situation (i.e., pressure gradient between southern and northern parts of the Indonesian Seas) achieves a balance again.

As a consequence of these seasonally varying flow patterns in addition to the blocking character of the Sunda Island chain, the Indonesian Seas, particularly the southeastern regions, act as a buffer for ocean water. At the beginning of the boreal winter season, water masses pile up. The volume transport through the inflow straits—at this time essentially the Karimata Strait, Makassar Strait, Lifamatola Passage, and the Halmahera Sea—does not match the outflow volume transport passing essentially through the Malacca Strait, Sunda Strait (between Sumatra and Java), Lombok Strait, Ombai Strait, and Timor Passage.

### 2.3.2 Seasonality of temperature and salinity

The monsoon-induced seasonal variability is also visible in the horizontal pattern and variability of salinity and temperature in the Indonesian Seas. In Fig. 2.3, lower panels, time series of simulated sea surface salinity (SSS) and temperature (SST) are presented as spatial and monthly averages for different subregions in the Indonesia Seas. Mayer et al. (2018) verified these data using satellite-derived data. Strong seasonal signals are obvious in both SST and SSS time series for all regions. The SST variation with two peaks per year results mainly from the sun's zenith point crossing this region twice a year on its way from the southern Tropic of Capricorn in December to the northern Tropic of Cancer in June and back. This is also obvious from the horizontal SST distributions in the same figure, where the northern half of the Indonesian Seas is warmer than the southern one during northern summer and cooler during northern winter.

The SSS behavior is dominated by the seasonally varying circulation in connection with freshwater input from precipitation and rivers. For the Java Sea, fresher surface water passes from SCS in the west during northern monsoon (January) and saltier North Pacific water from the Makassar Strait during southern monsoon (July). This produces the largest variation compared with other regions in the Indonesian Seas. Second largest variation is visible in the Malacca Strait, where we also have a clear seasonality in the flow direction and strength but more influence of freshwater input. During the boreal summer months, the flow stagnates, and there is negligible water exchange with the adjacent saltier ocean regions like the Andaman Sea in the northwest and the SCS in the southeast. Due to the high rates of freshwater input from the atmosphere and from rivers (see Section 2.5) compared with freshwater losses (evaporation), we observe a freshening during this season. When the flow from SCS toward the Andaman Sea is accelerated during winter monsoon (period of the strongest throughflow in the Malacca Strait), comparably salty SCS water is transported into the strait. This is visible in the horizontal SSS distributions as well as in the time series, where the SSS values increase in this region at the end of each year.

### 2.3.3 Long-term development of sea surface temperature and sea surface salinity

Climate change has been detected already in the Indonesian Seas from observations and from model simulations over the past decades. Liu et al. (2015) investigated the geostrophic transport of ITF during 1984–2013 by analyzing repeated expendable bathythermograph measurements along a transect between the Sunda Strait (Indonesia) and Fremantle (Western Australia). They found an increase of the volume transport by about 1 Sv per decade and explain this with an increase of the trade winds over the Pacific Ocean. Furthermore, long-term changes in SST and SSS have been recognized from quasi-realistic numerical hindcast simulation results (Mayer et al., 2018) produced with the circulation model HAMSOM mentioned in the Introduction section. The trends for different subdomains within the Indonesian Seas were investigated for the period of 1965–2014 using their spatially averaged properties. They show an SST increase of 1.05, 1.47, 1.08, and 0.96°C per century for the inner Sunda Shelf, Gulf of Thailand, Malacca Strait, and Java Sea, respectively, and an SSS decrease of 1.34, 0.96, 2.82, and 1.79 psu per century for the same subdomains, respectively. Table 2.1 lists also these values.

Noteworthy, hindcast model results show already signals due to climate change. Obviously, the long-term development of temperature and salinity can deviate enormously among different regions and consequently from an overall (global) average.

## 2.4 Water residence times

For the investigation of marine environments and ecosystems, it is advantageous to know about exchange rates or residence times of water in the region of interest. This gives an idea about the regional distribution of time scales for a renewal of the corresponding water bodies.

One way of studying time scales of exchange of ocean water in certain regions is the application of a Lagrangian model, if measured data about transport rates are not available. With this type of model, we trace virtual particles within the four-dimensional model space using the simulated flow field. By this means, it is possible to follow them and determine the time they need to leave a certain region.

For the Indonesian Seas, there are hardly any estimations for the renewal of water so far, and none of them aims at gaining information on the water renewal for certain regions. Stansfield and Garrett (1997) applied a simple model and based their estimations additionally on observed salinity data. They found the Gulf of Thailand flushing rate to be in the range of half a year. Nugrahadi et al. (2013) estimated exchange rates for the Madura Strait solely using numerical models.

Here, we apply also two numerical models according to Mayer et al. (2015): the first step is the regional hydrodynamical model HAMSOM mentioned in the Introduction section. For the second step, the 3D daily averaged model results of HAMSOM were averaged for each month of all years 2000–12 to gain a mean velocity field typical for the particular month. For February and August as representatives for the NW monsoon and SE monsoon seasons, simulations with a Lagrangian tracer model were performed for 2 years to investigate theoretical time scales of water renewal on the entire Sunda Shelf. Fig. 2.5 presents the model results for tracers, which started at different depth levels, giving the residence times in days: i.e., the duration of each tracer (or water parcel) necessary to leave the Sunda Shelf area. Each tracer simulation was utilized with one particle for each model grid box.

In general, residence times show values of less than 30 days to more than 2 years. This depends on the magnitude of the velocity and on the distance to the exit of the region. Naturally, it takes more time to leave an area for particles starting far away than close to the

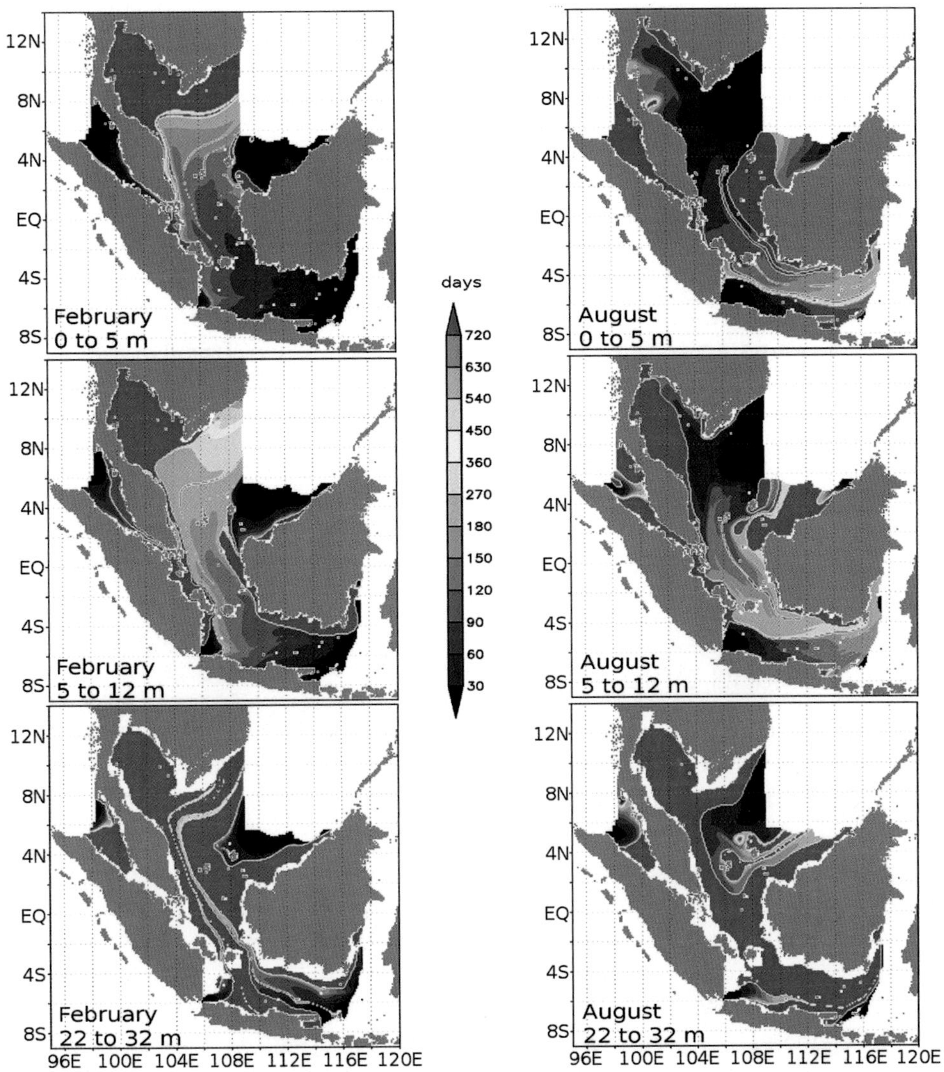


**FIGURE 2.5** Residence times in days of ocean water for the entire Sunda Shelf Sea: Residence times were calculated from tracer model results for the Sunda Shelf, forced with monthly mean velocity fields for February and August from the hydrodynamical model HAMSOM (Mayer et al., 2015), which were averaged for the period 2000 to 2012. February and August represent the ocean circulation during NW and SE monsoon seasons, respectively.

area's open boundary. Moreover, it is visible that the deepest layer, which is closest to the bottom, shows longer residence times because of slower velocities due to bottom friction.

For February, there is obviously a slow inflow into the southern SCS, off the Vietnamese coast, where we see long residence times. This inflow is moving mainly south- and eastward through the Java Sea, which shows faster exchange rates

(= decreasing residence times) toward its eastern boundary. A part of the SCS inflow, however, is retroflected and exiting the southern SCS, off the Borneo coast, where we see very short residence times. During this season of the year, the Gulf of Thailand is flushed very slowly, because the main flow passes the Gulf without entering it. In contrast, the Malacca Strait shows quite short residence times at least near the surface (bottom friction slows down the current speed again).

In August, the circulation is clearly different and with this, the distributions of the residence times as well. In the Java Sea, the main flow is directed from the east to the west leaving the Sunda Shelf at the location, where we detected an inflow in February. This is obvious from the very short residence times in the SCS off the Vietnamese coast. The Gulf of Thailand is also much more effected and flushed, because the northward flow in the SCS intrudes the Gulf region from the south. As in February, the main current in the Malacca Strait is directed northward with much lower magnitudes; hence, the surface layer seems to hardly move at all.

There are some thin lines of long residence times visible within areas of short residence times. Note, they are caused by tracers, which are transported directly to and around islands, where they are caught at the lee side (current shadow) extending their time to leave the region significantly.

## 2.5 Sources and sinks of freshwater

In the previous sections, we saw already that the subdomains of the Indonesian Seas are subject to different physical influences resulting in different oceanic regimes as can be seen, for example, from the different SSS levels (Fig. 2.3). In this section, we would like to add another aspect underlining the subdomains' various characteristics and an obvious signal of climate change: the strength of sources and sinks of freshwater in the regions and their long-term trends in the past 50 years (1965–2014).

The most important processes adding or removing freshwater to or from marine environments are rainfall (precipitation of freshwater), coastal river discharges, and evaporation: the first two being sources, and the latter being a sink of freshwater. Increasing or decreasing rainfall is usually accompanied with increasing or decreasing river discharge, which might, however, exhibit a certain delay depending on the properties of the land in the rivers' catchment area. Transport of higher or lower saline ocean water into a region can also be considered as an indirect sink or source of freshwater, but this will be ignored here, since it is only of secondary importance for our region of interest.

We have estimated the sums of these sources and sinks for the different subdomains over a long period. Precipitation data were directly taken from NCEP/NCAR reanalysis data (Kalnay et al., 1996). Evaporation is calculated according to the hydrodynamical model HAMSOM, where the evaporation rate is a function of SST, SSS, wind speed, air temperature, and relative (or alternatively specific) humidity. The data are indirectly validated by verification of the simulated SSS (Mayer et al., 2018). River discharge data were taken from the verified global hydrological model WaterGAP (Müller Schmied et al., 2016). There are no large-scale observational data available concerning evaporation rates or coastal river runoff. Therefore, we depend on these numerical model results for our estimations. Because of the performed verification, we can assume that they are realistic.

The freshwater balance can be expressed as freshwater volume transport in $m^3s^{-1}$. They are displayed in Fig. 2.6 as time series for the different regions for the period 1965–2014. At the right end of the graphs, the overall averages are marked. Note: The scales of the volume transport (y-axis) are varying with 0–300,000 $m^3\ s^{-1}$ for the Sunda Shelf Sea, 0–40,000 $m^3\ s^{-1}$ for the Malacca Strait, and 0–80,000 $m^3\ s^{-1}$ for the Gulf of Thailand and Java Sea.

It is interesting to compare the sources and the sink within each subdomain. For the Malacca Strait region, which sums up about half or less of the amounts of the other two subdomains Gulf of Thailand and Java Sea, the river discharge corresponds approximately to the precipitation rates, while the evaporation rates sum up to less than half of this. This region has a comparably small surface but a long coastline. Therefore, evaporation plays only a minor role. In contrast, evaporation nearly cancels out river runoff in the Gulf of Thailand and the Java Sea.

Long-term trends for all three constituents are clearly visible and result from an already ongoing change of climate. Straight lines in each graph in Fig. 2.6 display the linear trends; their values are listed in the legends just above the graphs and in Table 2.1. Most outstanding is the increase of precipitation during the years 1965–2014: For the entire inner Sunda Shelf, the Gulf of Thailand, the Malacca Strait, and the Java Sea, the trend in 100 years amounts to about 60%, 31%, 61%, and 118% of the average value, respectively, which corresponds to the value in 1990. Evaporation decreases by 34%, 10%, 23%, and 46% in 100 years, while the freshwater riverine input increases by 14%, 6%, 40%, and 15% of the corresponding average value. We see the strongest relative change toward freshening of the marine environment in the Java Sea, where the average sum from the respective three contributions amounts to a freshwater gain of 4.3 L $day^{-1}$ per $km^2$, the trend to +4.7 L $day^{-1}$ per $km^2$ per century. This is more than doubling, mainly due to an increase in precipitation, which is also more than doubling. As can be seen from the simulated long-term SSS time series in Fig. 2.3, lowest panel, this increasing freshwater input does not affect very much the SSS and, due to quite well mixed water, the salinity in the Java Sea, because this sea is a throughflow region with permanent ocean water sources from neighboring seas.

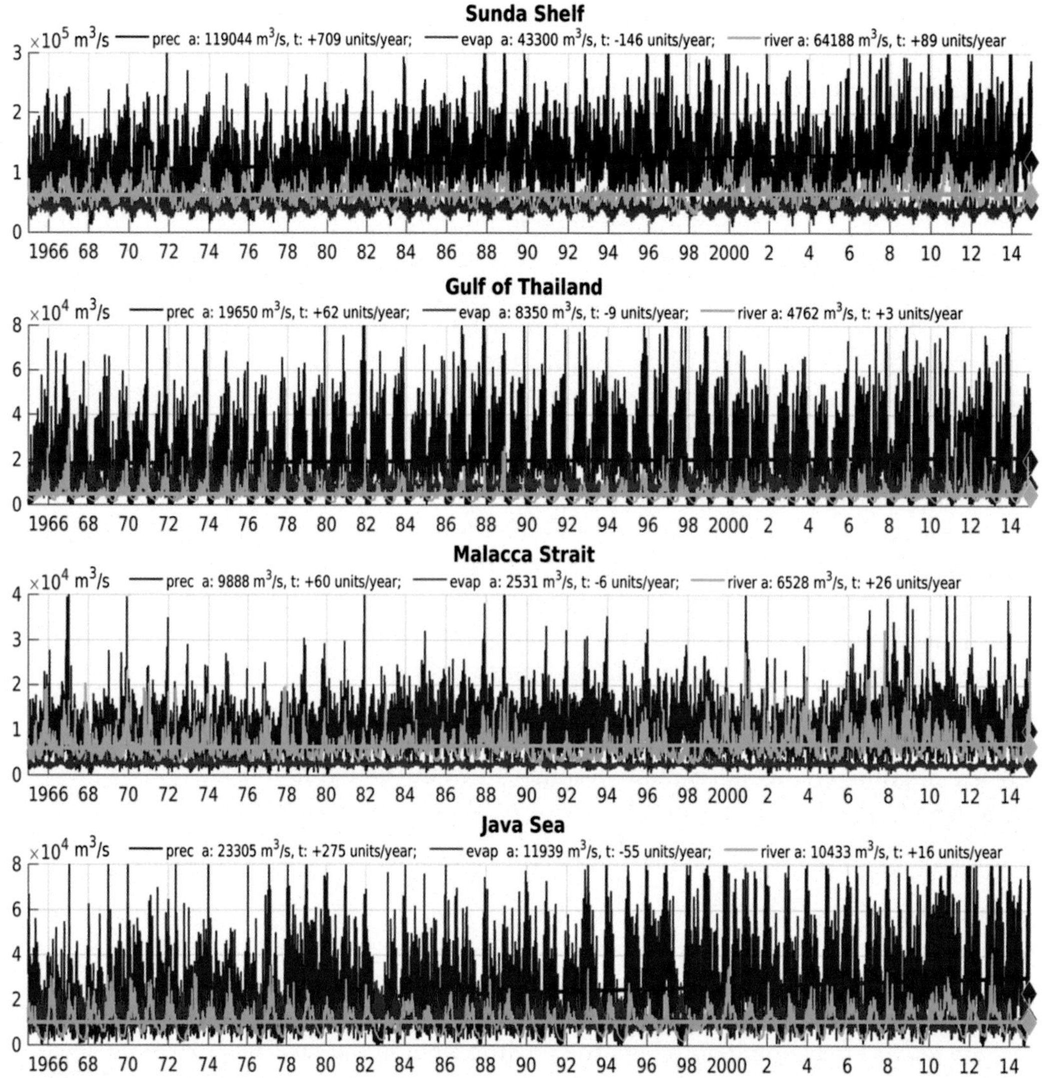


**FIGURE 2.6** Time series of total precipitation, evaporation, and river discharge for different subregions of the Indonesian Seas: In the legend line, "a" denotes the average value for 1965–2014, "t" denotes the linear trend calculated from these time series. Precipitation data were taken from NCEP/NCAR (Kalnay et al., 1996). Evaporation rates were calculated as done by the hydrodynamical model HAMSOM, where the rate is a function of SST, SSS, wind speed, air temperature, or relative or specific humidity. River discharge data were taken from the global hydrological model WaterGAP (Müller Schmied et al., 2016). Note: Scales for the freshwater volume flow are varying in the panels.

## 2.6 Remote sensing methods applied in coastal process studies

### 2.6.1 Available satellite data

In dynamically highly variable coastal areas, satellite data provide detailed synoptic information for coastal discharge studies. Sensors operating in the visible and infrared spectral range with different spatial and spectral resolutions receive electromagnetic radiation caused by watercolor or thermal emission modified by the atmosphere. From these radiometric quantities, different parameters are derived using atmospheric corrections and special algorithms. Examples are atmospheric corrected reflectance, concentration of optically active water constituent, true color images, or SST.

Ocean color sensors are the Sea-viewing Wide Field-of-view Sensor (SeaWiFS), the Moderate-resolution Imaging Spectro-radiometer (MODIS), Medium-Resolution Imaging Spectrometer (MERIS), Ocean and Land Color Instrument (OLCI), and the Visible Infrared Imaging Radiometer Suite (VIIRS). They require for global application wide paths (1000–2000 km) and several narrow spectral bands in the visible spectral range. NASA (National Aeronautics and Space Administration) operates the actually most important sensor MODIS on board the satellites, Terra (EOS AM, since 1999), and Aqua (EOS PM, since 2002), covering the earth every 1–3 days, both. MODIS collects data in 36 spectral bands between 0.4 and 14.4 μm with spatial resolutions of 250, 500, and 1000 m. Standard MODIS level 2 products with a spatial resolution of 1 km are atmospheric corrected water-leaving radiances, chlorophyll *a* (Chla) concentration, diffuse attenuation coefficient K490, and SST. Two channels with 250 m resolution are used to derive true color images (RGB) as a product. MODIS level 3 products with a 4 km × 4 km resolution were included for general applications. SeaWiFS operated with similar performance on board the OrbView-2-spacecraft (1997–2011). MERIS was operated from March 2002 until April 2012 by the European Space Agency (ESA) on board the Environmental Satellite (Envisat) covering the earth in 3 days. Full (300 m) and reduced resolution (1200 m) data with more spectral channels led to more level 2 products such as SST, Chla, suspended particulate matter (SPM), and colored dissolved organic matter (CDOM) absorption. The VIIRS sensor operating on Suomi National Polar-orbiting Partnership spacecraft (Suomi NPP) combines different sensors including the Advanced Very-High-Resolution Radiometer (AVHRR) of NOAA weather satellites for SST and MODIS. OLCI on Sentinel-3 A continues the MERIS application since February 2016. OLCI has a swath of 1270 km, a spatial resolution of 300 m, 21 spectral channels (400–1020 nm).

Landsat and SPOT with high spatial resolution developed for land applications were implemented for detailed investigations. Landsat series exists since 1982. Landsat 5 TM (Thematic Mapper), Landsat 7 ETM+ (Enhanced Thematic Mapper Plus), and Landsat 8 OLI (Operational Land Imager) are still operated by US Geological Survey (USGS). Landsat TM has 3 visible channels, 1 near infrared, 2 short infrared with 30 m resolution, and a thermal infrared channel with 120 m. ETM+ is additionally equipped with panchromatic band 0.5–0.9 μm (15 m) and thermal band (60 m). Landsat 8 OLI has four channels in the visible (30 m) and a panchromatic channel (15 m) und the thermal infrared sensor (TIRS) with 2 channels in the infrared (100 m). The swath width is 185 km and repeating rate 16 days. The commercial SPOT (Satellite Pour l'Observation de la Terre) satellites operated by French space agency CNES (Centre national d'études spatiales) since 1986 have spatial resolution of 8–20 m and a repetition rate of 23 days. Sentinel-2 is operated by ESA since 2015 with spatial resolution of 10 or 20 m in 10 spectral channels between 443 and 2190 nm.

Wind data of Topex/Poseidon (1992–2006), TMI (TRMM Microwave Imager, TRMM—Tropical Rainfall Measuring Mission, 1997–2015), and AMSR-E (Advanced Microwave Scanning Radiometer—Earth Observing System) on MODIS Aqua or QuikSCAT (Quick Scatterometer, 1999–2009) support climate studies. Furthermore, satellite-derived SSS (2009 composition of this book) is provided by NASA's Soil Moisture Active Passive (SMAP) mission and by ESA's Soil Moisture and Ocean Salinity (SMOS) mission (Bao et al., 2019; Ferster and Subrahmanyam, 2018).

### 2.6.2 Ocean color and its variation in Indonesian coastal waters

The Indonesian coastal waters are characterized by strong coastal discharge driven by high precipitation throughout the year. Monsoon and tides form a dynamical system highly variable in space and time. Therefore, satellite remote sensing is the only method to acquire synoptic observational information of larger areas. The large rivers of SE-Sumatra belong to the major tropical carbon and sediment sources for the world ocean. Knowledge about carbon sources is important because of their potential impact on coastal ecosystems and on climate change. Drainage of peatlands enhances the concentration of dissolved organic matter (DOM), and strong tidal currents increase the SPM load in rivers/estuaries reducing photosynthetically available radiation in the water column. On the other hand, rivers transport inorganic nutrients into the coastal regions inducing phytoplankton development expressed as Chla concentration.

The different water constituents and their absorption and scattering properties influence the watercolor, the basic information source for ocean color satellite applications. Satellite data of different spatial and temporal resolution allow the identification of sources of different water masses, the investigation of river discharge, and coastal transport processes in different spatial and temporal scales in relation to tidal and monsoon phases and to El Niño–Southern Oscillation (ENSO).

Watercolor results from the interaction between incident sun radiation and natural water and contains the information for the application of satellite data in the visible spectral range modified by the atmosphere. Color in terms of total internal spectral reflectance $\Re(\lambda)$ is the ratio of spectral upward radiance $L_u(\lambda)$ just beneath and spectral downward irradiance $E_s(\lambda)$ above the sea surface (Siegel et al., 2005; Siegel et al., 2009) according to following equation:

$$\Re(\lambda) = \frac{\pi L_u(\lambda)}{E_s(\lambda)} = \frac{t_d(R_w(\lambda))}{1 - r_u(R_w(\lambda))} \tag{2.2}$$

$\Re(\lambda)$ is proportional to water reflectance $R_w(\lambda)$, which is the ratio of total backscattering $b_b(\lambda)$ and total absorption coefficients $a(\lambda)$.

$$R_w(\lambda) = \frac{0.33\, b_b(\lambda)}{(a(\lambda) + b_b(\lambda))} \tag{2.3}$$

Total spectral absorption and backscattering coefficients consist of contributions of water and optically active water constituents.

$$b_b(\lambda) = b_{bw}(\lambda) + b_{bc}(\lambda) + b_{bp}(\lambda) \tag{2.4}$$

$$a(\lambda) = a_w(\lambda) + a_{ph}(\lambda) + a_d(\lambda) + a_y(\lambda) + a_s(\lambda) \tag{2.5}$$

The total backscattering coefficient $b_b(\lambda)$ consists of contributions by w (water molecules), c (chlorophyllous), and p (nonchlorophyll particles). The total absorption coefficient $a(\lambda)$ includes contribution of w (water), ph (phytoplankton), d (detritus), y (CDOM), and s (inorganic suspended matter). Particle scattering increases the reflectance in the entire spectral range and makes water turbid and bright. Watercolor is dominated by spectrally selective absorption of pure water, Chla, and CDOM (Siegel et al., 2005; Siegel et al., 2018). Examples are presented in Fig. 2.7. Absorption of pure seawater has a minimum in the blue spectral range and a maximum in the red leading to dark blue color in clear ocean water. Chla absorption has maxima near 443 and 670 nm. Absorption of CDOM increases exponentially to shorter wavelength. Each contribution consists of specific spectral absorption (Fig. 2.7, left diagram) multiplied by concentration of corresponding water constituent.

Chla, SPM, and CDOM are measured from water samples taken just beneath the sea surface and filtered through Whatman GFF filters. SPM is determined gravimetrically (Doerffer, 2002). Filters were weighted after filtration through preweighted filters. After filtration, Chla was extracted from the filters by ethanol and measured according to Lorenzen and Jeffrey (1980). CDOM absorption was estimated on filtered water (Højerslev, 1980). Photometric measurements were performed on site using the PC-Spec Photometer (380–700 nm). Transparency was determined in terms of Secchi disc depth.

Rivers of SE-Sumatra transport high amounts of dissolved carbon and sediment into the ocean, leading to high variable concentration and composition of optically active water constituents. Biooptical in situ measurements were performed in the period 2004–13 in five major rivers of SE Sumatra (Rokan, Siak, Kampar, Indragiri, and Musi). These large rivers are 350–600 km long and in river mouth areas a few kilometers wide

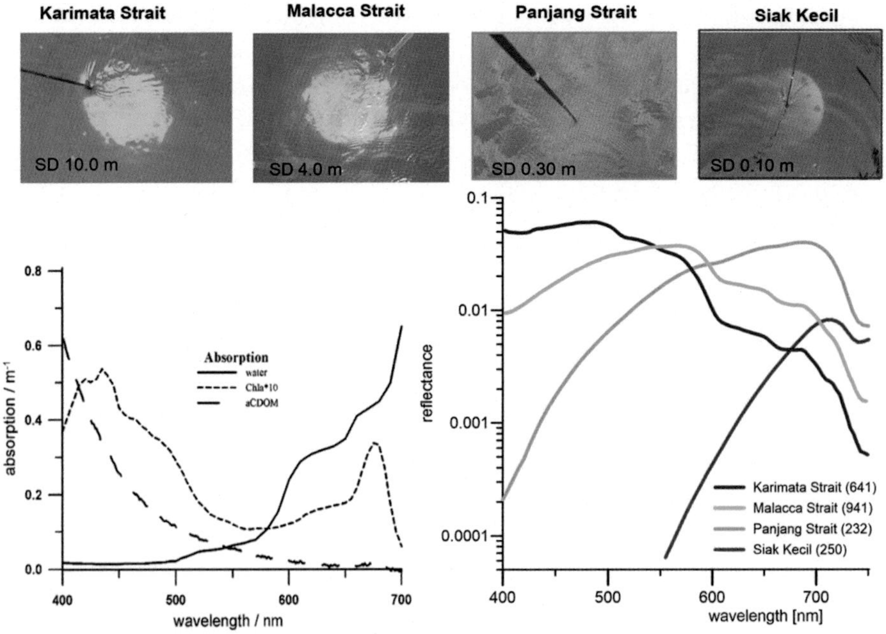


**FIGURE 2.7** Light properties of four different water masses in the Indonesian Seas: Specific absorption of pure water, Chla and CDOM (left diagram), photographs (upper panel), and reflectance (right diagram) of four different water masses (Karimata, Malacca, and Panjang Straits, and Siak Kecil). CDOM: colored dissolved organic matter; Chla: chlorophyll a. *Modified after Siegel et al. (2018).*

(Siegel et al., 2018). The rivers drain different landscapes from highlands to coastal plain with peatlands. The northern rivers, Rokan, Siak, and Kampar, drain into the Malacca Strait and southern rivers from Indragiri to Musi into the Karimata Strait. Rokan and Kampar are particularly exposed to tides because of the orientation of their shallow river estuaries. The Siak River System is the main DOC source in the region, particularly the source river Tapung Kanan and the tributaries Mandau, Siak Kecil, and Bukit Batu (Hendiarti et al., 2005, p. 5; Siegel et al., 2009).

Chla, SPM, and CDOM distributions of SE-Sumatra are described in Fig. 3 and Table 1 in Siegel et al. (2018). They are separated for rivers (salinity = 0), estuaries, and adjacent seas with mean values and variation range. Chla is an indicator for phytoplankton activity, SPM provides information on erosion or soil leaching, and CDOM loads about catchment areas. In the freshwater part of the rivers, Chla varied between 19.3 mg $m^{-3}$ in the Rokan River and 0.1 mg $m^{-3}$ in the industrial area of the Siak. In the mixing zone in front of the river mouth areas, nutrients increase Chla with maxima of 77.6 and 13.9 mg $m^{-3}$ off Rokan and Musi. Highest SPM content occurred in Rokan with 776 g $m^{-3}$. In the estuary of Rokan, 2200 g $m^{-3}$ was reached, 348 g $m^{-3}$ in Kampar, and 216 g $m^{-3}$ in Indragiri. CDOM, the absorbing part of DOM, was detected by CDOM absorption at 440 nm (aCDOM440). Highest mean CDOM absorption of 33.6 $m^{-1}$ occurred in the Siak

System followed by Rokan (14.8 $m^{-1}$), and Indragiri (13.1 $m^{-1}$). Maximum value of 51.8 $m^{-1}$ in Siak Kecil exceeded the known global maximum (Siegel et al., 2009) underlining the global importance as DOC sources. The lowest concentrations were observed in the open Karimata Strait with values partly in the range of ocean water.

Spectral reflectance of four different regions, Karimata Strait, Malacca Strait, Panjang Strait, and Siak Kecil from the Siak River System in Fig. 2.7 (right diagram), represents the high variation range (Siegel et al., 2005, 2018). Photographs of Secchi discs illustrate the variation of watercolor. Karimata Strait represents rather blue water with low concentrations and a reflectance maximum between 450 and 500 nm. The maximum shifted to 530–550 nm in the Malacca Strait. In the Panjang Strait, high SPM content and background CDOM led to highest reflectance between 550 and 680 nm and a strong slope in the short wavelength range reflecting turbid yellow-brownish water. In peat-draining rivers, very high CDOM absorption shifted the reflectance maximum to 700 nm, which made them so-called black water rivers. Transparency varied between 10 m in the Karimata Strait and 10 cm in the river.

In summary, extremely high in situ measured CDOM absorption of more than 50 $m^{-1}$ in the Siak River System identified them as black water rivers. This fact and extremely high SPM content (up to 2200 g $m^{-3}$ in the Rokan estuary) verify that the rivers of SE Sumatra belong to the major carbon and sediment sources for the world ocean. The northern rivers Rokan, Siak, and Kampar discharge into the Malacca Strait, and Indragiri, Batang Hari, and Musi into the western Karimata Strait. Different catchment areas like peatland and soil leaching areas determine the composition of the discharged water producing ocean color from clear black to turbid, milky waters.

## 2.6.3 Satellite-based studies of phytoplankton and coastal processes

### *2.6.3.1 Distribution of phytoplankton*

Studies of phytoplankton development in the Malacca Strait were performed by Tan et al. (2006) using SeaWiFS data. They compared satellite-derived chlorophyll with in situ measurements and verified an overestimation.

Hendiarti et al. (2004) investigated the throughflow through the Sunda Strait (southeast Sumatra) and coastal discharge into the western Java Sea using SeaWiFS data and the relation of both to Chla distribution and fish catches. Particularly around Java, the relation to the Monsoon phases was derived. During typical years, upwelling events were observed along the southern coast of Java during the southeast monsoon (June to October) with maximum in September. In the upwelling area, Chla and SST are 0.6–1.5 mg $m^{-3}$ and 25–28°C, respectively, and fronts coincided. The cold water transported nutrients into the euphotic zone and initiated phytoplankton growth. During El Niño in summer 1997, very strong upwelling occurred lasting until October–November with concentrations of 4 mg $m^{-3}$ (Susanto et al., 2006). In 1998, a year after the El Niño, intensity and extent of upwelling were much smaller than in normal years.

Satellite-derived Chla distributions off Southeast Sumatra show elevated concentrations along the coastline and in front of river plumes (Siegel et al., 2018).

### *2.6.3.2 Coastal discharge and influence of tidal and monsoon phases*

Coastal discharge of SE Sumatra is driven by heavy rain throughout the year, and transport processes are forced by monsoon and superimposed by tides. Tidal influence was studied in detail in the Siak River System (Siegel et al., 2009, 2018), where the rivers Siak, Siak Kecil, and Bukit Batu discharge into the Bengkalis Strait, the connection to the Malacca Strait. During low tide, Siak water propagates northward, mixes with Siak Kecil and Bukit Batu, and enters the Malacca Strait as shown in the quasi-true color image of Landsat 8 OLI from July 12, 2016 (Fig. 2.8, upper left panel). During increasing tide, Siak discharge stops, estuarine water may enter the river, and water of the last ebb phase is

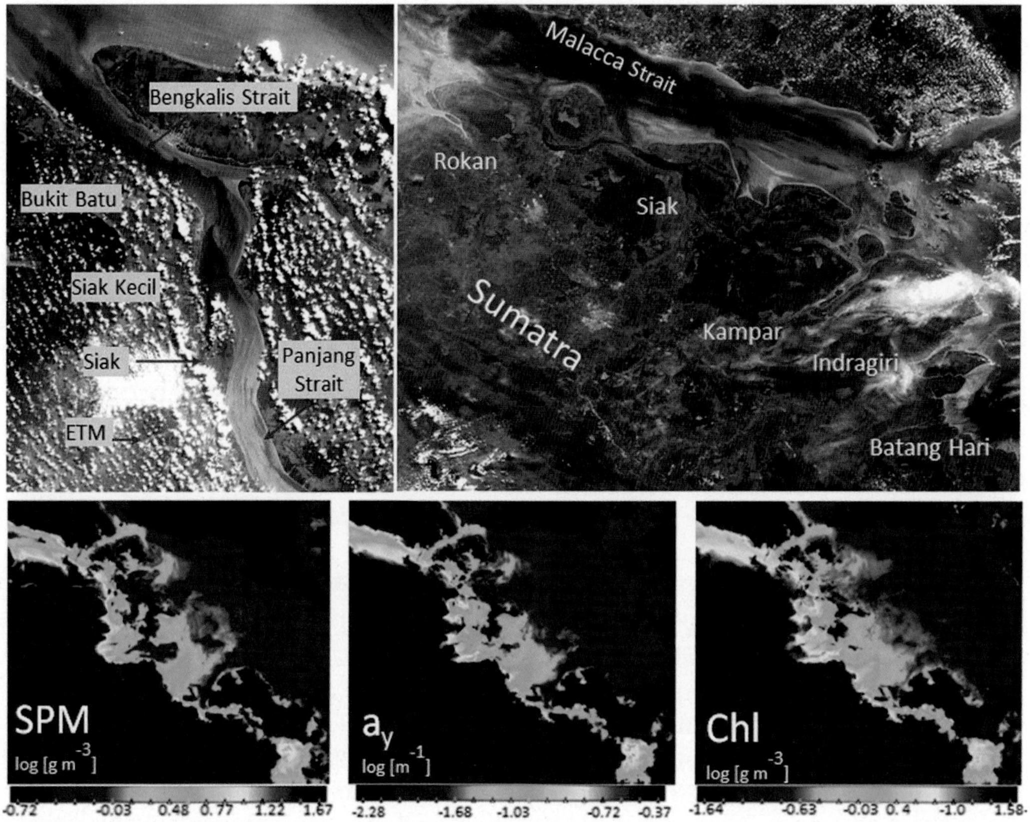


**FIGURE 2.8** Satellite images of riverine discharge: Upper left panel: Landsat 8 OLI quasi-true color image on July 12, 2016, of the Siak River System, the estuary and the Malacca Strait, showing the river plumes of Siak, Siak Kecil, and Bukit Batu. Upper right panel: MERIS true color image on May 8, 2011, showing the Rivers from Rokan to Batang Hari. Lower panels: MERIS-derived Chla, SPM, and aCDOM440, on November 4, 2008 (NW monsoon, IOW algorithm). aCDOM: colored dissolved organic matter absorption; Chla: chlorophyll a; OLI: Operational Land Imager; SPM: suspended particulate matter.

transported further southward into the Panjang Strait. This was documented by an image showing two plumes in the Panjang Strait with a distance of about 80–85 km. Satellite scenes combined with in situ observations verified the occurrence of a permanent estuarine turbidity maximum (ETM) zone in the Siak River. During both tidal phases, the tidal current in the entire water column initiated SPM resuspension at the saline front and consequently higher spectral reflectance. The ETM is described in detail in Siegel et al. (2009), but also indicated in Fig. 2.8 (upper left panel).

During all monsoon phases, satellite images confirm that the water of the northern rivers (Rokan, Siak, and Kampar) discharging into the Malacca Strait follows the general northwestward transport (Siegel et al., 2018; Wyrtki, 1961). The baroclinic pressure gradient between both entrances of the Malacca Strait initiates the transport supported by monsoon phases. Reduced sea level in the Andaman Sea (north of Malacca Strait) during NE winds and SE winds strengthen the northwest transport. Clear water of the Karimata Strait enters the Malacca Strait and improves the water quality there. Spreading of southern rivers (Indragiri, Musi) in the Karimata Strait is strongly related to monsoon phases. During northerly winds, Indragiri water is transported southward and during southeasterly winds northward. In the Musi River region, northerly winds guide small rivers south- and eastward. During southerly winds, the discharge including the Musi water spreads northward.

MERIS-derived products were used to compare the distribution of Chla, SPM, CDOM absorption, and DOC. CDOM and DOC were generated on the basis of own empirical algorithms. During northeasterly winds, strong gradients occurred particularly in CDOM with very high values in the river mouth area and in a coastal strip along the northern rivers and low values in the coastal area near the Musi river mouth. Distribution of SPM is different to the distribution of Chla, aCDOM440, and DOC. The southern rivers do not carry high SPM content.

MERIS products showed during both monsoon phases high CDOM, Chla, and TSM in the coastal strip of Sumatra with a larger lateral spread during the SE monsoon (Siegel et al., 2018, Fig. 10). Year-round, cells of clear water occur in the basin between Kampar, Indragiri, and offshore islands. This has also been detected in measured concentrations.

VIIRS-derived RGB and level 2 products are presented in Fig. 4 in Siegel et al. (2018). SST shows strong heating in shallow areas independent on the river plumes. Chla is rather high in contrast to measurements, which seems to be a problem of atmospheric correction or influence of high SPM and CDOM in the Chla algorithms.

#### *2.6.3.3 Climatological aspects*

Based on satellite-derived wind, precipitation, and SST, a comparison between the Malacca and Karimata Straits was performed for the period 2004–09 (Siegel et al., 2018, Fig. 6). NW monsoon occurs from December to March, SE monsoon from May to September, and intermonsoon periods in April and October/November. In the Malacca Strait, wind speed has no significant seasonal cycle, and slight maxima occurs at end of the intermonsoon period and beginning of the NW monsoon in November/December.

Monthly precipitation of both regions indicates dryer SE and wetter NW monsoonal seasons. The maximum correlates with the transition to NW monsoon and increasing winds. The minimum appears during the SE monsoon. In the Malacca Strait, the seasonal cycle of SST consists of a minimum in December/January and a maximum in March/April. The increase from minimum (28°C) to maximum (32°C) occurs quickly during low wind in January/February. After the maximum, SST decreases slowly during SE monsoon and stronger than in the following intermonsoon period of maximum wind speed. The maxima are higher than in the Karimata Strait, where a correlation exists between decreasing wind and increasing SST. In the Karimata Strait, minima in wind speed occur in March/April and October/November in the intermonsoon periods and maxima in December/January and July/August in the monsoon phases. Precipitation has a minimum during SE monsoon. SST minima occur in December/January and August/September and maxima in April/May and October/November. With increasing wind speed, wind mixing increases and SST decreases. The NW monsoon transports cold water from the SCS into the Karimata Strait leading to SST minimum.

Chla and SST of both regions have an influence on the ecosystem. Chla maxima occur in December/January and minima in March to August in the Malacca Strait. The maxima correspond to the SST minimum and the SST maxima in March/April to the Chla minima. In the Karimata Strait, the annual cycles of SST consist of two periods, and Chla follows in the opposite direction. High wind speed generates strong mixing with decreasing SST and transport of nutrients into the euphotic zone inducing phytoplankton growth. Low wind means little mixing, strong stratification, high SST, nutrient depletion in the surface layer, and reduced Chla.

In the Malacca Strait, the concentration gradients are more pronounced during NW than during SE monsoon (Siegel et al., 2018, Fig. 11). The concentrations are lower on the Malaysian than on the Indonesian side, because the NW monsoon transports the surface water toward the Indonesian coast and creates upwelling on the northern side. During the SE monsoon, the nutrient-rich coastal runoff of Sumatra is more distributed in the Malacca Strait, and higher concentrations are observed, especially in Chla. Seasonal means of MODIS-derived absorption of Gelbstoff (CDOM) + detritus, Chla, and SST are compared for SE and NW monsoon. Monthly precipitation in SE Sumatra indicates dryer SE and wetter NW monsoon season. Higher precipitation enhances coastal discharge. During NW monsoon, currents prevent offshore propagation of Indonesian coastal discharge. During SE monsoon, weaker winds support their offshore transport. During SE monsoon, coral reefs or touristic areas at Riau Islands between Malacca und Karimata Straits are more influenced by coastal discharge.

The archipelago of Indonesia plays a central role as an atmospheric source of heat in the coupled ocean–atmosphere circulation system of the tropical Pacific, ENSO. ENSO has three phases: the normal state, El Niño, and La Niña. El Niño reduces the rainfall, and La Niña brings stronger Pacific trade winds, higher SST, and precipitation. Higher precipitation strengthens transport of CDOM into the sea. MODIS derived CDOM absorption for the boreal summer (SE monsoon) of La Niña year 2010 and El Niño year

2015. SE winds force an offshore transport of coastal discharge in the Malacca Strait. Particularly, the higher precipitation during the La Niña year 2010 leads to higher CDOM values in the Malacca Strait.

Knowledge gaps and directions of future research

- The seasonally varying amount of river discharge is estimated only from model simulations. There are hardly measurements about the rivers' runoff, which influences the coastal ocean circulation.
- The physical processes driving coastal and near-coastal upwelling in the Indonesian Seas are not well understood and require further studies. Our previous results for the SCS have shown that the traditional view of a purely wind driven upwelling cannot fully explain the performance of near-coastal upwelling systems.
- In aquaculture areas (onshore and offshore), it is essential to know the local ocean circulation and water exchange rates to protect the natural marine environment from high loads of nutrients and other chemical substances.
- The regional effects of climate change may deviate enormously from global averages. Particularly for the Indonesian Seas, there is still a large uncertainty about future changes of the onset time and intensity of the winter and summer monsoons, and the effects on local conditions, namely the local seawater level, the frequency and intensity of flood events, the ocean circulation as well as the local temperature and salinity.

Implications/recommendations for policy and society

A sustainable management of coastal waters and their resources such as the maintenance of water quality or the assessment of productivity and fish stocks require an improved understanding of the local coastal circulation and physical upwelling processes. Therefore, the following points are suggested:

- An operational monitoring system is required to obtain reliable amounts of the seasonally varying freshwater discharge of the most relevant Indonesian rivers. The data should be available online in real time to the scientific community and organizations involved in the management of coastal zones.
- A scientific investigation of the underlying physical processes is needed to allow a prediction of the existence and strength of upwelling phenomena on a monthly or seasonal scale.
- An operational hydrodynamical (and wave) model system (ocean forecasting system) of the entire Indonesian Seas should be established. In a later stage, it should include ecosystem dynamics. It allows the short-term prediction of flood events and a consequent tracing of oil and other hazardous substances released from ships, oil platforms, and land.
- A permanently operating network of oceanographic moorings at certain key positions also within the ITF routes should be installed for the observation of physical parameters.

*Continued*

**Implications/recommendations for policy and society—cont'd**

The data are needed to validate and improve the operational model system for the ocean interior complementing surface information provided by satellite images.

- The results of global climate models (coupled ocean–atmosphere models) need to be translated into regional information ("regionalization"). Regional fine resolution climate models shall be driven by global climate model results to provide estimates of the regional and local sea level rise, of the increase in the strength and frequency of storm surges and flood events, and of the increase of water temperature.

# Acknowledgments

The authors like to thank scientists and students of University of Riau in Pekanbaru and Palembang, of the Hasanuddin University in Makassar, and of the Institute of Technology (ITB) in Bandung as well as scientists of the Indonesian Ministry for Research (RISTEK) and for Marine Affairs and Fisheries (KKP) for their great support before, during, and after our research cruises. The studies were funded by the Federal German Ministry for Education and Research under grants no. 03F0392E-IOW, 03F0392D, 03F0473E-IOW, 03F0473D, 03F0642C-IOW, and 03F0642D in the frame of the German–Indonesian cooperation SPICE.

# References

van Aken, H.M., 2005. Dutch oceanographic research in Indonesia in colonial times. Oceanography 18 (4), 30–41.

van Aken, H.M., Brodjonegoro, I.S., Jaya, I., 2009. The deep-water motion through the Lifamatola Passage and its contribution to the Indonesian throughflow. Deep-Sea Research I 56, 1203–1216. https://doi.org/10.1016/j.dsr.2009.02.001.

Bao, S., Wang, H., Zhang, R., Yan, H., Chen, J., 2019. Comparison of satellite-derived sea surface salinity products from SMOS, Aquarius, and SMAP. Journal of Geophysical Research: Oceans 124 (3), 1932–1944. https://doi.org/10.1029/2019JC014937. https://agupubs.onlinelibrary.wiley.com/doi/pdf/10.1029/2019JC014937.

Doerffer, R., 2002. Protocols for the validation of MERIS water products. In: Delwart, S., Huot, J.P. (Eds.), PO-TNMEL-GS-0043. ESA publications.

Egbert, G.D., Erofeeva, S.Y., 2002. Efficient inverse modeling of barotropic ocean tides. Journal of Atmospheric and Oceanic Technology 19 (2), 183–204. https://doi.org/10.1175/1520-0426(2002)019(0183:EIMOBO)2.0.CO;2.

Fang, G., Susanto, R.D., Wirasantosa, S., Qiao, F., Supangat, A., Fan, B., et al., 2010. Volume, heat, and freshwater transports from the South China Sea to Indonesian seas in the boreal winter of 2007-2008. Journal of Geophysical Research: Oceans 115 (C12). https://doi.org/10.1029/2010JC006225. https://agupubs.onlinelibrary.wiley.com/doi/pdf/10.1029/2010JC006225. https://agupubs.onlinelibrary.wiley.com/doi/abs/10.1029/2010JC006225.

Fang, G., Wang, Y., Wei, Z., Fang, Y., Qiao, F., Hu, X., 2009. Interocean circulation and heat and freshwater budgets of the South China Sea based on a numerical model. Dynamics of Atmospheres and Oceans 47 (1), 55–72. https://doi.org/10.1016/j.dynatmoce.2008.09.003. ISSN 0377-0265. http://www.sciencedirect.com/science/article/pii/S0377026508000596 (The South China Sea and its impact on climate).

Feng, X., Liu, H.L., Wang, F.C., Yu, Y.Q., Yuan, D.L., 2013. Indonesian throughflow in an eddy-resolving ocean model. Chinese Science Bulletin 58 (35), 4504–4514. https://doi.org/10.1007/s11434-013-5988-7.

Ferster, B.S., Subrahmanyam, B., June 2018. A comparison of satellite-derived sea surface salinity and salt fluxes in the southern ocean. Remote Sensing in Earth Systems Sciences 1 (1), 1–13. https://doi.org/10.1007/s41976-018-0001-5. ISSN 2520-8209.

Godfrey, J.S., 1996. The effect of the Indonesian throughflow on ocean circulation and heat exchange with the atmosphere: a review. Journal of Geophysical Research 101 (C5), 12217–12237.

Gordon, A.L., 2005. Oceanography of the Indonesian seas and their throughflow. Oceanography 18 (4), 14–27.

Gordon, A.L., Huber, B.A., Metzger, E.J., Susanto, R.D., Hurlburt, H.E., Adi, T.R., 2012. South China Sea throughflow impact on the Indonesian throughflow. Geophysical Research Letters 39 (11). https://doi.org/10.1029/2012GL052021.L11602.

Gordon, A.L., Sprintall, J., van Aken, H.M., Susanto, D., Wijffels, S., Molcard, R., et al., 2010. The Indonesian throughflow during 2004-2006 as observed by the INSTANT program. Dynamics of Atmospheres and Oceans 50 (2), 115–128. https://doi.org/10.1016/j.dynatmoce.2009.12.002. ISSN 0377- 0265. http://www.sciencedirect.com/science/article/pii/S0377026509000724 (Modeling and Observing the Indonesian Throughflow).

Hendiarti, N., Siegel, H., Frederik, M.C., Reissmann, J., Andiastuti, R., 2005. Remote Sensing Investigation of Coastal Discharge of Siak Estuary and Nearby Jakarta Bay, Proceedings of the 16th APEC Workshop on Ocean Models and Information System for the APEC Region. Ho Chi Minh City. OMISAR Project Publication.

Hendiarti, N., Siegel, H., Ohde, T., 2004. Investigation of different coastal processes in Indonesian waters using SeaWiFS data. Deep Sea Research Part II: Topical Studies in Oceanography 51 (1–3), 85–97.

Højerslev, N.K., 1980. On the Origin of Yellow Substances in marine Environments. Rep. No 42. Institute of Physical Oceanography.

Kalnay, E., Kanamitsu, M., Kistler, R., Collins, W., Deaven, D., Gandin, L., et al., 1996. The NCEP/NCAR reanalysis 40-year project. Bulletin American Meteorology Society 77, 437–471.

Koch-Larrouy, A., Lengaigne, M., Terray, P., Madec, G., Masson, S., 2010. Tidal mixing in the Indonesian Seas and its effect on the tropical climate system. Climate Dynamics 34 (6), 891–904. https://doi.org/10.1007/s00382-009-0642-4. ISSN 1432-0894.

Liu, Q.Y., Feng, M., Wang, D., Wijffels, S., 2015. Interannual variability of the Indonesian throughflow transport: a revisit based on 30 year expendable bathythermograph data. Journal of Geophysical Research: Oceans 120 (12), 8270–8282. https://doi.org/10.1002/2015JC011351. ISSN 2169-9291.

Lorenzen, C., Jeffrey, S., 1980. Determination of chlorophyll in seawater. UNESCO Technical Papers in Marine Science 35 (1), 1–20.

Mayer, B., Rixen, T., Pohlmann, T., 2018. The spatial and temporal variability of air-sea $CO_2$ fluxes and the effect of net coral reef calcification in the Indonesian seas: a numerical sensitivity study. Frontiers in Marine Science 5. https://doi.org/10.3389/fmars.2018.00116. https://www.frontiersin.org/article/10.3389/fmars.2018.00116.

Mayer, B., Stacke, T., Stottmeister, I., Pohlmann, T., 2015. Sunda shelf seas: flushing rates and residence times. Ocean Science Discussions 12, 863–895. https://doi.org/10.5194/osd-12-863-2015. https://www.ocean-sci-discuss.net/12/863/2015/.

Müller Schmied, H., Adam, L., Eisner, S., Fink, G., Flörke, M., Kim, H., et al., 2016. Variations of global and continental water balance components as impacted by climate forcing uncertainty and human water use. Hydrology and Earth System Sciences 20 (7), 2877–2898. https://doi.org/10.5194/hess-20-2877-2016. https://www.hydrol-earth-syst-sci.net/20/2877/2016/.

Nagai, T., Hibiya, T., 2015. Internal tides and associated vertical mixing in the Indonesian Archipelago. Journal of Geophysical Research: Oceans 120 (5), 3373–3390. https://doi.org/10.1002/2014JC010592. ISSN 2169-9291.

Nagai, T., Hibiya, T., Bouruet-Aubertot, P., 2017. Nonhydrostatic simulations of tide-induced mixing in the Halmahera Sea: a possible role in the transformation of the Indonesian throughflow waters. Journal of Geophysical Research: Oceans 122 (11), 8933–8943. https://doi.org/10.1002/2017JC013381. ISSN 2169-9291.

Nugrahadi, M.S., Duwe, K., Schroeder, F., Goldmann, D., 2013. Seasonal variability of the water residence time in the Madura Strait, East Java, Indonesia. Asian Journal of Water, Environment and Pollution 10 (1), 117–128.

Nugroho, D., Koch-Larrouy, A., Gaspar, P., Lyard, F., Reffray, G., Tranchant, B., 2017. Modelling explicit tides in the Indonesian seas: an important process for surface seawater properties. Marine Pollution Bulletin. https://doi.org/10.1016/j.marpolbul.2017.06.033. http://www.sciencedirect.com/science/article/pii/S0025326X17305131. ISSN 0025-326X.

van Oldenborgh, G., Collins, M., Arblaster, J., Christensen, J.H., Marotzke, J., Power, S., et al., 2013. Annex I: atlas of global and regional climate projections. Climate Change 1311–1393.

Oppo, D.W., Rosenthal, Y., 2010. The great Indo-Pacific Communicator. Science 328 (5985), 1492–1494. https://doi.org/10.1126/science.1187273. http://science.sciencemag.org/content/328/5985/1492.full.pdf. http://science.sciencemag.org/content/328/5985/1492. ISSN 0036-8075.

Pariwono, J.I., Ilahude, A.G., Hutomo, M., 2005. Oceanography of the Indonesian seas. Oceanography 18 (4), 42–49.

Ray, R.D., Egbert, G.D., Erofeeva, S.Y., 2005. A brief overview of tides in the Indonesian seas. Oceanography 18 (4), 74–79.

Ray, R.D., Susanto, R.D., 2016. Tidal mixing signatures in the Indonesian seas from high-resolution sea surface temperature data. Geophysical Research Letters 43 (15), 8115¢8123. https://doi.org/10.1002/2016GL069485. https://agupubs.onlinelibrary.wiley.com/doi/pdf/10.1002/2016GL069485. https://agupubs.onlinelibrary.wiley.com/doi/abs/10.1002/2016GL069485.

Siegel, H., Gerth, M., Ohde, T., Heene, T., 2005. Ocean colour remote sensing relevant water constituents and optical properties of the Baltic Sea. International Journal of Remote Sensing 26 (2), 315–330.

Siegel, H., Stottmeister, I., Gerth, M., Baum, A., Samiaji, J., 2018. Remote sensing of coastal discharge of SE Sumatra (Indonesia). In: Barale, V., Gade, M. (Eds.), Remote Sensing of the Asian Seas. Springer.

Siegel, H., Stottmeister, I., Reißmann, J., Gerth, M., Jose, C., Samiaji, J., 2009. Siak River system? East-Sumatra: characterisation of sources, estuarine processes, and discharge into the Malacca Strait. Journal of Marine Systems 77 (1–2), 148–159.

Sprintall, J., Gordon, A.L., Koch-Larrouy, A., Lee, T., Potemra, J.T., Pujiana, K., et al., 2014. The Indonesian seas and their role in the coupled ocean–climate system. Nature Geoscience 7, 487. https://doi.org/10.1038/ngeo2188.

Sprintall, J., Wijffels, S., Gordon, A.L., Ffield, A., Molcard, R., Susanto, R.D., et al., 2004. Instant: New international array to measure the Indonesian throughflow. Eos, Transactions – American Geophysical Union 85 (39), 369–376.

Sprintall, J., Wijffels, S.E., Molcard, R., 2009. Direct estimates of the Indonesian throughflow entering the Indian ocean: 2004–2006. Journal of Geophysical Research 114 (C07001), 1–58.

Stansfield, K., Garrett, C., 1997. Implications of the salt and heat budgets of the Gulf of Thailand. Journal of Marine Research 55 (5), 935–963. https://doi.org/10.1357/0022240973224184. ISSN 0022-2402. http://tinyurl.sfx.mpg.de/u2ve.

Storto, A., Masina, S., 2016a. C-GLORSv5: an improved multipurpose global ocean eddy-permitting physical reanalysis. Earth System Science Data 8 (2), 679–696. https://doi.org/10.5194/essd-8-679-2016. https://www.earth-syst-sci-data.net/8/679/2016/.

Storto, A., Masina, S., 2016b. The CMCC Eddy-Permitting Global Ocean Physical Reanalysis (C-GLORS v5, 1980-2014). https://doi.org/10.1594/PANGAEA.857995.

Susanto, R.D., Gordon, A.L., Zheng, Q., 2001. Upwelling along the coasts of Java and Sumatra and its relation to ENSO. Geophysical Research Letters 28 (8), 1599–1602. https://doi.org/10.1029/2000GL011844. https://agupubs.onlinelibrary.wiley.com/doi/pdf/10.1029/2000GL011844. https://agupubs.onlinelibrary.wiley.com/doi/abs/10.1029/2000GL011844.

Susanto, R.D., Marra, J., 2005. Effect of the 1997/98 El Niño on chlorophyll a variability along the southern coasts of Java and Sumatra. Oceanography 18. https://doi.org/10.5670/oceanog.2005.13.

Susanto, R.D., Moore, T., Marra, J., 2006. Ocean color variability in the Indonesian Seas during the SeaWiFS era. Geochemistry, Geophysics, Geosystems 7 (5). https://doi.org/10.1029/2005GC001009. https://agupubs.onlinelibrary.wiley.com/doi/pdf/10.1029/2005GC001009. https://agupubs.onlinelibrary.wiley.com/doi/abs/10.1029/2005GC001009.

Susanto, R.D., Wei, Z., Adi, R.T., Fan, B., Li, S., Fang, G., 2013. Observations of the Karimata Strait througflow from December 2007 to November 2008. Acta Oceanologica Sinica 32 (5), 1–6. https://doi.org/10.1007/s13131-013-0307-3.

Susanto, R.D., Wei, Z., Adi, T.R., Zheng, Q., Fang, F., Fan, B., et al., 2016. Oceanography surrounding Krakatau volcano in the Sunda Strait, Indonesia. Oceanography 29. https://doi.org/10.5670/oceanog.2016.31.

Tan, C.K., Ishizaka, J., Matsumura, S., Yusoff, F.M., Mohamed, M.I.H., 2006. Seasonal variability of SeaWiFS chlorophyll a in the Malacca straits in relation to Asian monsoon. Continental Shelf Research 26 (2), 168–178.

Wyrtki, K., 1961. Physical Oceanography of the Southeast Asian Waters. Scientific Results of Marine Investigation as of the South China Sea and the Gulf of Thailand. NAGA Rept. 2. Univ. Calif.

3

# Human interventions in rivers and estuaries of Java and Sumatra

Tim C. Jennerjahn[1,2], Antje Baum[1], Ario Damar[3], Michael Flitner[4], Jill Heyde[4], Ingo Jänen[1], Martin C. Lukas[1,4], Muhammad Lukman[5], Mochamad Saleh Nugrahadi[6], Tim Rixen[1], Joko Samiaji[7], Friedhelm Schroeder[8]

[1]LEIBNIZ CENTRE FOR TROPICAL MARINE RESEARCH (ZMT), BREMEN, GERMANY; [2]FACULTY OF GEOSCIENCE, UNIVERSITY OF BREMEN, BREMEN, GERMANY; [3]CENTER FOR COASTAL AND MARINE RESOURCES STUDIES, BOGOR AGRICULTURAL UNIVERSITY (IPB UNIVERSITY), INDONESIA; [4]SUSTAINABILITY RESEARCH CENTER (ARTEC), UNIVERSITY OF BREMEN, GERMANY; [5]DEPARTMENT OF MARINE SCIENCE, HASANUDDIN UNIVERSITY (UNHAS), MAKASSAR, INDONESIA; [6]AGENCY FOR THE ASSESSMENT AND APPLICATION OF TECHNOLOGY (BPPT), JAKARTA, INDONESIA; [7]UNIVERSITY OF RIAU, PEKANBARU, INDONESIA; [8]INSTITUTE FOR COASTAL RESEARCH, HELMHOLTZ CENTRE GEESTHACHT, GERMANY

## Abstract

*Indonesia's rivers are of global importance in terms of their dissolved and particulate fluxes as well as in terms of the controlling natural factors and the human interventions in their catchments. The rivers of the volcanic islands of Java and Sumatra have been affected by different kinds of anthropogenic environmental transformation. While on Java river damming, urbanization and agriculture under irrigation have strongly affected water quality and dissolved and particulate river loads since the late 19th century, the recent degradation of peatlands and conversion to plantations strongly affect water quality, dissolved organic carbon fluxes, and dissolved oxygen in rivers on Sumatra. In this chapter, the Brantas River on Java and the blackwater river Siak on Sumatra serve as the main examples illustrating the interplay between natural control factors and human interventions in watersheds and the related socioeconomic and governance issues. The last part of this chapter provides directions for future research and recommendations for policy and society.*

## Abstrak

*Sungai-sungai di Indonesia memegang peranan penting dalam hal aliran muatan padatan terlarut dan tersuspensi, sebagaimana juga dalam hal pengaturan berbagai pengaruh alamiah dan aktifitas manusia di daerah tangkapannya. Pulau Jawa dan Sumatera menjadi fokus perhatian khususnya karena laju perubahan tata guna lahan yang tinggi. Masalah utama di Pulau Jawa yang terkait dengan sungai adalah pembendungan sungai, laju urbanisasi yang tinggi dan irigasi pertanian yang berakibat kepada perubahan beban padatan tersuspensi dan terlarut. Sementara fokus di Pulau Sumatera adalah tentang perubahan rawa gambut menjadi lahan perkebunan yang berakibat kepada perubahan kualitas air, seperti aliran karbon organik terlarut, dan kandungan oksigen terlarut. Sungai Brantas di Jawa dan Sungai Siak di Sumatera dapat menjadi contoh dari interaksi antara pengaruh faktor alamiah dan manusia di daerah aliran sungai dengan berbagai*

Science for the Protection of Indonesian Coastal Ecosystems (SPICE). https://doi.org/10.1016/B978-0-12-815050-4.00002-X

*faktor sosial ekonomi dan isu pengelolaan yang berujung pada isu pemanfaatan berlebih dan kerusakan lingkungan. Akhirnya, penelitian ini juga memberikan arahan untuk topik-topik penelitian prioritas dan rekomendasi bagi berbagai kebijakan yang dapat dilakukan di masa datang.*

**Chapter outline**

**3.1 Introduction** .......... **46**
**3.2 Drivers of environmental change affecting river fluxes** .......... **48**
**3.3 Natural factors, human interventions, and extreme events controlling river fluxes** .......... **49**
3.3.1 The Brantas River, Java, as an example of high suspended matter rivers .......... 50
*3.3.1.1 Variations in sources, composition, and fate of nutrients* .......... 53
*3.3.1.2 Variations in sources, composition, and fate of suspended sediments and particulate organic matter* .......... 57
*3.3.1.3 Effects on phytoplankton abundance and community composition* .......... 60
*3.3.1.4 Effects on the dissolved oxygen regime of the lower Brantas* .......... 61
3.3.2 The Siak River, Sumatra, as an example of blackwater rivers .......... 65
*3.3.2.1 Variations in dissolved organic carbon and dissolved oxygen* .......... 66
*3.3.2.2 Sources and fate of nutrients* .......... 69
**3.4 Governance and management programs** .......... **71**
**Acknowledgments** .......... **74**
**References** .......... **74**

## 3.1 Introduction

Water is essential for all life on Earth, and rivers are the life veins of continents. They collect and redistribute water and all its dissolved and particulate ingredients on their way to the coast, where they ultimately empty their loads into the ocean. The estuaries of rivers are transition zones where the land-derived load of substances nourishes the coastal seas. They are rich in natural resources and have provided goods and services to humans for a long time. Consequently, they are among the most densely populated regions of the world. This, in turn, makes them particularly vulnerable to global environmental change. Tropical coastal zones receive the major part of the annual riverine inputs of freshwater and dissolved and particulate substances (Milliman and Farnsworth, 2011). They harbor some of the most productive and diverse ecosystems on Earth, such as mangrove forests, seagrass meadows, and coral reefs. River fluxes of water, sediment, and other dissolved and particulate substances are therefore a major determinant of the structure and functions of coastal ecosystems, of the economic potential of coastal zones, and hence of the livelihoods of the coastal population.

In general, natural factors such as geology, climate, hydrology, and vegetation are the first-order controls of river flows. More than 4500 years ago, early civilizations developed near rivers and used their goods and services (Giosan et al., 2012; Macklin and Lewin, 2015). Since then, human interventions in river catchments have altered the flow of rivers and the amount and composition of their load, culminating in the Anthropocene during which land use change and alterations of hydrology in concert with human-induced climate change became the first-order controls of river fluxes in some regions (Syvitski et al., 2005).

Indonesia's rivers are of global importance in terms of their fluxes as well as in terms of the controlling natural factors and human interventions (e.g., Milliman et al., 1999; Milliman and Farnsworth, 2011; Syvitski et al., 2014). Indonesia is located in a zone of high tectonic activity, which is a major driver of physical denudation, providing mineral surface area for chemical weathering, which is also high because of high precipitation and high runoff temperature (Gaillardet et al., 1999). In many places, dissolved and particulate river loads are high (e.g., Jennerjahn et al., 2013; Milliman and Farnsworth, 2011). However, Indonesia stretches over a very large area with a multitude of landforms, including vast low-lying areas in which vegetation has a more prominent role in determining river fluxes. Many of these areas are peatland swamps, and Indonesia contributes nearly 50% of the total tropical peatland swamp area (Joosten, 2010; Page et al., 2011). The vast peatland systems in Indonesia are concentrated in eastern Sumatra (mainly Riau, Jambi, and South Sumatra provinces), in Borneo, and in the southern part of Papua (Hooijer et al., 2010). In those regions, the dissolved river loads are much higher than the particulate ones and dominated by organic substances (Alkhatib et al., 2007; Baum et al., 2007; Moore et al., 2013; Wit et al., 2015). Those rivers are called "blackwater" rivers.

To guarantee the future supply of estuarine and coastal ecosystem services, it is of utmost importance to understand the effects of human interventions in river catchments and their interaction with outcomes of human-induced climate change and natural controls on the amount and composition of dissolved and particulate river fluxes. Identifying cause–effect relationships and quantifying respective fluxes is a prerequisite for developing measures toward mitigating undesired consequences of human activities in river catchments. This was investigated in the clusters "Coastal Ecosystem Health" and "Marine Geology and Biogeochemistry" of the Indonesian–German research and education program "Science for the Protection of Indonesian Coastal Marine Ecosystems" (SPICE). The effects of land use and regulations of hydrology on the dissolved and particulate river loads, the underlying processes, and the socioeconomic boundary conditions were investigated with a multi- and interdisciplinary approach mainly on the two most populous Indonesian islands, in the high suspended matter load river Brantas on Java and in the blackwater river Siak on Sumatra. Furthermore, land use changes and sediment sources, their drivers, and watershed and coastal governance were investigated within the cluster "Governance und Management of Social-Ecological

Coastal Systems" (SPICE II) and the subproject "Upstream-Downstream Linkages and New Instruments in Coastal and Watershed Governance" (SPICE III) with a main regional focus on the Citanduy River and the Segara Anakan lagoon in Java and analyses of Payments for Environmental Services pilot schemes in the catchments of the Brantas in Java, the Way Besay in Sumatra, and the Kapuas in West Kalimantan. This chapter synthesizes the findings of these SPICE clusters.

## 3.2 Drivers of environmental change affecting river fluxes

The major drivers of environmental change affecting river fluxes of dissolved and particulate matter are human activities in river catchments and the coastal zone in combination with natural hazards and outcomes of climate change. Of the latter, sea level rise in combination with land subsidence can contribute to increased flooding of low-lying coastal areas and as such affect processes and fluxes in river estuaries. This is of particular relevance along the northern coast of Java where a number of regions are affected by land subsidence with rates of up to 22 cm $year^{-1}$, mostly because of groundwater and gas extraction (Chaussard et al., 2013). In some areas, flooding is frequently observed, for example, in the Greater Jakarta metropolitan area (Simarmata, 2018; Steinberg, 2007).

More important in this context are possible changes in atmospheric moisture distribution and in the frequency and intensity of extreme weather events. All IPCC scenarios (representative concentration pathways: RCP2.6, RCP4.5, RCP6.0, RCP8.5) project an increase in precipitation over Indonesia until the year 2100 (van Oldenborgh et al., 2013) with some uncertainty on the southernmost islands of the Indonesian island arc, where precipitation may also decrease.

While the more than 17,000 islands forming Indonesia are spread over a wide area with large differences in tectonic activity, the islands of Sumatra and Java are located on an active continental margin, which belongs to the tectonically most active regions in the world (Lowman et al., 1999). As those two islands with approximately 184 million inhabitants are holding 77% of the total Indonesian population (Badan Pusat Statistik, 2019), the natural hazards of volcanic eruptions, earthquakes, and landslides are also major threats to the population. During the past two decades, the Indonesian island arc has been very active as documented by, for example, the major 2004 tsunami resulting from a major submarine earthquake, the Yogyakarta earthquake in 2006, the Merapi eruption in 2010, and the Sunda Strait tsunami in 2018. Moreover, floods and landslides are also commonly observed consequences of extreme precipitation events (e.g., Cepeda et al., 2010).

Human uses of the land and regulations of hydrology are manifold all over Indonesia, and they are an important control of river fluxes on Sumatra and Java, where human impact is strongest due to high population and intensive use. Agriculture and plantations are the major forms of land use on both islands, with Java producing 50% of the nation's rice, and oil palm plantations on Sumatra and Kalimantan producing 50% of the global palm oil (Ramdani and Hino, 2013). The conversion of natural forests to other land uses largely contributes to increasing river fluxes, whereas the damming of rivers may lead to a reduction of river fluxes of sediments, carbon, nutrients, and other elements. While the damming of rivers is not such a big issue on Sumatra, >35 dams have been constructed on Java mainly for purposes of flood control, irrigation, power generation, and drinking water supply, with consequences for the amount and composition of river fluxes (Whitten et al., 1996). The Cirata, Saguling, and Jatiluhur Dams in the Citarum River watershed in West Java provide a good example on how multiple damming of a river has affected sediment and nutrient inputs into its estuary. It has been observed that damming of the Citarum has reduced sediment input into the Java Sea significantly. The Jatiluhur Dam retained $1.2 \times 10^6$ t year$^{-1}$ of sediment in its reservoir, which would otherwise have contributed substantially to the Citarum sediment discharge of $1.8 \times 10^6$ t year$^{-1}$ (Paryono et al., 2017).

Peatland area burning and its conversion to other land uses are the main factors contributing to the dynamics of river fluxes in the respective areas in Sumatra, Borneo, and Papua. The Riau province, in particular, which hosts more than 30% of the peatland area in Indonesia, has experienced severe peatland degradation in the past few decades (Vetrita and Cochrane, 2020).

## 3.3 Natural factors, human interventions, and extreme events controlling river fluxes

In terms of natural control factors, Indonesia is a global hot spot of weathering and erosion. Intensive physical weathering and erosion, high runoff, and high temperatures are the most important preconditions for high chemical weathering and ultimately for high river loads of dissolved and particulate matter (Gaillardet et al., 1999). In Indonesia, and in particular on the islands of Sumatra and Java, which are located directly on the Java trench where the Australian plate is subducted under the Eurasian plate, tectonic uplift rates and volcanic activity are high (Lowman et al., 1999). Consequently, physical denudation and the supply of mineral particles are high. Precipitation rates of 2000–6000 mm year$^{-1}$ are among the highest in the world (Global Precipitation Climatology Centre, 2018), and temperature is also high. As a result, Indonesian rivers generally have very high yields of dissolved and even more of particulate matter (Milliman and Farnsworth, 2011). However, there are also large low-lying areas covered by extensive peat swamp forests, in particular on the island of Sumatra. Rivers draining these areas have generally low particulate matter loads, but instead contain large amounts of dissolved organic matter that gives them a dark color, the so-called "blackwater" rivers.

### 3.3.1 The Brantas River, Java, as an example of high suspended matter rivers

The Brantas is the second largest river of Java and is located in its eastern part. It is a medium-sized river of 320 km length that has its origin near the volcano Mount Arjuna and drains an area of 11,050 km$^2$. It empties into the coastal waters of the shallow Madura Strait through two major branches, the Wonokromo and the Porong. The climate is driven by monsoon with a dry season from June to September and a wet season from November to April. The latter is also the time during which 80%–90% of the annual river discharge occurs, with the Porong responsible for approximately 80% of it (Jennerjahn et al., 2013).

The climate is dominated by the monsoons, but in contrast to the western part of the island, eastern Java experiences only one wet season during the months November to April with an average annual rainfall of 2220 mm. This is also the period of peak river discharge, which has been in the order of 217 m$^3$ s$^{-1}$ on an annual average during the period of 1991–96 (Jennerjahn et al., 2004). Approximately 40 km upstream from Surabaya City the river branches in two, the northeastward flowing Surabaya River and the eastward-directed Porong River being the major transporting agent for water and sediment. During the wet season, when almost 80% of the water supplied by the Brantas is diverted to the Porong, its average discharge can be about 600 m$^3$ s$^{-1}$, which may double to 1200 m$^3$ s$^{-1}$ in extremely wet years (Hoekstra, 1989). During El Niño years, however, discharge can be much lower due to a decrease in precipitation. Approximately 10 km before Surabaya City, the Surabaya River then branches in two, the Kali Mas with a low volume of water passing Surabaya City and the eastward-directed Wonokromo River, which discharges into Madura Strait 30 km north of the Porong River. Due to high sediment loads, particularly during the wet season, the Porong has a strongly prograding delta. During the dry season, flow is mainly diverted to the town of Surabaya, and discharge of the Porong is extremely low (Hoekstra, 1989). The Madura Strait is characterized by a mixed diurnal–semidiurnal tide in the micro- to mesotidal range (Hoekstra, Nolting and van der Sloot, 1989). An analysis of five decades of rainfall over the Brantas River catchment in East Java since the 1950s showed a decrease of accumulated rainfall, an increasing intensity of wet season rainfall and a prolonged dry period (Aldrian and Djamil, 2008).

Approaches to watershed governance in Indonesia have over time reflected a myriad of shifting political and socioeconomic interests (e.g., Lukas and Flitner, 2019), and the situation is no different in the Brantas area. In the early days of Indonesian independence in the 1950s, river basin development was specified as a national priority, with the Brantas River basin identified as an area of top concern (Fujimoto, 2013). Over the following four decades, development of the different phases of the Brantas Master Plan was supported by the Government of Japan through grant and loan funding and technical cooperation, initially as part of postwar reparations (JICA, 1998). The implementation of the different phases was supported by funding from donors such as Japan, the Asian Development Bank (ADB), the International Bank for Reconstruction and Development (World Bank), and the Austrian government (Fujimoto, 2013). As with other watersheds in Indonesia, in the prereform period, developments were led by the central government through the Ministry of Public Works (Bhat et al., 2005). The four phases of the Brantas Master Plan shown below reflect shifting national priorities and realities on the ground (see Adi et al., 2013 and references therein):

- Phase I (1962–1972): A flood control program was initiated, dams were constructed, and river channel capacity improved in the upper portion of the catchment.
- Phase II (1973–1984): An irrigation development program was implemented; reservoirs, barrages, and technical irrigation systems were installed, with one of the primary goals being to support the government's drive to achieve rice self-sufficiency (Ramu, 2004).
- Phase III (1985–1999): A water supply program for domestic and industrial use was implemented.
- Phase IV (2000–2020): Integrated water resource management was implemented for water resources conservation and water quality.

During that time, the hydrological regime and land use changed dramatically, forest cover decreased to around 10%, and rice cultivation and plantations (mainly maize and sugar cane) increased to >60% of the total catchment area (Adi et al., 2013 and references therein; Fig. 3.1). The regulations of hydrology had the following major benefits: (1) mitigation of flood discharge (50 years return period), (2) generation of electric hydropower on the order of 233 MW per year, (3) supply of water for irrigation of 345,000 ha of paddy field, (4) supply of raw water for domestic and industrial use on the order of 300 million $m^3$ per year, and (5) supply of fresh water at a rate of 13.5 $m^3\ s^{-1}$ for an area of 11,000 ha of brackish aquaculture in the Brantas lowlands (Ramu, 2004). The total wetland area of the lowlands was estimated to 18,000 ha in 2015 and consisted almost exclusively of aquaculture ponds. Those have mainly replaced mangroves, which covered an area of only 1700 ha in 2015 (Maryantika and Lin, 2017).

The main water structures are eight dams and respective reservoirs that were constructed between 1970 and 2001 with initial gross and effective capacities of 647 and 479 million $m^3$, respectively. High erosion rates and corresponding sediment inputs decreased the gross and effective capacities to 405 million $m^3$ (63%) and 343 millon $m^3$ (72%), respectively (Ramu, 2004). The largest reservoir, Sutami (also called Karangkates), located in the upstream portion of the catchment, experienced the largest decrease of effective capacity down to 57% after 31 years of operation. Besides inputs related to land use, the high sedimentation in reservoirs is also strongly affected by ejected material from the active volcanoes Mt. Kelud and Mt. Semeru. For example, an estimated 100–300 million $m^3$ of volcanic material per eruption can raise the river bed by about 2.5–7.5 m. In 1990, the eruption of Mt. Kelud filled the Wlingi reservoir with sediment, which then had to be dredged (Omachi and Musiake, 2003).

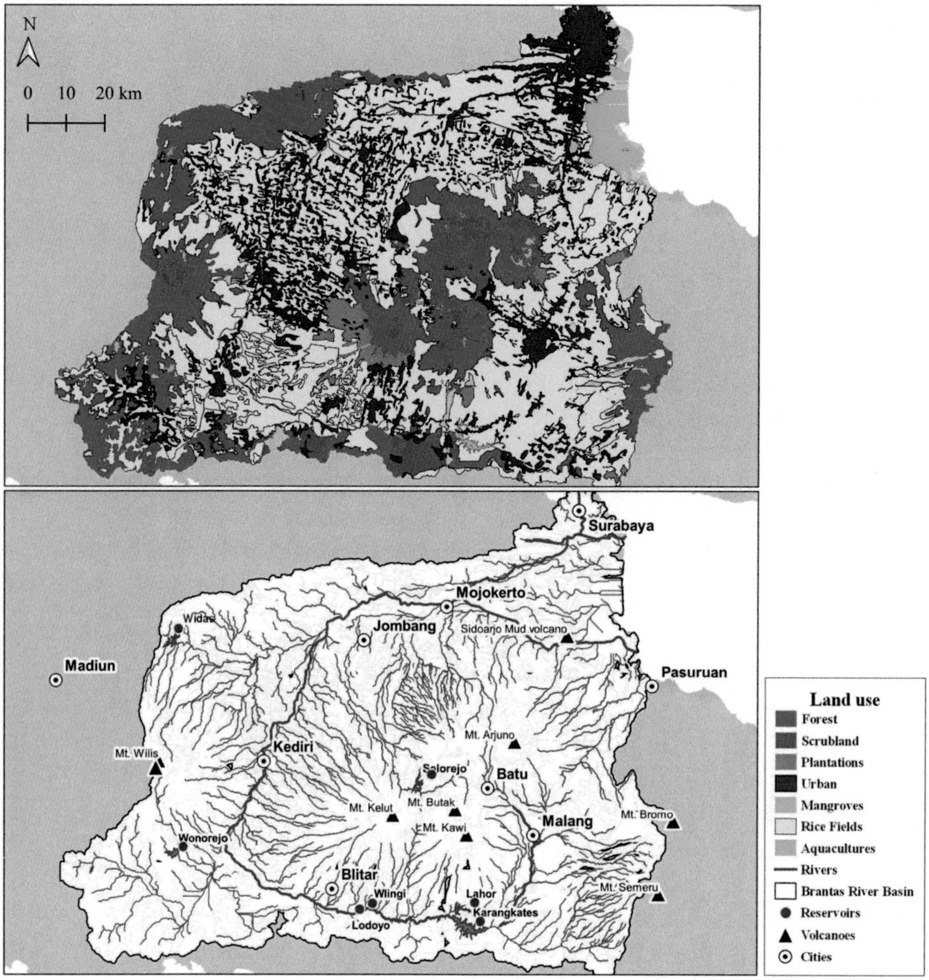


**FIGURE 3.1** Map of the Brantas River catchment with major land use/cover (upper panel) and the river network, major dams and reservoirs, volcanoes, and major cities (lower panel). Maps were generated by Ingo Jänen using QGIS 3.14.1-Pi (https://qgis.org/de/site/). The drainage basin map was created with the DEM ASTER (USGS, https://glovis.usgs.gov/app). *The land cover map is based on data from 1990 to 2015 retrieved from http://webgis.menlhk.go.id:8080/pl/pl.htm.*

Of particular relevance for the river fluxes of dissolved and particulate substances was the eruption of the mud volcano "Lusi" in eastern Java in the downstream portion of the Brantas River in 2006. Mud volcanoes are natural disasters known from Java, but it was debated if gas drilling operations triggered the "Lusi" eruption starting on May 29, 2006 (Davies et al., 2010; Jänen et al., 2013; Sawolo et al., 2009; Tingay et al., 2015). The mud flow reached early maximum rates of up to 180,000 $m^3$ $day^{-1}$ (Mazzini et al., 2007) and in February 2017 still had a rate of 80,000 $m^3$ $day^{-1}$ (Fallahi et al., 2017). The mud covered large areas (ca. 7 $km^2$) and led to the displacement of almost 60,000 people (Davies et al., 2010; Mazzini et al., 2009; Richards, 2011). Part of the mud was directed into the nearby Porong River, one of the two major branches of the Brantas River that discharges into Madura Strait, approximately 20 km from the river mouth.

#### *3.3.1.1 Variations in sources, composition, and fate of nutrients*

The concentrations of dissolved inorganic nutrients are overall high in the Brantas River and its tributaries and reservoirs, despite some seasonal and spatial variations (Jänen et al., 2013; Jennerjahn et al., 2004; Jennerjahn and Klöpper, 2013). In the upstream portion of the river near Mount Arjuna, the concentration varied between 200-250 μM for nitrate, 6–11 μM for phosphate, and 800–1200 μM for silicate; the former two were a little lower in the midstream portion of the catchment (Ponnurangam, 2011). These values are high when compared with those of forested, almost unused headwaters of rivers like, for example, in Scotland and Japan. Nitrate was in the range of 10 μM and phosphate around 0.1 μM in the upstream portion of the Scottish Dee River, mainly montane with moorland and peat cover, and only increased further downstream with increasing agriculture (Edwards et al., 2000). In 18 streams draining forested watersheds in Japan, nitrate varied between $<10$ and 50 μM except for two streams with elevated atmospheric N deposition (Shibata et al., 2001). Nitrate concentrations between 4-23 μM were also measured in streams in forested US watersheds (Ohte et al., 2001). In contrast, rivers draining highly used watersheds, i.e., mainly for agriculture, can have much higher nitrate concentrations from tens to hundreds of micromoles like, for example, in the German Weser River, which had nitrate concentrations from 150 μM to several hundreds of micromoles throughout the whole catchment, which is densely populated and mainly used for agriculture (Hirt et al., 2012). Nitrate concentrations in the Mississippi River varied between 70 and 500 μM at several stations in the agriculture-dominated catchment between 1980 and 2008 (Sprague et al., 2011). Major land use in the Brantas River catchment is agriculture with about 60%, the majority of which is rice cultivation under irrigation, also in the upstream portion of the river, while forest cover is on the order of 10–20% (Pristiwati, 2015, p. 64; Yoshino et al., 2017). Following a classification scheme of trophic status by Smith et al. (1999), the Brantas River water is highly eutrophic even in the upstream portion of the catchment as a result of the naturally high weathering rates and the inputs from the agriculture-dominated hinterland.

The effect of these high natural and anthropogenic inputs can also be observed in the upstream reservoirs. The Sutami reservoir in the mainstem, being the largest in the Brantas River, and the Selorejo reservoir in the Konto River, a tributary of the Brantas, displayed clear signs of eutrophication with nitrate concentrations of up to 150 μM and excessive growth of aquatic plants. Huge amounts of phytoplankton ($4-5 \times 10^3$ cells $L^{-1}$) forming a green "carpet" were recorded in the Sutami reservoir, while about one half of the Selorejo reservoir was covered with water hyacinths (Fig. 3.2; Jennerjahn and Klöpper, 2013), both of which are indicators of high nutrient concentrations (Lung'ayia et al., 2001). Only a small portion of the excess nutrients coming from the surrounding

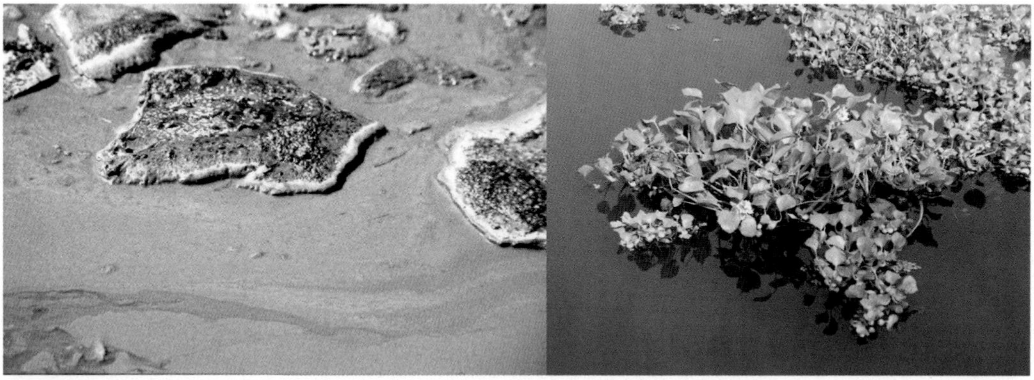

**FIGURE 3.2** The visual result of eutrophication in the Sutami and Selorejo reservoirs, a green "carpet" of millions of phytoplankton cells per liter in the former (left) and a large cover of water hyacinths in the latter (right). The gray-brown patch (ca. 30 × 20 cm) in the left photo is organic tissue/sludge, which was released after an accident in a nearby tapioca factory. The decomposition of this organic matter also contributed to oxygen consumption and nutrient release. *Photos taken by Tim Jennerjahn.*

weathering and unused fertilizer are consumed in the turbulent waters of the river. When river flow is being slowed down by a dam, its turbulence decreases. The increasing light penetration then allows photosynthesizing aquatic plants to consume part of the nutrients to form their biomass, i.e., the aforementioned algae and water hyacinths as well as nitrogen-fixing cyanobacteria in the case of the Brantas River (Jennerjahn and Klöpper, 2013). High primary production entails the formation of large and rapidly settling particles with high organic matter (OM) content. This, in turn, leads to intensive microbial decomposition of the sinking organic matter as documented by low concentrations of dissolved oxygen in deeper parts of the reservoirs (1.4–4.5 mg $L^{-1}$, 19%–65% saturation; (Jennerjahn and Klöpper, 2013)). These low oxygen concentrations form unfavorable conditions for fish and benthic organisms and can be even lethal for some of them (Vaquer-Sunyer and Duarte, 2008).

In the midstream portion of the river 50–100 km from the coast and before its diversion into Porong and Wonokromo, the concentration of dissolved inorganic nitrogen (DIN = nitrate + nitrite + ammonium) was up to 180 μM in the wet season and up to 150 μM in the dry season; it consisted almost exclusively of nitrate, and the N/P ratio was in the range of 25–40. Downstream in the Porong and Wonokromo, the amount and composition of nutrients were different and varied markedly between seasons. The Wonokromo, which transports about 20% of the total Brantas discharge to the coast, displayed high DIN concentrations consisting almost exclusively of nitrate during the wet season on its way to the coast. The river plume extends several kilometers out into Madura Strait as demonstrated by DIN concentrations and a salinity of 10 (Fig. 3.3; Jänen et al., 2013). During the dry season, the very low discharge is responsible for slightly higher DIN concentrations, with ammonium making up half of it. The strongly increasing salinity several kilometers before the river mouth indicates that the

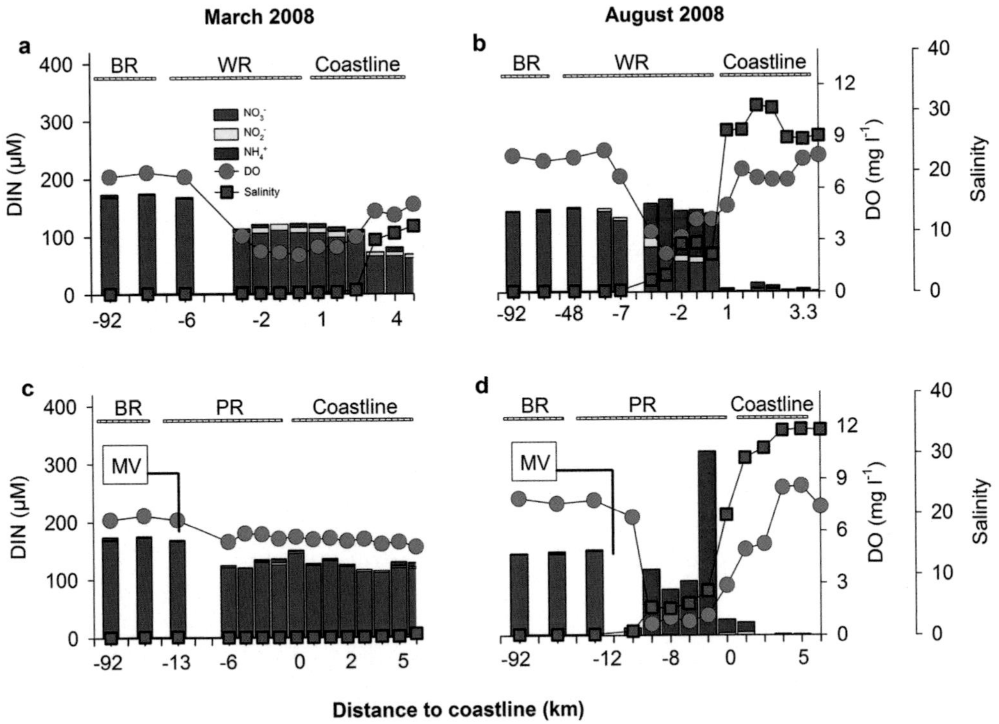


**FIGURE 3.3** Distribution of dissolved inorganic nitrogen (DIN) species, dissolved oxygen (DO), and salinity in the Wonokromo transect during the wet (A) and dry season (B) and in the Porong transect during the wet (C) and the dry season (D). BR, Brantas; Coastline, Madura Strait coastal waters; *MV*, mud volcano; *PR*, Porong; *WR*, Wonokromo. *Adapted from Jänen et al. (2013).*

estuarine mixing zone geographically moved inside the river. Consequently, DIN is $< 5$ μM in coastal waters. While the majority of the dissolved nutrients in the Wonokromo comes from the agriculture-dominated hinterland, it receives additional input when passing through the city of Surabaya, which discharges untreated sewage into the river, and from the downstream brackish water aquaculture ponds, which also release untreated effluents into the river. Because of the low discharge, the portions of the urban and aquaculture contributions are larger during the dry season (Jänen et al., 2013). For comparison, concentrations of several hundreds of micromoles of DIN, the majority of which being ammonium, were also observed in the Taiwanese Danshuei River that receives effluents from the city of Taipei (Wen et al., 2008). High concentrations of ammonium originating from the release of untreated aquaculture effluents were also observed in coastal waters of the Chinese island of Hainan, where large areas along the coast are covered by brackish water aquaculture ponds. There, in the absence of rivers, shrimp and fish pond water had average DIN concentrations of 55 μM, in peaks of up to 400 μM, with about three quarters of it being ammonium. In back-reef areas in coastal waters DIN still amounted to 5–15 μM (Herbeck et al., 2013).

The seasonal variations are even stronger in the Porong, which transports ca. 80% of the total Brantas discharge, but can become almost stagnant during the dry season, because almost all the water is directed into the Wonokromo to wash out the dirt of Surabaya. To avoid flooding of the lowlands by the usually high sediment load of the river, the Porong is diked and therefore does not come into touch with the surrounding lowlands, which are mainly covered by aquaculture ponds. The high nitrogen load coming from the hinterland is rapidly transported into coastal waters as demonstrated by a high DIN concentration and zero salinity 5 km off the river mouth. The only downstream input comes from the mud volcano. Its input is negligible during the wet season, but leads to high ammonium concentrations in the Porong during the dry season, because the porewater of the mud contained >4000 μM of dissolved ammonium (Jänen et al., 2013; Jennerjahn et al., 2013). Nevertheless, as in the Wonokromo, the estuarine mixing zone moves geographically into the river in the dry season; hence, nutrients become rapidly depleted in coastal waters (Fig. 3.3; Jänen et al., 2013). The high DIN concentrations are in the range reported from other rivers, which are strongly affected by human uses like, for example, in several European rivers (Middelburg and Nieuwenhuize, 2000) and in the Mississippi where the links to agriculture and other human uses in the river and the consequences for coastal eutrophication have been extensively demonstrated (Goolsby et al., 2000; Turner and Rabalais, 2003). High nutrient inputs and respective consequences in terms of eutrophication have also been reported for other Indonesian rivers, in particular on the intensively used island of Java. Only moderate to low concentrations of nitrate and phosphate were found in the Citanduy River debouching into the Segara Anakan Lagoon in south central Java despite intensive agriculture in parts of the hinterland, most probably because of the short flushing time of 1–3 days on average in most parts of the lagoon (Holtermann et al., 2009; Jennerjahn et al., 2009). In contrast, high concentrations of total nitrogen and total phosphorus including high DIN and phosphate were reported from rivers discharging into Jakarta Bay (van der Wulp et al., 2016), where they lead to eutrophication (Damar et al., 2012). Long-term monitoring in Jakarta Bay from 2001 to 2013 shows that this estuarine system has also experienced elevated eutrophication levels (Damar et al., 2019).

Monsoonal variability similarly affects the riverine nutrient loads into Jakarta Bay, which received as much as $21.3 \times 10^3$ t DIN year$^{-1}$ and $6.7 \times 10^3$ t P year$^{-1}$, respectively, in 2001, the majority of which was discharged into the bay during the rainy season (Damar et al., 2012). However, the level of eutrophication does not correlate linearly with the increasing nutrient input during the rainy period, since the growth of phytoplankton also depends on light availability. Long-term monitoring of Jakarta Bay between 2001 and 2013 has shown that increasing turbidity during the rainy period also increased light limitation, and an elevated phytoplankton biomass was observed during the dry period when waters were calmer (Damar et al., 2019).

A global analysis of inorganic nitrogen fluxes into the ocean showed that high DIN fluxes into coastal waters are correlated with high precipitation and high population density (Smith et al., 2003). Budget calculations show that the amount of nutrients discharged into Madura Strait coastal waters is moderate to high on a global scale. With a yield of $91 \times 10^3$ Mol N $km^2$ $year^{-1}$, the Brantas River falls into the lower end of the highest category of N yields ($85–170 \times 10^3$ Mol N $km^2$ $year^{-1}$; Smith et al., 2003).

#### *3.3.1.2 Variations in sources, composition, and fate of suspended sediments and particulate organic matter*

The concentrations and loads of total suspended matter (TSM) and particulate organic carbon (POC) in the Brantas River display large spatial and seasonal variations related to the combination of the various natural and anthropogenic controls. The TSM concentration varied between a few mg $L^{-1}$ in the reservoirs and >1000 mg $L^{-1}$ in the estuarine portion of the Porong after the "Lusi" input (Jennerjahn and Klöpper, 2013; Jennerjahn et al., 2004, 2013). While concentrations of particulate OM were generally low in the dry season, a major part of it was of autochthonous origin, i.e., phytoplankton, as indicated by C/N ratios and stable carbon and nitrogen isotope composition. In contrast, during the wet season, particulate OM concentrations were much higher than during the dry season and the majority of the material was of terrestrial origin, mainly derived from the dominant crop plants and soils, rice, and sugarcane (Fig. 3.4; Jennerjahn et al., 2004; Aldrian et al., 2008; Jennerjahn and Klöpper, 2013).

In the estuarine portion of the river, downstream sources added to the material transported to and deposited in Madura Strait coastal waters. In both, the Porong and the Wonokromo, fringing estuarine mangroves contribute to the POC export. In the Wonokromo sewage release from the city of Surabaya and input of aquaculture pond effluents contribute to the POC export, while the Porong does not receive effluents from the surrounding aquaculture ponds. Because of the usually very high wet season sediment load and substantial deposition in the river, the Porong riverbed is elevated over the surrounding land and is diked to protect the lowlands from flooding during the wet season. Therefore, the Porong does not receive lowland inputs except for the "Lusi" mud, a large part of which is deposited in the estuarine portion of the river (Jennerjahn et al., 2013). While this contributes to a further elevation of the riverbed, sandmining for construction purposes contributes to a destabilization of the river banks and bed (Fig. 3.5).

TSM and POC also displayed seasonal variations in the downstream portions of the Brantas as well as in coastal waters. While TSM and POC were about an order of magnitude higher in the Porong and Wonokromo in the wet than in the dry season, the seasonal difference was much smaller in coastal waters with only slightly elevated concentrations in the wet season. Because of the river damming, the Brantas generally has much lower transport energy and capacity than many other rivers. Also, rapid energy dissipation when the plume reaches coastal waters and high bottom friction in the very shallow (1–2 m) nearshore waters promotes a rapid settling of larger particles close to the river mouth. The high sediment accretion, in turn, contributed also to the progradation of the Porong delta and promoted the growth of mangroves. However, the extremely high rates of sediment accretion of approximately 20 cm $year^{-1}$ in the wet season can even hamper the growth of mangrove trees (Sidik et al., 2016).

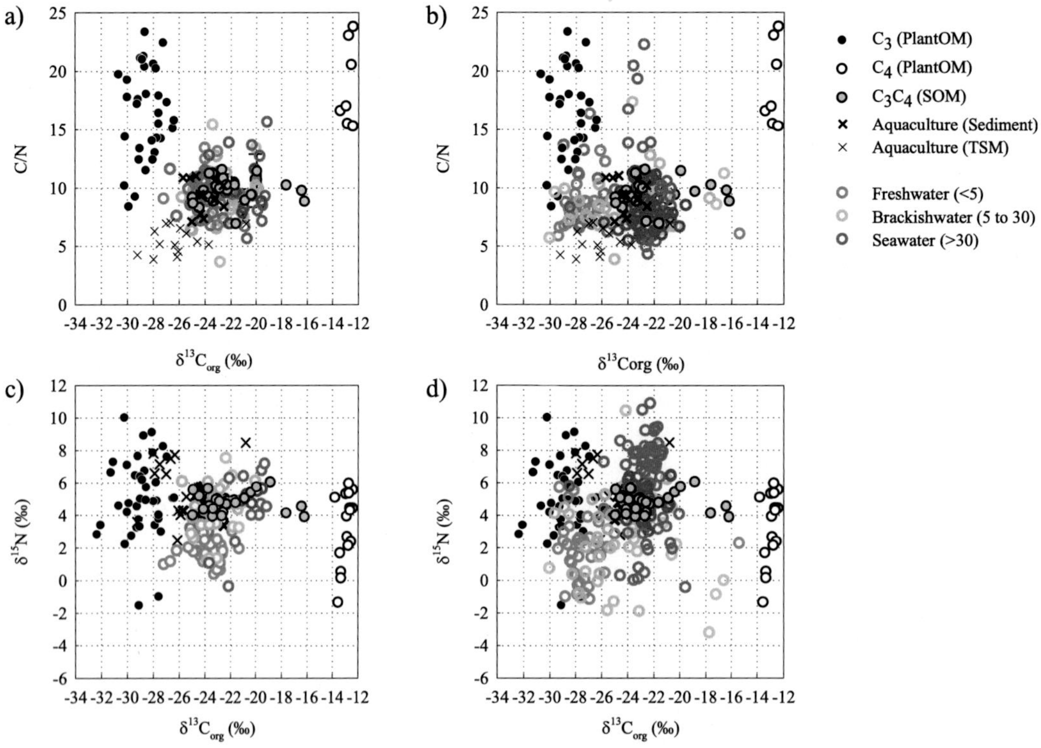


**FIGURE 3.4** The organic matter composition of riverine–estuarine TSM during wet (A, C) and dry (B, D) seasons in 2007 and 2008 (colored symbols) compared with that of quantitatively important sources (plants: C3 PlantOM, C4 PlantOM; soils: C3C4SOM). TSM samples are divided into three groups by salinity (freshwater <5, brackish water 5–30, seawater >30). *TSM*, total suspended matter.

**FIGURE 3.5** Photographs displaying (left) the introduction of part of the "Lusi" mud into the Porong 20 km before the mouth and (right) the sandmining activities in the downstream portion of the Porong. *Photos taken by Tim Jennerjahn.*

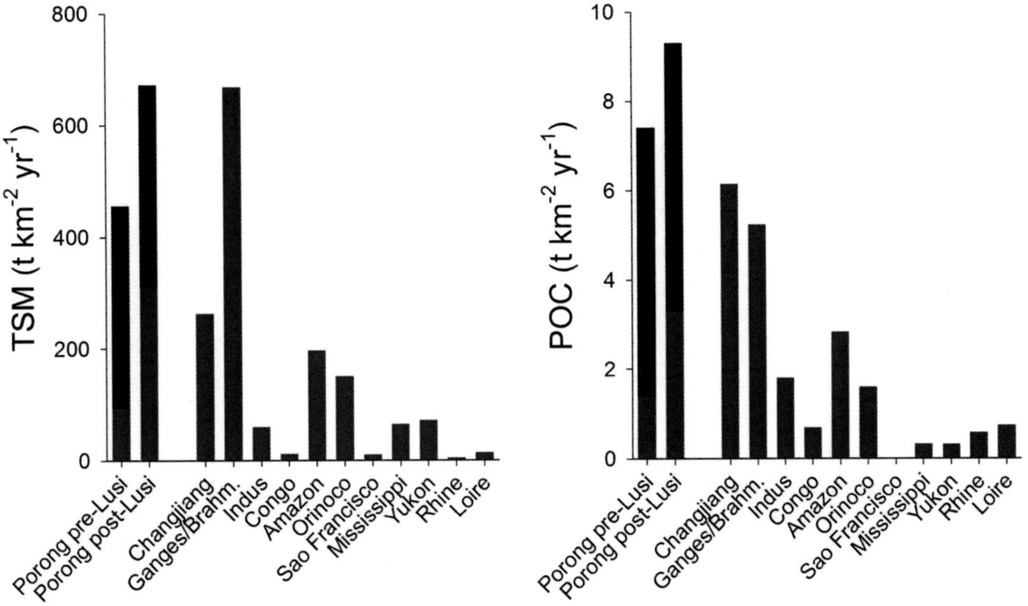


**FIGURE 3.6** Global comparison of (left) TSM and (right) POC yields of selected rivers (blue bars). The Porong yields include amounts retained in reservoirs behind dams (brown bars). Note that the other rivers are also affected by dams, but retained sediment amounts are not known. *POC*, particulate organic carbon; *TSM*, total suspended matter.

On a global scale, the sediment yield of the Brantas of 123 t $km^{-2}$ $year^{-1}$ is only moderate despite being located in an area of naturally high weathering and erosion (Fig. 3.6). Even with the "Lusi" additions, which is just a downstream temporary point source, the yield of 406 t $km^{-2}$ $year^{-1}$ does not reach the yield of major world rivers such as the Ganges/Brahmaputra (Jennerjahn et al., 2013). This is due to the multitude of human interventions in the catchment. Deforestation and land conversion to agriculture on Java culminated in the 1990s and increased erosion rates from about 2 to 10–14 mm $year^{-1}$ in the Brantas River catchment, which is much higher than natural erosion rates (Lavigne and Gunnell, 2006; Montogomery, 2007). This is counteracted by the construction of large dams and reservoirs, which started in the 1970s (Adi et al., 2013). Sediment deposition reduced the initial gross storage capacity of 647 $\times$ $10^6$ $m^3$ of the eight major reservoirs to 390 $\times$ $10^6$ $m^3$ (60%) by 2010 (Subijanto, 2010). In contrast, the calculated POC yield of 1.8 t $km^{-2}$ $year^{-1}$ is at the upper end of the range for major world rivers, probably mainly because of the erosion of agricultural soils, which are particularly rich in organic matter. With the addition of the "Lusi" input, the POC yield increases to 4.3 t $km^{-2}$ $year^{-1}$, which is much higher than for most major world rivers except for the Changjiang, Huanghe, and Ganges/Brahmaputra that drain some very erodible regions (Degens et al., 1991; Jennerjahn et al., 2013; Ludwig et al., 1996).

The aforementioned differences in downstream human interventions also affect the final fate of sediment and OM deposition in Madura Strait coastal waters. Due to its much higher river load and its downstream disconnection from the surrounding lowland, the Porong mainly serves as a conduit for hinterland-derived weathering and erosion products that lead to high deposition of refractory terrestrial OM diluted by high amounts of lithogenic material in coastal waters. In contrast, the much lower Wonokromo input and the higher portion of autochthonous OM promote the accumulation of more easily degradable OM in coastal sediments (Jennerjahn et al., 2004; Propp et al., 2013).

#### *3.3.1.3 Effects on phytoplankton abundance and community composition*

Changes in the amount and community composition of phytoplankton in coastal waters related to land use change and regulations of hydrology in river catchments are well known; prominent examples of it are the Mississippi and the Gulf of Mexico in North America and the Danube and the northwestern Black Sea in Europe (e.g., Humborg et al., 1997; Rabalais et al., 2000). The construction of the "Iron Gate" dams in the river Danube, approximately 800 km from the coast, led to nutrient depletion of river water leaving the reservoir. Decreasing river flow and turbulence in the reservoir increased light penetration and allowed for phytoplankton blooms consuming major part of the nitrogen, phosphorus and silicon. On the way toward the coast, nutrient-depleted waters were filled up with nitrate and phosphate originating from the effluents of agriculture, industry, and households downstream, but received only little silicate, which stems exclusively from natural weathering. As a consequence of the river damming and the growing industrialization and agriculture in Romania from the 1970s on, nitrate and phosphate inputs into the NW Black Sea increased and the silicate input decreased. The increased nutrient input and changed stoichiometry entailed excessive algal blooms, a shift in phytoplankton community composition from diatom dominated to more nonbiomineralizing species and hypoxic coastal bottom waters (Humborg et al., 1997).

In the Brantas reservoirs, the phytoplankton community composition is usually dominated by diatoms. However, during times of high nutrient enrichment, the proportion of cyanobacteria increases strongly (Jennerjahn and Klöpper, 2013). Despite the similarities with the Danube—NW Black Sea and the Mississippi—Gulf of Mexico in terms of land use change and river damming, the respective ecological response of Madura Strait coastal waters is different. Nitrogen and phosphorus inputs from the Brantas River are high, but the silicate input is also high. Therefore, despite seasonal variations in nutrient input, the phytoplankton community is at all times strongly dominated by diatoms (Damar, 2012; Jennerjahn et al., 2004). The major reason for this is the extraordinary high silicate supply, because the Brantas River catchment is located in an area of extremely high mechanical and chemical weathering and erosion (Jennerjahn et al., 2006). Despite the damming effect, the river supply of dissolved silicate is so high that diatoms can thrive in Madura Strait coastal waters. Seasonal variations in phytoplankton abundance are mainly due to dilution and light limitation.

Phytoplankton abundance is typically lower during the wet than during the dry season because of dilution in the much larger amount of water transported and because of the much higher suspended load and much larger extension of the river plume reducing light penetration and hence photosynthesis (Damar, 2012).

In both, the Porong and Wonokromo estuaries, the optimum combination of underwater light and nutrient availability for phytoplankton growth were observed during the dry season. However, during the rainy season when nutrient concentrations are an order of magnitude higher (Fig. 3.3), the also much higher sediment load (Jennerjahn et al., 2013) causes light limitation. Therefore, optimum growth conditions with reduced turbidity leading to higher phytoplankton biomass are achieved in Madura Strait coastal waters in 3–5 km distance of the river mouths during that time (Damar, 2012). A similar seasonal variation in phytoplankton biomass has also been observed in Jakarta Bay (Damar et al., 2019).

Owing to the generally much higher nutrient input, the maximum biomass as depicted by the chlorophyll *a* (Chl *a*) concentration off the Porong River was with 22 $\mu g\ L^{-1}$ three times higher than off the Wonokromo River (7 $\mu g\ L^{-1}$). In both estuaries, the phytoplankton community was dominated by diatoms (Skeletonema spp. and Pseudonitzschia spp.); dinoflagellates (Ceratium spp. and Dinophysis spp.) made a minor contribution only (Damar, 2012).

Phytoplankton productivity and biomass display considerable spatial variability in response to varying nutrient content as observed in several studies conducted in Java's and Sumatra's estuaries (Table 3.1). In particular, the nutrient-rich estuaries of the eutrophied northern coast of Java promote higher primary production and phytoplankton biomass.

#### *3.3.1.4 Effects on the dissolved oxygen regime of the lower Brantas*

The most critical effects on the water quality in the lower Brantas are generated by biogeochemical processes; most important is oxygen consumption by mineralization of particulate organic matter. Other oxygen-consuming processes, such as nitrification and oxygen consumption within the sediment, are of minor importance for the overall water quality (Schroeder et al., 2004; Schroeder and Knauth, 2013). In particular in the lower Brantas, in the vicinity of the city of Surabaya, critically low dissolved oxygen concentrations can be found (Fig. 3.7); they result in frequent fish kills. However, since the oxygen concentrations are highly variable, with large differences between rainy and dry season and day and night, other methods with higher time resolution are required for a better understanding of the underlying processes. The data shown in Fig. 3.7 originate from automated online stations, which have a sampling rate of 1–3 h (Siregar et al., 2004).

To understand the biogeochemical processes that are involved in the oxygen consumption, the MERMAID station, a new high-tech water quality online station, had been installed in the lower Brantas River (Schroeder et al., 2004), at the beginning of the Wonokromo River in 2002. All data presented and discussed in the following originate from the MERMAID station (Schroeder and Knauth, 2013; Schroeder et al., 2013).

**Table 3.1** Variability of primary production and phytoplankton biomass in some estuaries in Java and Sumatra subjected to different levels of nutrient input.

| Area | Annual production (g C $m^{-2}$ $year^{-1}$) | Phytoplankton biomass (μg Chl *a* $L^{-1}$) | Sources |
|---|---|---|---|
| Banten Bay, Java Sea | 163 | – | Alianto et al. (2010) |
| Porong Estuary, Java Sea | – | 4.75 | Damar (2012) |
| Wonokromo Estuary, Java Sea | – | 2.51 | Damar (2012) |
| **Lampung Bay, Java** | | | |
| - inner part (Hurun coast) 1999 | 70 | – | Tambaru (2000) |
| - inner part (Hurun coast) 2000 | 152 | – | Sunarto (2001) |
| - inner part 2000–01 | 196 | 5.77 | Damar (2003) |
| - middle part 2000–01 | 40 | 2.22 | Damar (2003) |
| - outer part 2000–01 | 31 | 0.78 | Damar (2003) |
| **Jakarta Bay, Java** | | | |
| - inner part 1991 | 166–214 | – | Kaswadji et al. (1993) |
| - inner part 2000–01 | 503 | 31.4 | Damar (2003) |
| - middle part 2000–01 | 119 | 15.8 | Damar (2003) |
| - outer part 2000–01 | 47 | 2.20 | Damar (2003) |
| **Semangka Bay, Sumatra** | | | |
| - inner part (river plume) 2000–01 | 14 | 7.11 | Damar (2003) |
| - middle part 2000–01 | 40 | 1.21 | Damar (2003) |
| - outer part 2000–01 | 22 | 0.44 | Damar (2003) |

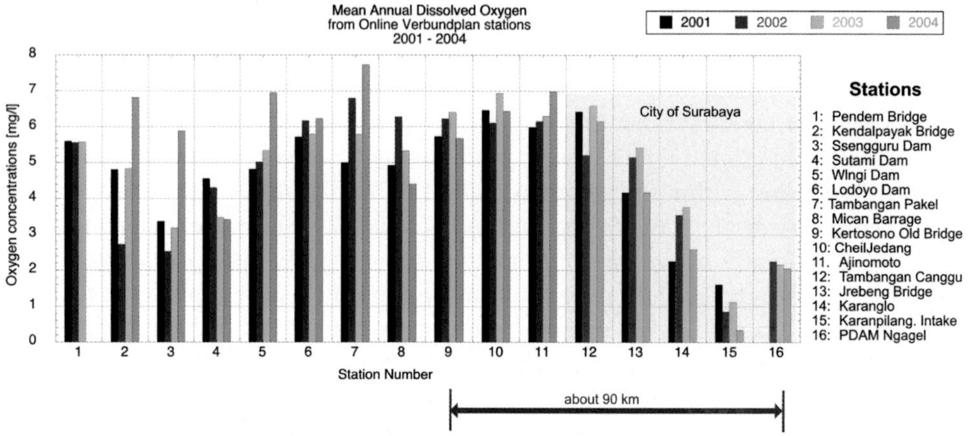


**FIGURE 3.7** Mean annual dissolved oxygen concentrations in the years 2001–04. *Data were retrieved from Jasa Tirta Public Company (PJT, pers. communication).*

In general, dissolved oxygen concentrations are lower and display higher daily fluctuations during the dry than during the rainy season. Corresponding with low river discharge and low turbidity, the median oxygen concentrations vary between 0 and 2 mg $L^{-1}$ during the dry season, while they vary between 3 and 5 mg $L^{-1}$ during the rainy season (Fig. 3.8). The more pronounced daily fluctuations during the dry season are likely the result of the diurnal variation of autochthonous primary production during the day and its remineralization during the night. Photosynthetic production results in high

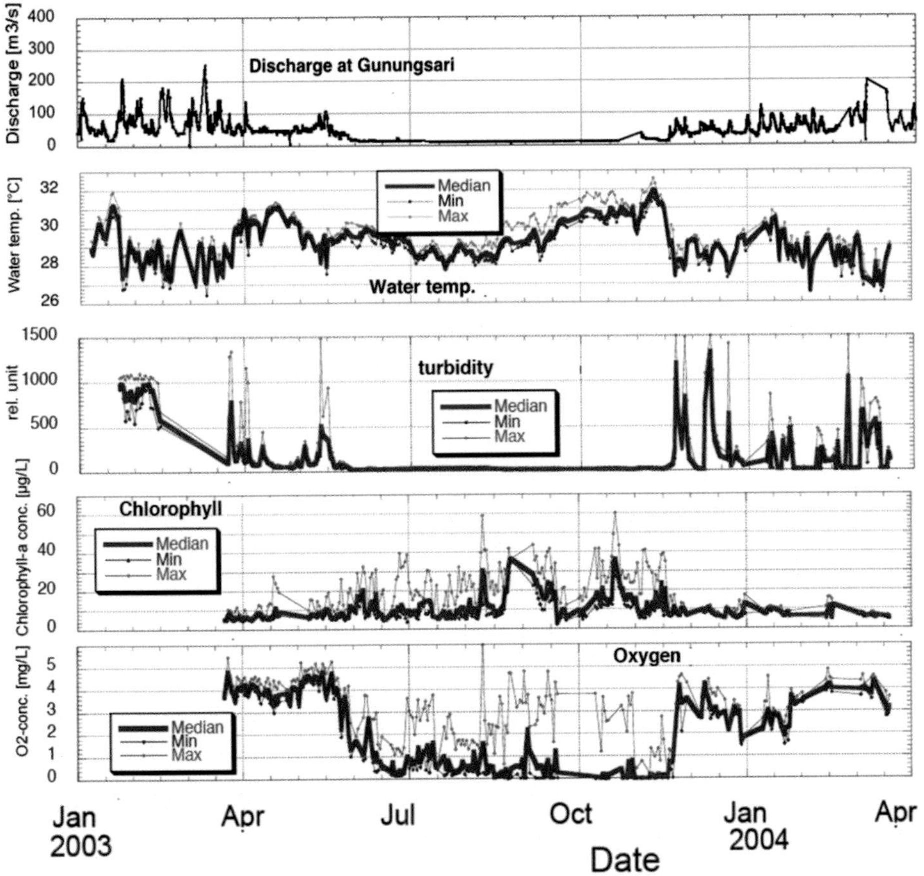


**FIGURE 3.8** MERMAID time series of several physicochemical and biological data collected from January 2003 until April 2004. From top to bottom: fresh water discharge, water temperature, turbidity, chlorophyll a fluorescence, and dissolved oxygen. *Adapted from Schroeder and Knauth (2013).*

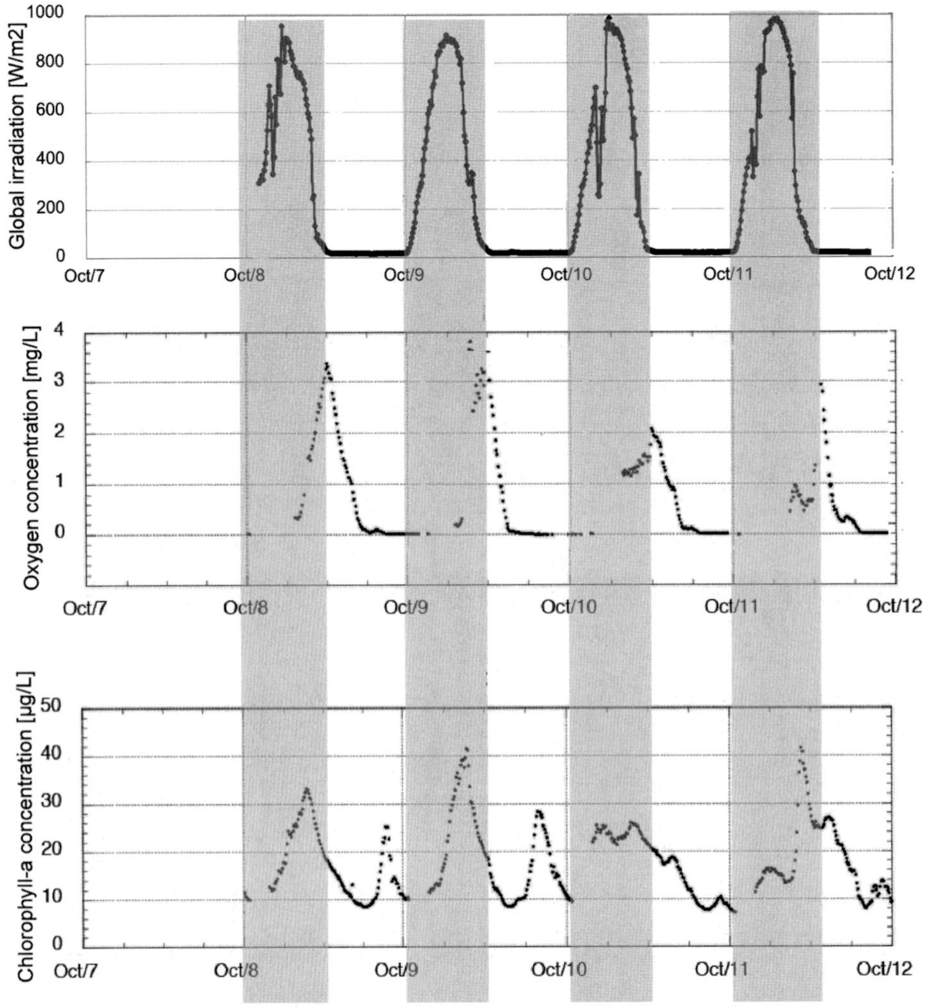


**FIGURE 3.9** MERMAID time series data for October 7–12. From top to bottom: Global irradiation, dissolved oxygen, chlorophyll a. Daylight times are indicated by gray bars. *Modified from Schroeder and Knauth (2013).*

Chl *a* and oxygen concentrations until sunset, and remineralization consumes it almost completely during the night (Fig. 3.9). Additional Chl *a* peaks during the night are probably the result of flow irregularities due to operation of upstream weirs (Schroeder and Knauth, 2013).

These high-resolution time series data allow to derive important process parameters, such as oxygen exchange with the atmosphere, microbial respiration, and the net ecosystem production (Odum, 1956; Schroeder and Knauth, 2013; Seeley, 1969). With these, the total respiration of the lower Brantas section from km 260 to km 300 is estimated to be on the order of 70 t $O_2$ day$^{-1}$, which nearly matches the input of ca. 100 t day$^{-1}$ BOD (biological oxygen demand) equivalent as reported by the monitoring authority BAPPEDAL (Badan Perencana Pembangunan Daerah = regional body for planning and development). To keep the oxygen values in this section of the Brantas above the "hypoxia" threshold of 2 mg $L^{-1}$ (Ekau et al., 2010; Vaquer-Sunyer and Duarte, 2008) under all meteorological and hydrological conditions, a maximum organic carbon input from domestic and industrial discharge was estimated to 33 t $day^{-1}$ of BOD, only one third of the existing discharge. This is quite a political and administrative challenge considering that domestic sewage accounts for about two thirds of the total organic carbon discharge and that there is no central sewage system in Surabaya that could be connected to a purification plant. It demonstrates that the Indonesian water quality management is not yet sufficiently integrated into river basin management, which mainly focuses on water quantity (Ertsen et al., 2018). In addition, the monitoring agencies of East Java lack a proper data management and do not have a complete inventory of the industries it had licensed and should monitor (Fatimah, 2017). However, there are prospects of an improvement. For example, the regulatory initiative "Water Patrol" in East Java takes care of regular inspections (D'Hondt, 2019).

### 3.3.2 The Siak River, Sumatra, as an example of blackwater rivers

The Siak is one of the main rivers draining the province of Riau in central Sumatra, Indonesia. It originates from the confluence of the headstreams Sungai (= river) Tapung Kanan and Sungai Tapung Kiri (km ~155) and passes through the adjacent lowlands with the cities of Pekanbaru (km 180), Perawang (km 220), and Siaksriindrapura (km 285) before discharging into the Strait of Malacca (km 370, Fig. 3.10). The Mandau River is the main tributary and cuts like the S. Tapung Kanan through peat swamps, which cover 22% of the Siak River catchment area of 10.423 $km^2$.

Tropical peat swamps are forested and separated from the groundwater flow of the hinterland and the underlying rocks. Since weathering of minerals is the main nutrient source in terrestrial ecosystems, peat swamps are nutrient-deficient ecosystems. They rely on precipitation as the main water and nutrient source. Even though peat swamps and their rivers host a number of species, except freshwater fish species, only very few of them are strongly associated with peat swamp forests (Posa et al., 2011). Due to their physical and chemical properties, peat soils were long spared from cultivation, but this has changed (Posa et al., 2011). Growing demand increased prices, in particular for pulp and fibers. Over time, oil palm and other plantations on peat soils have become profitable even though mineral soils are more suitable for farming. This, together with the rampant granting of concession licenses by district governors since political decentralization, has led to the large-scale conversion of peatland forests. The evaluation of satellite data showed that peatland deforestation in Indonesia increased from 2210 $km^2$ $year^{-1}$ between 2000 and 2010 to 2253 $km^2$ $year^{-1}$ between 2010 and 2015 (Miettinen et al., 2016). In 2015, only 6% of the original pristine peat swamp forest was left behind, and in the province of Riau degraded peat swamp forests, shrubs and secondary forests as well as managed lands covered 16%, 5%, and 66% of the peatlands,

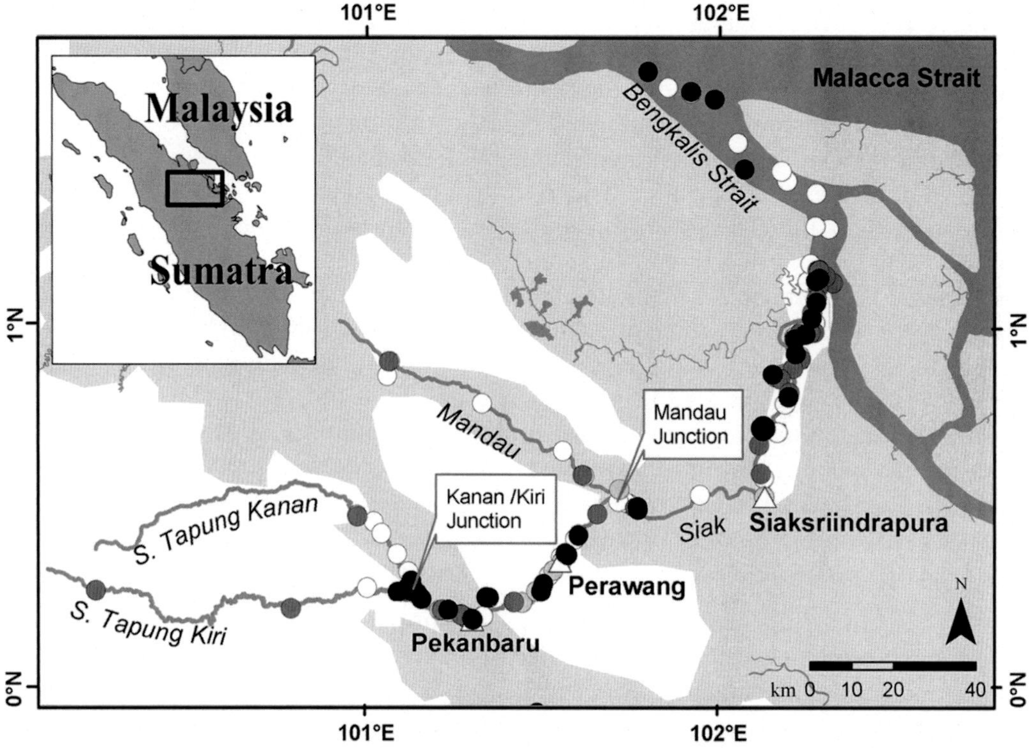


**FIGURE 3.10** The Siak River catchment area with its headstreams S. Tapung Kanan and S. Tapung Kiri and its major tributary Mandau River. The white triangles mark the location of the main cities along the Siak. The gray-shaded area shows the distribution of peat soils as obtained from FAO/UNESCO (2003). The black, white, and gray circles indicate the sampling sites of seven expeditions between March 2004 and October 2009. *Modified from Rixen et al. (2010).*

respectively (Miettinen et al., 2016). This land use and land cover change was accompanied by rapid population growth. According to projections of local authorities, the population of the capital Pekanbaru will increase from 671,777 inhabitants in 2006 to up to 1.3 million in the year 2031.

#### *3.3.2.1 Variations in dissolved organic carbon and dissolved oxygen*

The Siak owes its dark brown color to the high dissolved organic carbon (DOC) concentrations. DOC originates from peat soil leaching. Since the S. Tapung Kanan and the Mandau are the main peat draining tributaries, the DOC concentrations increased from approximately 500 to 1300 μM and from 1300 to 1900 μM, respectively, around the Kanan/Kiri and Mandau junctions in the Siak (Fig. 3.11A). To better understand the interaction between DOC, dissolved oxygen, and dissolved inorganic nutrients, a DOC decomposition experiment was conducted, and a box diffusion model was developed. While 27% of the DOC was degradable within a 2-week period, the remaining 73% of the DOC appeared to be refractory on timescales of days to months. Moreover, the DOC

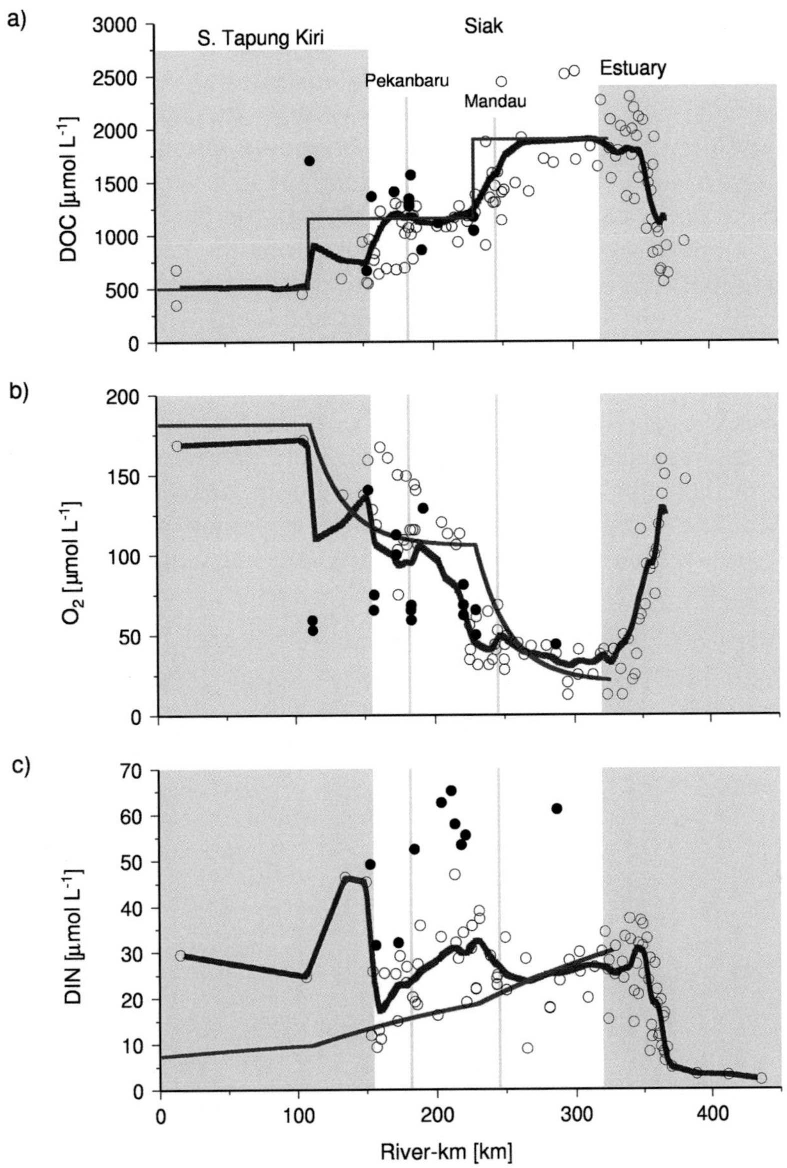


**FIGURE 3.11** DOC (A), oxygen (B), and DIN concentrations (C) measured in a water depth of 1 m versus river km (circles). Black circles indicate data measured during an expedition in March 2004. The blue lines show the DOC, oxygen, and DIN concentrations averaged for all expeditions. The red line in figure (A) indicates the DOC concentrations used to run the model. The red lines in figures (B) and (C) show the resulting oxygen and DIN concentrations. The vertical lines mark the locations of Pekanbaru and the Mandau junction. *DIN*, dissolved inorganic nitrogen; *DOC*, dissolved organic carbon. *Data taken from Rixen et al. (2010).*

decomposition experiment suggests an exponential decay and a half-life of 43 h for the most labile fraction (Rixen et al., 2008). In the estuary, DOC concentrations decrease due to mixing between DOC-rich river and DOC-poor ocean waters.

In the Siak, the high DOC concentrations reduced the light penetration depth to 15–20 cm and prevented photosynthesis to become a dominant factor in the river (Baum et al., 2007; Siegel et al., 2009). Photosynthesis is an oxygen source, which strongly influences the diurnal oxygen cycle in rivers. During the day, it increases the oxygen concentrations, which are reaching a maximum during the late afternoon prior to the sunset. During the night when the lack of light prevents photosynthesis, the decomposition of organic matter lowers the oxygen concentrations (Fig. 3.12). The absence of such a pronounced diurnal cycle and a mean Chl *a* concentration of 0.3 $\mu g\ L^{-1}$, which is far below concentrations measured elsewhere (Baum and Rixen, 2014), suggest that photosynthesis is negligible as an oxygen source in the Siak River.

However, the oxygen concentrations, which were measured along the river, were inversely correlated with the DOC concentrations (Fig. 3.11A and B). They decreased from 170 μM in the S. Tapung Kiri to 12 μM at the beginning of the Siak estuary. In the ocean, fish species start to suffer from oxygen deficiency at oxygen concentrations

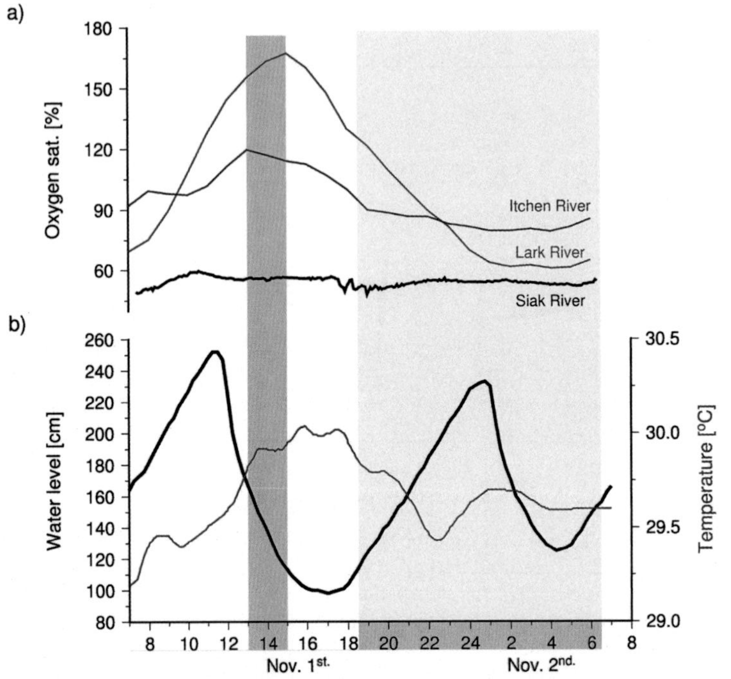


**FIGURE 3.12** Oxygen saturation (A) as well as water level and water temperature in the Siak versus time (B). The Siak data were measured at Pelita Pantai in Pekanbaru during an expedition in November 2008 and cover 1 day (~24 h) as explained in more detail by Rixen et al. (2010). Oxygen saturation from the Itchen and Lark rivers was obtained from Butcher et al. (1927) and was plotted against the hours of a day.

of <133 μM, and an oxygen concentration of 60 μM is suggested as threshold defining the upper limit of hypoxia in fisheries (Ekau et al., 2010). In view of these thresholds, oxygen deficiency appears to be a critical environmental vector in the Siak. It explains the adaptation of freshwater fish species to peat draining rivers as mentioned before. However, it suggests also that oxygen depletion might have caused the reported periodical occurrence of mass fish mortalities in the Siak, whereas other causes such as water discharges cannot be ruled out. The inverse correlation between oxygen and DOC concentrations suggests that peat soil leaching and the subsequent DOC decomposition in the river are main factors controlling the oxygen concentration in the Siak.

The box diffusion model was used to project possible consequences of further peat soil leaching related to land conversion. Driven by DOC concentrations of 520 μM as initially measured in the S. Tapung Kiri and a two-step increase at river km 105 and km 215, the resulting calculated oxygen concentrations match the measured oxygen quite well (Fig. 3.11). Model calculations suggest that an increase of DOC concentrations by 15% is sufficient to cause anoxia in the Siak River. Considering that land use and land cover changes enhanced peat soil leaching by 70% already, this suggests that small changes in land management can easily lead to anoxic events and mass mortality of fishes in the Siak.

### *3.3.2.2 Sources and fate of nutrients*

In the Siak River, DIN concentrations ranged between 2-54 μM during all expeditions except in March 2004, where they reached values of up to 65 μM (Fig. 3.13). Ignoring the

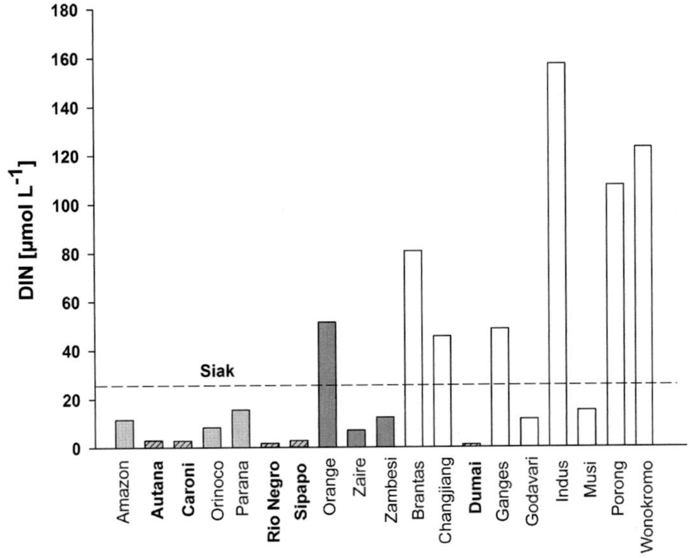


**FIGURE 3.13** Mean DIN concentration of the Siak (dashed line) compared with that of other tropical rivers in South America (light gray bars), Africa (dark gray bars), and Asia (white bars). Striped bars and bold labels indicate blackwater rivers. *DIN*, dissolved inorganic nitrogen. *Figure taken from Baum and Rixen (2014).*

exceptional period in 2004, the mean DIN concentration in the Siak River mainstream amounted to 25 μM (Fig. 3.13), which is much lower than in other Asian rivers, but it exceeds by far DIN concentrations measured in other blackwater rivers. Since blackwater rivers mostly drain peatlands, they are nutrient-poor and have a mean DIN concentration <3 μM (Castillo et al., 2004; Lewis and Weibezahn, 1981).

In addition to nutrient leaching from soils, the decomposition of organic matter is the main nutrient source in rivers. POC concentrations in the Siak are generally low (Baum et al., 2007), indicating that the decomposition of dissolved organic matter is the main nutrient source in the Siak. Simulations with the box diffusion model displaying an increasing DIN concentration toward the Siak estuary corroborate this (Fig. 3.11C). The decomposition of dissolved organic matter continuously releases nutrients, which remain unutilized in the absence of photosynthesis. Accordingly, nutrients accumulate in the river. However, while the calculated DIN concentrations in the estuary agree quite well with the measured ones, those between the S. Tapung Kiri and estuary are much lower than the measured ones. This discrepancy is probably related to the fact that the model does not account for other important nutrient sources such as soil leaching and wastewater discharge.

Industrial plants (e.g., petroleum, gas, rubber, paper, and fiber processing industries) as well as rubber, oil palm, and fiber plantations are located along the Siak. Wastewaters from rubber and paper processing industries are known to be heavily enriched in DIN (Agamuthu, 1999; Korhonen et al., 2004). Because of a lack of sewage purification, industrial and domestic wastewaters are discharged untreated into the Siak via wastewater channels (Rixen et al., 2010). Wastewater channels draining Pekanbaru had very high mean DIN (1170 μM in March 2004, 548 μM in September 2004) concentrations, much higher than the mean DIN concentration of 25 μM in the Siak River (Baum and Rixen, 2014). Phosphate concentrations in the Siak River were mostly <5 μM and generally followed the DIN trend, except for September 2004, when concentrations of up to 197 μM were observed at single stations along the river, which were attributed to wastewater discharges. However, mass balance calculations indicate that the amount of wastewater discharge is not sufficient to sustain the high DIN concentration in the Siak (Baum and Rixen, 2014).

March is usually the time when oil palm plantations are fertilized (Sinarmas, plantation manager, pers. comm.). In general, each palm tree receives 2.5 kg N fertilizer per year, which is five times higher than N fertilization at other oil palm estates in Indonesia and elsewhere (FAO, 1977; FAO, 2005; Omoti et al., 1983; Schroth et al., 2000). This intensive fertilization is required to meet the nitrogen demand of oil palms, as they are growing on the nutrient-poor peat soils. According to our model calculations, the washout of <1% of the N fertilizer is sufficient to explain the high DIN concentration observed in the Siak River in March 2004 (Baum and Rixen, 2014). Since the application of fertilizer generally increases the microbial activity in soils, it is conceivable that it also contributed to the high DOC and low oxygen concentrations in the Siak (Fig. 3.11A and B).

## 3.4 Governance and management programs

River, river basin, and watershed management became issues of national political importance in Indonesia in the course of the 20th century, in particular on the island of Java, where a naturally highly dynamic environment was profoundly transformed by humans early on. Following an era of intensive anthropogenic environmental transformations through forest exploitation and plantation establishment under colonial rule and agricultural expansion in the 19th and early 20th century, concerns over upland degradation and their potential impacts on dry season water flows and flood frequency in the lowlands were raised in the 1920/30s (Galudra and Sirait, 2009; Lukas, 2015).

From the 1960s onward, state-led river, river basin, and watershed management interventions have served as state building and development strategies and linked environmental concerns with politics over forest control (Galudra and Sirait, 2009; Lukas, 2015). The political focus on lowlands and uplands, the framing of the environment and environmental issues, and the kinds of interventions thereby shifted over time in line with changing national and international political interests and discourses (Lukas and Flitner 2019). From the late 1960s onward, some of Java's major river basins, including the Brantas, Citanduy, and Solo River basins, were targeted by internationally and nationally funded river basin development projects (Fujimoto, 2013; Lukas, 2015). These projects, which served economic development, food security, and state building goals, focused on agricultural expansion, reservoir construction, irrigation, and flood protection in the lowlands and the establishment of corresponding river basin management authorities under the Ministry of Public Works (Fujimoto, 2013; Lukas and Flitner, 2019). The land use changes and river regulations pushed in the frame of these projects substantially altered many of Java's riverscapes.

Political attention then gradually turned to the uplands in the 1970/80s (Lukas and Flitner, 2019). Substantial international and national funds have since been invested into watershed conservation, with a focus on agricultural extension, field terracing, and tree planting on upland farmers' private plots, aiming to reduce sediment input into rivers, reservoirs, irrigation schemes, and coastal waters (Lukas, 2015, 2017a). In this context, watershed management authorities were established under the Ministry of Forestry.

The limited effects of these watershed conservation efforts on river sediment loads were first attributed to various procedural shortcomings in program design and implementation (Purwanto, 1999; USAID, 1984; USAID, 1985; ADB, 1996). Our recent research in the Citanduy River watershed has shown that the one-sided political focus of watershed discourses and interventions on upland farmers' private lands and the corresponding neglect of a large range of other sediment sources have limited the effectiveness of watershed management (Lukas, 2015). In addition to farmers' rainfed agricultural plots, the anthropogenic causes of high river sediment loads that have been largely neglected to date include various land use transformations of the 19th and early 20th centuries, erosion on contested state forest and plantation lands, state forest management practices, slope cuts to enlarge agricultural fields in valley floors, agriculture in riparian zones, erosion from roads, trails, and settlements, and finally the river channel and floodplain modifications pushed in the frame of the earlier river basin development projects (Lukas, 2017b). Of particular relevance are historically rooted land conflicts and widespread conflicts over forest resources; many conflict lands are hot spots of erosion, contributing to high river sediment loads (Lukas, 2014, 2015). Through political and institutional entanglements of state-led watershed protection with the management and control of extensive forest tracts claimed by the state forest corporation, these conflicts have also contributed to one-sided watershed debates. In these debates, forester-dominated watershed agencies tend to put all blame on smallholders, while sidelining the roles of state forest management practices and the various other causes of erosion (Lukas, 2015). This has severely limited the effectiveness of watershed conservation (Lukas, 2017a).

While Java's rivers and watersheds were heavily transformed by humans by the 1980s already, anthropogenic impacts on many rivers and watersheds in Sumatra, Kalimantan, and other less densely populated areas of Indonesia have drastically intensified over the past few decades. Water flows and sediment loads have been affected by rapid expansions of various land and natural resources uses, including large-scale oil palm plantation development (e.g., Merten et al., 2016) and mining (e.g., Stapper, 2006).

The continuous commitment of the Indonesian state to watershed protection is reflected in the recurring declaration of "priority watersheds", watersheds with a large proportion of "critical", i.e. erosion-prone, lands prioritized for conservation efforts. The increase in their reported numbers from 22 in 1984 to 42 in 2000 (Nugroho et al., 2004) and to 118 by 2015 (Kurnia, 2015) indicates not only changes in watershed conditions and evaluation criteria, but also a clear political commitment to watershed conservation.

In the mid-1990s, the potential of watershed payment for environmental services (PES) schemes to address long-standing watershed management issues started to gain traction globally (Jack et al., 2008; Stanton et al., 2010). Such mechanisms, which cover a diverse range of approaches ranging from those that are rather market-oriented to those that support collective action, aim to incentivize land and resource practices that reduce negative environmental externalities through interventions that have positive environmental impacts. In Indonesia, interest in piloting such mechanisms started to become evident in the early 2000s (Heyde, 2016). By then, pilot PES schemes had been implemented in several watersheds across Indonesia. The initiatives tended to be rather small-scale, initiated by external nongovernmental actors and tended to have rather uncertain long-term prospects (Heyde, 2017; Heyde et al., 2012). As is frequently the case in watershed PES schemes, there were often challenges in showing environmental impacts (Ferraro, 2011; Heyde, 2017; Pattanayak et al., 2010; Sommerville et al., 2009).

In the mid-2000s, a small-scale PES initiative was piloted in the upper Brantas River watershed. The project was mainly funded by an external donor, with some support from the parastatal Brantas River Basin Operator (Perusahaan Umum Jasa Tirta 1, PJT1). The initiative aimed at linking small-scale farmers in two villages with PJT1 in an effort to explore options for reducing sediment input to a hydropower reservoir. Farmers' groups received support for tree planting and soil conservation activities, with the assumption that these would reduce the river sediment load. There was no attempt to directly quantify these expected environmental impacts to link them explicitly to rewards to farmers for their actions, something that is particularly challenging in multifunctional landscapes with many different land use actors (Heyde, 2017). The initiative was not continued by PJT1 once the pilot project had been completed.

While debate continues about the economic and environmental efficacy of watershed PES approaches as a way of achieving different benefits (Salzman et al., 2018; Wunder et al., 2018), a comprehensive analysis of sedimentation in the context of the Segara Anakan Lagoon and its watershed in southern Java sounded a note of caution. In that case, it was determined that PES would likely have little impact on reducing soil erosion in the watershed and sedimentation of the lagoon (Heyde et al., 2017).

The work from the Segara Anakan area highlighted the need to focus on resolving long-standing sociopolitical issues and conflicts rather than introducing new PES approaches. Core challenges in that area are also evident in other Indonesian watersheds. These include tenure conflicts, overlapping and conflicting government policies and programs, jurisdictional conflicts between government agencies, weak sharing of information, and, importantly, a deep and historically rooted lack of trust between small-scale resources users and state agencies. Under circumstances where these types of issues have not been adequately addressed, introducing PES mechanisms could in fact increase tensions rather than working to resolve environmentally damaging conflicts over resources (Heyde et al., 2017).

Knowledge gaps and directions of future research

- Nutrient and organic matter sources have to be better constrained and fluxes need to be quantified to assess the potential impact on coastal ecosystem health and to improve water quality management. Major contributions in the Brantas are as follows: (1) dissolved inorganic nitrogen (nitrate) and particulate organic matter are mainly derived from the agriculture-dominated hinterland, (2) dissolved organic matter and dissolved inorganic nitrogen (ammonium) are resulting from downstream sewage inputs of Surabaya and other cities, and (3) dissolved inorganic nitrogen (ammonium) and dissolved and particulate organic matter are released from lowland aquaculture ponds. Implementing targeted management measures requires prior source identification and flux quantification.
- Sediment budgets need to take into account the entire range of sediment sources and the reductions of sediment flux related to river damming.
- The societal causes of soil erosion and land use patterns and changes need to be better understood in many river catchments, based on rigorous, open-ended empirical research. Such research needs to take into account historical trajectories, link material and social dynamics, and cover processes, structures, and actors across all relevant scales within and beyond the respective watersheds.

**Implications/recommendations for policy and society**

- A reduction of nutrient, organic matter, and untreated sewage inputs from agriculture, aquaculture, and urban settlements is highly recommended to maintain a water quality that does not impair the health of ecosystems and humans.
- Initiatives to reduce river sediment loads need to consider a broader range of sediment sources than has been the case to date. In addition to upland smallholders' agricultural plots, which have been the focus of watershed debates and interventions to date, a large number of other sediment sources need to be taken into account and addressed, including, for example, erosion-prone state forest and plantation lands, slope cuts in valley floors, agriculturally used riparian zones, as well as roads, trails, and settlements.
- The strongly reduced river sediment load by damming causes sediment starvation at the coast and makes it more vulnerable to increasing erosion and habitat loss under the accelerating sea level rise. This should be taken into account when decisions on regulations of hydrology are made.
- Watershed debates and management have been dominated by state forest institutions to date. The corresponding entanglement of interests of forest production and control with watershed policies has impeded open-ended debates over the entire range of watershed issues and their causes. Watershed management should thus be disentangled from state forest management institutionally.
- Market-based natural resource management instruments such as PES need to be considered with caution and only after being sure that other governance challenges (unclear land and resource tenure, unclear or overlapping policies and programs on the ground, and lack of trust between small-scale farmers and state agencies) do not stand in the way of the potential success of a PES approach.

# Acknowledgments

We thank the large number of enthusiastic students from Indonesia, Germany, and other parts of the world, who invested a lot of time and effort in conducting thesis projects or internships. We are also grateful for the assistance provided by numerous technicians from the participating organizations. The research reported here was made possible by the continuing financial and administrative support of the German Federal Ministry of Education and Research (Grant Nos. 03F0301, 03F0456, 03F0644), the Indonesian Ministry for Research and Technology (RISTEK), the Indonesian Ministry for Maritime Affairs and Fisheries (KKP), and the German Academic Exchange Service (DAAD).

# References

ADB, 1996. Report and Recommendation of the President to the Board of Directors on Proposed Loans and Technical Assistance grant to the Republic of Indonesia for the Segara Anakan Conservation and Development Project. Asian Development Bank.

Adi, S., Jänen, I., Jennerjahn, T.C., 2013. History of development and attendant environmental changes in the Brantas River basin, Java, Indonesia, since 1970. Asian Journal of Water, Environment and Pollution 10, 5–15.

Agamuthu, P., 1999. Specific biogas production and role of packing medium in the treatment of rubber thread manufacturing industry wastewater. Bioprocess Engineering 21, 151–155.

Aldrian, E., Chen, C.-T.A., Adi, S., Prihartanto, S.N., Nugroho, S.P., 2008. Spatial and seasonal dynamics of riverine carbon fluxes of the Brantas catchment in East Java. Journal of Geophysical Research 113, G03029. https://doi.org/10.1029/2007JG000626.

Aldrian, E., Djamil, Y.S., 2008. Spatio-temporal climatic change of rainfall in East Java, Indonesia. International Journal of Climatology 28, 435–448.

Alianto, Adiwilaga, E.M., Damar, A., Harris, E., 2010. Estimates on small pelagic fish potential based on the primary production approach in sea waters. In: Proceedings of the 6th Annual National Seminar of the Results of Fisheries and Marine Research. University of Gajah Mada Yogyakarta, ISBN 978-979-19942-0-0.

Alkhatib, M., Jennerjahn, T.C., Samiaji, J., 2007. Biogeochemistry of the Dumai River estuary, Sumatra, Indonesia, a tropical blackwater river. Limnology and Oceanography 52, 2410–2417.

Badan Pusat Statistik, 2019. Statistics of Indonesia. https://www.bps.go.id.

Baum, A., Rixen, T., 2014. Dissolved inorganic nitrogen and phosphate in the human affected blackwater river Siak, central Sumatra, Indonesia. Asian Journal of Water Environment and Pollution 11, 13–24.

Baum, A., Rixen, T., Samiaji, J., 2007. Relevance of peat draining rivers in Central Sumatra for the riverine input of dissolved organic carbon into the ocean. Estuarine, Coastal and Shelf Science 73, 563–570.

Bhat, A., Ramu, K., Kemper, K., 2005. Institutional and Policy Analysis of River basin Management: The Brantas River, East Java, Indonesia. World Bank Policy Research Working. Paper no. 3611.

Butcher, R.W., Pentelow, F.T.K., Woodley, J.W.A., 1927. Diurnal variations of the gaseous contents of river waters. Biochemistry Journal 21, 945–957.

Castillo, M.M., Allan, J.D., Sinsabaugh, R.L., Kling, G.W., 2004. Seasonal and interannual variation of bacterial production in lowland rivers of the Orinoco basin. Freshwater Biology 49, 1400–1414.

Cepeda, J., Smebye, H., Vangelsten, B., Nadim, F., Muslim, D., 2010. Landslide Risk in Indonesia. Global Assessment Report on Disaster Risk Reduction. United Nations (UN) International Strategy for Disaster Reduction (ISDR). Retrieved from http://www.preventionweb.net/english/hyogo/gar/2011/en/bgdocs/Cepeda_et_al._2010.pdf.

Chaussard, E., Amelung, F., Abidin, H., Hong, S.H., 2013. Sinking cities in Indonesia: ALOS PALSAR detects rapid subsidence due to groundwater and gas extraction. Remote Sensing of Environment 128, 150–161.

D'Hondt, L.Y., 2019. Addressing Industrial Pollution in Indonesia: The Nexus between Regulation and Redress Seeking. Dissertation University of Leiden, ISBN 978 94 028 1697 6.

Damar, A., 2003. Effects of enrichment on nutrient dynamics, phytoplankton dynamics and productivity in Indonesian tropical waters: a comparison between Jakarta bay, Lampung bay and Semangka bay. In: Berichte aus dem Forschungs- und Technologiezentrum Westküste der Universität Kiel, Nr. 29. Büsum.

Damar, A., 2012. Net phytoplankton community structure and its biomass dynamics in the Brantas River estuary, Java, Indonesia. In: Subramanian, V. (Ed.), Coastal Environments: Focus on Asian Regions. Springer, Capital Publishing Company, Dordrecht, Heidelberg, New York, London, New Delhi, pp. 173–189.

Damar, A., Colijn, F., Hesse, K.-J., Wardiatno, Y., 2012. The eutrophication states of Jakarta, Lampung and Semangka Bays: nutrient and phytoplankton dynamics in Indonesian tropical waters. Journal of Tropical Biology and Conservation 9 (1), 61–81.

Damar, A., Hesse, K.-J., Colijn, F., Vitner, Y., 2019. The eutrophication states of the Indonesian sea large marine ecosystem: Jakarta Bay, 2001–2013. Deep-Sea Research Part II 163, 72–86.

Davies, R., Manga, M., Tingay, M., Lusianga, S., Swarbrick, R.R., Sawolo, et al., 2010. (2009) the Lusi mud volcano controversy: was it caused by drilling? Marine and Petroleum Geology 27, 1651–1657.

Degens, E.T., Kempe, S., Richey, J.E., 1991. Summary: biogeochemistry of major world rivers. In: Degens, E.T., Kempe, S., Richey, J.E. (Eds.), Biogeochemistry of Major World Rivers, SCOPE 42. Wiley, Chichester, pp. 323–347.

Edwards, A.C., Cook, Y., Smart, R., Wade, A.J., 2000. Concentrations of nitrogen and phosphorus in streams draining the mixed land-use Dee Catchment, north-east Scotland. Journal of Applied Ecology 37, 159–170.

Ekau, W., Auel, H., Pörtner, H.O., Gilbert, D., 2010. Impacts of hypoxia on the structure and processes in pelagic communities (zooplankton, macro-invertebrates and fish). Biogeosciences 7, 1669–1699.

Ertsen, M., Setyorini, D., Nooy, C., van Beusekom, M., Boogaard, F., Arisandi, P., et al., 2018. Mainstreaming Water Quality in River basin Management in the Brantas River Basin, Indonesia. International Conference Water Science for Impact, Wageningen, Netherlands (Abstract).

Fallahi, M.J., Obermann, A., Lupi, M., Karyono, K., Mazzini, A., 2017. The plumbing system feeding the Lusi eruption revealed by ambient noise tomography. Journal of Geophysical Research – Solid Earth 122, 8200–8213.

FAO, 1977. The Oil Palm, FAO Economic and Social Development Series, Better Farming Series 24. Rome.

FAO, 2005. Fertilizer Use by Crop in Indonesia. FAO, Food and Agriculture Organization of the United Nations, Rome.

FAO/UNESCO, 2003. Digital Soil Map of the World and Derived Soil Properties.

Fatimah, I., 2017. Performance of Local Governments in Regulating Industrial Water Pollution: An Empirical Study on Norm-Setting, Monitoring and Enforcement by the Environmental Agencies of East Java Province, and the Districts Gresik and Mojokerto (Summary in English).

Ferraro, P.J., 2011. The future of payments for environmental services. Conservation Biology 25 (6), 1134–1138. https://doi.org/10.1111/j.1523-1739.2011.01791.x.

Fujimoto, K., 2013. Brantas river basin development plan of Indonesia. In: Aid as Handmaiden for the Development of Institutions. Springer, pp. 161–194.

Gaillardet, J., Dupre, B., Louvat, P., Allegre, C.J., 1999. Global silicate weathering and $CO_2$ consumption rates deduced from the chemistry of large rivers. Chemical Geology 159, 3–30.

Galudra, G., Sirait, M., 2009. A discourse on Dutch colonial forest policy and science in Indonesia at the beginning of the 20th century. International Forestry Review 11, 524–533. https://www.jstor.org/stable/43739830.

Giosan, L., Clift, P.D., Macklin, M.G., Fuller, D.Q., Constantinescu, S., Durcan, J.A., et al., 2012. Fluvial landscapes of the Harappan civilization. Proceedings of the National Academy of Sciences 109, E1688–E1694.

Global Precipitation Climatology Centre, 2018. www.dwd.de/EN/ourservices/gpcc/gpcc.html.

Goolsby, D.A., Battaglin, W.A., Aulenbach, B.T., Hooper, R.P., 2000. Nitrogen flux and sources in the Mississippi River basin. The Science of the Total Environment 248, 75–86.

Herbeck, L.S., Unger, D., Wu, Y., Jennerjahn, T.C., 2013. Effluent, nutrient and organic matter export from shrimp and fish ponds causing eutrophication in coastal and back-reef waters of NE Hainan, tropical China. Continental Shelf Research 57, 92–104. https://doi.org/10.1016/j.csr.2012.05.006.

Heyde, J., 2016. Environmental governance and resource tenure in times of change: experience from Indonesia. In: Dissertation in Fulfilment of the Requirements of the Doctoral Committee. Dr.rer.pol.) of the University of Bremen (University of Bremen). Retrieved from. http://elib.suub.uni-bremen.de/edocs/00105588.pdf.

Heyde, J., 2017. Conditionality in practice: experience from Indonesia. In: Namirembe, S., Leimona, B., van Noordwijk, M., Minang, P.A. (Eds.), Co-investment in Ecosystem Services: Global Lessons from Payment and Incentive Schemes. Retrieved from. http://www.worldagroforestry.org/sites/default/files/Ch26_ConditionalityInPractice_ebook.pdf.

Heyde, J., Lukas, M.C., Flitner, M., 2012. Payments for Environmental Services in Indonesia: A Review of Watershed-Related Schemes. Artec paper Nr. 186.

Heyde, J., Lukas, M.C., Flitner, M., 2017. Payments for Environmental Services: A New Instrument to Address Long-Standing Watershed and Coastal Issues in Indonesia? Policy Paper. Artec paper Nr. 213.

Hirt, U., Kreins, P., Kuhn, U., Mahnkopf, J., Venohr, M., Wendland, F., 2012. Management options to reduce future nitrogen emissions into rivers: a case study of the Weser River basin, Germany. Agricultural Water Management 115, 118–131.

Hoekstra, P., 1989. Hydrodynamics and depositional processes of the Solo and Porong deltas, East Java, Indonesia. In: Proceedings of the KNGMG Symposium Coastal Lowlands, Geology and Geotechnology. Kluwer, Dordrecht, The Netherlands, pp. 161–173.

Hoekstra, P., Nolting, R.F., van der Sloot, H.A., 1989. Supply and dispersion of water and suspended matter of the rivers Solo and Brantas into the coastal waters of east Java, Indonesia. Netherlands Journal of Sea Research 23, 501–515.

Holtermann, P., Burchard, H., Jennerjahn, T., 2009. Hydrodynamics of the Segara Anakan lagoon. Regional Environmental Change 9, 245–258.

Hooijer, A., Page, S., Canadell, J.G., Silvius, M., Kwadijk, J., Wösten, H., et al., 2010. Current and future $CO_2$ emissions from drained peatlands in Southeast Asia. Biogeosciences 7, 1505–1514.

Humborg, C., Ittekkot, V., Cociasu, A., von Bodungen, B., 1997. Effect of Danube River dam on Black Sea biogeochemistry and ecosystem structure. Nature 386, 385–388.

Jack, B.K., Kousky, C., Sims, K.R.E., 2008. Designing payments for ecosystem services: lessons from previous experience with incentive-based mechanisms. Proceedings of the National Academy of Sciences 105, 9465–9470. https://doi.org/10.1073/pnas.0705503104.

Jänen, I., Adi, S., Jennerjahn, T.C., 2013. Spatio-temporal variations in nutrient supply of the Brantas River to Madura Strait coastal waters, Java, Indonesia, related to human alterations in the catchment and a mud volcano. Asian Journal for Water, Environment and Pollution 10, 73–94.

Jennerjahn, T.C., Ittekkot, V., Klöpper, S., Adi, S., Nugroho, S.P., Sudiana, N., et al., 2004. Biogeochemistry of a tropical river affected by human activities in its catchment: Brantas River estuary and coastal waters of Madura Strait, Java, Indonesia. Estuarine, Coastal and Shelf Science 60, 503–514.

Jennerjahn, T.C., Jänen, I., Propp, C., Adi, S., Nugroho, S.P., 2013. Environmental impact of mud volcano inputs on the anthropogenically altered Porong River and Madura Strait coastal waters, Java, Indonesia. Estuarine, Coastal and Shelf Science 130, 152–160. https://doi.org/10.1016/j.ecss.2013.04.007.

Jennerjahn, T.C., Klöpper, S., 2013. Does high silicate supply control phytoplankton and particulate organic matter composition in two eutrophied reservoirs in the Brantas River catchment, Java, Indonesia? Asian Journal for Water, Environment and Pollution 10, 41–53.

Jennerjahn, T.C., Knoppers, B.A., Souza, W.F.L., Brunskill, G.J., Silva, E.I.L., Adi, S., 2006. Factors controlling dissolved silica in tropical rivers. In: Ittekkot, V., Unger, D., Humborg, C., Tac An, N. (Eds.), The Silicon Cycle – Human Perturbations and Impacts on Aquatic Systems, SCOPE 66. Island Press, Washington, Covelo, London, pp. 29–51.

Jennerjahn, T.C., Nasir, B., Pohlenga, I., 2009. Spatio-temporal variation of dissolved inorganic nutrients in the Segara Anakan lagoon, Java, Indonesia. Regional Environmental Change 9, 259–274.

JICA, 1998. Development of the Brantas River Basin: Cooperation of Japan and Indonesia. Japan International Cooperation Agency, Tokyo.

Joosten, H., 2010. The Global Peatland $CO_2$ Picture: peatland status and drainage related emissions in all countries of the world. Wetlands International. Ede. www.wetlands.org.

Kaswadji, R.F., Widjaja, F., Wardiatno, Y., 1993. Produktifitas primer dan laju pertumbuhan fitoplankton di perairan pantai Bekasi (Phytoplankton primary productivity and growth rate in the coastal waters of Bekasi Regency). Jurnal Ilmu-Ilmu Perairan Dan Perikanan Indonesia 1 (2), 1–15.

Korhonen, K., Liukkonen, T., Ahrens, W., Astrakianakis, G., Boffetta, B., Burdorf, A., et al., 2004. Occupational exposure to chemical agents in the paper industry. International Archives of Occupational and Environmental Health 77, 451–460.

Kurnia, K., 2015. 118 Sungai besar di Indonesia kritis, siap-siap darurat banjir. Retrieved February 6, 2017, from http://www.galamedianews.com/nasional/39175/reviews-all.html.

Lavigne, F., Gunnell, Y., 2006. Land cover change and abrupt environmental impacts on Javan volcanoes, Indonesia: a long-term perspective on recent events. Regional Environmental Change 6, 86–100.

Lewis Jr., W.M., Weibezahn, F., 1981. The chemistry and phytoplankton of the Orinoco and Caroni rivers, Venezuela. Archiv Fur Hydrobiologie 91, 521–528.

Lowman, P., Yates, J., Masuoka, P., Montgomery, B., O'Leary, J., Salisbury, D., 1999. A digital tectonic activity map of the earth. Journal of Geoscience Education 47, 428–437.

Ludwig, W., Probst, J.-L., Kempe, S., 1996. Predicting the oceanic input of organic carbon by continental erosion. Global Biogeochemical Cycles 10, 23–41.

Lukas, M.C., 2014. Eroding battlefields: land degradation in Java reconsidered. Geoforum 56, 87–100. https://doi.org/10.1016/j.geoforum.2014.06.010.

Lukas, M.C., 2015. Reconstructing Contested Landscapes. Dynamics, Drivers and Political Framings of Land Use and Land Cover Change, Watershed Transformations and Coastal Sedimentation in Java, Indonesia. Doctoral thesis. University of Bremen, Germany. Available at: https://elib.suub.uni-bremen.de/edocs/00106383-1.pdf.

Lukas, M.C., 2017a. Konservasi daerah aliran sungai di Pulau Jawa, Indonesia: Terjebak dalam konflik sumberdaya hutan. University of Bremen. Artec-Paper 212, Sustainability Research Center (artec). https://www.uni-bremen.de/fileadmin/user_upload/sites/artec/Publikationen/artec_Paper/212_paper.pdf.

Lukas, M.C., 2017b. Widening the scope: linking coastal sedimentation with watershed dynamics in Java, Indonesia. Regional Environmental Change 17 (3), 901–914. https://doi.org/10.1007/s10113-016-1058-4.

Lukas, M.C., Flitner, M., 2019. Scalar fixes of environmental management in Java, Indonesia. Environment and Planning E: Nature and Space 2 (3), 565–589. https://doi.org/10.1177/2514848619844769.

Lung'ayia, H., Sitoki, L., Kenyanya, M., 2001. The nutrient enrichment of Lake Victoria (Kenyan waters). Hydrobiologia 458, 75–82.

Macklin, M.G., Lewin, J., 2015. The rivers of civilization. Quaternary Science Reviews 114, 228–244.

Maryantika, N., Lin, C., 2017. Exploring changes of land use and mangrove distribution in the economic area of Sidoarjo District, East Java using multi-temporal Landsat images. Information Processing in Agriculture 4, 321–332.

Mazzini, A., Nermoen, A., Krotkiewski, M., Podladchikov, Y., Planke, S., Svensen, H., 2009. Strike-slip faulting as a trigger mechanism for overpressure release through piercement structures. Implications for the Lusi mud volcano, Indonesia. Marine and Petroleum Geology 26, 1751–1765.

Mazzini, A., Svensen, H., Akhmanov, G.G., Aloisi, G., Planke, S., Malthe-Sørenssen, A., et al., 2007. Triggering and dynamic evolution of the LUSI mud volcano, Indonesia. Earth and Planetary Science Letters 261, 375–388.

Merten, J., Röll, A., Guillaume, T., Meijide, A., Tarigan, S., Agusta, H., et al., 2016. Water scarcity and oil palm expansion: social views and environmental processes. Ecology and Society 21 (2), 5. https://doi.org/10.5751/ES-08214-210205.

Middelburg, J.J., Nieuwenhuize, J., 2000. Uptake of dissolved inorganic nitrogen in turbid, tidal estuaries. Marine Ecology Progress Series 192, 79–88.

Miettinen, J., Shi, C., Liew, S.C., 2016. Land cover distribution in the peatlands of Peninsular Malaysia, Sumatra and Borneo in 2015 with changes since 1990. Global Ecology and Conservation 6, 67–78.

Milliman, J.D., Farnsworth, K.L., 2011. River Discharge to the Coastal Ocean – A Global Synthesis. Cambridge University Press, Cambridge.

Milliman, J.D., Farnsworth, K.L., Albertin, C.S., 1999. Flux and fate of fluvial sediments leaving large islands in the East Indies. Journal of Sea Research 41, 97–107.

Montgomery, D.R., 2007. Soil erosion and agricultural sustainability. Proceedings of the National Academy of Sciences 104, 13268–13272.

Moore, S., Evans, C.D., Page, S.E., Garnett, M.H., Jones, T.G., Freeman, C., et al., 2013. Deep instability of deforested tropical peatlands revealed by fluvial organic carbon fluxes. Nature 493, 660–663.

Nugroho, C., Priyono, S., Cahyono, S.A., 2004. Teknologi pengelolaan daerah aliran sungai: Cakupan, permasalahan, dan upaya penerapannya. Watershed management Technology: coverage, problems, and efforts for implementation. Balai Penelitian dan Pengembangan Teknologi Pengelolaan DAS, IBB. https://balittanah.litbang.pertanian.go.id/ind/dokumentasi/prosiding/mflp2004/nugroho.pdf.

Odum, H.T., 1956. Primary production in flowing waters. Limnology and Oceanography 1, 102–117. https://doi.org/10.4319/lo.1956.1.2.0102.

Ohte, H., Mitchell, M.J., Shibata, H., Tokuchi, N., Toda, H., Iwatsubo, G., 2001. Comparative evaluation on nitrogen saturation of forest catchments in Japan and northeastern United States. Water, Air and Soil Pollution 130, 649–654.

Omachi, T., Musiake, K., 2003. Changes of runoff mechanism of the Brantas River over the past 30 years. In: Proceedings of 2nd Conference of the Asia Pacific Association of Hydrology and Water Resources. Singapore. http://rwes.dpri.kyoto-u.ac.jp/~tanaka/APHW/APHW2004/proceedings/APHW2004proc.htm.

Omoti, U., Ataga, D.O., Isenmila, A.E., 1983. Leaching losses of nutrients in oil palm plantations determined by tension lysimeters. Plant and Soil 73, 365–376.

Page, S.E., Rieley, J.O., Banks, C.J., 2011. Global and regional importance of the tropical peatland carbon pool. Global Change Biology 17, 798–818.

Paryono, D.A., Susilo, S.B., Dahuri, R., Susenno, H., 2017. Sedimentasi delta sungai Citarum, kecamatan Muara Gembong, kabupaten Bekasi (sedimentation at delta of Citarum River, Muara Gembong district, Bekasi regency). Jurnal Penelitian Pengelolaan Daerah Aliran Sungai (Journal of Watershed Management Research) 1, 15–26.

Pattanayak, S.K., Wunder, S., Ferraro, P.J., 2010. Show me the money: do payments supply environmental services in developing countries? Review of Environmental Economics and Policy 4 (2), 254–274. https://doi.org/10.1093/reep/req006.

Ponnurangam, A.E., 2011. Biogeochemical Characterization of the Sources of Carbon and Nutrient Input in the Brantas River, Estuary and Coastal Waters of Madura Strait, Java, Indonesia. MSc Thesis. University of Bremen.

Posa, M.R.C., Wijedasa, L.S., Corlett, R.T., 2011. Biodiversity and conservation of tropical peat swamp forests. BioScience 61, 49–57.

Pristiwati, G.A., 2015. Assessment of the Effects of Land Use Changes on Spatio-Temporal Dynamics in the Flood Exposure: Case Study: The Brantas Watershed, East Java, Indonesia. MSc Thesis. Wageningen University. http://edepot.wur.nl/356177.

Propp, C., Jänen, I., Jennerjahn, T., 2013. Sources and degradation of sedimentary organic matter in coastal waters off the Brantas River estuary, Java, Indonesia. Asian Journal for Water, Environment and Pollution 10, 95–115.

Purwanto, E., 1999. Erosion, Sediment Delivery and Soil Conservation in an upland Agricultural Catchment in West Java, Indonesia. A Hydrological Approach in a Socio-Economic Context. Doctoral thesis. Faculty of Earth Sciences, Vrije Universiteit, Amsterdam.

Rabalais, N.N., Turner, R.E., Justic, D., Dortch, Q., Wiseman Jr., W.J., Sen Gupta, B.K., 2000. Gulf of Mexico biological system responses to nutrient changes in the Mississippi River. In: Hobbie, J. (Ed.), Estuarine Science: A Synthetic Approach to Research and Practice. Island Press, Washington, pp. 241–268.

Ramdani, F., Hino, M., 2013. Land use changes and GHG emissions from tropical forest conversion by oil palm plantations in Riau province, Indonesia. PLoS One 8 (7), e70323. https://doi.org/10.1371/journal.pone.0070323.

Ramu, K., 2004. Brantas River basin case study Indonesia. Background paper. In: Integrated River basin and the Principle of Managing Water Resources at the Lowest Approriate Level – when and Why Does it (Not) Work in Practice? World Bank. Http://Siteresources.Worldbank.Org/INTSAREGTOPWATRES/Resources/Indonesia_BrantasBasinFINAL.Pdf.

Richards, J.R., 2011. Report into the Past, Present and Future Social Impacts of Lumpur Sidoarjo. Technical Report. Humanitus Sidoarjo Fund.

Rixen, T., Baum, A., Pohlmann, T., Balzer, W., Samiaji, J., Jose, C., 2008. The Siak, a tropical black water river in central Sumatra on the verge of anoxia. Biogeochemistry 90, 129–140.

Rixen, T., Baum, A., Sepryani, H., Pohlmann, T., Jose, C., Samiaji, J., 2010. Dissolved oxygen and its response to eutrophication in a tropical black water river. Journal of Environmental Management 91, 1730–1737.

Salzman, J., Bennett, G., Carroll, N., Goldstein, A., Jenkins, M., 2018. The global status and trends of payments for ecosystem services. Nature Sustainability 1 (3), 136.

Sawolo, N., Sutriono, E., Istadi, B.P., Darmoyo, A.B., 2009. The LUSI mud volcano triggering controversy: was it caused by drilling? Marine and Petroleum Geology 26, 1766–1784.

Schroeder, F., Boer, M., Wijanarko, D.A., 2013. Development and application of the MERMAID water quality monitoring station in the Brantas River, Java, Indonesia. Asian Journal of Water, Environment and Pollution 10, 25–39.

Schroeder, F., Knauth, H.-D., 2013. Water quality time-series data of the lower Brantas River, East Java, Indonesia: results from an automated water quality monitoring station. Asian Journal of Water, Environment and Pollution 10, 55–72.

Schroeder, F., Knauth, H.-D., Pfeiffer, K., Nöhren, I., Duwe, K., Jennerjahn, T., et al., 2004. Water quality monitoring of the BRANTAS Estuary, Indonesia. In: Oceans '04 MTS/IEEE Techno-Ocean '04. Kobe. IEEE Cat. No.04CH37600), pp. 115–120.

Schroth, G., Rodrigues, M.R.L., D'Angelo, S.A., 2000. Spatial patterns of nitrogen mineralization, fertilizer distribution and roots explain nitrate leaching from mature Amazonian oil palm plantation. Soil Use and Management 16, 222–229.

Seeley, C.M., 1969. The diurnal curve in estimates of primary productivity. Chesapeake Science 10, 322–326.

Shibata, H., Kuraji, K., Toda, H., Sasa, K., 2001. Regional comparison of nitrogen export to Japanese forest streams. The Scientific World 1, 572–580.

Sidik, F., Neil, D., Lovelock, C.E., 2016. Effect of high sedimentation rates on surface sediment dynamics and mangrove growth in the Porong River, Indonesia. Marine Pollution Bulletin 107, 355–363.

Siegel, H., Stottmeister, I., Reißmann, J., Gerth, M., Jose, C., Samiaji, J., 2009. Siak River system – East-Sumatra: characterisation of sources, estuarine processes, and discharge into the Malacca strait. Journal of Marine Systems 77, 148–159.

Simarmata, H., 2018. Phenomenology in Adaptation Planning. An Empirical Study of Flood-Affected People in Kampung Muara Baru Jakarta (Singapore).

Siregar, M.R.T., Hiskia, Wahyu, Y., Wiranto, G., Mashari, I., 2004. On-line water quality monitoring on Brantas river East Java Indonesia. In: IEEE International Conference on Semiconductor Electronics. Kuala Lumpur, ISBN 0-7803-8658-2, p. 5. https://doi.org/10.1109/SMELEC.2004.1620825.

Smith, S.V., Swaney, D.P., Talaue-McManus, L., Bartley, J.D., Sandhei, P.T., McLaughlin, C.J., et al., 2003. Humans, hydrology, and the distribution of inorganic nitrogen loading to the ocean. BioScience 53, 235–245.

Smith, V.H., Tilman, G.D., Nekola, J.C., 1999. Eutrophication: impacts of excess nutrient inputs on freshwater, marine, and terrestrial ecosystems. Environmental Pollution 100, 179–196.

Sommerville, M.M., Jones, J.P.G., Milner-Gulland, E.J., 2009. A revised conceptual framework for payments for environmental services. Ecology and Society 14 (2), 34.

Sprague, L.A., Hirsch, R.M., Aulenbach, B.T., 2011. Nitrate in the Mississippi River and its tributaries, 1980 to 2008: are we making progress? Environmental Science and Technology 45, 7209–7216.

Stanton, T., Echavarria, M., Hamilton, K., Ott, C., 2010. State of Watershed Payments: An Emerging Marketplace. Ecosystem Marketplace. Available online: http://www.forest-trends.org/documents/files/doc_2438.pdf.

Stapper, D., 2006. Artisanal Gold Mining, Mercury and Sediment in central Kalimantan, Indonesia. Canada: Master thesis, University of Victoria, Victoria.

Steinberg, F., 2007. Jakarta: environmental problems and sustainability. Habitat International 31 (3), 354–365.

Subijanto, T.W., 2010. Integrated water resource management and water governance to improve water security in the Brantas River basin, Indonesia. Proceedings River basin Study Visit. Spain. www.tecniberia.es.

Sunarto, 2001. Pola hubungan intensitas cahaya dan nutrient dengan produktivitas primer phytoplankton di Teluk Hurun Lampung. M.Sc. Thesis. Graduate school of Bogor Agricultural University.

Syvitski, J.P.M., Cohen, S., Kettner, A.J., Brakenridge, G.R., 2014. How important and different are tropical rivers? — an overview. Geomorphology 227, 5–17.

Syvitski, J.P.M., Vörösmarty, C.J., Kettner, A.J., Green, P., 2005. Impact of humans on the flux of terrestrial sediment to the global coastal ocean. Science 308, 376–380.

Tambaru, R., 2000. Pengaruh intensitas cahaya pada berbagai waktu inkubasi terhadap produktivitas primer fitoplankton di perairan Teluk Hurun (The influence of incubation time on phytoplankton primary productivity in Hurun Bay, Indonesia). M.Sc. Thesis. Graduate school of Bogor Agricultural University.

Tingay, M.R.P., Rudolph, M.L., Manga, M., Davies, R.J., Wang, C.-Y., 2015. Initiation of the Lusi mudflow disaster. Nature Geoscience 8, 493–494.

Turner, R.E., Rabalais, N.N., 2003. Linking landscape and water quality in the Mississippi River basin for 200 years. BioScience 53, 563–572.

USAID, 1984. Citanduy II – Organizational Difficulties Hinder Implementation of an Ambitious Integrated Rural Development Project. Audit Report No. 2-497-84-04. Manila: Regional Inspector General for Audit.

USAID, 1985. Citanduy II Assessment. Special Evaluation (Washington).

Vacquer-Sunyer, R., Duarte, C.M., 2008. Thresholds of hypoxia for marine biodiversity. Proceedings of the National Academy of Sciences 40, 15452–15457.

Van Oldenborgh, G.J., Collins, M., Arblaster, J., Christensen, J.H., Marotzke, J., Power, S.B., et al., 2013. Annex I: atlas of global and regional climate projections. In: Stocker, T.F., Qin, D., Plattner, G.-K., Tignor, M., Allen, S.K., Boschung, J., et al. (Eds.), Climate Change 2013: The Physical Science Basis. Contribution of Working Group I to the Fifth Assessment Report of the Intergovernmental Panel on Climate Change. Cambridge University Press, Cambridge, United Kingdom and New York, NY, USA.

Van der Wulp, S.A., Damar, A., Ladwig, N., Hesse, K.-J., 2016. Numerical simulations of river discharges, nutrient flux and nutrient dispersal in Jakarta Bay, Indonesia. Marine Pollution Bulletin 110, 675–685.

Vetrita, Y., Cochrane, M.A., 2020. Fire frequency and related land-use and land-cover changes in Indonesia's peatlands. Remote Sensing 12, 5. https://doi.org/10.3390/rs12010005.

Wen, L.-S., Jiann, K.-T., Liu, K.-K., 2008. Seasonal variation and flux of dissolved nutrients in the Danshuei estuary, Taiwan: a hypoxic subtropical mountain river. Estuarine, Coastal and Shelf Science 78, 694–704.

Whitten, T., Soeriaatmadja, R.E., Afiff, S.A., 1996. The Ecology of Java and Bali. The Ecology of Indonesia Series Vol. II. Periplus Editions, Dalhousie University, Halifax, p. 1028.

Wit, F., Müller, D., Baum, A., Warneke, T., Pranowo, W.S., Müller, M., et al., 2015. The impact of disturbed peatlands on river outgassing in Southeast Asia. Nature Communications 6, 10155.

Wunder, S., Brouwer, R., Engel, S., Ezzine-de-Blas, D., Muradian, R., Pascual, U., et al., 2018. From principles to practice in paying for nature's services. Nature Sustainability 1 (3), 145.

Yoshino, K., Setiawan, Y., Shima, E., 2017. Land use analysis using time series of vegetation Index derived from satellite Remote Sensing in Brantas River watershed, East Java, Indonesia. Geoplanning: Journal of Geomatics and Planning 4, 109–120. https://doi.org/10.14710/geoplanning.4.2.109-120.

4

# Carbon cycle in tropical peatlands and coastal seas

Tim Rixen[1,7], Francisca Wit[1], Andreas A. Hutahaean[2], Achim Schlüter[1], Antje Baum[1], Alexandra Klemme[3], Moritz Müller[4], Widodo Setiyo Pranowo[5], Joko Samiaji[6], Thorsten Warneke[3]

[1]LEIBNIZ CENTRE FOR TROPICAL MARINE RESEARCH (ZMT), BREMEN, GERMANY; [2]MINISTRY OF MARITIME AFFAIRS, JAKARTA, INDONESIA; [3]INSTITUTE FOR ENVIRONMENTAL PHYSICS, UNIVERSITY OF BREMEN, GERMANY; [4]SWINBURNE UNIVERSITY OF TECHNOLOGY, SARAWAK CAMPUS, KUCHING, SARAWAK, MALAYSIA; [5]RESEARCH & DEVELOPMENT CENTER FOR MARINE & COASTAL RESOURCES (P3SDLP), JAKARTA, INDONESIA; [6]UNIVERSITY OF RIAU, PEKANBARU, INDONESIA; [7]INSTITUTE OF GEOLOGY, UNIVERSITÄT HAMBURG, GERMANY

## Abstract

*This chapter provides background information on peat and more specifically on Indonesian peatlands and their role in the global carbon cycle and summarizes information on human-induced $CO_2$ emissions caused by peat oxidation and fires. In contrast to these so-called on-site $CO_2$ emissions, not much was known about off-site $CO_2$ emissions prior to the joint Indonesian–German project Science for the Protection of Indonesian Coastal Ecosystems (SPICE). Off-site $CO_2$ emissions are $CO_2$ emissions caused by the mobilization of peat carbon along the land–ocean continuum, which were studied by us in the framework of SPICE. Our results allowed us to establish comprehensive carbon budgets showing that peatland preservation and restoration are crucial measures to combat global warming and mitigate climate change impacts caused. e.g., by sea level rise. Furthermore, we conducted socioeconomic experiments and used the established carbon budget to demonstrate the economic conflict that arises between restoration and transformation of peatlands into plantations.*

## Abstrak

*Bab ini memberikan informasi mengenai gambut dan lebih khusus lagi tentang lahan gambut di Indonesia dan peranannya dalam siklus karbon global, serta merangkum informasi mengenai emisi $CO_2$ yang disebabkan oleh aktivitas manusia sehingga mengakibatkan terjadinya oksidasi dan kebakaran lahan gambut. Terkait emisi $CO_2$ yang terdapat di tempat ini telah banyak di ketahui, namun hal yang kontras dengan informasi terkait emisi $CO_2$ di luar lokasi masih kurang diketahui sebelum program kerjasama riset SPICE antara Indonesia dan Jerman dilakukan. Emisi $CO_2$ di luar lokasi merupakan emisi $CO_2$ yang disebabkan oleh mobilisasi karbon yang berasal dari gambut di sepanjang kontinum darat hingga laut, yang diteliti dalam kerangka kerja program SPICE. Riset ini menunjukkan hasil pengukuran karbon budget yang komprehensif dimana aspek pelestarian dan restorasi lahan gambut merupakan upaya yang sangat penting untuk memerangi*

Science for the Protection of Indonesian Coastal Ecosystems (SPICE). https://doi.org/10.1016/B978-0-12-815050-4.00011-0 

*pemanasan global dan mitigasi perubahan iklim yang ditimbulkan, misalnya akibat kenaikan permukaan laut dan meningkatnya emisi $CO_2$. Lebih lanjut, dalam riset SPICE ini dilakukan juga eksperimen sosial ekonomi dan menggunakan pengukuran karbon budget yang telah ditetapkan untuk menunjukkan potensi konflik ekonomi yang mungkin terjadi antara aktivitas restorasi dan transformasi lahan gambut menjadi perkebunan, pertanian atau tujuan lainnya.*

## Chapter outline

**4.1 Introduction** ........ **85**
**4.2 Background information** ........ **86**
4.2.1 Peat ........ 86
4.2.2 Peatland types ........ 87
4.2.3 Vegetation and biodiversity ........ 88
4.2.4 Peatland distribution and carbon storage ........ 88
**4.3 Indonesian peatlands** ........ **89**
4.3.1 History of Indonesian peat swamps ........ 89
4.3.2 Peat properties ........ 90
4.3.3 Peat carbon accumulation ........ 91
4.3.4 Land use and cover changes in Indonesia ........ 93
4.3.5 The hydrological cycle of Indonesian peatlands ........ 94
**4.4 Peat carbon losses** ........ **97**
4.4.1 $CO_2$ emissions caused by peat and forest fires ........ 97
4.4.2 $CO_2$ emissions caused by peat soil oxidations ........ 98
4.4.3 Off-site $CO_2$ emission ........ 98
**4.5 Land–ocean continuum** ........ **99**
4.5.1 SPICE study area ........ 102
4.5.2 Dissolved organic carbon ........ 104
4.5.3 Dissolved organic carbon yields ........ 105
4.5.4 $CO_2$ emission from rivers ........ 106
4.5.5 Dissolved inorganic carbon yields ........ 109
4.5.6 Leaching and erosion ........ 109
4.5.7 Priming ........ 111
**4.6 Estuaries and the ocean** ........ **112**
4.6.1 Dissolved organic carbon ........ 112
*4.6.1.1 The microbial organic carbon pump in the ocean* ........ 112
*4.6.1.2 Dissolved organic carbon discharges into the ocean* ........ 113
*4.6.1.3 The fate of dissolved organic carbon in the ocean* ........ 114
4.6.2 $CO_2$ emissions from the coastal ocean ........ 115
4.6.3 Organic carbon burial ........ 116
4.6.4 The invisible carbon footprint ........ 117

4.6.5 The marine peat carbon budget ........ 118
4.6.6 Emission factors ........ 119
**4.7 Ecosystem $CO_2$ emissions ........ 120**
4.7.1 Net on-site ecosystem $CO_2$ exchange ........ 120
4.7.2 $CO_2$ emission from pristine peat swamps ........ 121
4.7.3 $CO_2$ emission from disturbed peatlands ........ 121
**4.8 Evaluation of $CO_2$ emissions ........ 123**
4.8.1 Climate response to cumulative emissions of $CO_2$ ........ 123
4.8.2 $CO_2$ reduction potential ........ 123
4.8.3 $CO_2$ emissions and land losses ........ 125
4.8.4 Climate pledges and gaps ........ 126
**4.9 Socioeconomic implications ........ 127**
4.9.1 REDD+ ........ 127
4.9.2 SPICE field experiments ........ 129
**4.10 Outlook ........ 131**
**References ........ 132**

## 4.1 Introduction

Peat swamps are carbon-rich and extremely fragile ecosystems. They occur in tropical, temperate, and boreal regions, play an important role in the global carbon cycle, and thus influence climate today as well as in the geological past. Today, tropical peat swamps are in the center of the sustainability conflict (WCED, 1987) that arises between the necessity to feed and increase the welfare of a growing population while preserving natural resources for the next generation. Due to the increasing demand for timber, pulp, fibers, palm oil, and other goods, plantations on peat became profitable even though peat is a problematic soil for agriculture. Therefore, it has long been spared from cultivation with the consequence that peatlands became one of the last few retreats for wildlife (Posa et al., 2011). Accordingly, more than one-third of all birds and mammals recorded in peat swamp forests have an IUCN Red List status, which means they are threatened, vulnerable, or endangered. Among them are, for instance, orangutans.

Drainage of peat swamps also increased the vulnerability of peat to fires (Turetsky et al., 2014). In 1997, peat and forest fires were running out of control, and acrid smoke from these fires covered Indonesia and neighboring countries in Southeast (SE) Asia (BBC News, 1998; Jim, 1999). The resulting mass mortalities of coral reefs and especially the severe health problems of the affected people caused international outcries (Abram et al., 2003; BBC News, 1998), which lead to the very first idea of the joint German–Indonesian project Science for the Protection of Indonesian Coastal Ecosystems (SPICE). In 2003, SPICE was initiated and covered a variety of environmental problems as shown in this book. This chapter addresses the issue of peat carbon and more specific impacts of peatland degradation on rivers and the coastal ocean. The global relevance of peat drainage and fire as $CO_2$ source to the atmosphere has long been recognized (Hooijer et al., 2006; Page et al., 2002), whereas impacts of peatland degradation on rivers and the coastal ocean were hardly studied before SPICE. To establish a comprehensive carbon budget including carbon fluxes along the land–ocean continuum, we compiled data on $CO_2$ emissions caused by drainage and fire and studied the fate of peat carbon in rivers and the ocean. The results will be evaluated in the context of international climate agreements and Indonesian climate pledges and with respect to their relevance for coastal protection. Furthermore, socioeconomical experiments were carried out to better understand problems associated with the implementation of imposed climate and environmental goals. Prior to discussing these topics, we will provide some background information on peat, its formation, and distribution as well as on some particularities of tropical peat in comparisons to peat swamps at higher latitudes.

## 4.2 Background information

### 4.2.1 Peat

Peat is an organic-rich material, which accumulates to form peat soils (Fig. 4.1). The organic carbon content, the thickness of the peat layer, and the period during which peat is water-saturated are main characteristics used to define peat soils that are also called "organic soils" and "histosols" (Joosten and Clarke, 2002; USDA, 1999).

Even though the characteristics used to define peat soils differ, e.g., in terms of soil thickness and periods during which peat is water-saturated, they agree on the soil organic carbon content, which must exceed 12%. Instead of organic carbon, also the organic matter content is used to define peat soils where at commonly a factor of 1.7 is

**FIGURE 4.1** Peat layers at the banks of the Siak river in the central Sumatran province Riau, Indonesia.

used to convert organic carbon into organic matter content. Both the organic matter and organic carbon content refer to the dry weight (Andriesse, 1988). In addition to carbon (12%–60%), other major organic matter–composing elements are oxygen (30%–40%), hydrogen (2%–6%), and nitrogen (3%–4%). However, peat is largely a leftover of organic matter, which was produced by plants and could not fully be respired by microbes and fungi in soils. This incomplete decomposition results from the chemical composition of organic matter produced by plants and specific environmental conditions developing in peat soils.

Plants contribute ~80% to the living biomass of the Earth's biosphere (Bar-On et al., 2018), and the chemical composition of organic matter produced by terrestrial plants differs from that produced by marine organisms (Cragg et al., 2020). Vascular plants, which are the most dominant clade among plants, produce high amounts of lignin, which forms a key structural element in wood and bark. Vascular plants and their derivate peat contain 5%–35% and 5.6%–25% lignin, respectively (Gandois et al., 2014; Smith et al., 1958). Chemically, lignin is a complex phenolic polymer, which is resistant against microbial decomposition. Stress caused by, e.g., herbivory or pathogen attack favors the production of phenolic compounds in plants to strengthen their secondary (defense) metabolisms (Yule et al., 2018). Since in nutrient-poor peat swamps, plants depend, in particular, on such a carbon-based defense mechanism, plants growing in peat swamps are even enriched in phenolic compounds (Yule et al., 2018). Nevertheless, there are enzymatic reactions capable of degrading lignin, which evolved in fungal and bacterial lineages (Nelsen et al., 2016).

Phenol oxidase is one of the most import lignin-degrading enzymes, and its activity is considered as a key factor controlling the preservation of terrestrial organic matter in soils and sediments (e.g., Pind et al., 1994). The phenol oxidase activity depends on the availability of oxygen and the pH, whereas low oxygen concentrations and acidic conditions lower the activity of phenol oxidase (Kang et al., 2018; Pind et al., 1994). Accordingly, the phenol oxidase activity is low in swamps, where pore waters hinder the penetration of oxygen from the atmosphere into soils and organic acids lower the pH to values ranging approximately between three and five (Andriesse, 1988; Dasgupta et al., 2015; Smith et al., 1958). The resulting low phenol oxidase activity reduces the decomposition and favors the preservation of organic matter, which finally leads to the formation of peat in swamps.

### 4.2.2 Peatland types

Peatlands are areas in which peat occurs whereby mires are peatlands where peat is actively formed. If peat is formed in a depression, it is called fen, and if peat rises above the surrounding and builds a dome, it is called bog (Joosten and Clarke, 2002). Doming happens because peat behaves almost like a sponge taking-up and holding water by capillary forces. Bogs are referred to as ombrogenous mires if they are cut off from groundwater flow from the underlying rocks and rely on precipitation as the only water source. Geogenous mires are in turn fed by groundwater that was in contact with bedrocks. Since groundwater carries nutrients from weathering of mineral soils, the separation from the groundwater flow strongly reduces the nutrient supply. Hence, at such condition, atmospheric wet and dry depositions are significant nutrient sources, but since these sources are poor in comparison with groundwater nutrients inputs, peat bogs are nutrient deficient sites.

### 4.2.3 Vegetation and biodiversity

In contrast to peatlands at higher latitudes where grass-like plants such as sedges, fens and sphagnum mosses are the dominant plants in bogs, tropical peatlands are forested and reveal higher biodiversity in terms of plants (1524), mammals (123), birds (268), reptiles (75), amphibians (27), and freshwater fish (219) species (Posa et al., 2011). However, being bogs, tropical peat swamps are nutrient-poor sites and therefore support a lower species diversity than tropical rainforests growing on mineral soils (Posa et al., 2011).

Tropical peat bogs reveal a concentric succession of specific tree-dominated plant communities (Anderson and Muller, 1975). In general, the number of tree species, the tree height, and girth decrease from the perimeter toward the center of the bogs. Dominant tree species are, e.g., *Gonystylus bancanus* and *Campnosperma coriacea* at the rim, which are replaced by *Shorea albida* until *Tristania obovata* and *Combretocarpus rotundatus* dominate in the center of the bogs (Anderson and Muller, 1975).

### 4.2.4 Peatland distribution and carbon storage

Peatlands develop in geomorphological depressions in the hinterland (inland peatlands) and in coastal flat plains (coastal peatlands) where the penetration of seawater hinders the groundwater outflow and favors the formation of wetlands. The term wetland includes all areas that are inundated and/or saturated by water and in that the vegetation is adapted for life in water-saturated soils (Joosten and Clarke, 2002). In addition to peatlands, also swamps, marshes, and mangroves are considered as wetlands, but to convert a wetland into a peatland, an organic-rich peat layer has to be formed. This takes time and is favored by the accumulation specific clay minerals restricting the groundwater flow (Gastaldo, 2010) and a high water supply preventing the drying up of the wetland during dry seasons. Nevertheless, peat soils are not always inundated and saturated by water. During the dry season, a decreasing water level allows air to penetrate into the upper soil horizon. This periodically ventilated upper part of the soil is termed acrotelm in contrast to catotelm, which is the constantly saturated deeper part of the peat soil (Holden, 2005).

Globally, peatlands cover an area of circa $4.4 \times 10^{12}$ $m^2$ (Xu et al., 2018; Yu et al., 2010) and hold approximately 650 Pg C Fig. 4.2 (Green and Page, 2017; Joosten, 2010; Wang et al., 2009; Yu et al., 2010). Considering the global soil organic carbon storage of 3000 Pg C (Köchy et al., 2015), this means that peat, which covers only 3.0% of land surface area, holds 21.6% of the soil organic carbon stock. This emphasizes the role of peat swamps in

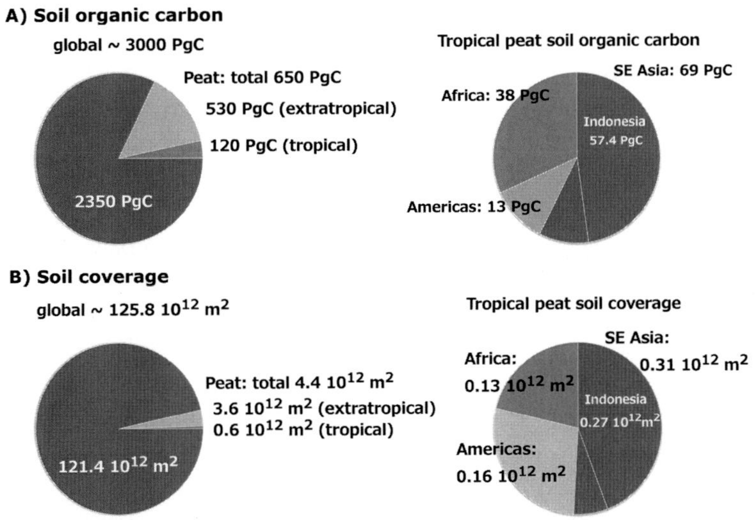


**FIGURE 4.2** Distribution of soil organic matter (A) and soil coverage (B).

the global carbon cycle, as it implies that changes in this comparatively small peatland area could have a large impact on the soil organic carbon reservoirs.

## 4.3 Indonesian peatlands

With a peatland area of $0.27 \times 10^{12}$ $m^2$ (Joosten, 2010) and a peat carbon stock of 57.4 Pg C (Page et al., 2011), Indonesia hosts about 50% of the tropical peatlands and holds 48% of the tropical peat carbon (Fig. 4.2).

### 4.3.1 History of Indonesian peat swamps

In Indonesia peat swamps, forests have existed for 20 million years without any significant change in their floristic composition (Anderson and Muller, 1975), but their more recent history is strongly linked to the sea level rise after the last glacial maximum at circa 18,000 years BP (BP = before present, where present means 1950, Dommain et al., 2011; Dommain et al., 2014). At this time, the glacio-eustatic drop of sea level by about 120 m exposed the Sunda Shelf and created a large landmass. This landmass is known as the Sundaland connecting the SE Asian continent with the islands of Borneo, Sumatra, and Java (Molengraaff, 1921). It was similar in size to present-day Europe, and many of the current rivers were tributaries of much larger river systems. This giant tropical lowland hosted peatlands, which are today buried below marine sediments.

It was proposed that the glacio-eustatic sea level rise forced coastal peat swamps on the Sundaland to move inland with the consequence that the areal extent of SE Asian peatlands hardly changed during the past 10,000 years (Abrams et al., 2018). An alternative interpretation is that the cold climate reduced the peat formation during the glacial period and that the sea level rise controlled the expansion of coastal peatlands thereafter (Dommain et al., 2011, 2014). The postglacial development of coastal peatlands was divided into three phases: During the first phase, the fast rising sea level impeded the expansion of peatlands until circa 5000–7000 years ago. At this time, the sea level rise slowed down and introduced the second phase with a sea level rise of 2.4 mm $yr^{-1}$. This sea level rise favored the development of coastal peatlands because coral reefs (Ridgwell et al., 2003), and mangroves (Woodroffe et al., 1985) could keep up with this slow sea level rise and stabilized the coast. During the past 5000 years, the sea level rise reversed, and a slowly decreasing sea level set the stage for the third phase. During this phase, coastal peatlands expanded exponentially due to mangroves and coral reefs, which migrated offshore and made space for peat swamps to expand (Dommain et al., 2014).

Today, global warming and the resulting thermal expansion of ocean waters and the melting of glaciers again reverted the trend and caused a global mean sea level rise of 3.6 mm $yr^{-1}$ between 1993 and 2010 (Church et al., 2013). This is introducing the fourth phase, but in contrast to the past three phases, humans instead of sea level changes exert the main control on the fate of peatlands by disturbing the peat carbon cycle.

## 4.3.2 Peat properties

The bulk composition of peat and physical peat properties are important parameters to characterize the carbon cycle in peatlands. In addition to organic matter, water and minerals are the bulk components of peat. Minerals, which we consider as lithogenic matter, are solid components, whereas organic matter occurs as a solid peat component and as dissolved organic matter in pore water. Solid organic matter is also considered as particulate organic matter (POM) or particulate organic carbon (POC) and organic matter dissolved in pore water is termed DOM or dissolved organic carbon (DOC).

The bulk density is one of the most important physical parameters used to describe peat as it reflects the peat composition (Andriesse, 1988). However, one has to distinguish between the density of the individual components (water, POM, lithogenic matter), the density of peat without water (dry bulk density) and with water (bulk density), and the peat carbon density (PCD). Densities are expressed in g $cm^{-3}$. The density of organic and lithogenic matter is about 1.1 and 2.6 g $cm^{-3}$, respectively, and the density of freshwater is $\sim$1 g $cm^{-3}$ (Rixen et al., 2019 and references therein). Adding up the densities of the solids (POM and lithogenic matter) results in the dry bulk density, whereas the bulk density includes additionally the weight of water. The PCD defines the weight of carbon in peat. It is expressed in g carbon (C) $cm^{-3}$, depends on the water content (Fig. 4.3), and is often used to establish carbon budgets, as we will see in the next chapter.

In Indonesia, the mean peat organic carbon content is 51.3% and by far exceeds the lower threshold value of 12% for peat soils (Warren et al., 2012). This organic carbon content corresponds to an organic matter content of 87.2% (51.3% $\times$ 1.7) and suggests a mineral contribution of 12.8%. The mean PCD is about 0.05 g C $cm^{-3}$ (Murdiyarso et al.,

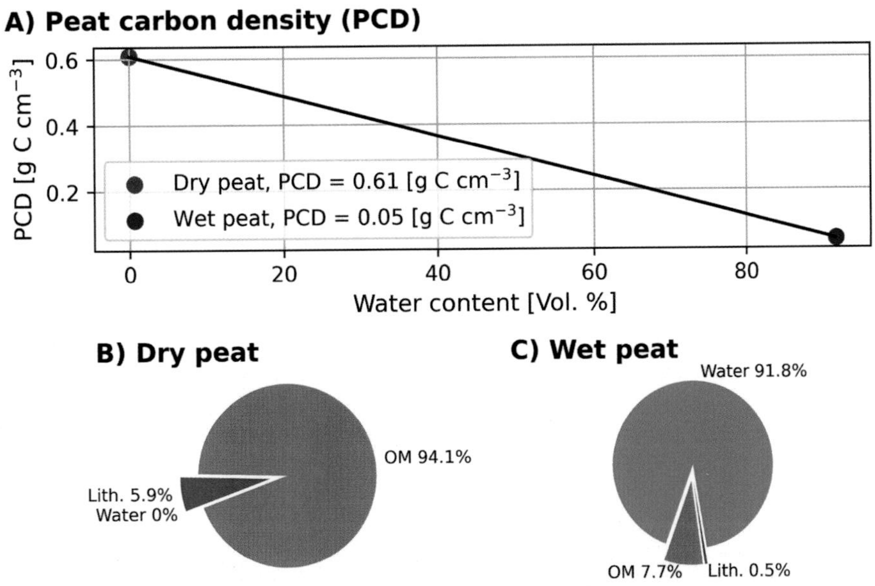


**FIGURE 4.3** (A) Water content versus peat carbon density (PCD). Composition of dry (B) and wet peat (C). *OM*, organic matter; *Lith.*, lithogenic matter. The numbers refer to the percent by volume, not weight.

2010), which means that 1 $cm^{-3}$ of wet peat contains 0.05 g organic carbon or 0.085 g organic matter (= 0.05 g × 1.7). If 0.085 g organic matter contributes 87.2% to the dry weight, lithogenic matter must contribute the remaining 12.08%, which equals 0.012 g. This amounts to a dry bulk density of 0.097 g $cm^{-3}$ (=0.085 g OM + 0.012 g Lith. per $cm^3$), which agrees quite well with earlier results (Andriesse, 1988) suggesting mean dry bulk densities ranging between 0.09 and 0.12 g $cm^{-3}$.

To estimate the mean water content, the volume of the solid components has to be subtracted from 1 $cm^3$. The volume of the solid components can be obtained by dividing the weight by the density of organic and lithogenic matter. Accordingly, 0.085 g organic matter and 0.012 g lithogenic matter occupy a volume of 0.083 $cm^3$ and suggest that 1 $cm^3$ of peat contains on average 0.918 $cm^3$ water (1−0.083 $cm^3$). If this peat dries and loses its water, the mean PCD of 0.05 g $cm^{-3}$ increases to 0.61 g C $cm^{-3}$ due to the shrinking peat volume and increasing contribution of organic matter to the peat (Fig. 4.3).

## 4.3.3 Peat carbon accumulation

The mean PCD and vertical peat growth rates are used to calculate organic carbon accumulation in peat soils, which represents the amount of $CO_2$ that is sequestered from the atmosphere in a given time. Compared with other ecosystems (Table 4.1), it allows to estimate the role of peatlands in the global carbon cycle. The area-normalized peat carbon accumulation is the product of the PCD and the peat growth rate, expressed in $g\ C\ m^{-2}\ yr^{-1}$. Accordingly, a mean PCD of $0.05\ g\ C\ cm^{-3}$ and mean vertical growth rates of $1–2\ mm\ yr^{-1}$ (Murdiyarso et al., 2010) imply peat carbon accumulation rates of $50–100\ g\ C\ m^{-2}\ yr^{-1}$.

**Table 4.1** Areal expansion and organic carbon accumulation rates in soils on land, in coastal ecosystems, and in the ocean.

| | Area ($10^{12}\ m^2$) | | Organic carbon accumulation ($TgCyr^{-1}$) | | |
|---|---|---|---|---|---|
| **Ecosystem** | **From** | **To** | **From** | **To** | **Mean** |
| Soil (without peat) | 121 | 125 | 50[t] | 50[t] | 50[t] |
| Freshwater (lakes without reservoirs) | | | 30[s] | 70[s] | 50 |
| Total land | | | 80 | 120 | 100 |
| Saltmarsh | 0.20[a] | 0.40[a] | 5[m] | 97[a] | 51 |
| Sea grass | 0.17[a] | 0.60[a] | 10[b] | 308[b] | 159 |
| Peat (temperate, polar) | 1.19[d] | 3.79[d] | 20[c] | 121[e] | 71 |
| Peat (tropical) | 0.37[f] | 0.61[g] | 28 | 46 | 37 |
| Mangroves | 0.08[i] | 0.14[h] | 5[b] | 34[p,q,r] | 20 |
| Coral reef | 0.28[j] | 0.60[a] | | | |
| Estuary | 1.10[l] | 1.80[k] | 70[a,n] | 81[k] | 76 |
| Total coastal ecosystems (total) | 3.39 | 8.00 | 138 | 687 | 413 |
| Shelf | | 26.00[l] | 45[k] | 75[n] | 60 |
| Ocean (slope and deep sea) | | 355.00[l] | 15[n] | 17[o] | 16 |
| Total ocean (shelf and ocean) | | | 60 | 92 | 76 |
| Total (land, coastal, and ocean) | | | 278 | 899 | 589 |

[a]Alongi (2014) and references therein.
[b]Duarte (2017) and references therein.
[c]Wang et al. (2009).
[d]Joosten (2010), Xu et al. (2018), Yu et al. (2010).
[e]Yu (2012).
[f]Yu et al. (2010).
[g]Joosten (2010).
[h]Giri et al. (2011).
[i]Hamilton and Casey (2016).
[j]Spalding et al. (2001).
[k]Duarte et al. (2005) and references therein.
[l]Cai (2011).
[m]Duarte et al. (2013) and references therein.
[n]Hedges and Keil (1995).
[o]Cartapanis et al. (2016).
[p]McLeod et al. (2011) and references therein.
[q]Iglesias-Rodriguez et al. (2002).
[r]Extrapolated from data obtained from Saderne et al. (2018).
[s]Cole et al. (2007).
[t]Regnier et al. (2013), Isson et al. (2020)

Another way to calculate organic carbon accumulation rates is to take cores from peatlands, date them, and estimate the amount of carbon that accumulated over time within the core (Page et al., 2004). According to such paleostudies, the carbon accumulation rates varied during the Holocene and were highest between 5000 and 10,000 BP (Dommain et al., 2011). As of 5000 BP, organic carbon accumulation rates decreased and reached modern values of about 65–77 g C $m^{-2}$ $yr^{-1}$ in coastal lowlands and 20–31 g C $m^{-2}$ $yr^{-1}$ in inland peat swamps. Considering that approximately 80% of the peatlands are located at the coast, and only 20% occur inland, Dommain et al. (2011, 2014) suggest a peat carbon accumulation rate of 56–68 g C $m^{-2}$ $yr^{-1}$. This is well within the range of those accumulation rates, which were derived from the PCD and vertical peat growth rates (50 – 100 g C $m^{-2}$ $yr^{-1}$) and only slight fall below their mean of 75 g C $m^{-2}$ $yr^{-1}$.

Peat carbon accumulation rates of 50–100 g $m^{-2}$ $yr^{-1}$ multiplied by the modern Indonesian peatland area of $0.27 \times 10^{12}$ $m^2$ result in a total Indonesian peat carbon accumulation of 14–27 Tg C $yr^{-1}$, which agrees quite well with previous estimates of 10–30 Tg C $yr^{-1}$ (Sorensen, 1993). From both estimates, one can surmise a mean Indonesian peat carbon accumulation of 20 Tg C $yr^{-1}$, and considering a peatland area of $0.27 \times 10^{12}$ $m^2$, one can surmise a mean area-normalized peat carbon accumulation of 75 g C $m^{-2}$ $yr^{-1}$. Assuming that the Indonesian peat carbon accumulation rate is representative for the whole tropics, and considering a global tropical peatland area of $0.37–0.61 \times 10^{12}$ $m^2$, one can suggest a tropical peat carbon accumulation of 28–46 Tg C $yr^{-1}$ (Table 4.1).

### 4.3.4 Land use and cover changes in Indonesia

In Indonesia, deforestation on an industrial level started around 1960 (Leifeld et al., 2019). In 2001, the Indonesian Ministry of Forestry in cooperation with the European Union jointly developed the plan to convert huge areas of peatland into oil palm plantations to meet the growing market for palm oil (Hooijer et al., 2006; Sargeant, 2001). Ironically, the production of biodiesel, which increased the demand for palm oil, was intended as a measure to decrease $CO_2$ emission into the atmosphere.

Already in 2007, about 90% of the peatlands of Sumatra and Borneo were disturbed and no longer in their original state (Miettinen and Liew, 2010). Despite the Indonesian government imposing a moratorium on the clearing of primary forests and conversion of peatlands in 2010 (Indonesia, 2009; UNEP, 2019), peatland deforestation continued. Satellite data showed that peatland deforestation increased from 2210 $km^2$ $yr^{-1}$ between 2000 and 2010 to 2253 $km^2$ $yr^{-1}$ between 2010 and 2015 (Miettinen et al., 2016). In 2015, only 6% of the original pristine peat swamp forest was left untouched. Furthermore, it is doubtful whether the remaining fraction of peat swamp forest is still pristine if one considers the relevance of atmospheric deposition for peat bogs and the pollution of the atmosphere with, e.g., bioavailable nitrogen.

### 4.3.5 The hydrological cycle of Indonesian peatlands

Precipitation rates over SE Asia are among the highest reported worldwide and are strongly influenced by the Australian–Indonesian monsoon (Wang, 2009). The Australian–Indonesian monsoon is in turn affected by anomalies of the coupled circulation in the ocean and atmosphere such as the Madden–Julian Oscillation (MJO), the Indian Ocean Dipole (IOD), and in particular El Niño–Southern Oscillation (ENSO; Fig. 4.4, Madden and Julian, 1994; Saji et al., 1999; Webster et al., 1999). The periodicity of the MJO is $< 1$ year, while the IOD and ENSO operate on timescales of several years. Negative ENSO phases, referred to as La Niña, are associated with precipitation rates above normal. Positive ENSO excursions are called El Niño and reduce precipitation rates over Indonesia, as positive IOD phases do. The IOD describes differences between sea surface temperatures anomalies in the eastern and western Indian Ocean, and the Dipole Mode Index (DMI) indicates its strength. The bimonthly Multivariate ENSO Index (MEI) is one among many indices showing the strength of ENSO.

To quantify precipitation rates, we used data obtained from the Global Precipitation Climatology Centre (GPCC, Schneider et al., 2011). The GPCC produces global analyses of monthly mean precipitation on the Earth's land surface based on *in situ* rain-gauge data. The data cover the period from 1983 onward and reveal a spatial resolution of $1 \times 1$ degree. According to this data, the monthly mean predication rate over Indonesia, in the borders between 12°S and 8°N and 95°E – 152°E, is $212 \pm 34.6$ mm. Due to the monsoon-driven seasonality, the monthly mean precipitation rates are lowest in August

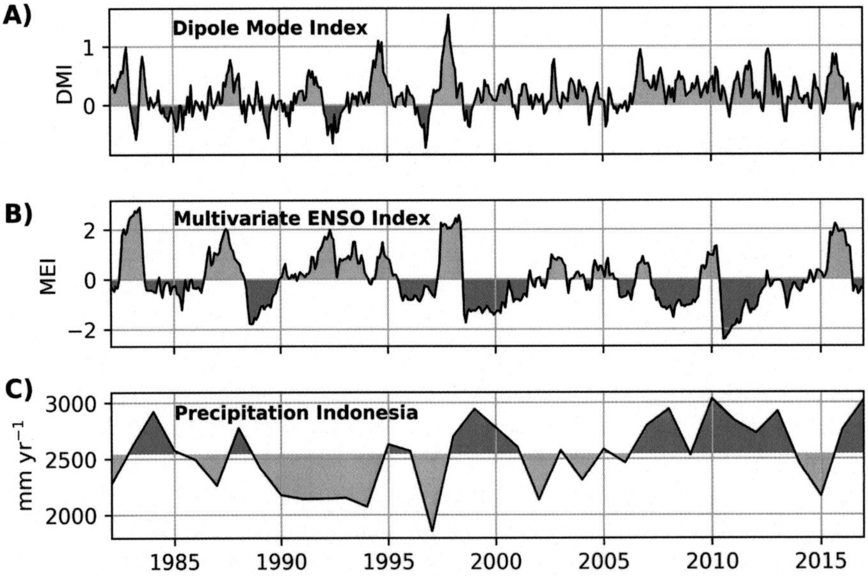


**FIGURE 4.4** Indices describing anomalies of the coupled circulation in the ocean and atmosphere and the all over Indonesian precipitation rates. *The data were obtained from https://www.esrl.noaa.gov/psd/gcos_wgsp/Timeseries/DMI/ (May 2020) https://www.esrl.noaa.gov/psd/enso/mei (May 2020), and Schneider et al. (2011).*

(160 mm) and highest in December/January (265 mm). The annual mean precipitation rate is 2544 ± 587 mm (= mm $yr^{-1}$ = L $m^{-2}$ $yr^{-1}$).

According to the global "Simulated Topological Network" (STN-30p), the Indonesian river discharge amounts to 1420 mm, contributing 11% to the global mean river discharge (Syvitski et al., 2005). Groundwater discharges into the ocean represent 0.01%–10% of the total freshwater discharge into the ocean (Moosdorf and Oehler, 2017 and references therein). Ignoring groundwater fluxes and changes of the groundwater storage allows us to calculate evapotranspiration by subtraction of river discharges from precipitation rates (= P − R). Accordingly, a mean precipitation rate of 2544 mm and mean river and groundwater discharge of 1420 mm amounts to a mean evapotranspiration of 1124 mm or 44%.

In SE Asia, local field studies determined evapotranspiration ranging from 37% to 72% (Kumagai et al., 2005; Moore et al., 2013), whereas evapotranspiration decreases with an increasing perturbation of the pristine peat swamp forests. For example, in central Borneo, evapotranspiration rates decrease from 69% in a nearly pristine peat swamp forest to 56% in a disturbed and fire-affected peat swamps forest, down to 38% in an area covered by shrubs (Hirano et al., 2015; Moore et al., 2013). On managed lands such as oil palm plantations, evaporation rates of 44% have been measured, but they increase with the age of the plantation. Mature oil palm plantations reveal evapotranspiration rates similar to those measured in pristine forests (Comte et al., 2012; Manoli et al., 2018).

To estimate evapotranspiration of oil palm plantations on disturbed peatlands, we linked land cover–specific evapotranspiration rates and biomasses estimates (Table 4.2).

The data indicate that evapotranspiration rates decrease with a decreasing biomass due to an increasing degree of forest degradation (Fig. 4.5). However, evapotranspiration rates level off at a biomass density of >10 kg $m^{-2}$. A decreasing water vapor pressure deficit (VPD) is assumed to limit evapotranspiration rates at high biomass density in less degraded and pristine swamp forests. VPD is the difference between the water vapor pressure at saturation and the actual water vapor pressure. It is inversely related to

**Table 4.2** Land covers and evapotranspiration rates from peatlands as obtained from Hirano et al. (2015) and Moore et al. (2013) and as discussed in the text as well as biomass estimates (total, as well as above and below ground).

| Land cover | | ET | Biomass | | |
|---|---|---|---|---|---|
| | | | Total | Above | Below |
| | | [%] | | [kg $m^{-2}$] | |
| Pristine PSF | | 69 | 20.7 ± 5.3 | 18.2 ± 2.6 | 2.5 ± 1.2 |
| Degraded PSF | Logged forest | 63 | 9.7 ± 4.2 | 8.5 ± 2.4 | 1.2 ± 0.6 |
| Secondary forest | Fire damaged | 56 | 6.4 ± 3.1 | 5.6 ± 1.7 | 0.8 ± 0.4 |
| Ferns and shrubs | Crop and shrubs lands | 38 | 1.5 ± 0.5 | 1.2 ± 0.2 | 0.3 ± 0.1 |
| Managed land | Oil palm plantation | 48 | 3.0 ± 1.6 | 2.4 ± 0.8 | 0.6 ± 0.4 |

Obtained from Hergoualc'h et al. (2011).

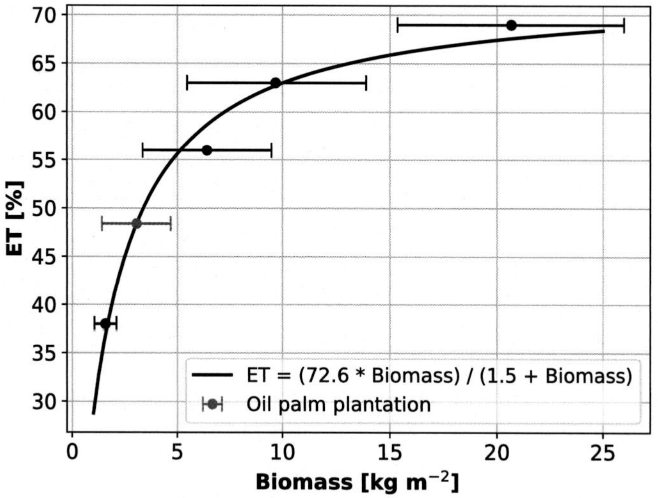


**FIGURE 4.5** Total biomass versus evapotranspiration rates. Table 4.4 shows the data and Eq. (4.1) shows the function, which was used to calculate the evapotranspiration rates from oil palm plantations as indicated by the *red circle*.

relative humidity and terminates transpiration if it approaches zero (Damour et al., 2010 and references therein).

The Michaelis–Menten function is an often-used equation to describe dependencies approaching an upper limit, which is given as $V_{max}$. Adapted to the link between evapotranspiration (ET) and biomass, it can be written as follows:

$$ET = \frac{V_{max} \times Biomass}{K_M + Biomass} \tag{4.1}$$

$V_{max}$ is the maximum ET, and $K_M$ is the half-reaction rate. We applied a least-squares optimization to determine $K_M$ and $V_{max}$. The resulting $V_{max}$ of 72.6% suggests that the evapotranspiration of 72%, as measured by Kumagai (2005) in Sarawak, Malaysia, was close its maximum. Nevertheless, for establishing a carbon budget in Indonesia, we will use an evapotranspiration rate of 69% for pristine peat swamp because of two reasons. First, it was measured in central Borneo, and second, it accounts for the biomass variability within pristine swamps forests, which decreases toward the center of the bogs. However, the obtained $K_M$ and $V_{max}$ values suggest an evapotranspiration rate of 48% for oil palm plantations with a mean biomass density of 3.0 kg $m^{-2}$.

To compute a mean evapotranspiration rate for Indonesian disturbed peatlands, we weighted the land cover–specific evapotranspiration rates with the areal extension of the respective land covers as obtained from Miettinen et al. (2016) (Table 4.3). This results in a mean evapotranspiration rate of 51%. Since this value is relatively close to the evapotranspiration rates of 44% as derived from the mean all over Indonesian precipitation rates and river discharges, we assume that a mean evapotranspiration rate of 51% for disturbed Indonesian peatland is reliable.

**Table 4.3** Land cover, evapotranspiration rates, land cover distribution, and weighted evapotranspiration. The latter is calculated as followed: (ET × land cover distribution)/100. The sum of the weighted evapotranspiration rates (51%) is assumed to be representative for Indonesian peatlands.

| Land cover | ET [%] | Land cover distribution [%] | Weighted ET [%] |
|---|---|---|---|
| Pristine PSF | 69 | 6.4 | 4.4 |
| Degraded PSF | 63 | 22.8 | 14.4 |
| Secondary forest | 56 | 11.1 | 6.2 |
| Ferns and shrubs | 38 | 5.4 | 2.1 |
| Managed land | 48 | 49.8 | 23.9 |
| Total | | 95.5 | 51.0 |

## 4.4 Peat carbon losses

As already mentioned before, the global relevance of peat carbon loss due to drainage and fire has been recognized prior to and during the very early phase of SPICE (Hooijer et al., 2006; Page et al., 2002). The respective benchmark studies initiated a number of follow-up projects aiming at enhancing the data density and improving estimates. Furthermore, in 2014, the IPPC published an update of its guideline to quantify $CO_2$ emission from peat draining rivers (IPCC, 2014). Within the following three chapters, we compile data on peat carbon losses from the literature and evaluate the IPCC guideline.

### 4.4.1 $CO_2$ emissions caused by peat and forest fires

One year prior to the initiation of SPICE, the first estimate on carbon losses caused by the ENSO-favored peat and forest fires in 1997/98 was published (Page et al., 2002). According to this estimate, 810–2570 Tg C was released as $CO_2$ during this ENSO-driven fire event. Inverse modeling technique and the evaluation of satellite data, in turn, suggest carbon losses of about 1000–1069 Tg C (Rödenbeck et al., 2003; van der Werf et al., 2010). During the 2015/16 ENSO event, again forest and peat forest fires were running out of control with fire-induced carbon emissions of about 200–300 Tg C (Boer et al., 2018; Heymann et al., 2017; Huijnen et al., 2016; Lohberger et al., 2018). The radiocarbon ($^{14}C$) content measured in carbonaceous aerosol samples collected in Singapore during the 2015 fire events revealed a mean carbon age of 800 ± 420 years, which implies that 85% of the burned biomass was peat (Wiggins et al., 2018).

According to the second Indonesian Biennial Update Report (BUR) to the United Nations Framework Convention on climate change in 2018 (Boer et al., 2018), the average annual mean peat fire emission was 66.4 ± 54.4 Tg C $yr^{-1}$ between 2000 and 2016 (Boer et al., 2018). Other estimates suggested annual mean SE Asian fire-induced carbon emission of 18–1110 Tg C $yr^{-1}$ with an average of 173 Tg C $yr^{-1}$ between 1997 and 2016 (van der Werf et al., 2017). Peat fires were assumed to contribute 42.8% to the total SE Asian fire emissions suggesting a mean peat fire emission of 74 Tg C $yr^{-1}$ (van der Werf et al., 2008, 2017). Considering that Indonesia holds 87% of the SE Asian peatlands (Fig. 4.2) further reduces this estimate to 64 Tg C $yr^{-1}$. This, in turn, agrees quite well with those of the BUR. Dividing the latter by the Indonesian peat area of about $0.27 \times 10^{12}$ $m^2$ suggests an Indonesian peat fire emission of 245.9 g C $m^{-2}$ $yr^{-1}$.

### 4.4.2 $CO_2$ emissions caused by peat soil oxidations

While we were evaluating our first SPICE data, "Delft Hydraulics" in cooperation with "Wetland International" and "Alterra Wageningen" published in 2006 the peat $CO_2$ report (Hooijer et al., 2006). Based on drainage depth and its correlation with $CO_2$ emissions, $CO_2$ emission from degraded peat soils in SE Asia was estimated to range between 97 and 233 Tg C $yr^{-1}$. Approximately 82% of these emissions were originated in Indonesia resulting in an Indonesian $CO_2$ emission caused by peat oxidation of 89–233 Tg C $yr^{-1}$ (Hooijer et al., 2006, 2010). An update, which used the 2015 land cover data from Miettinen et al. (2016) and peat oxidation values obtained from the Intergovernmental Panel on Climate Change (IPCC, 2014), suggested a mean SE Asian peat carbon emission of 132–159 Tg C $yr^{-1}$ (Miettinen et al., 2017). Assuming again that 82% of this emission is from Indonesia results in an Indonesian peat carbon emission of 108–130 Tg C $yr^{-1}$. According to the BUR, the $CO_2$ emission rate caused by peat oxidation in Indonesia amounts to $83 \pm 9$ Tg C $yr^{-1}$ (Boer et al., 2018). This estimate falls below of those obtained from Miettinen et al. (2017) and reveals that estimates of $CO_2$ emission range from 74 ($=83 \pm 9$) to 130 Tg C $yr^{-1}$, which on average suggest peat oxidations rate of about 102 Tg C $yr^{-1}$. Taking the Indonesian peatland area into account, this results in a mean peat carbon oxidation rate of about 374 g C $m^{-2}$ $yr^{-1}$.

### 4.4.3 Off-site $CO_2$ emission

The methods which the IPCC (IPCC, 2014) suggest to estimate carbon loss from disturbed peatland includes a procedure to calculate $CO_2$ emission caused by soil leaching and erosion and the subsequent oxidation of the mobilized peat carbon. This procedure is called "off-site $CO_2$ emission due to DOC loss from drained organic soils" ($CO_2$-$C_{DOC}$) in contrast to on-site emission from peat oxidation and fires. Since the data available to constrain the emission factors were assumed to be insufficient, Miettinen et al. (2017) and also the BUR ignored $CO_2$-$C_{DOC}$ in their peat carbon budget. $CO_2$-$C_{DOC}$ can be calculated as shown by Eqs. (4.2) and (4.3).

$$CO_2 - C_{DOC}\left[tC\ yr^{-1}\right] = A \cdot EF_{DOC} \tag{4.2}$$

$$EF_{DOC}\left[tC\ ha\ yr^{-1}\right] = DOC_{Flux-natural} \cdot \left(1 + \Delta DOC_{Drainage}\right) \cdot Frac_{DOC-CO_2} \tag{4.3}$$

where A, drained land area [ha]; $EF_{DOC}$, emission factor for annual $CO_2$ emissions due to DOC losses; $DOC_{Flux\text{-}natural}$, DOC flux rate from undisturbed peat soils [tC $ha^{-1}$ $yr^{-1}$]; $\Delta DOC_{Drainage}$, proportional increase of the DOC leaching rate relative to $DOC_{Flux\text{-}natural}$; $Frac_{DOC\text{-}CO2}$, fraction of leached DOC, which is oxidized to $CO_2$.

**Table 4.4** DOC emission factors.

| DOC emission factors | Units | IPCC (2014) Min. | Max. | Mean |
|---|---|---|---|---|
| $EF_{DOC}$ | $tC\ ha^{-1}\ yr^{-1}$ | 0.56 | 1.14 | 0.82 |
| $DOC_{Flux\text{-}natural}$ | $tC\ ha^{-1}\ yr^{-1}$ | 0.49 | 0.64 | 0.57 |
| $\Delta DOC_{Drainage}$ | | 0.43 | 0.78 | 0.60 |
| $Frac_{DOC\text{-}CO_2}$ | | 0.8 | 1.00 | 0.90 |

Table 4.4 shows the values of these "DOC emission factors" for rivers draining disturbed tropical peatlands as given in Tier 1 of the IPCC report in 2014.

The term "$DOC_{Flux\text{-}natural}$" forms the core for the calculation of $CO_2$-$C_{DOC}$. It is the product of the measured DOC concentration in the river and the water discharge. Thereby, it equals the DOC yield. Accordingly, the term "$Frac_{DOC\text{-}CO2}$" quantifies the fraction of the DOC that is oxidized after the DOC leaves the considered section of the river. In general, DOC yields are used to estimate riverine DOC discharges into the ocean. Applied to rivers draining into the ocean, this means that $EF_{DOC}$ quantifies the $CO_2$ production caused by the oxidation of DOC, which was discharged by rivers into the ocean.

However, using the given equations and the numbers shown in Tables 4.5 and 4.6 results in mean DOC yield (=$DOC_{Flux\text{-}natural}$) of 91 and 56 g C $m^{-2}$ $yr^{-1}$ from rivers draining disturbed and pristine peat swamps, respectively. Since it is assumed that 90% of these discharges is respired and emitted, this causes a $CO_2$ emission from the ocean in the atmosphere of 82.1 and 51.3 g C $m^{-2}$ $yr^{-1}$, respectively. Applied to the Indonesian peatland with and area of $0.27 \times 10^{12}$ $m^2$, these numbers amount to total off-site $CO_2$ emission of 22.1 Tg C $yr^{-1}$ for disturbed peatlands and of 13.8 Tg C $yr^{-1}$ prior to peatland degradation.

According to our SPICE results, these fluxes are too low as they are based on the measured DOC concentrations and ignore DOC respiration in the rivers. The latter feeds the $CO_2$ emissions from rivers into the atmosphere and the discharges of dissolved inorganic carbon (DIC) into the ocean. Within the framework of SPICE, we studied these processes, and in the following chapter, we will introduce our results and revise the IPCC concept.

## 4.5 Land–ocean continuum

The land–ocean aquatic continuum defines the transition zone between terrestrial ecosystems and the open ocean (Billen et al., 1991; Ward et al., 2017). Water is the moving agent, and precipitation drives the water movement. Leaching and erosion are the processes describing the associated carbon transport from the terrestrial ecosystems into the rivers. Leaching is the transport of dissolved carbon by the groundwater flow and, in case of flooding, by the surface run-off into the river. Carbon erosion addresses the physical erosion and transport of particulate carbon into rivers.

**Table 4.5** River overview.

| | Catch-ment | Peat Cover. | Prec. [mm | R[b] [%] | ET | POC Mean | POC std | DOC Mean | DOC std | $pCO_2$ Mean | $pCO_2$ std | pH | DIC |
|---|---|---|---|---|---|---|---|---|---|---|---|---|---|
| | [$km^2$] | [%] | $yr^{-1}$] | | [%] | [$\mu$mol $L^{-1}$] | | | | [$\mu$atm] | | | [$\mu$mol $L^{-1}$] |
| Musi | 56,931 | 4 | 2718 | 0.43 | 51 | 143 | 51 | 303 | 61 | 4316 | 928 | 6.85 | 929 |
| B.-Hari | 44,890 | 5 | 2432 | 0.60 | 51 | 109 | 0 | 311 | 39 | 2400 | 18 | 7.07 | 809 |
| Indragiri | 17,968 | 12 | 2410 | 0.97 | 51 | 482 | 113 | 757 | 99 | 5777 | 527 | 6.30 | 480 |
| Kampar | 26,195 | 22 | 2868 | 0.80 | 51 | 156 | 39 | 1280 | 63 | | | 6.40 | |
| Siak | 10,423 | 22 | 2664 | 0.78 | 51 | 499 | 109 | 1900 | 640 | 8555 | 528 | 5.13 | 298 |
| Rokan | 19,258 | 30 | 2827 | 0.80 | 51 | 1034 | 18 | 781 | 53 | | | 6.50 | |
| Maludam | 91 | 91 | 3798 | 0.38 | 69 | | | 3802 | 805 | 8098 | 800 | 4.80 | 267 |
| Lupar[a] | 6541 | 0 | 3798 | 0.12 | 51 | | | 148 | | 1274 | 198 | 6.90 | 303 |
| Rajang | 52,010 | 2 | 3771 | 0.39 | 51 | | | 171 | 26 | 2445 | 250 | 6.70 | 395 |
| Sebuyau | 451 | 61 | 2854 | 0.94 | 51 | | | 3026 | 1047 | 8834 | 1050 | 4.20 | 281 |
| PSF1 | 34 | 100 | 2749 | 0.62 | 69 | 117 | 3 | 5667 | 42 | | | | |
| PSF2 | 13 | 100 | 2749 | 0.62 | 51 | 442 | 75 | 4583 | 83 | | | | |
| PSF3 | 64 | 100 | 2749 | | 51 | 300 | 175 | 4025 | 183 | | | | |

[a]Peat soil coverage of the Lupar was set to zero because DOC concentrations were measured upstream the peatland.
[b]River area in % of the catchment.

**Table 4.6** Yields and fluxes as well as the contribution of erosion and leaching to the total carbon loss by these processes, the contribution of $CO_2$ yields, and river discharges to the total carbon loss and the contribution of DOC, DIC, and POC yields to the river discharges. "Δ" shows the difference between disturbed and pristine peatlands. The carbon fluxes refer to areal extent of the Indonesian peatland of $0.27 \times 10^{12}$ $m^2$.

| | | Disturbed peat lands | | | Pristine (Maludam) peat lands | | | Δ |
|---|---|---|---|---|---|---|---|---|
| | | Yield | Flux | | Yield | Flux | | |
| | | [g C $m^{-2}$ $yr^{-1}$] | [Tg C $yr^{-1}$] | [%] | [g C $m^{-2}$ $yr^{-1}$] | [Tg C $yr^{-1}$] | [%] | [%] |
| 1 | Leaching and erosion | 148.2 | 40.0 | | 84.5 | 22.8 | | 75 |
| 2 | Erosion (POC yield) | 5.9 | 1.6 | 4.0 | 1.2 | 0.3 | 1.3 | 433 |
| 3 | Leaching | 142.3 | 38.4 | 96.0 | 83.3 | 22.5 | 98.7 | 71 |
| | | | | 100 | | | | |
| 4 | $CO_2$ yield[a] | 66.0 | 17.8 | 44.5 | 21.5 | 5.8 | 25.4 | 207 |
| 5 | River discharge[b] | 82.2 | 22.2 | 55.5 | 63.0 | 17.0 | 74.6 | 31 |
| | | | | 100.0 | | | 100.0 | |
| 6 | DOC yield | 71.2 | 19.2 | 86.6 | 58.0 | 15.4 | 92.3 | 22 |
| 7 | DIC yield | 5.1 | 1.4 | 6.2 | 3.8 | 1.0 | 5.9 | 38 |
| 8 | POC yield | 5.9 | 1.6 | 7.2 | 1.2 | 0.03 | 1.8 | 431 |
| | | | | 100 | | | | |

*DIC*, dissolved inorganic carbon; *DOC*, dissolved organic carbon; *POC*, particulate organic carbon.
[a]The sum of $CO_2$ yield and river discharge equals leaching and erosion.
[b]River discharge is sum of DOC, DIC, and POC.

In peatlands, leaching and erosion mostly refer to the mobilization and transport of DOC and POC, whereas leaching includes also DIC formed during the respiration of DOC and POC. Particulate inorganic (carbonate) carbon (PIC) and DIC derived from its dissolution play only a minor role in peatlands (Wit et al., 2018). DOC and POC are decomposed along the entire land–ocean continuum, and produced DIC is emitted as $CO_2$ into the atmosphere or discharged via estuaries into the ocean. In estuaries and the ocean, DOC and POC respiration as well as DIC inputs interact with the marine carbonate system.

## The marine carbonate system

The marine carbonate system is a series of temperature and salinity-dependent equilibrium reactions controlling the distribution of the individual carbonate species in the ocean ($CO_2 \Leftrightarrow HCO_3^- \Leftrightarrow CO_3^{2-}$) and thereby the amount of carbon stored by the ocean.

The sum of the carbonate species is the dissolved inorganic carbon (DIC), and the sum of the charges carried by the carbonates species is the carbonate alkalinity. Jointly with charges carried by other weak acids (mostly boric acid), it forms the total alkalinity. It strongly interacts with pH, whereby an increasing pH lowers the $CO_2$ concentration and increases that of $CO_3^{2-}$ by shifting the equilibriums reactions toward the right side. This favors the precipitation of carbonate as it increases the saturation state of carbonate minerals, Ω (Morse et al., 2007; Mucci, 1986).

$$\Omega = \frac{[Ca^{2+}] \times [CO_3^{2-}]}{K_{sp}(T, S)} \quad (4.4)$$

where $K_{sp}$ is the temperature and salinity-dependent solubility product of the carbonate minerals, of which calcite and aragonite are the dominant carbonate minerals in the ocean. However, a saturation state of $\Omega < 1$ indicates that the solution is undersaturated, which means carbonate minerals do not precipitate and those which are added to the solution dissolve. Vice versa, in an oversaturated solution ($\Omega > 1$), minerals precipitate. Although the ocean is three to five times oversaturated with respect to calcium carbonate, inhibitors such as sulfate ions ($SO_4^{2-}$) prevent the inorganic precipitations of calcium carbonate. To enforce the precipitation of calcium carbonate, calcifying organisms oversaturate waters in their calcifying fluid 10 to 15 times (McCulloch et al., 2017).

A decreasing pH shifts the carbonate system toward the left. This lowers Ω and enhances $CO_2$ concentrations. Since the $CO_2$ partial pressure ($pCO_2$) results from the $CO_2$ concentration and its solubility, shifts of the carbonate system toward $CO_2$ increase the emission of $CO_2$ from the ocean into the atmosphere. The $pCO_2$ difference between the ocean and the atmosphere ($\Delta pCO_2 = pCO_2\text{-ocean} - pCO_2\text{-atmosphere}$) determines $CO_2$ fluxes (F) between these two reservoirs, which can be calculated as follows:

$$F = k \times \alpha \times \Delta pCO_2 \quad (4.5)$$

where "k" is the $CO_2$ gas transfer velocity and "α" is the solubility of $CO_2$ in seawater. "α" converts the partial pressures $pCO_2$ into mole C per liter. "k" strongly depends on wind speeds and increases with increasing winds speed (Wanninkhof, 2014).

*Continued*

**The marine carbonate system—cont'd**

Since the current mean seawater pH is about 8.00, only ~1% of the carbon stored in the ocean occurs as $CO_2$. The vast majority is stored as $HCO_3-$, which does not affect the $pCO_2$ in the ocean. Accordingly, the ocean stores approximately 60 times more carbon than the atmosphere while keeping a $\Delta pCO_2$ close to zero during preindustrial times. Today, the ocean absorbs about 26.9% of the anthropogenic $CO_2$ (Table 4.3) because human-caused $CO_2$ emissions increase the $pCO_2$ in the atmosphere and therewith the $CO_2$ flux into the ocean.

On an ecosystem level, leaching and erosion cannot be measured directly but are derived from DOC, DIC, and POC discharges into the ocean as well as from $CO_2$ emissions from rivers into the atmosphere. After a short introduction to our SPICE working area, we will quantify these fluxes and thereby the off-site $CO_2$ emission from disturbed and pristine peatlands.

### 4.5.1 SPICE study area

During the three phases of SPICE, we increased our working area. We started to work at the Siak river, included almost all major rivers draining central Sumatra during the second phase, and focused on the coastal ocean off Sumatra during the third phase of SPICE. During this phase, SPICE-associated projects were initiated, which aimed at studying peat-draining rivers in Sarawak, Malaysia (Fig. 4.6, Table 4.5).

The Siak is one of the main rivers draining the province of Riau in central Sumatra, Indonesia. It originates at the confluence of the two headstreams Sungai Tapung Kanan and Sungai Tapung Kiri, passes through adjacent lowlands, and discharges after 370 km into the Bengkalis Strait, which is part of the Malacca Strait. According to Miettinen and Liew (2010), 86% of the Sumatran peatlands were degraded in 2008, and rivers were used to export felled trees (Fig. 4.7).

All investigated Indonesian rivers originate in the western Sumatran high lands, pass through the adjacent lowlands, and discharge into the Malacca Strait and the Java Sea, respectively. On their way to the ocean, the rivers cut through peat soils, which are located in the coastal flat plains and cover 3.5% to 30.2% of their catchments.

Sarawak is the largest among the 13 states of Malaysia and comprises the north-western part of the island of Borneo. It holds the largest share of Malaysia's peatlands, which cover about 14,659 $km^2$ or 12% of the state's area (Joosten, 2010). Among the main studied peat draining rivers are the Maludam, the Lupar, the Rajang, and the Sebuyau. Peat soil coverage of these rivers ranged from 2% to 91%. The Maludam with a peat soil coverage of 91% drains the Maludam National Park. This park holds a protected peat swamp forest growing on an up to 10-m-thick peat dome. Hence, the data obtained from the Maludam are considered to be representative for a river draining a pristine peat

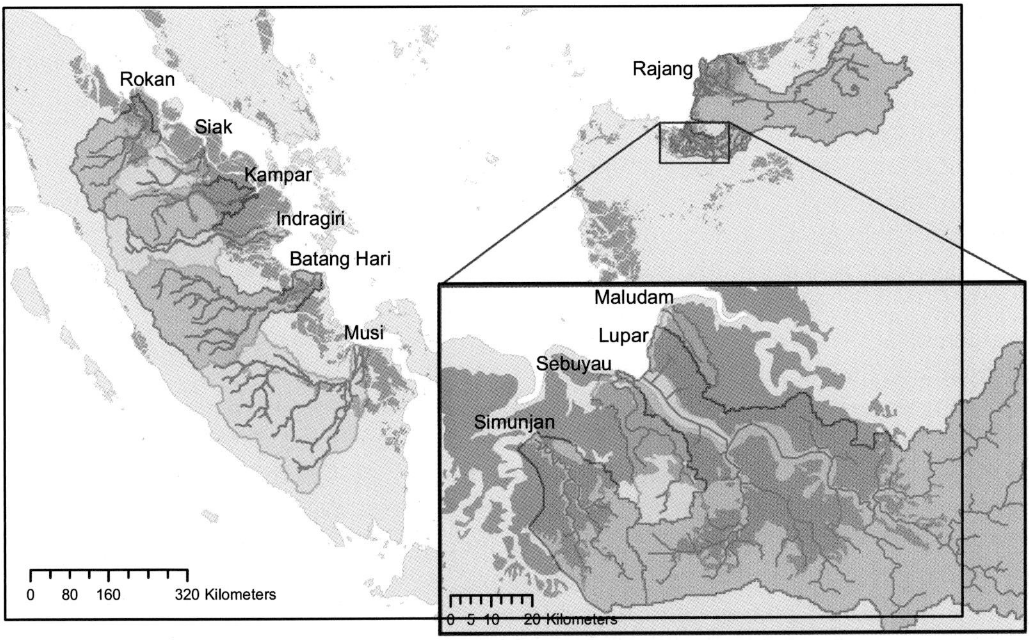


**FIGURE 4.6** Working area: *Blue lines* indicate the studied rivers, and the blue shaded and the brown areas show the river catchment basins and peatlands, respectively. Catchment size was derived from Hydro-SHEDS at 15 s resolution, using Esri's ArcMap 10.5, and the peat maps were downloaded from www.globalforestwatch.org for Indonesia and Malaysia.

**FIGURE 4.7** Timber rafts (left) and black water (right) at the bank of a Siak tributary in 2005.

swamp forest. Furthermore, we integrated data obtained from a study by Moore et al. (2013) into our analysis. This study was conducted in central Borneo, where small rivers (PSF1) and two channel systems (PSF2 and PSF3) drain a pristine and a disturbed peat swamp, respectively.

### 4.5.2 Dissolved organic carbon

DOC is the main component in the peat carbon cycle along the land–ocean continuum. Its concentration determines the watercolor, and the higher the DOC concentration is, the darker appears the water due to the increasing light absorption. Since at high DOC concentrations, the water looks almost like tea, peat-draining rivers are also named "black-water rivers" (Fig. 4.7). The dark color lowers the penetration depth of sunlight, which in addition to low nutrient concentrations strongly reduces the growth of phytoplankton in black-water rivers (Baum et al., 2014; Rixen et al., 2008; Siegel et al., 2009). In the following paragraphs, we will discuss how DOC concentrations are used to characterize rivers and calculate DOC discharges into the ocean and the respiration rates of organic matter in the investigated rivers.

To characterize rivers that drain peatlands, DOC concentrations measured along a river are averaged, and/or DOC end-members are determined. The latter is based on DOC concentrations, which decrease in the estuary of the rivers due to mixing of DOC-rich river water and DOC-poor ocean water. Vice versa, the mixing increases the salinity from zero (100% river water) in the inner estuaries to typical ocean values of 33–35 (100% ocean waters) in the outer estuaries. A linear correlation between DOC concentrations and salinity indicates a conservative mixing. This means that DOC sinks and sources such as respiration and DOC inputs from tributaries are negligible within the mixing zone. At such circumstances, the zero salinity y-intercept defines the riverine DOC end-member concentration.

During the six expeditions to the Siak river between 2004 and 2013, the DOC end-member concentrations increased with rising precipitation rates until the precipitation rates exceed 310 mm (Rixen et al., 2016). At this precipitation rate, the DOC end-member concentration decreased due to an acrotelm overflow and a resulting enhanced surface run-off. Since the DOC concentrations within the surface run-off are much lower than in groundwater, an increased surface run-off dilutes and lowers DOC concentrations in rivers. The mean DOC concentration in the Siak was $1900 \pm 640$ $\mu$mol $L^{-1}$ (Table 4.7). The mean DOC concentrations measured in the other rivers are lower and reveal a positive correlation with peat soil coverage (Table 4.5, Fig. 4.8).

**Table 4.7** Parameters used to calculate $CO_2$ emission from estuaries and the coastal ocean into the atmosphere.

| Parameter | Units | Estuaries | Coastal ocean |
|---|---|---|---|
| Area | $km^2$ | 10,818 | 12,7674 |
| Wind speed | $m\ s^{-1}$ | 5.6 | 5.6 |
| Temperature | °C | 29 | 29 |
| Salinity | | 10 | 30 |
| k | $cm\ hr^{-1}$ | 9.8 | 9.5 |
| $pCO_2$ | $\mu$atm | 2038 | 554 |
| $CO_2$ yield | $g\ C\ m^{-2}\ yr^{-1}$ | 493 | 43 |
| $CO_2$ flux | $Tg\ C\ yr^{-1}$ | 5.3 | 5.5 |

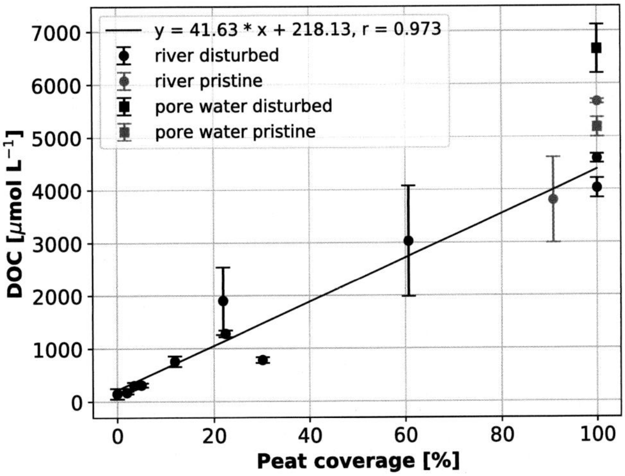


**FIGURE 4.8** Peatland coverage versus DOC concentrations measured in the rivers and pore waters. *DOC*, dissolved organic carbon.

DOC concentrations that increase with increasing peat soil coverage indicate first of all that peatlands are the main source of DOC in the rivers. Secondly, the resulting regression equations indicate a concentration of about 4316 μmol L$^{-1}$ in rivers draining disturbed peat soils (peat soil coverage = 100%). The mean pore water DOC concentration in disturbed peat soils was 6685 ± 458 μmol L$^{-1}$ (Gandois et al., 2013). This exceeds by far DOC concentrations in rivers, which indicates that DOC decomposition is an important process in rivers draining disturbed peatlands.

In pristine peat swamps, the situation appears to be more complex. Here, the pore water DOC concentration of 5183 ± 183 μmol L$^{-1}$ (Gandois et al., 2013) falls below those measured in PSF1 (5667 μmol L$^{-1}$) and exceeds those measured in the Maludam (3802 μmol L$^{-1}$, Fig. 4.8). In a study prior to the Maludam expedition (Rixen et al., 2016), it was concluded that DOC decomposition in rivers draining pristine peat swamp is negligible because pore water DOC concentrations were almost equal to those measured in the river waters of PSF1. In contrast, the comparably low Maludam DOC concentration implies that DOC decomposition is also of relevance in rivers draining pristine peat swamps. This difference between the Maludam and PSF1 is difficult to explain and could be an indication that the Maludam is not as pristine as we thought. However, we still consider PSF1 and the Maludam as rivers representative for pristine peat swamps because human impacts in their catchments are still low in comparison with Sumatran river catchments, which are heavily impacted by land use and cover changes (LULCC).

### 4.5.3 Dissolved organic carbon yields

River discharges, in addition to DOC end-member concentrations, are required to calculate DOC yields, and thereby, DOC discharges into the ocean (see Eqs. 4.7 and 4.8). Since river discharges into the ocean are difficult to measure, they are often derived from precipitation and evapotranspiration rates, as shown in Eq. (4.6)

$$\text{DOC yield}\left[\text{g Cm}^{-2}\text{yr}^{-1}\right] = \left(\text{Precipitation} \times \left(\frac{100 - \text{Evapotranspiration}}{100}\right)\right) \times \text{DOC}_{\text{end-member}} \quad (4.6)$$

$$\text{DOC discharge}\left[\text{Tg Cyr}^{-1}\right] = \frac{\text{DOC}_{\text{yield}} \times \text{Catchment area}}{10^{12}} \quad (4.7)$$

Within these equations, the catchment area is given in $m^2$, "Precipitation" is the mean annual precipitation rate over the catchment area in mm $yr^{-1}$, and "Evapotranspiration" is given in percentage. DOC end-member concentration, which is given in μmol $L^{-1}$, needs to be multiplied by the molecular mass of C, 12.01 and $10^{-6}$, to be converted into g $L^{-1}$.

Precipitation rates obtained from different sources and covering varying periods as well as the selection of evapotranspiration rates strongly influence the calculation of river discharges and DOC yields. To reduce such effects, we used the 1 × 1 degree gridded GPCC rainfall data (Schneider et al., 2011) and a mean evapotranspiration rate of 51% (see Table 4.3) to calculate discharges for all rivers except for the Maludam and PSF1. Since these two rivers drain nearly pristine peat swamps, we used the determined evapotranspiration rate of 69%. Table 4.7 shows the precipitation and evapotranspiration rates as well as the used DOC concentrations for each river, and Fig. 4.9B displays the DOC yields. Due to the newly derived evapotranspiration rates and the GPCC rainfall data, the obtained DOC yields differ slightly from those published in previous papers (e.g., Moore et al., 2013; Müller et al., 2015; Müller et al., 2016; Müller-Dum et al., 2019; Rixen et al., 2016; Wit et al., 2018).

Nevertheless, the recalculated data, as the previously published ones, show that DOC yields and DOC concentrations increase with an increasing peat soil coverage (Fig. 4.9B). The resulting regression equation suggests a DOC yield of 71.2 g C $m^{-2}$ $yr^{-1}$ for rivers draining disturbed peatlands (peat soils coverage = 100%, Table 4.6). DOC yields obtained from pristine PSF1 and the Maludam are lower with values of 53 and 60 g C $m^{-2}$ $yr^{-1}$ (mean = 58.0, Table 4.6).

For rivers draining pristine peat swamps, DOC yields obtained by applying IPCC methods (= $DOC_{Flux\text{-}natural}$) and using our data result in a similar DOC yield of about 56 g C $m^{-2}$ $yr^{-1}$. This was expected because the data from PSF1 were used to determine the DOC emission factors (see, e.g., Table 2A.3 in IPCC, 2014). The IPCC-derived mean DOC yield of 91 g C $m^{-2}$ $yr^{-1}$ for disturbed peatlands exceeds those calculated by us (71 g C $m^{-2}$ $yr^{-1}$). This deviation results most likely from the larger data set used by us. However, multiplied by a peatland coverage of $0.27 \times 10^{12}$ $m^2$, DOC yields of 71 g C $m^{-2}$ $yr^{-1}$ and 58 g C $m^{-2}$ $yr^{-1}$ result in DOC discharges of 19.2 and 15.4 Tg C $yr^{-1}$ from disturbed and pristine rivers into the ocean, respectively (Table 4.6).

## 4.5.4 $CO_2$ emission from rivers

$CO_2$ emissions from rivers into the atmosphere are calculated as those from the ocean in the atmosphere (Eq. 4.5). What differs is the parameterization of the $CO_2$ exchange velocity "k." In principle, "k" is controlled by the turbulence in the aquatic boundary layer. In the ocean, wind speeds control "k," which increases with increasing wind speeds.

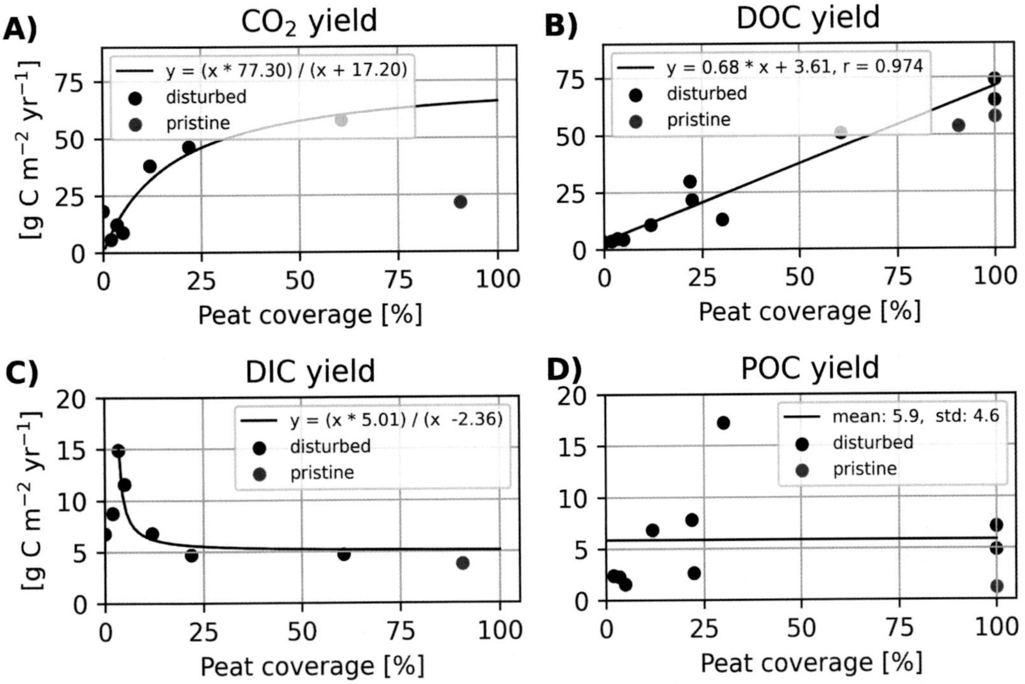


**FIGURE 4.9** Peat coverage versus $CO_2$ (A), DOC (B), DIC (C), and POC (D) yields. *DIC*, dissolved inorganic carbon; *DOC*, dissolved organic carbon; *POC*, particulate organic carbon.

Even though the main driver is known, there are a number of equations used to calculate "k" (e.g., Wanninkhof, 1992, 2014). According to these equations, a moderate breeze with a wind speeds of up to 10 m $s^{-1}$ causes "k" values of about 25 cm $h^{-1}$. In rivers, the turbulence in the aquatic boundary layer strongly depends on the bottom friction, which increases with decreasing water depth and increasing current velocity (e.g., Borges et al., 2004; Raymond and Cole, 2001; Raymond et al., 2013). Common parameterization of "k" for rivers in SE Asia results in values of up to 64 cm $h^{-1}$ (Raymond et al., 2013). However, results of measurements made in the Amazon river, by determining Radon ($^{222}Rn$) accumulation in free-floating chambers and oxygen balance calculation in the Siak river, indicate "k" values of 22–25 cm $h^{-1}$ (Devol et al., 1987; Rixen et al., 2008). These lower "k" values match with results obtained by floating chamber experiments in Malaysian rivers (Müller et al., 2015, 2016; Wit et al., 2015). In these rivers, "k" values of up 28.5 cm $h^{-1}$ were determined with mean values ranging between 17 and 26 cm $h^{-1}$. Nevertheless, "k" remains poorly constrained. To prevent a transfer of uncertainties associated with the determination of "k" to our $CO_2$ flux calculation, we used a "k" value based on the mean measured in Malaysian rivers of 22 cm $h^{-1}$ to calculate $CO_2$ fluxes from all studied rivers.

The $pCO_2$ was measured during five expeditions, which were carried out between 2009 and 2016 (Müller et al., 2016; Müller-Dum et al., 2019; Wit et al., 2015, 2018). In the studied rivers, the mean measured $pCO_2$ ranged between 1100 and 8555 μatm. In a global comparison, a $pCO_2 > 5000$ μatm is among the highest so far reported values (Lauerwald et al., 2015).

In addition to the $pCO_2$ in rivers, the atmospheric $pCO_2$ concentration is also required to calculate $CO_2$ fluxes across the water–air interface (Eq. 4.5). In Indonesia at the atmospheric $pCO_2$ observation site at Bukit Kototabang, the annual mean $pCO_2$ increased from 377.5 to 408.5 between 2004 and 2018 (Dlugokencky et al., 2020). In comparison with the $CO_2$ concentrations in rivers, which ranged between 1100 and 8555 ppm, atmospheric $CO_2$ concentrations of 393 ± 15.5 ppm are relatively low. However, since $pCO_2$ in rivers by far exceeds those in the atmosphere, it largely controls the $CO_2$ emission from rivers into the atmosphere. Changes of the atmospheric concentrations are, in turn, negligible during the period of our observations. With this in mind, we used an atmospheric $CO_2$ concentration of 393 ppm to calculate $CO_2$ emission from rivers.

Riverine $CO_2$ fluxes were based on the mean measured $pCO_2$ in the rivers and calculated in mmol C $cm^{-2}$ $hr^{-1}$ (see Eq. 4.5). This flux was multiplied by a factor of $1.0512 \times 10^6$ to be converted into g $m^{-2}$ $yr^{-1}$ and subsequently multiplied by the river surface area to obtain $CO_2$ fluxes across river water interface into the atmosphere.

The fraction of the Earth surface that is covered by rivers is small and poorly constrained. A recent analysis suggests a global river surface area of $0.77 \pm 0.01 \times 10^{12}$ $m^2$ (Allen and Pavelsky, 2018). This represents 0.6% of the global area covered by soils and almost equals the areal extent of tropical peat soils (Fig. 4.1). The Landsat (GRWL) Database (Allen and Pavelsky, 2018) and information on river length and width as obtained from Google Earth were used to estimate the river surface areas at our study sites (Wit et al., 2015). Thereby, changes in the river surface areas caused by temporal variations in the hydrological cycle were ignored. On average, the river surface area represents 0.65 ± 0.27% of the catchment area (Table 4.5), which corresponds to the global mean. To compare $CO_2$ fluxes from the river into the atmosphere to DOC yields and obtain their ratio, $CO_2$ fluxes were converted into $CO_2$ yields by dividing the $CO_2$ fluxes by the river catchment area.

Similar to DOC concentrations and yields, $CO_2$ yields also increase with increasing peat soil coverage but level off at peat soil coverage larger than 30% (Fig. 4.9A). This is assumed to be a pH effect, since a low pH limits the activity of the phenol oxidase and therewith the DOC decomposition in rivers, which are characterized by a high peat soil coverage and low pH values (Klemme et al., 2021). To address this pH effect, we fitted the $CO_2$ yields to the peat soil coverage by using the Michaelis–Menten equation as shown in Eq. (4.8):

$$\mathbf{CO_2\ yield} = \frac{\mathbf{V_{max} \times Peat\ soil\ coverage}}{\mathbf{K_M + Peat\ soil\ coverage}} \tag{4.8}$$

where $V_{max}$ equals maximal $CO_2$ yields. A least square optimization was applied to determine $K_M$ and $V_{max}$. The resulting $K_M$ and $V_{max}$ values of 77.3 and 17.2 g C $m^{-2}$ $yr^{-1}$ and a peat soil coverage of 100% suggest a $CO_2$ yield from disturbed peatlands of 66 g C $m^{-2}$ $yr^{-1}$ (Table 4.6, Fig. 4.9A). The $CO_2$ yield derived from the Maludam is with 22.5 g C $m^{-2}$ $yr^{-1}$ much lower and indicates that the perturbation of peatlands increases the

$CO_2$ yield from 22.5 to 66 g C $m^{-2}$ $yr^{-1}$. Multiplied by the Indonesian peatland area of $0.27 \times 10^{12}$ $m^2$, these $CO_2$ yields result in $CO_2$ emissions of 5.8 and 17.8 Tg C $yr^{-1}$ from rivers draining pristine and disturbed peatlands, respectively (Table 4.6). This indicates an increase of 207% due to peatland degradation.

### 4.5.5 Dissolved inorganic carbon yields

Although DOC discharges and $CO_2$ emissions largely balance DOC leaching, it ignores the part of the respired DOC that remains as DIC in the river (Table 4.6). Hence, DIC discharges into the ocean have to be quantified to calculate leaching as the sum of DIC and DOC discharges and $CO_2$ emission.

DIC is part of the carbonate system and can be calculated based on the measured pH and $pCO_2$ by using the $CO_2$sys program (Humphreys et al., 2020; Lewis and Wallace, 1998), but this approach ignores the role of organic acids on the carbonate system. PHREEQC (Parkhurst and Appelo, 2013) is another program offering the possibility to integrate the impact of organic acids into calculations of the carbonate system (Dasgupta et al., 2015). However, we used the $CO_2$sys program since acid strength of peat-derived organic acids depends on the vegetation type (Dasgupta et al., 2015), and no data from tropical peat areas were available.

DIC yields were calculated like DOC yields (see Eqs. 4.7 and 4.8), except that instead of DOC, the DIC concentrations were used to calculate the yields. In contrast to DOC yields, DIC yields decrease with an increasing peat soil coverage, which is most likely a pH effect (Fig. 4.9C). The pH decreases with increasing peat soil coverage (Table 4.6) and shifts the carbonate system toward the left. This enhances the $CO_2$ concentrations and emissions, which in turn decreases the DIC concentrations and yields in the river.

### 4.5.6 Leaching and erosion

To calculate leaching, we used the equations shown in Fig. 4.9. They describe the impact of peat soil coverage on the $CO_2$, DIC, and DOC yields. As leaching is the sum of these yields, they also describe the influence of peat soil coverage on leaching (Fig. 4.10). Due to the comparably low DIC yields, its effect is negligible, and the calculated leaching rate follows the trend shown by $CO_2$ and DOC yields and increases with an increasing peat soil coverage. The leaching rate at a peat coverage of 100% is with 142.3 g C $m^{-2}$ $yr^{-1}$ lower than the one previously estimated of 183 g C $m^{-2}$ $yr^{-1}$ (Rixen et al., 2016). The higher leaching rate was derived from a numerical model, which included an exponential DOC decay function. The only recently discovered pH limitation of the phenol oxidase activity in rivers (Klemme et al., 2021) was ignored.

The Maludam is the only investigated river, which drains a more or less pristine peat swamp forest, that almost (91%) covers the entire river catchment and in which pH as well as DOC and $CO_2$ concentrations were measured (Müller et al., 2015). The Maludam leaching rate was 79 g C $m^{-2}$ $yr^{-1}$, and as discussed before, the underlying $CO_2$ and DIC yields are assumed to be representative for rivers draining pristine peatlands.

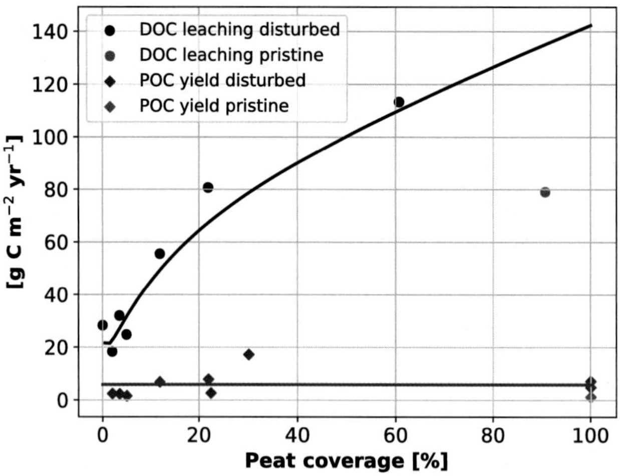


**FIGURE 4.10** Peat coverage versus leaching and POC yield. *POC*, particulate organic matter.

However, its DOC yield was lower than those derived from PSF1 due to the higher peat soils coverage of PSF1 (100%, see Fig. 4.9B). Accordingly, we used the DOC yield for the PSF1 as well as the $CO_2$ and DIC yields from the Maludam to calculate the leaching rate from pristine peat swamps forests (= peat soil coverage 100%). Considering that DIC yields are low and $CO_2$ yields are leveling off, the resulting error is assumed to be low. However, the leaching rate of 83.3 g C m$^{-2}$ yr$^{-1}$ is slightly higher than those derived for the Maludam (Table 4.6).

In comparison with a leaching rate of 142.3 g C m$^{-2}$ yr$^{-1}$ from disturbed peatlands, a leaching rate of 83.3 g C m$^{-2}$ yr$^{-1}$ implies that peat soil degradation increases leaching by 71% (Table 4.6). The influence of peatland degradation on the DOC supply and the hydrological cycle could cause this difference. To disentangle the impact of these two processes, we recalculated leaching rates but considered an evapotranspiration rate of 69% instead of 51%, also for rivers draining disturbed peatlands. Due to resulting reduced yields, the calculated leaching rate decreases from 142.3 to 115.2 g C m$^{-2}$ yr$^{-1}$, which means that changes in the hydrological cycle increase leaching by 23%. Consequently, the supply of DOC seems to have increased the leaching by 38% from 83.3 to 115 g C m$^{-2}$ yr$^{-1}$.

In addition to leaching, erosion is the second main process via which carbon is transferred from soils into rivers. Where leaching is associated with DOC, erosion refers to the transport of POC, whereby its yields are considered to be representative for erosion rates. This approach ignores POC decomposition in the river, but since $CO_2$ yields include the release of $CO_2$ from respired POC, this underestimate does not affect the estimate of the total carbon loss from soils. Accordingly, leaching and erosion add to the total loss and input of carbon from soils and into rivers.

Contrary to leaching, the POC yields obtained from rivers draining disturbed peatlands do not correlate with the peat soil coverage and reveal a mean of 5.9 g C m$^{-2}$ yr$^{-1}$

(Figs. 4.9D and 4.10). This low yield illustrates that erosion contributes only 4% to the total carbon loss from soils, whereas leaching is the main process via which peat carbon is introduced into rivers (Table 4.6).

The data from central Borneo show that POC yields from disturbed peatlands were much higher than those from pristine peatlands, the latter of which amount to ~1.2 g C $m^{-2}$ $yr^{-1}$. In comparison with the mean POC yield of disturbed rivers (5.9 g C $m^{-2}$ $yr^{-1}$), this represents an increase of 433% due to peatland degradation (Table 4.6). Such an increase is imaginable considering the impact of deforestation on soils, but relatively speaking, the low POC yield hardly affects the total carbon loss from peatlands.

Up to now, we have estimated the total carbon losses from soils, $CO_2$ emissions from rivers into the atmosphere and carbon discharges into the ocean. Now, it is interesting to know how peatland degradation influences these fluxes. We have found that in disturbed peatlands, leaching and erosion are much higher (75%), and in contrast to pristine peatlands, a smaller fraction of the leached and eroded carbon is discharged into the ocean (Table 4.6). This implies that peatland degradation preferentially increases DOC respiration and the resulting $CO_2$ emissions into the atmosphere. The underlying mechanisms are, as discussed before, the decrease of the evapotranspiration rates and the increased peat decomposition. The decrease of the evapotranspiration rates is a response to the decrease in biomass caused by peatland degradation (Fig. 4.5). Mechanisms through which peatland degradation increases the peat decomposition remain to be discussed.

### 4.5.7 Priming

Priming is a collective term subsuming a variety of reactions in response to perturbations, which increase the release of carbon and nutrients from soils. Originally, Kuzyakov et al. (2000) defined priming as a strong short-term change in the turnover of soil organic matter caused by comparatively moderate treatments of soil. However, applied to peatlands, deforestation and drainage of peat swamps, as well as fertilization of plantations, and the regrowth of secondary vegetation are perturbations, which could cause priming. Underlying mechanisms are an enhanced microbial activity due to an increased food supply.

Deforestation and drainage increase the food supply because they enhance the supply of oxygen and raise the pH and temperature in soils (Kobayashi, 2016). Considering that in general enzyme activities increase with increasing temperatures, all three responses to deforestation and drainage foster the activity of the peat decomposing enzyme phenol oxidase (Kang et al., 2018; Klemme et al., 2021). Consequences are an enhanced supply of digestible DOC, which further favors the microbial activity and the leaching of DIC and especially DOC from soils. Accordingly, drainage depth and soil temperature are also considered as prime factors controlling on-site $CO_2$ emission from peat soils (Holden, 2005; Hooijer et al., 2006, 2010).

Fertilization, which is a common practice on plantations (e.g., Baum and Rixen, 2014) and, as discussed before, a consequence of atmospheric pollution, increases the food supply more indirectly by raising the metabolism of plants. Secondary forest plant species reinforce the associated enhanced food supply to the heterotopic microbes by providing labile leaf organic matter. Results obtained from leaf litter leaching experiments revealed, e.g., that leaching and decomposition rates of leaves of endemic peat forest plants are much lower than those of leaves of secondary forest species (Yule and Gomes, 2009). Hence, priming and the enhanced formation of more labile DOC could explain the preferentially increase of DOC respiration and $CO_2$ emissions into the atmosphere in response to peatland degradation.

## 4.6 Estuaries and the ocean

Even though priming seems to favor DOC respiration, more than half (55.5%) of the leached and eroded peat carbon is introduced into the ocean (Table 4.6). This poses the question: what happens to peat carbon in the marine realm? In principle, there are three answers to this question: (1) It can be emitted from the ocean into the atmosphere, (2) it is transferred into the marine sediments, or (3) it remains in the oceanic water column. The fate of the peat carbon influences the carbon cycle on different time scales. Terrestrial carbon that is emitted into the atmosphere is an immediate $CO_2$ source to the atmosphere. Carbon that is buried in marine sediments and absorbed by the ocean could potentially be removed from the atmosphere for hundreds to thousands of years. Since the marine biosphere is quite small and short-lived, there are only two main carbon reservoirs, which are sufficiently large to take up significant amounts of terrestrial carbon, and this is the marine DIC and DOC pool. In the following paragraphs, we will discuss the potential uptake of peat carbon by the marine DOC pool, its burial in marine sediments, and finally the uptake by the DIC pool.

### 4.6.1 Dissolved organic carbon

#### *4.6.1.1 The microbial organic carbon pump in the ocean*

DOC represents the largest organic carbon pool in the ocean. With a reservoir size of 662 Pg C, it exceeds by far those of the living biomass in the ocean (6 Pg C) and the amount of carbon held by the preindustrial atmosphere (589 Pg C, Bar-On et al., 2018; Ciais et al., 2013; Hansell, 2013; Hansell et al., 2009). Hence, changes in the reservoir size of the marine DOC pool could influence the atmospheric $CO_2$ concentration significantly. The degradation of marine organic matter and the associated formation of persistent refractory DOC are referred to as the microbial organic carbon pump (Jiao et al., 2010). Due to the varying chemical composition of organic matter produced by plants and plankton as discussed before, terrestrial DOC is often assumed to be refractory, which increases its survival in the marine realm (Cao et al., 2018; Cragg et al., 2020; Zigah et al., 2017). An opposing opinion is that independently of its origin and nature, DOC is decomposed until its concentration falls below a threshold of about 40–50 $\mu mol\ L^{-1}$, which corresponds to the DOC concentration in the deep ocean (Arrieta et al., 2015; Middelburg, 2015). At such low concentrations, the probability that a microbe meets a

DOC molecule approaches zero, and under such circumstances, DOC remains preserved. Vice versa, this means that whenever DOC is supplied and the DOC concentration exceeds the threshold concentration, it is decomposed. Hence, the DOC pool and, thereby, the $CO_2$ storage within the DOC pool remain more or less constant. To estimate the potential relevance of Indonesian rivers for the marine DOC pool, we will quantify their DOC discharges into the ocean and use a remote sensing technique to trace DOC in the coastal seas.

#### *4.6.1.2 Dissolved organic carbon discharges into the ocean*

The overall Indonesian peat soil coverage is about 14% (Joosten, 2010). According to the regression equation obtained from the correlation between DOC yield and peat soil coverage (Fig. 4.9B), this results in a DOC yield of 13 g C $m^{-2}$ $yr^{-1}$. Multiplied by the total Indonesian land area of $1.9 \times 10^{12}$ $m^2$, this DOC yield amounts to a total DOC discharge into the ocean of 25 Tg C $yr^{-1}$. The comparison of this DOC discharge to those from Indonesian peatlands of 19.2 Tg C $yr^{-1}$ (Table 4.6) suggests that 73% of the Indonesian DOC originates from peatlands. However, DOC discharges from the numerous small Indonesian peat draining rivers are as high as those of the world's largest river, the Amazon, which has a DOC discharge of about 22 Tg C $yr^{-1}$ (Richey et al., 1990).

Global estimates suggest a riverine DOC discharge of 170–250 Tg C $yr^{-1}$ and indicate wetlands within the river catchments as the main DOC source (Dai et al., 2012; Harrison et al., 2005; Ludwig et al., 1996). According to these estimates, Indonesia contributes 15%–22% to the global riverine DOC discharge into the ocean (Fig. 4.11) while contributing only 11% to the global mean river discharge. Compared with the river discharge, these high DOC discharges emphasize the role of Indonesian peat swamps and priming as DOC sources to the ocean.

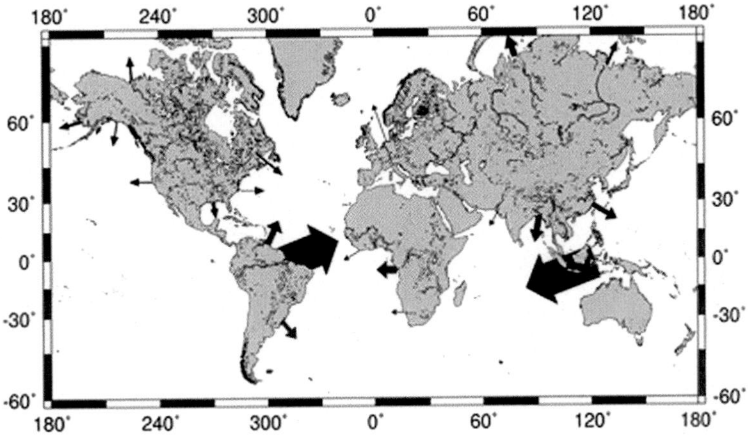


**FIGURE 4.11** Riverine DOC discharges. The widths of the arrows are proportional to the discharge. *DOC*, dissolved organic carbon. *Data source: Harrison et al. (2005) and Baum et al. (2007).*

### *4.6.1.3 The fate of dissolved organic carbon in the ocean*

To trace the terrestrial DOC in the ocean, we developed an algorithm that allowed us to translate satellite data into DOC concentrations (Siegel et al., 2009; Siegel et al., 2019). The algorithm is an empirical band ratio algorithm that estimates colored dissolved organic matter (CDOM) absorption from the reflectance ratio at two wavelengths and then further calculates DOC concentration with an empirical relationship to CDOM.

The evaluation of the satellite-derived DOC concentrations showed high DOC concentrations along the coast, which decreased with increasing distance to the coast and reached the 40 μmol $L^{-1}$ threshold concentration in open waters (Fig. 4.12). This supports the assertion that in the Indonesian Seas, the vast majority of the land-derived organic matter is decomposed. However, the algorithm is based only on in situ data from coastal waters and probably underestimates DOC because of, i.e., photobleaching of CDOM, which occurs in the surface layer of the ocean (Weyhenmeyer et al., 2012).

Results obtained from a numerical ocean circulation model study (Mayer et al., 2018) can also be used to further support the hypothesis. The results of the circulation model showed that on average, water from the South China Sea flushes the coastal ocean between Sumatra and Borneo and continues into the Java Sea. The transport rate from the South China Sea into the Java Sea was approximately $0.8 \times 10^6\ m^3\ s^{-1}$. River discharges into this area can be estimated by using the mean Indonesian precipitation rate of 2544 mm, an evapotranspiration rate of 51%, and a relevant land area, comprised of Sumatra and half of Borneo ($\sim 0.85 \times 10^{12} m^2$). The resulting river discharge is about

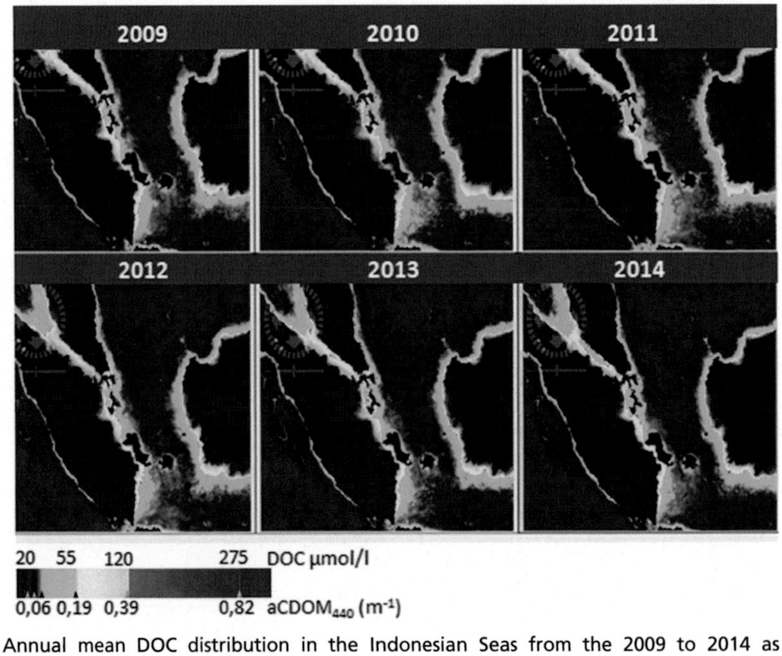


**FIGURE 4.12** Annual mean DOC distribution in the Indonesian Seas from the 2009 to 2014 as derived from satellite data. *DOC*, dissolved organic carbon.

$0.07 \times 10^6$ $m^3$ $s^{-1}$. Assuming a peat coverage of about 15% for Sumatra and Borneo (Miettinen et al., 2016) suggests in turn a riverine DOC concentration of ~781 μmol $L^{-1}$ (Fig. 4.8). In the South China Sea and the Java Sea, far-out from the coast, DOC concentrations in the surface waters were close to the 40 μmol $L^{-1}$ threshold concentration (Fig. 4.12). If one assumes a mixing between ocean ($0.8 \times 10^6$ $m^3$ $s^{-1}$, 40 μmol $L^{-1}$) and river waters ($0.07 \times 10^6$ $m^3$ $s^{-1}$, 781 μmol $L^{-1}$), the riverine DOC discharges could increase DOC concentrations from 40 to 100 μmol $L^{-1}$. Since, despite the potential to increase DOC concentrations significantly, DOC concentrations are close the 40 μmol $L^{-1}$ threshold concentration in the open Java Sea, we expect that the vast majority of the DOC supplied by Indonesian rivers is already decomposed in the coastal ocean. This is our best guess, which is based on the available data and needs to be evaluated in future studies. However, if DOC is decomposed it either enters the DIC pool or is emitted into the atmosphere.

### 4.6.2 $CO_2$ emissions from the coastal ocean

To estimate $CO_2$ emissions from the coastal ocean off Sumatra into the atmosphere, research cruises were carried out in October 2009, October 2012, and April 2013. During these cruises, an underway system for the determination of $CO_2$ concentrations and fluxes was installed onboard of the used vessels. Wind speeds required to calculate "k" were obtained from the Quick Scatterometer (Ricciardulli and Wentz, 2015). Furthermore, total alkalinity (TA) and DIC concentrations were determined in discrete samples, and individual parameters of the carbonate system were computed (Humphreys et al., 2020; Lewis and Wallace, 1998). These parameters also include the saturation state of aragonite ($\Omega_{AR}$) and calcite ($\Omega_{CA}$), which influence the precipitation and dissolution of the respective minerals.

As described by Wit et al. (2018), salinity thresholds are often used to determine the spatial extent of estuaries and the coastal ocean. Off the coast of Sumatra, the border between estuaries and the coastal ocean is predetermined at a salinity equal to 25. By correlating the salinity and distance to shore, this border is found at an approximate distance of 3 km. Based on this distance, the surface area of the estuaries is estimated using ArcGIS 10.4 and, assuming that estuaries influence the entire coastline, amounts to 10,818 $km^2$. The perimeter of the coastal ocean is based on a salinity of circa 32.8. This coincided with a distance of circa 67 km and resulted in a surface area of 127,674 $km^2$. Wit et al. (2018) showed the geographical extent of estuaries and the coastal ocean off Sumatra. In contrast to Wit et al. (2018) who used the Nightingale et al. (2000) "k" parameterization, we used here those from Wanninkhof (2014) and an atmospheric $CO_2$ concentration of 393 ppm as for the river studies. Accordingly, our results deviate slightly from those presented by Wit et al. (2018). Table 4.7 summarizes the data used to calculate the $CO_2$ yields and fluxes.

The mean $pCO_2$ of 2038 and 554 μatm in the estuaries and the coastal ocean are low in comparison with those measured in the rivers but high in comparison with $pCO_2$

measured in the ocean beyond shelves where they rarely exceed 1000 μatm (Emeis et al., 2018; Takahashi et al., 2014). Due to a lower turbulence in the aquatic boundary layer, the $CO_2$ exchange velocity "k" is with about ~9.6 cm $h^{-1}$ much lower in the estuaries and the coastal ocean than in the studied rivers where we used a "k" of 22 cm $h^{-1}$. In the ocean, a "k" of 22 cm $h^{-1}$ corresponds to wind speeds of ~8.35 m $s^{-1}$. Such a strong breeze causes white capes to occur but, at least to our data and experiences, not constantly. Nevertheless, one has to admit that uncertainties associated with the determination of "k" affect $CO_2$ flux calculation significantly. Even though it appears that "k" is better constrained in the ocean than in rivers, the use of the older "k" parameterization of Wanninkhof (1992) instead of those from Wanninkhof (2014) raised "k" from 10 to 19 cm $h^{-1}$ at a wind speed of 5.6 m $s^{-1}$. This increase of "k" would almost double our calculated $CO_2$ emission from estuaries and the coastal ocean into the atmosphere. Using a "k" of 63 cm $h^{-1}$ as also suggested to be suitable for rivers in a comparative study (see Wit et al., 2015 and references therein) would increase the calculated $CO_2$ emission from rivers by a factor of almost three. However, as discussed before, a "k" of 22 appears to be the most appropriate choice for the studied rivers, and ~9.6 cm $h^{-1}$ was derived from an updated version of the wind-depending "k" parameterization.

The calculated $CO_2$ yields from the estuaries and the coastal ocean amount to 493 and 43 g C $m^{-2}$ $yr^{-1}$, respectively. Calculated into fluxes, the $CO_2$ emission from estuaries of 5.3 Tg C $yr^{-1}$ almost equals that in the coastal ocean of 5.5 Tg C $yr^{-1}$ due to the larger spatial extension of the coastal ocean. Since the data refer to the coastal ocean of Sumatra, they have to be compared with $CO_2$ emissions from Sumatran rivers and discharges of Sumatran rivers into the coastal ocean.

To estimate $CO_2$ emissions and carbon discharges from Sumatran rivers into the estuaries and the coastal ocean, we used the equations shown in Fig. 4.9 and the Sumatran peat coverage of 15.6%. This amounts to a total $CO_2$ emission of 17.1 Tg C $yr^{-1}$ from rivers into the atmosphere and a river discharge into the ocean of 12.0 Tg C $yr^{-1}$ (Table 4.8). This river discharge exceeds the total degassing of $CO_2$ from estuaries and the coastal ocean of 10.8 Tg C $yr^{-1}$, which implies that carbon is buried in sediments and/or enters the marine DIC pool.

### 4.6.3 Organic carbon burial

Organic burial has not been measured directly but downscaled from global mean values (Wit et al., 2018). Hence, the following estimate is associated with large uncertainties but the best we can do under the given circumstances. On a global scale, the areal extent of estuaries ($1.45 \times 10^{12}$ $m^2$) and shelves ($26 \times 10^{12}$ $m^2$) in combination with their organic carbon burial rates results in area-normalized organic carbon burial rates of approximately 48 and 1.7 g C $m^2$ $yr^{-1}$, respectively. Multiplied by the areal extent of estuaries and the coastal ocean off Sumatra, this amounts to organic carbon burial rates of 0.52 and 0.23 Tg C $yr^{-1}$, respectively, and a total organic carbon burial of 0.74 Tg C $yr^{-1}$.

**Table 4.8** Carbon fluxes from Sumatra in comparison with those from disturbed and pristine peatlands and the difference between those from disturbed and pristine peatlands (Δ). $CO_2$ emission from the ocean, the carbon uptake of the ocean, and sedimentation, which refers to disturbed peat soils, were derived from river discharges and the contribution of the fluxes to river discharges as obtained from Sumatra.

| | Sumatra | | Peat land | | |
|---|---|---|---|---|---|
| | | | Disturb. | Pristine | Δ |
| | [Tg C $yr^{-1}$] | [%] | [Tg C $yr^{-1}$] | | [%] |
| Leaching and erosion | 29.1 | 100 | 40.0 | 22.8 | 75 |
| Leaching | 26.4 | 91 | 38.4 | 22.5 | 71 |
| Erosion | 2.7 | 9 | 1.6 | 0.3 | 433 |
| $CO_2$ emission river | 17.1 | 59 | 17.8 | 5.8 | 207 |
| River discharges | 12.0 | 41 | 22.2 | 17.0 | 31 |
| $CO_2$ emission ocean | 10.9 | 91 | 20.1 | 15.4 | 31 |
| Ocean uptake | 0.8 | 7 | 1.9 | 1.6 | 19 |
| Sedimentation | 0.3 | 2 | 0.18 | 0.03 | 500 |

A contribution of terrestrial organic matter to the total organic carbon of 44% (Burdige, 2005) indicates a burial rate of terrestrial organic carbon of 0.3 Tg C $yr^{-1}$ (Table 4.8). An organic carbon burial of 0.3 Tg C $yr^{-1}$ in addition to a $CO_2$ degassing of 10.9 Tg C $yr^{-1}$ still falls below the total discharge of carbon form rivers in the ocean (12.0 Tg C $yr^{-1}$). This imbalance calls for an additional carbon sink of about 0.8 Tg C $yr^{-1}$, which is assumed to be the uptake of terrestrial carbon by the marine DIC pool.

### 4.6.4 The invisible carbon footprint

The absorption of riverine carbon inputs by the marine DIC pool is called the invisible carbon footprint (Wit et al., 2018). Mechanistically, it is a feedback mechanism on the marine $CO_2$ emissions caused by a reduced carbonate production due to riverine carbon inputs, which lower the seawater pH. As discussed in the box. "The marine carbonate system", the pH is a key variable, which controls $CO_2$ fluxes into the atmosphere and via $\Omega$ also the precipitation of carbonates. Since riverine carbon inputs decrease the pH and shift the carbonate system to the left, they enhance $CO_2$ concentrations and lower $\Omega$. In turn, a reduced $\Omega$ favors the carbonate dissolution and lowers the carbonate production, which again increases the pH and reduces $CO_2$ concentrations and emission into the atmosphere.

Due to the respiration of riverine DOC and POC as well as the discharges of DIC, we determined $\Omega$ of the carbonate mineral aragonite ($\Omega_{AR}$) to be $<1$ in estuaries (Wit et al., 2018). It increased to about 3.5 at the outer rim of the coastal ocean at a salinity of $\sim$30. In coral reefs, an $\Omega_{AR}$ of $<4.5$ already hinders the coral growth, and a $\Omega_{AR}$ of $\sim$3 is considered as the threshold below which reefs turn from a net accretion into a net dissolution (Eyre et al., 2018; Langdon and Atkinson, 2005). Accordingly, the low $\Omega_{AR}$ off the coast of Sumatra documents the strong impact of river discharges on the marine carbonate system and suggests carbonate dissolution in estuaries and reduced carbonate production in the coastal ocean. However, prior to dissolved carbonate, it must have been supplied or produced.

Wit et al. (2018) estimated a riverine discharge of particulate inorganic carbon (PIC) of 0.23 Tg C $yr^{-1}$. According to our SPICE data, sediments off the coast of Sumatra contain 0.63% organic and 1.73% carbonate carbon (Michel et al., 2015), which suggests a carbonate carbon burial rate of 2.3 Tg C $yr^{-1}$. Since the carbonate carbon burial exceeds PIC inputs from land, carbonate must also be produced in the ocean. As indicated by our results, mollusks are the marine carbonate producers in sediments off the coast of Sumatra, while coral reefs develop only outside the region influenced by rivers discharges (Michel et al., 2015). This means that in addition to PIC from land, also carbonate shells provide carbonate whose dissolution lowers the $CO_2$ concentrations in the water. Considering that peatland degradation increases carbon discharge by 31% (Table 4.6), we believe that the low measured $\Omega_{AR}$ was already an effect of peatland degradation. This implies that peatland degradation lowered carbonate production and increased the dissolution of carbonate, which accumulated in marine sediments prior to the onset of peatland degradation.

To quantify the uptake of terrestrial carbon by the marine DIC pool, we calculated the difference between river inputs, $CO_2$ emissions, and burial rates. Off Sumatra, this amounts to an ocean uptake of 0.8 Tg C $yr^{-1}$ (Table 4.10). In summary, this means that off the coast of Sumatra 91% of the carbon discharged from rivers into the ocean was emitted into the atmosphere, 7% absorbed by the ocean, and 2% buried in marine sediments (Table 4.8).

### 4.6.5 The marine peat carbon budget

So far, we have discussed the fate of carbon that was introduced by Sumatran rivers into the coastal ocean off the coast of Sumatra. However, to complete the off-site $CO_2$ emission budget from peatlands, we apply results obtained from the study off the coast of Sumatra to the entire Indonesian peatlands. Of interest are here the survival rate of POC (11%) that is discharged into the ocean and the fraction of the total riverine carbon discharge that is emitted into the atmosphere (91%).

The POC survival rate of 11% in addition to the POC discharge from rivers draining pristine and disturbed peatlands can be used to estimate burial rates of peat carbon. POC discharges from rivers draining pristine and disturbed peatlands amount to 0.3 and 1.6 Tg C $yr^{-1}$, as discussed before. Assuming that 11% of this riverine POC discharge is buried in sediments suggests a peat organic carbon burial rate of 0.18 and 0.03 Tg C $yr^{-1}$ in pristine and disturbed peatlands, respectively (Table 4.8).

The fraction of the total riverine carbon discharge that is emitted into the atmosphere allows us to estimate the fraction of the peat-derived river discharge that is emitted from the ocean into the atmosphere. Considering that approximately 91% of the riverine carbon discharge is emitted into the atmosphere, the $CO_2$ emission results in 20.1 and 15.4 Tg C $yr^{-1}$ from disturbed and pristine peatlands, respectively (Table 4.8).

Since the riverine carbon discharge, $CO_2$ emission, and burial rates are known, also the absorption by the DIC pool can be calculated as the difference between riverine inputs, burial rates, and $CO_2$ emission. This results in a DIC absorption of 1.9 and 1.6 Tg C $yr^{-1}$ for disturbed and pristine peatlands, respectively (Table 4.8).

Thereby, we have an estimate of the fate of peat carbon that enters the ocean, which shows that the enhanced riverine carbon discharges due to peatland degradation are largely balanced by increased $CO_2$ emissions from the ocean into the atmosphere.

### 4.6.6 Emission factors

The off-site $CO_2$ emission from peatlands is the sum of $CO_2$ emission from rivers (17.8 Tg C $yr^{-1}$) and the ocean (20.1 Tg C $yr^{-1}$), and both so far have been quantified (Table 4.8). The total off-site $CO_2$ emission from disturbed peatlands in comparison with the leaching and erosion rate shows that 94.8% of the eroded and leached carbon is emitted from rivers and the ocean into the atmosphere (Table 4.9). Divided by the peatland area, this results in area-normalized off-site $CO_2$ emission of 140.5 g C $m^{-2}$ $yr^{-1}$, which exceeds those derived from the IPCC of 82.1 g C $m^{-2}$ $yr^{-1}$ by 71%.

Similar to our results (94.8%), the IPCC also assumes that 90%–100% of the carbon that is discharged into the ocean is emitted as $CO_2$ into the atmosphere. However, instead of carbon leaching and erosion, it uses the DOC yield to quantify carbon losses from peat soils. Accordingly, the respiration of DOC in rivers as well as those of POC in rivers and the ocean is ignored. This is a conceptual problem, which is reinforced by peatland degradation as this preferentially supplies degradable DOC and increases $CO_2$ emissions from rivers.

To account for these missing carbon fluxes, the term that describes the DOC yield and its increase due to peatland degradation in Eq. (4.3) has to be replaced by leaching and erosion and their rise to peatland degradation as shown in Eq. (4.9).

$$\begin{aligned}\mathrm{EF_{aquatic}}\left[\mathrm{gC\,m^{-2}g^{-1}yr^{-1}}\right] = \Big[&\left(\mathrm{leaching_{natural}} \times \left(1 + \Delta \mathrm{leaching_{perturbation}}\right)\right) \\ &+ \left(\mathrm{erosion_{natural}} \times \left(1 + \Delta \mathrm{erosion_{pertubation}}\right)\right)\Big] \times \mathrm{Frac_{emission}}\end{aligned} \tag{4.9}$$

**Table 4.9** Fate of carbon mobilized by carbon leaching and erosion from disturbed peatlands.

| | [Tg C $yr^{-1}$] | [g C $m^{-2}$ $yr^{-1}$] | [%] |
|---|---|---|---|
| Leaching and erosion | 40.0 | 148.2 | |
| $CO_2$ emission to the atmosphere (ocean and rivers) | 37.9 | 140.5 | 94.8 |
| Uptake by the ocean | 1.9 | 7.0 | 4.8 |
| Sediments | 0.18 | 0.7 | 0.4 |

**Table 4.10** SPICE emission factors.

| Factors | Units | Value |
|---|---|---|
| $Erosion_{natural}$ | [g C $m^{-2}$ $yr^{-1}$] | 1.2 |
| $\Delta Erosion_{perturbation}$ | | 4.33 |
| $Leaching_{natural}$ | [g C $m^{-2}$ $yr^{-1}$] | 83.3 |
| $\Delta Leaching_{perturbation}$ | | 0.71 |
| $Frac_{emisison}$ | | 0.948 |

where $EF_{aquatic}$ = emission factor for annual $CO_2$ emissions along the land–ocean continuum; $leaching_{natural}$ = leaching rate from undisturbed peat soils [g C $m^{-2}$ $yr^{-1}$]; $\Delta leaching_{perturbation}$ = proportional increase of the leaching rate relative to natural leaching rates; $erosion_{natural}$ = POC yield from undisturbed peat soils [g C $m^{-2}$ $yr^{-1}$]; $\Delta erosion_{perturbation}$ = proportional increase of the POC yield relative to POC yield form pristine peatlands; $Frac_{emission}$ = fraction of leached and eroded carbon, which is oxidized to $CO_2$ along the land–ocean continuum. The individual emission factors are listed in Table 4.10.

Since $EF_{aquatic}$ represents the area-normalized off-site $CO_2$ emission from disturbed peatlands, it has to be multiplied by the peatland area (A) to obtain off-site $CO_2$ emission in Tg C $yr^{-1}$ as shown by Eq. (4.10).

$$CO_{2-off-site}\left[Tg\ C\ yr^{-1}\right] = A \cdot EF_{aquatic} \quad (4.10)$$

## 4.7 Ecosystem $CO_2$ emissions

Off-site and on-site $CO_2$ emissions from peat soils represent the total $CO_2$ loss from peat soils. However, if one is interested in the $CO_2$ emission from the ecosystem "peat swamp" into the atmosphere, one has to also consider the vegetation (living biomass). It strongly affects the impact of on-site $CO_2$ emissions from soils on the on-site $CO_2$ emission from the ecosystem "peat swamp" into the atmosphere. For instance, trees growing by the utilization of $CO_2$ that is emitted from soils lower the impact of on-site $CO_2$ emissions from soils on those from peatlands into the atmosphere. The net ecosystem $CO_2$ exchange (NEE) is a parameter, which takes into account the influence of the vegetation on the $CO_2$ emission from the ecosystem.

### 4.7.1 Net on-site ecosystem $CO_2$ exchange

Eddy covariance is a method which allows to measure $CO_2$ fluxes above the treetops at the canopy and thus to quantify on-site $CO_2$ emissions from peatlands into the atmosphere. Net fluxes reflect changes in the amount of carbon stored in the biomass due to imbalances between the formation and respiration of organic matter. The IPCC distinguishes between six ecosystem immanent organic carbon reservoirs: above- (AB) and belowground biomass (BB), deadwood (DW), litter (LI), harvested wood products (HWPs), and soils (SO). Changes in the soil reservoir (SO) are represented by off-site and on-site $CO_2$ emissions from peat soils as discussed before. Since off-site $CO_2$ emissions from soils occur in rivers and the ocean, they are not or at least hardly captured in eddy covariance measurements in the peat swamps. Vice versa, leaching and erosion appear even as an increase of peat biomass.

$$\text{Net flux}\left[\frac{\text{gC}}{\text{m}^{2*}\text{yr}}\right] = [\Delta\text{Vegetation} + \Delta\text{Peat} + \text{Leaching}_{\text{DOC}} + \text{Erosion}_{\text{POC}}] \tag{4.11}$$

If 'Vegetation' is considered as one carbon reservoir, which includes AB, BB, DW, HWP and LI, net fluxes result from changes in the amount of carbon stored in vegetation (ΔVegetation), peat soils (ΔPeat) as well as from leaching and erosion as shown in Eq. (4.11).

## 4.7.2 $CO_2$ emission from pristine peat swamps

Since pristine peat swamp forests reveal a climax condition, it is assumed that the biomass is nearly constant (ΔVegetaion = 0). Their peat carbon accumulation rate of 75 g C $m^{-2}$ $yr^{-1}$ increases ΔPeat, and leaching and erosion from pristine peat swamps amount to 84.5 g C $m^{-2}yr^{-1}$ (Table 4.6). Inserting these numbers into Eq. (4.11) results into a net on-site ecosystem $CO_2$ flux of 159.5 g C $m^{-2}$ $yr^{-1}$.

To estimate off-site $CO_2$ emissions, one has to consider that in pristine peat swamps, instead of 94.8%, only 93% of the leached and eroded peat carbon is respired and emitted as $CO_2$ into the atmosphere (Table 4.6). Hence a leaching and erosion rate of 84.5 g C $m^{-2}yr^{-1}$ results into an off-site $CO_2$ emission of 78.5 g C $m^{-2}yr^{-1}$. Multiplied by the peatland area of $0.27 \times 10^{12}$ $m^2$, these numbers suggest an on-site and off-site $CO_2$ emission of about 43.0 and −21.2 Tg $Cyr^{-1}$, respectively, which results in a total net ecosystem $CO_2$ flux (NEE) of 21.8 Tg C $yr^{-1}$ (Fig. 4.13A).

## 4.7.3 $CO_2$ emission from disturbed peatlands

So far, we are aware of seven peatland sites in SE Asia at which on-site $CO_2$ emission was determined. The annual mean on-site $CO_2$ emission varied between −532 and 499 g C $m^{-2}$ $yr^{-1}$, showing that peatlands switched from $CO_2$ sinks (positive NEE) to $CO_2$ sources (negative NEE, Suzuki et al., 1999). One study, which was conducted in peatlands from Thailand, suggested that peat formation and the regrowth of secondary forests caused the $CO_2$ uptake by a pristine and disturbed peat swamp, respectively (Suzuki et al., 1999). However, observations from Sarawak showed that the Maludam forest acted as a $CO_2$ source to the atmosphere with a mean on-site $CO_2$ emission of −420 g C $m^{-2}$ $yr^{-1}$ (Tang Che Ing, 2017), while a much more disturbed site in the vicinity of the park absorbed $CO_2$ with an on-site $CO_2$ emission of 136 g C $m^{-2}$ $yr^{-1}$ (Kiew et al., 2018). A more linear link between peatland degradation and on-site $CO_2$ emission was obtained by a long-term

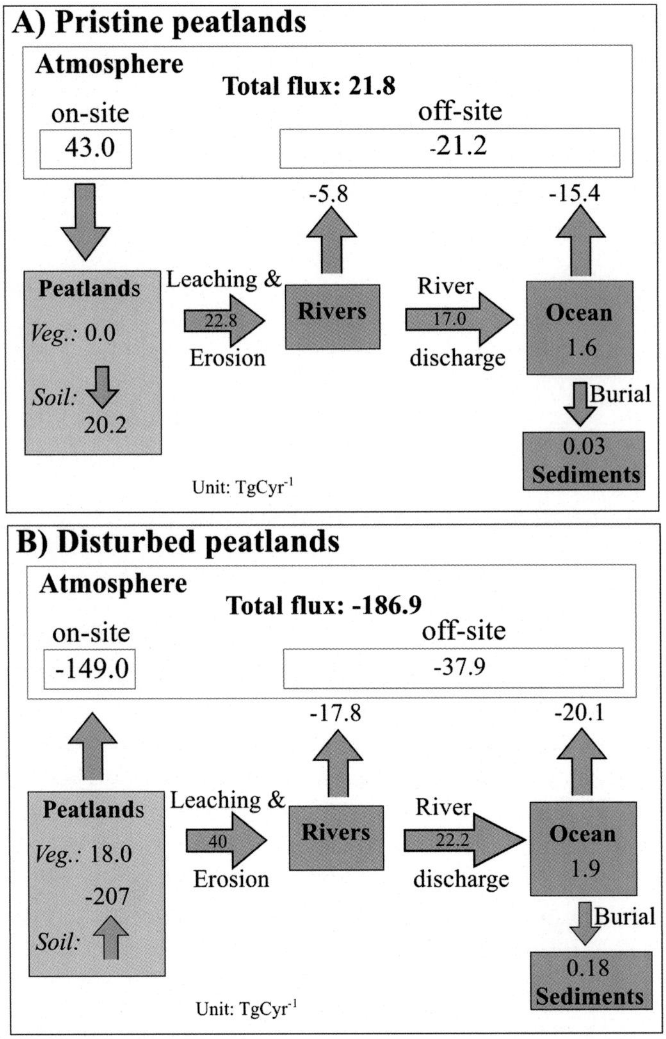


**FIGURE 4.13** Carbon fluxes along the land–ocean continuum at Indonesian peatlands in their natural (A) and disturbed state (B). "Veg." indicates vegetation in contrast to soil, which includes the dead biomass.

study carried out in a disturbed peat swamp in central Borneo (Hirano et al., 2007, 2012). Here the on-site $CO_2$ emission decreased from $-174$ to $-499$ g C $m^{-2}$ $yr^{-1}$ with drainage depth and increasing degree of forest degradation.

An alternative method to estimate on-site $CO_2$ emission is suggested by the IPCC (2014). It is based on the quantification of changes in the reservoir sizes of all the six organic carbon reservoirs. The BUR applied the IPCC methodology (Boer et al., 2018) and suggested a carbon loss of $149 \pm 68$ Tg $Cyr^{-1}$, which divided by peatland area of $0.27 \times 10^{12}$ $m^2$ results in an on-site $CO_2$ emission of $-552 \pm 252$ g C $m^{-2}$ $yr^{-1}$. This is within the range of those measured in SE Asia but much lower than $-920$ g C $m^{-2}$ $yr^{-1}$ as obtained in recent model study by Günther et al. (2020). Our mean on-site $CO_2$ emission from soil of 167 Tg C $yr^{-1}$ (620 g C $m^{-2}$ $yr^{-1}$ × 0.27 × $10^{12}$ $m^2$) due to peat oxidation (374 g C $m^{-2}$ $yr^{-1}$) and fires (246 g C $m^{-2}$ $yr^{-1}$) exceeds the BUR estimate 149 ± 68 Tg $Cyr^{-1}$, which implies that a growing vegetation might act as $CO_2$ sink (18 Tg C $yr^{-1}$ = 167−149). Considering additionally our leaching and erosion of 40 Tg C $yr^{-1}$ amounts to total carbon loss from 207 Tg C $yr^{-1}$. In contrast to pristine peat swamps where leaching and erosion is balanced by a positive on-site $CO_2$ emission, it is balanced by the decomposition of peat (−ΔPeat) in disturbed peat soils. However, a mean on-site $CO_2$ emission of 149 Tg $Cyr^{-1}$ in addition to an off-site emission of 37.9 Tg C $yr^{-1}$ (Table 4.9) amounts to a total net ecosystem $CO_2$ flux (NEE) from disturbed Indonesian peatlands of 186.9 Tg C $yr^{-1}$ (Fig. 4.13). Accordingly, ignoring off-site $CO_2$ emissions means to underestimate the NEE from disturbed peatlands by 20%.

## 4.8 Evaluation of $CO_2$ emissions

Far-reaching decisions toward sustainability were taken in 2015 during the United Nations Climate Change Conference (COP 21) in Paris. There the international community decreed to reduce $CO_2$ emissions to keep global warming below 1.5–2°C, which, as mentioned before, calls upon a rapid transition to net zero greenhouse gas (GHG) emissions unlit 2050 (Figueres et al., 2017). The total amount of $CO_2$ that can be emitted into the atmosphere while keeping this goal is called the "remaining carbon budget" (Rogelj et al., 2019). This amount shrinks with each molecule of $CO_2$ that is emitted into the atmosphere. One way to evaluate the $CO_2$ emission caused by peatland degradation is to quantify their contribution to the shrinking "remaining carbon budget" (Fig. 4.14).

### 4.8.1 Climate response to cumulative emissions of $CO_2$

There are various ways to calculate the remaining carbon budget and results differ (Rogelj et al., 2019). Here, we used the cumulative $CO_2$ emissions caused by burning fossils fuel, global mean $CO_2$ concentrations in the atmosphere and global mean temperature anomalies to calculate the remaining carbon budget (Fig. 4.14). All three data sets are publicly available, and the reader can recalculate and update the remaining carbon budget, which is necessary as it shrinks from year to year.

### 4.8.2 $CO_2$ reduction potential

High $CO_2$ emissions imply, in turn, a high potential to reduce $CO_2$ emission by restoring peat swamp forests. Based on carbon emission rates from rewetted areas of about 52 g C $m^{-2}$ $yr^{-1}$ (Günther et al., 2020), the $CO_2$ reduction potential due to peatland restoration can be estimated. Using the rate of 52 g C $m^{-2}$ $yr^{-1}$ and assuming that the entire Indonesian peatlands are rewetted suggest an on-site $CO_2$ emission of 14 Tg C $yr^{-1}$. This falls well below the quota of 31.5 Tg C $yr^{-1}$ and represents a reduction of the on-site emission of 135 Tg C $yr^{-1}$ (149 -14 Tg C $yr^{-1}$). However, such reduction would be associated with three main problems: time, management efforts, and emissions of other GHGs such as methane.

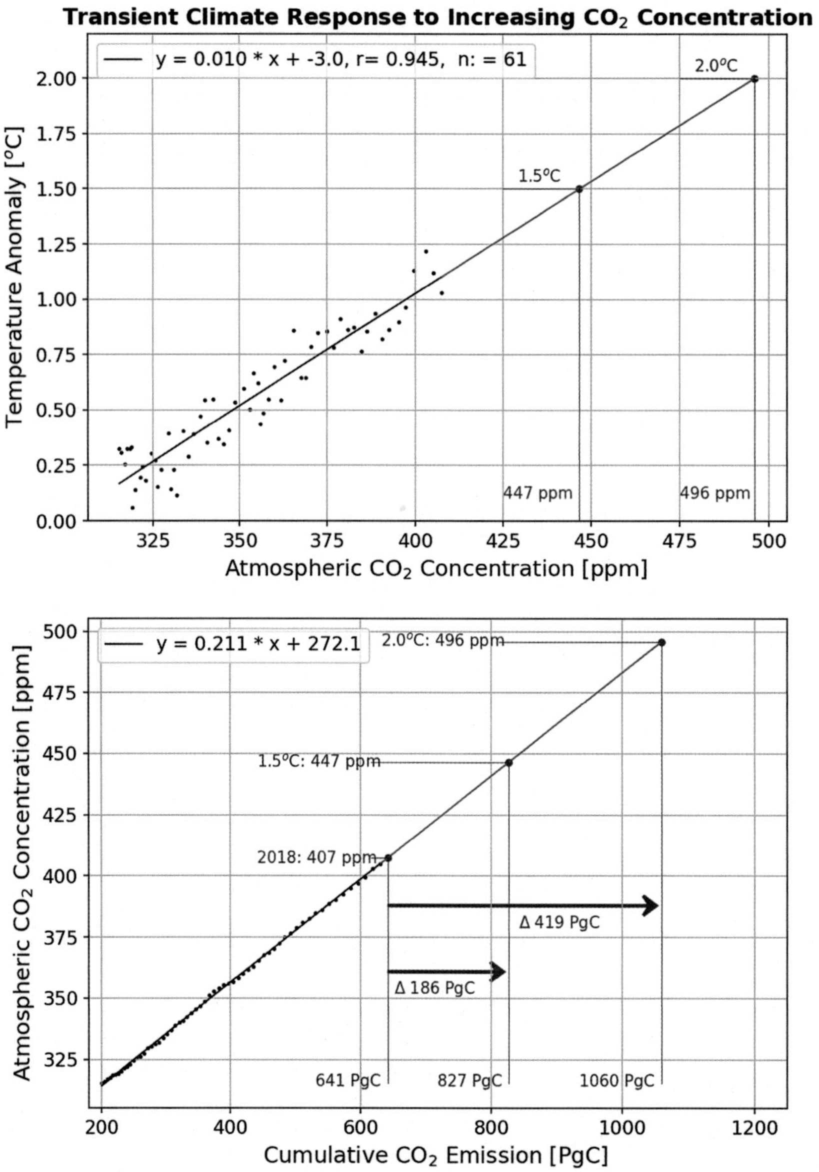


**FIGURE 4.14** Atmospheric $CO_2$ concentrations versus global mean temperature anomalies (upper panel) and cumulative $CO_2$ emissions versus atmospheric $CO_2$ concentrations. *The data were obtained from Friedlingstein et al. (2019), Lenssen et al. (2019), NOAA and Dlugokencky (2016), Osborn and Jones (2014), and Smith et al.(2008).*

The global society aims at reaching net zero GHG emissions within the upcoming 30 years (Figueres et al., 2017), but since it takes 60–170 years for a peat swamp forest to regrow (Hapsari et al., 2018), the reduction of $CO_2$ emission through peatlands restoration is a long-term strategy. Furthermore, approximately only 10% of the peatlands are protected (Posa et al., 2011), and it is a long way to increase this to 100% as we will discuss later, and finally rewetting favors the formation of methane. This bears the risk that enhanced emissions of methane outweigh the reduction of $CO_2$ on the greenhouse effect. Methane has a global warming potential of at least 28 times the one of $CO_2$ over 100 years and has contributed 20% to the current global warming (Myhre et al., 2013 and references therein). However, the concept behind the global warming potential is debated, and methane sources are still poorly constrained, which also includes methane emissions related to palm oil plantations (Saunois et al., 2020). Taking this into account, model calculations carried out by Günther et al. (2020) suggest that enhanced methane emissions due to rewetting are insufficient to outbalance the impact of $CO_2$ on global warming. Hence, peatland restoration seems to remain an important measure to reduce GHG emissions even though its impact on the latter is slow and knowledge gaps regarding the cycle of methane have to be closed in future studies.

### 4.8.3 $CO_2$ emissions and land losses

Subsidence is another major issue in peatlands. Results derived from field studies show a subsidence of 1420 mm within the first 5 years after drainage and of 50 mm $yr^{-1}$, thereafter (Hooijer et al., 2012). In line with a former local study (Wosten et al., 1997), satellite-derived data showed a mean subsidence rate of 22 mm $yr^{-1}$ with no significant difference between plantations and degraded peat swamp forests between 2007 and 2011 (Hoyt et al., 2020). Due to compaction and shrinking, the PCD increases from 0.05 g C $cm^{-3}$ in pristine peat swamps (Murdiyarso et al., 2010) to about 0.09 g C $cm^{-3}$ in plantations (Hooijer et al., 2012). This PCD in addition to carbon losses from peat soils can be used to estimate the subsidence caused by the latter. An on-site $CO_2$ emission from soils (167 Tg C $yr^{-1}$) in addition to leaching and erosion of 40 Tg C $yr^{-1}$ (Fig. 4.13) amounts to a mean peat carbon loss of 207 Tg C $yr^{-1}$. This corresponds to a carbon loss rate of 766.7 g C $m^{-2}$ $yr^{-1}$ (= 207 Tg C$yr^{-1}$/($0.27 \times 10^{12}$ $m^2$), which divided by the PCD of 0.09 g C $cm^{-3}$ results into a peat erosion rate of 8.5 mm $yr^{-1}$.

The sea level rise in the Indonesian Sea resembles the global mean (https://uhslc.soest.hawaii.edu/). It rose, as mentioned before, by 3.6 mm $yr^{-1}$ between 1993 and 2010 (Church et al., 2013) and is assumed to increase to 10 mm $yr^{-1}$ until the year 2100 (Church et al., 2013; Vermeer and Rahmstorf, 2009). In addition to the erosion rate of 8.5 mm $yr^{-1}$ caused by carbon losses from peat soils, a sea level rise of 3.6 mm $yr^{-1}$ results in relative sea level rise of 12.1 mm $yr^{-1}$ (8.5 + 3.6 mm $yr^{-1}$). Deltas with a relative sea level rise of 6.2 mm $yr^{-1}$ are already classified as deltas in peril or even in great peril (Syvitski et al., 2009). Since about 80% of the Indonesian peatlands are located within coastal flood plains and cover 11% of the Indonesian landmass, subsidence increases the vulnerability of Indonesian coast to sea level rise. More specifically, it enhances the risk to lose a significant area of land by enhancing coastal erosion (Rixen et al., 2016; Whittle and Gallego-Sala, 2016). The Dutch and German coast as well as many other places (Hooijer et al., 2012 and references therein) including the famous Venice Lagoon, in Italy (Fornasiero et al., 2020), are examples showing land losses due to peatland degradation.

### 4.8.4 Climate pledges and gaps

In 2009, during the G-20 meeting in Pittsburgh, United States, Indonesia pledged to reduce $CO_2$ emissions by 26% and 41% with support from international communities until 2020 from the Business as Usual (BAU) scenario with domestic resources and support from international communities, respectively (MoEF, 2015). With this statement, Indonesia was one of the first developing countries to announce an ambitious emission reduction target. Even though the BAU level was not specified (Fei and Shuang-Qing, 2012), Indonesia developed with the Presidential Regulation No. 61/2011 a National Action Plan on GHG reduction (RAN-GRK, Indonesia, 2009; UNFCCC, 2011). The RAN-GRK included, as mentioned before, a moratorium on the clearing of primary forests and conversion of peatlands from 2010 onward and listed the intended reduction of $CO_2$ emission according to sectors (Indonesia, 2009; Indrarto et al., 2012; UNEP, 2019). In 2016, Indonesia submitted the INDC report (Act No. 16/2016) with the commitment to reduce GHG emission by 29% and 41% relative to BAU until the year 2030. Table 4.11 shows the data according to BUR (Boer et al., 2018).

The unit "$CO_2e$" used in Table 4.3 means $CO_2$ equivalent. It expresses the GHG effect of $CO_2$ and other GHGs such as methane and $N_2O$, wherein $CO_2$ contributes on average 82% to GHG emission in Indonesia (Boer et al., 2018).

The data presented in Table 4.11 illustrate that the pledged reduction is actually an increase of the $CO_2$ emission by 52.5% in comparison with 2010, but this is an increase, which is lower as it could be at BAU conditions. The respective measures applied to achieve this goal is a moderate increase of $CO_2$ emissions by burning fossil fuel and a reduction of the $CO_2$ emission caused by LULCC. In comparison with 2010, the latter amounts to 56% (116 Tg $CO_2e$ $yr^{-1}$) from 206.5 Tg $CO_2e$ $yr^{-1}$ in 2010 to 89.7 Tg $CO_2e$ $yr^{-1}$ in 2030 (Table 4.3). With an estimated on-site emission reduction potential of 135 Tg C $yr^{-1}$, conservation and restoration of peat swamp forest could be a significant contribution to reach this goal, but as mentioned before, there are management issue,

**Table 4.11** Indonesian greenhouse gas emission (GHG-E) in 2010, GHG-E as expected according to the business as usual (BAU) scenario, intended GHG reduction, reduction in comparison with BAU, pursued GHG-E in 2030, and the pursued GHG-E in 2030 in comparison with those in 2010.

| | | 2030 | | | | |
|---|---|---|---|---|---|---|
| | **2010** | **BAU** | **Reduction** | **Reduction** | **Pursued** | **2010/2030** |
| Sector | Tg C− $CO_2e$ | Tg C− $CO_2e$ | | [%] | Tg C− $CO_2e$ | [%] |
| Energy | 123.5 | 455.2 | 85.6 | 18.8 | 369.5 | 199.1 |
| Industry | 9.8 | 19.0 | 0.8 | 4.0 | 18.2 | 85.7 |
| LULCC | 206.5 | 227.7 | 138.0 | 60.6 | 89.7 | −56.5 |
| Waste | 24.0 | 80.7 | 3.0 | 3.7 | 77.7 | 223.9 |
| Total | 363.8 | 782.5 | 227.5 | 29.1 | 555.0 | 52.5 |

The data are obtained from Boer et al. (2018).

which will be discussed in following section, and two ecosystem-related problems. The two ecosystem-related problems are the slow-growing biomass in peatlands and enhanced methane emission associated with the rewetting of disturbed peatlands. Therewith, peatland restoration is a long-term strategy with regard to emission goals, which in turn calls for additional measures to reach the short-term emission targets until 2030. To keep the Paris goals, these measures should include the reduction of $CO_2$ emissions caused by burning fuel by switching from fossil fuel to alternative, clean energy sources, such as wind, hydrogen, and solar energy. However, with regard to climate stability and the mitigation of climate change impacts caused, e.g., by the rising sea level, conservation and restoration are considered as urgent measures which implementation cannot be delayed.

## 4.9 Socioeconomic implications

In Indonesia, land classified as forest falls mostly under the jurisdiction of the Ministry of Forestry (MOF), while the National Land Agency and subnational authorities administer all other lands (Hein, 2013 and references therein). With a few exceptions, forestland belongs to the state. Private, and formal property exists only on land classified as non-forest. The purview of MOF includes >70% of the Indonesian landmass, and it has the right to assign land as state forest (Indrarto et al., 2012). So far, 10% of this land has been gazetted and thereby turned into state forests. In 2011 also, only 10% of the tropical swamps were classified as protected areas (Posa et al., 2011). Consequences are that local and indigenous people claim customary rights, and forests have been allocated for large management activities. The unclear tenure is seen as counterproductive in promoting sustainable forest management, and the REDD+ mechanism is, in turn, assumed to be a strategy that potentially could help to solve this dilemma.

### 4.9.1 REDD+

REDD+ refers to the "reducing emissions from deforestation and forest degradation in developing countries." It was first negotiated by UNFCC in 2005 with the goal to reduce GHG emissions through enhanced forest management and received substantial attention in 2007 during COP 13, in Bali, Indonesia. There, the "Bali Action Plan" was developed, which includes further requests. These are sustainable management of forests, conservation of forest carbon stocks, and enhancement of forest carbon stocks.

The REDD+ mechanism is designed as an international payment for ecosystem service (PES) scheme providing incentives to avoid deforestation and forest degradation through emission trading or result-based payments (Hein, 2013 and references therein). Carbon fixation services provided by forests are converted into comparable and exchangeable units, such as carbon credits. These are tradable certificates: GHG mitigation projects generate it, and emitters buy it. The revenue can be used to finance REDD+ projects. According to the European Union Emissions Trading System (EU ETS), the $CO_2$ price was about 20 €/t$CO_2$ in 2010. Thereby, a square meter ($m^2$) of land receives an economic value based on its carbon storage. Applied to peatlands, this means one has to add up the carbon density of living biomass and peat, including all dead organic matter and multiply this value with the $CO_2$ price.

For example, a mean peat dome thickness of 5 m and a PCD of 0.05 g C $cm^{-3}$ amounts to a peat carbon storage of 250 kg C $m^{-2}$. Considering the mean density of living biomass of about 20.7 kg $m^{-2}$ (Table 4.2) and that 1 g C corresponds to 3.66 g $CO_2$ the $CO_2$ storage results to 0.9925 t $CO_2$ $m^{-2}$ (= 270 kg C $m^{-2}$). Multiplied by a $CO_2$ price of 20 €/t$CO_2$, this results in an economic value of 19.8 € per 1 $m^2$ of peatland. Accordingly, the total value of the Indonesian peatlands with an area of $0.27 \times 10^{12} m^2$ is 5359 billion € (~5887 billion USD).

The human-induced $CO_2$ emission from Indonesian peatlands of 186.9 Tg C $yr^{-1}$ (Fig. 4.13) represents an annual loss of 13.7 billion € $yr^{-1}$, which amounts to a loss 0.05 € per $m^{-2}$ of disturbed peatlands per year. Just for comparison, an oil palm production of 3.3 t oil $ha^{-1}$ $yr^{-1}$ (Noleppa and Cartsburg, 2016, and references therein) and an oil palm price of about 600 € $t^{-1}$ (https://www.indexmundi.com/commodities/?commodity=palm-oil&currency=eur, excess date: May 2020) suggests an economic value of an oil palm plantation of 0.2 € $m^{-2}$.

We know that those values have to be considered with extreme caution, as the carbon and oil prices are highly volatile and the overall exercise is just a rough back of the envelope calculation. Nevertheless, it illustrates the main problem: The profit obtained from plantations could overcompensate the deterioration caused by $CO_2$ emissions. This fuels the conflict between conservation and restoration on the one hand and the transformation of peatlands into oil palm plantations on the other hand. Accordingly, the discussion about the $CO_2$ price reflects the difficulty of achieving a balance between the welfare of the current generation and sustainability. REDD+ can in turn be seen as an attempt to tackle this conflict, which has the potential for improvements.

Rimba Raya was one of the first Indonesian REDD+ projects. It is located approximately 200 km southwest of the river PSF1 in the Seruyan district in Central Kalimantan province on the island of Borneo (Indriatmoko et al., 2014). The project area is divided into so-called carbon accounting areas with the peat swamp forest and a management zone, which includes disturbed peatlands with villages and plantations. The aim of the project is to protect the peat swamp forest and to provide livelihood programs and employment for the communities with the help of the profit obtained by selling carbon credits. The project developer is "Infinite Earth," a Hong Kong based, and private company specialized in conservation. International companies such as Allianz and Microsoft are among the major clients.

In 2015, there were about 35 REDD+ activities distributed all over Indonesia (MoEF, 2015). However, the REDD+ policy options are controversially debated in the scientific literature (Engel and Palmer, 2008; Tata et al., 2014). Implementing such a scheme is difficult from various perspectives. According to general economic theory, it would be best to implement such a scheme on an individual basis: each landowner who does not clear cut, or even slash and burn their land, receives a payment for not choosing this individual ration path (Engel et al., 2009). This requires a certain property structure, i.e., clearly assigned property rights, and a proper monitoring and enforcement systems. Hence, transaction costs are often seen as prohibitive for such an individual scheme. Therefore, collective schemes, which make payments to a community, or which give a compensation for financing public goods, such schools or public amenities, are paid by such a scheme. Additionally, empirical evidence indicates that forest owners might not react to economic incentives as economic theory would predict (Lapeyre et al., 2015).

### 4.9.2 SPICE field experiments

To find out what type of PES, individual, group, or community payments would have the largest effect for forest conservation, we used framed field experiments within the framework of SPICE (Fine et al., 2008). In particular, we wanted to know if collective payments, which seem to be the only feasible option, would be able to reduce peatland degradation.

Using a within subject design playing various public goods experiments, in seven different villages in Riau (Sumatra/Indonesia) with 225 individuals, we investigated how different types of financial incentives could influence forest conservation. People, playing together in a group, get an endowment of 50.000 IDR (3.20€), which represents 10 units of standing timber. They now have to decide how much they want to harvest and how much they want to keep for conservation. Harvesting brings a secure income to the individual. The experimenter doubles investments into conservation. However, then those gains are equally split among all group members. Therefore, within this experiment, conservation is highly beneficial to the group, but it only pays to the investing individual, if other group members are doing the same. In the next step, we provide financial incentives to people, which should motivate them to invest more in conservation. We tested three types of payment schemes, based on real-life examples of PES, whereby incentives can be offered individually, collectively (as a group), or as a reinvestment in a public good, such as a school.

Following conventional economist wisdom, we find evidence that individual incentives are the most effective in yielding significant increases in forest conservation (see Table 4.12). The payment to the group even leads to a decrease in conservation efforts, which shows the negative expectation of people about the willingness to cooperate with the other members in the group, their fellow villagers. The treatment with the payment to the local school leads to higher contributions for conservation. However, this change is not significant. When asked about the preferences for one or the other payment scheme, it is clear that the school treatment is favored by most people (see Table 4.12).

In a controlled experimental setting, like in an art factual field experiment, there is no potential for any "cheating," but corruption is an aspect to be considered while establishing such a PES scheme. According to the corruption perception index of transparency international of 2019, which is a composite index of various international and national

**Table 4.12** (A) Contributions to the public good according to treatment, (B) stated preference for a treatment.

| (A) Treatment | Mean | Baseline mean (FPGE) | Difference |
|---|---|---|---|
| Individual | 26.00 (1.15) | 23.33 (1.14) | −2.667 (0.979) |
| School | 23.84 (1.16) | 23.33 (1.14) | −0.511 (1.046) |
| Group | 22.96 (1.14) | 23.33 (1.14) | 0.378 (1.074) |
| (B) Preferences in vote (1 = most preferred) | (1) Individual (percent) | (2) School | (3) Group |
| 1 | 87 (38.84) | 91 (40.62) | 46 (20.54) |
| 2 | 77 (34.38) | 80 (35.71) | 67 (29.91) |
| 3 | 60 (26.79) | 53 (23.66) | 111 (49.55) |
| Total | 224 | 224 | 224 |

Standard errors are in parentheses and indicate significance at 1% level, at 5% level, and at 10% level.

corruption perception indices, Indonesia gets on a scale between 0 (highly corrupt) to 100 (very clean) a score of 40, which ranks it on place 85 of 198 countries starting with the country being perceived as least corrupt (https://www.transparency.org/en/cpi#, data obtained Oct. 2020). This problem is recognized and the Indonesian government established a Corruption Eradication Commission (KPK) in 2002 (www.kpk.go.id). However, we also designed an experiment, which aims to learn more about corruption and its implication on the design on PES.

We ran the experiment with the same people in the same communities, where we ran the public good experiments. We did this with the intention of unveiling some of the complex ways in which corruption may affect individual incentives to conserve, particularly with respect to a forestry PES scheme. We used the framed experiment to gauge individual incentives of community members to engage in bribery over community forest property. Community members were asked to imagine a situation involving illegal loggers seizing community forest, whereby an individual would be elected to hire a legal service to mitigate the problem. We gave each individual the option to pick the firm, which ranged from not corrupt to very corrupt, by offering increasing levels of bribes for picking their respective service. By choosing the most corrupt firm, individuals are actually asking the community to pay a bribe so high that the value of the land would be lost on the community; thus, the legal service may have saved the standing forest but it is equal to the monetary value of the land. In framed field experiments, people make decisions in an artificially created situation (Fine et al., 2008). However, experimentalists use monetary incentives for the players, so that the choice each player makes has real monetary consequences for the person making a choice and the people they are interacting with (in our case the community). This shall increase the external validity of the experiment. Therefore, choosing a more corrupt firm meant a higher pay-off for the individual choosing the firm and lower pay-offs for the community as a whole. This artificial situation could mimic a real decision context in Indonesia since under the decentralization laws, communities have to jointly take a decision if they agree on the state giving a concession for their land to a harvesting company. However, we found that over 85% of our sample chose a corrupt firm, and nearly 18% of the individuals were willing to engage in a bribe that leaves fellow community members with nothing. Such a willingness to cheat their community decreases the reliability of estimates on how much of the $CO_2$ emissions could be mitigated due to the implementation of a PES scheme.

The open question remains if such a scheme should be based on individual or collective payments? Many REDD+ schemes rely on collective forms of payments. This is not only due to development goals, which are most of the time also tried to be achieved as a double dividend, but also due to high transaction costs, which particularly hinder the poor to participate in such programs and again favor powerful organizations (Gallemore et al., 2015). Even if experimental evidence indicated that an individual incentive scheme would work best, we would suggest that option due to the aforementioned arguments and due to autonomy and elbow rooms for villages, which are well maintaining their forest cover and therefore peatland. This would probably help to avoid some of the imbalances to the particularly highly perceived legitimacy and preference of the community. Thus, we believe that an individual incentive scheme, in favor of a scheme that helps to provide public goods as schools or community activities, is the preferred choice.

## 4.10 Outlook

During the past years, an opposing approach slowly becomes a reality (Barbier et al., 2020). Instead of getting paid for reducing $CO_2$ emissions, it is demanded that the emitter pays a carbon tax directly. A tax is a compulsory financial charge imposed upon a taxpayer by a governmental organization to fund public expenditures. Hence, a carbon tax can be seen as a compulsory financial charge imposed upon every legal body, which emits $CO_2$ to cover costs arising from consequences of rising $CO_2$ concentrations in the atmosphere. This does not solve the problem of imposing a price on $CO_2$ emission, but charging the polluter (e.g., plantation owner) provides advantages, as it allows outcompeting economic interests without the need to raise funds. Tax revenues can in turn be used to finance livelihood programs and provide employment for local communities. In practice, this could mean to continue and expand REDD+-like initiatives but on a tax-based funding scheme and with an improved security with respect to the enforcement of regulations and laws. An option that might be interesting is to combine reforestation and conservation with the production of, e.g., wind energy. The development and operation of wind parks in forests is possible and an increasing trend, because it generates income and is minimally invasive. Peat swamp forests could grow, stabilize the coast, and sequester $CO_2$ while the produced green energy lowers the demand for fossil fuel.

**Knowledge gaps and directions of future research**

The global relevance of peat drainage and fire as $CO_2$ sources to the atmosphere has long been recognized. In contrast, impacts of peatland degradation on rivers and the coastal ocean were hardly been studied before SPICE. Accordingly, $CO_2$ emissions along the land–ocean continuum, which are considered as off-site $CO_2$ emissions, in contrast on-site $CO_2$ emissions caused by peat oxidation and fires, are often ignored in carbon budgets. Comprehensive carbon budgets that encompass and even go beyond the aspects of climate and biodiversity are required to define climate goals and fully capture consequences of peatland degradation. Land that literally vanishes as $CO_2$ into the atmosphere strongly increases the vulnerability of coastal flood plains to erosion and sea level rise. Such consequences are poorly studied even though they imply that there is no other option than restoration and preservation if one wants to maintain the land. Furthermore, the fate of peat carbon transported into the ocean as well as its impact on life in the sea is still a mystery.

**Implications/recommendations for policy and society**

When peat oxidizes to $CO_2$, it vanishes into the air. This warms our climate, leaves habitats destroyed, and threatens the resource availability for future generations. Accordingly, peatland restoration is a crucial step to mitigate climate change impacts and toward sustainability even though its impact on the $CO_2$ emission is too slow to fulfill climate pledges in time. Therefore, it is recommended to strengthen efforts to reduce $CO_2$ emissions by switching from fossil fuel to alternative, clean energy sources, such as wind, hydrogen, and solar energy. Regarding climate stability, biodiversity, mitigation of climate change impacts caused, e.g., by the rising sea level, and therewith sustainability, conservation and restoration of peat swamp forests are considered as urgent measures which implementation should not be delayed. To achieve these goals, it is recommended to (1) turn peatlands into state forests and protected areas, (2) implement the moratorium on the clearing of primary forests and conversion of peatlands, (3) foster research to fill urgent knowledge gaps regarding, e.g., the on-site and off-site emission of methane in peatlands, and (4) introduce a $CO_2$ tax which is sufficiently high to outcompete agricultural interests and helps to finance restoration measures and livelihood programs as well as to improve management practices.

# References

Abram, N.J., Gagan, M.K., McCulloch, M.T., Chappell, J., Hantoro, W.S., 2003. Coral reef death during the 1997 Indian Ocean Dipole linked to Indonesian wildfires. Science 301, 952–955.

Abrams, J.F., Hohn, S., Rixen, T., Merico, A., 2018. Sundaland peat carbon dynamics and its contribution to the holocene atmospheric $CO_2$ concentration. Global Biogeochemical Cycles 32, 704–719.

Allen, G.H., Pavelsky, T.M., 2018. Global extent of rivers and streams. Science 361, 585.

Alongi, D.M., 2014. Carbon cycling and storage in mangrove forests. Annual Review of Marine Science 6, 195–219.

Anderson, J.A.R., Muller, J., 1975. Palynological study of a holocene peat and a miocene coal deposit from NW Borneo. Review of Palaeobotany and Palynology 19, 291–351.

Andriesse, J.P., 1988. Nature and Management of Tropical Peat Soils. Soil Resources, Management and Conservation Service FAO Land and Water Development Division. FAO - Food and Agriculture Organization of the United Nations, Rome, Italy.

Arrieta, J.M., Mayol, E., Hansman, R.L., Herndl, G.J., Dittmar, T., Duarte, C.M., 2015. Response to comment on "Dilution limits dissolved organic carbon utilization in the deep ocean". Science 350, 1483–1483.

Bar-On, Y.M., Phillips, R., Milo, R., 2018. The biomass distribution on Earth. Proceedings of the National Academy of Sciences 115, 6506.

Barbier, E.B., Lozano, R., Rodríguez, C.M., Troëng, S., 2020. Adopt a carbon tax to protect tropical forests. Nature 578, 213–216.

Baum, A., Rixen, T., 2014. Dissolved inorganic nitrogen and phosphate in the human affected blackwater river Siak, central Sumatra, Indonesia. Asian Journal of Water Environment and Pollution 11, 13–24.

Baum, A., Rixen, T., Samiaji, J., 2007. Relevance of peat draining rivers in central Sumatra for the riverine input of dissolved organic carbon into the ocean. Estuarine, Coastal and Shelf Science 73, 563–570.

BBC News, 1998. Haze – what Can Be Done?.

Billen, G., Lancelot, C., Meybeck, M., N, P., 1991. Si retention along the aquatic continuum from land to ocean. In: Mantoura, R.F.C., Martin, J.-M., Wollast, R. (Eds.), Ocean Margin Processes in Global Change. John Wileys & Sons.

Boer, R., Dewi, R.G., Ardiansyah, M., Siagian, U.W., 2018. Indonesia, Second Biennial Update Report under the United Nations Framework Convention on Climate Change. Directorate General of Climate Change, Ministry of Environment and Forestry, Jakarta, Indonesia.

Borges, A.V., Delille, B., Schiettecatte, L.-S., Gazeau, F., Abril, G., Frankignoulle, M., 2004. Gas transfer velocities of $CO_2$ in three European estuaries (Randers Fjord, Scheldt, and Thames). Limnology and Oceanography 49, 1630–1641.

Burdige, D.J., 2005. Burial of terrestrial organic matter in marine sediments: a re-assessment. Global Biogeochemical Cycles 19, GB4011.

Cai, W.-J., 2011. Estuarine and coastal ocean carbon paradox: $CO_2$ sinks or sites of terrestrial carbon incineration? Annual Review of Marine Science 3, 123–145.

Cao, X., Aiken, G.R., Butler, K.D., Huntington, T.G., Balch, W.M., Mao, J., et al., 2018. Evidence for major input of riverine organic matter into the ocean. Organic Geochemistry 116, 62–76.

Cartapanis, O., Bianchi, D., Jaccard, S.L., Galbraith, E.D., 2016. Global pulses of organic carbon burial in deep-sea sediments during glacial maxima. Nature Communications 7.

Church, J.A., Clark, P.U., Cazenave, A., Gregory, J.M., Jevrejeva, S., Levermann, A., et al., 2013. Sea level change. In: Stocker, T.F., Qin, D., Plattner, G.-K., Tignor, M., Allen, S.K., Boschung, J., et al. (Eds.), Climate Change 2013: The Physical Science Basis. Contribution of Working Group I to the Fifth Assessment Report of the Intergovernmental Panel on Climate Change. Cambridge University Press, Cambridge, United Kingdom and New York, NY, USA.

Ciais, P., Sabine, C., Bala, G., Bopp, L., Brovkin, V., Canadell, J., et al., 2013. Carbon and other biogeochemical cycles. In: Stocker, T.F., Qin, D., Plattner, G.-K., Tignor, M., Allen, S.K., Boschung, J., et al. (Eds.), Climate Change 2013: The Physical Science Basis. Contribution of Working Group I to the Fifth Assessment Report of the Intergovernmental Panel on Climate Change. Cambridge University Press, Cambridge, United Kingdom and New York, NY, USA.

Cole, J.J., Prairie, Y.T., Caraco, N.F., McDowell, W.H., Tranvik, L.J., Striegl, R.G., et al., 2007. Plumbing the global carbon cycle: integrating inland waters into the terrestrial carbon budget. Ecosystems 10, 171–184.

Comte, I., Colin, F., Whalen, J.K., Grünberger, O., Caliman, J.-P., 2012. Agricultural practices in oil palm plantations and their impact on hydrological changes, nutrient fluxes and water quality in Indonesia: a review. In: Sparks, D.L. (Ed.), Advances in Agronomy, vol. 116. Academic Press, Burlington, USA, pp. 71–124.

Cragg, S.M., Friess, D.A., Gillis, L.G., Trevathan-Tackett, S.M., Terrett, O.M., Watts, J.E.M., et al., 2020. Vascular plants are globally significant contributors to marine carbon fluxes and sinks. Annual Review of Marine Science 12, 469–497.

Dai, M., Yin, Z., Meng, F., Liu, Q., Cai, W.-J., 2012. Spatial distribution of riverine DOC inputs to the ocean: an updated global synthesis. Current Opinion in Environmental Sustainability 4, 170–178.

Damour, G., Simonneau, T., Cochard, H., Urban, L., 2010. An overview of models of stomatal conductance at the leaf level. Plant, Cell and Environment 33, 1419–1438.

Dasgupta, S., Siegel, D.I., Zhu, C., Chanton, J.P., Glaser, P.H., 2015. Geochemical mixing in peatland waters: the role of organic acids. Wetlands 35, 567–575.

Devol, A.H., Quay, P.D., Richey, J.E., Martinelli, L.A., 1987. The role of gas exchange in the inorganic carbon, oxygen, and 222Rn budgets of the Amazon River. Limnology and Oceanography 32, 235–248.

Dlugokencky, E.J., Mund, J.W., Crotwell, A.M., Crotwell, M.J., Thoning, K.W., 2020. Atmospheric Carbon Dioxide Dry Air Mole Fractions from the NOAA GML Carbon Cycle Cooperative Global Air Sampling Network, 1968–2019, Version: 2020-07.

Dommain, R., Couwenberg, J., Joosten, H., 2011. Development and carbon sequestration of tropical peat domes in south-east Asia: links to post-glacial sea-level changes and holocene climate variability. Quaternary Science Reviews 30, 999–1010.

Dommain, R., Couwenberg, J., Glaser, P.H., Joosten, H., Suryadiputra, I.N.N., 2014. Carbon storage and release in Indonesian peatlands since the last deglaciation. Quaternary Science Reviews 97, 1–32.

Duarte, C.M., 2017. Reviews and syntheses: hidden forests, the role of vegetated coastal habitats in the ocean carbon budget. Biogeosciences 14, 301–310.

Duarte, C.M., Middelburg, J.J., Caraco, N., 2005. Major role of marine vegetation on the oceanic carbon cycle. Biogeosciences 2, 1–8.

Duarte, C.M., Losada, I.J., Hendriks, I.E., Mazarrasa, I., Marba, N., 2013. The role of coastal plant communities for climate change mitigation and adaptation. Nature Climate Change 3, 961–968.

Emeis, K., Eggert, A., Flohr, A., Lahajnar, N., Nausch, G., Neumann, A., et al., 2018. Biogeochemical processes and turnover rates in the northern benguela upwelling system. Journal of Marine Systems 188, 63–80.

Engel, S., Palmer, C., 2008. Payments for environmental services as an alternative to logging under weak property rights: the case of Indonesia. Ecological Economics 65, 799–809.

Engel, A., Szlosek, J., Abramson, L., Liu, Z., Lee, C., 2009. Investigating the effect of ballasting by $CaCO_3$ in *Emiliania huxleyi*: I. Formation, settling velocities and physical properties of aggregates. Deep Sea Research Part II: Topical Studies in Oceanography 56, 1396–1407.

Eyre, B.D., Cyronak, T., Drupp, P., De Carlo, E.H., Sachs, J.P., Andersson, A.J., 2018. Coral reefs will transition to net dissolving before end of century. Science 359, 908.

Fei, T., Shuang-Qing, X., 2012. Definition of business as usual and its impacts on assessment of mitigation efforts. Advances in Climate Change Research 3, 212–219.

Figueres, C., Schellnhuber, H.J., Whiteman, G., Rockström, J., Hobley, A., Rahmstorf, S., 2017. Three years to safeguard our climate. Nature 546.

Fine, R.A., Smethie Jr., W.M., Bullister, J.L., Rhein, M., Min, D.-H., Warner, M.J., et al., 2008. Decadal ventilation and mixing of Indian Ocean waters. Deep Sea Research Part I: Oceanographic Research Papers 55, 20–37.

Fornasiero, A., Gambolati, G., Putti, M., Teatini, P., Ferraris, S., Pitacco, A., et al., 2020. Subsidence Due to Peat Soil Loss in the Zennare Basin (Italy): Design and Set-Up of the Field experiment.

Friedlingstein, P., Jones, M.W., O'Sullivan, M., Andrew, R.M., Hauck, J., Peters, G.P., et al., 2019. Global carbon budget 2019. Earth System Science Data 11, 1783–1838.

Gallemore, C., Di Gregorio, M., Moeliono, M., Brockhaus, M., Prasti, H.R.D., 2015. Transaction costs, power, and multi-level forest governance in Indonesia. Ecological Economics 114, 168–179.

Gandois, L., Cobb, A.R., Hei, I.C., Lim, L.B.L., Salim, K.A., Harvey, C.F., 2013. Impact of deforestation on solid and dissolved organic matter characteristics of tropical peat forests: implications for carbon release. Biogeochemistry 114, 183–199.

Gandois, L., Teisserenc, R., Cobb, A.R., Chieng, H.I., Lim, L.B.L., Kamariah, A.S., et al., 2014. Origin, composition, and transformation of dissolved organic matter in tropical peatlands. Geochimica et Cosmochimica Acta 137, 35–47.

Gastaldo, R.A., 2010. Peat or no peat: Why do the Rajang and Mahakam Deltas differ? International Journal of Coal Geology 83, 162–172.

Giri, C., Ochieng, E., Tieszen, L.L., Zhu, Z., Singh, A., Loveland, T., et al., 2011. Status and distribution of mangrove forests of the world using earth observation satellite data. Global Ecology and Biogeography 20, 154–159.

Green, S.M., Page, S., 2017. Tropical peatlands: current plight and the need for responsible management. Geology Today 33, 174–179.

Günther, A., Barthelmes, A., Huth, V., Joosten, H., Jurasinski, G., Koebsch, F., et al., 2020. Prompt rewetting of drained peatlands reduces climate warming despite methane emissions. Nature Communications 11, 1644.

Hamilton, S.E., Casey, D., 2016. Creation of a high spatio-temporal resolution global database of continuous mangrove forest cover for the 21st century (CGMFC-21). Global Ecology and Biogeography 25, 729–738.

Hansell, D.A., 2013. Recalcitrant dissolved organic carbon fractions. Annual Review of Marine Science 5, 421–445.

Hansell, D.A., Carlson, C.A., Repeta, D.J., Schlitzer, R., 2009. Dissolved organic matter in the ocean: a controversy stimulates new insights. Oceanography 22, 202–211.

Hapsari, K.A., Biagioni, S., Jennerjahn, T.C., Reimer, P., Saad, A., Sabiham, S., et al., 2018. Resilience of a peatland in Central Sumatra, Indonesia to past anthropogenic disturbance: improving conservation and restoration designs using palaeoecology. Journal of Ecology 106, 2473–2490.

Harrison, J.A., Caraco, N., Seitzinger, S., 2005. Global patterns and sources of dissolved organic matter export to the coastal zone: results from a spatially explicit, global model. Global Biogeochemical Cycles 19. https://doi.org/10.1029/2005GB002480.

Hedges, J.I., Keil, R.G., 1995. Sedimentary organic matter preservation: an assessment and speculative synthesis. Marine Chemistry 49, 81–115.

Hein, J., 2013. Reducing emissions from deforestation and forest degradation (REDD+), transnational conservation and aces to land in Jambi. In: Inodnesia. EFForTS Discussion Paper Series. University Goettingen, Goettingen, Germany, p. 28.

Hergoualc'h, K., Verchot, L.V., 2011. Stocks and fluxes of carbon associated with land use change in Southeast Asian tropical peatlands: a review. Global Biogeochemical Cycles 25, GB2001.

Heymann, J., Reuter, M., Buchwitz, M., Schneising, O., Bovensmann, H., Burrows, J.P., et al., 2017. $CO_2$ emission of Indonesian fires in 2015 estimated from satellite-derived atmospheric $CO_2$ concentrations. Geophysical Research Letters 44, 1537–1544.

Hirano, T., Segah, H., Harada, T., Limin, S., June, T., Hirata, R., et al., 2007. Carbon dioxide balance of a tropical peat swamp forest in Kalimantan, Indonesia. Global Change Biology 13, 412–425.

Hirano, T., Segah, H., Kusin, K., Limin, S., Takahashi, H., Osaki, M., 2012. Effects of disturbances on the carbon balance of tropical peat swamp forests. Global Change Biology 18, 3410–3422.

Hirano, T., Kusin, K., Limin, S., Osaki, M., 2015. Evapotranspiration of tropical peat swamp forests. Global Change Biology 21, 1914–1927.

Holden, J., 2005. Peat land hydrology and carbon release: Why small-scale process matters. Philosophical Transactions of the Royal Society 363, 2891–2913.

Hooijer, A., Silvius, M., Wösten, H., Page, S.E., 2006. Peat – $CO_2$, Assessment of $CO_2$ emissions from drained peat lands in SE Asia. In: Delft Hydraulic Report Q3943. WL Delft Hydraulics, Delft, p. 36.

Hooijer, A., Page, S., Canadell, J.G., Silvius, M., Kwadijk, J., Wösten, H., et al., 2010. Current and future $CO_2$ emissions from drained peatlands in Southeast Asia. Biogeosciences 7, 1505–1514.

Hooijer, A., Page, S., Jauhiainen, J., Lee, W.A., Lu, X.X., Idris, A., et al., 2012. Subsidence and carbon loss in drained tropical peatlands. Biogeosciences 9, 1053–1071.

Hoyt, A.M., Chaussard, E., Seppalainen, S.S., Harvey, C.F., 2020. Widespread subsidence and carbon emissions across Southeast Asian peatlands. Nature Geoscience 13, 435–440.

Huijnen, V., Wooster, M.J., Kaiser, J.W., Gaveau, D.L.A., Flemming, J., Parrington, M., et al., 2016. Fire carbon emissions over maritime southeast Asia in 2015 largest since 1997. Scientific Reports 6, 26886.

Humphreys, M.P., Gregor, L., Pierrot, D., van Heuven, S.M.A.C., Lewis, E.R., Wallace, D.W.R., 2020. PyCO2SYS: Marine Carbonate System Calculations in Python (Version 1.5.1). Zenodo.

Iglesias-Rodriguez, M.D., Armstrong, R.A., Feely, R.A., Hood, R.R., Kleypas, J., Milliman, J.D., et al., 2002. Progress made in study of ocean's calcium carbonate budget. EOS Transactions, American Geophysical Union 83, 365–375.

Indonesia, 2009. Intended Nationally Determined Contributions (INDC).

Indrarto, G.B., Murharjanti, P., Khatarina, J., Pulungan, I., Ivalerina, F., Rahman, J., et al., 2012. The context of REDD+ in Indonesia Drivers, agents and institutions. In: Working Paper, vol. 92. CIFOR, Bogor, Indonesia, p. 116.

Indriatmoko, Y., Atmadja, S., Ekaputri, A.D., Komalasari, M., 2014. Rimba Raya biodiversity reserve project, central Kalimantan, Indonesia. In: Sills, E.O., Atmadja, S.S., de Sassi, C., Duchelle, A.E., Kweka, D.L., Resosudarmo, I.A.P., et al. (Eds.), REDD+ on the Ground: A Case Book of Subnational Initiatives Across the globe. Bogor, Indonesia. Center for International Forestry Research, Bogor, Indonesia.

IPCC, 2014. In: Hiraishi, T., Krug, T., Tanabe, K., Srivastava, N., Baasansuren, J., Fukuda, M., et al. (Eds.), 2013 Supplement to the 2006 IPCC Guidelines for National Greenhouse Gas Inventories: Wetlands Methodological Guidance on Lands with Wet and Drained Soils, and Constructed Wetlands for Wastewater Treatment. IPCC, Switzerland.

Isson, T.T., Planavsky, N.J., Coogan, L.A., Stewart, E.M., Ague, J.J., Bolton, E.W., et al., 2020. Evolution of the global carbon cycle and climate regulation on earth. Global Biogeochemical Cycles 34 e2018GB006061.

Jiao, N., Herndl, G.J., Hansell, D.A., Benner, R., Kattner, G., Wilhelm, S.W., et al., 2010. Microbial production of recalcitrant dissolved organic matter: long-term carbon storage in the global ocean. Nature Reviews Microbiology 8, 593–599.

Jim, C.Y., 1999. The forest fires in Indonesia 1997–98: possible causes and pervasive consequences. Geography 84, 251–260.

Joosten, H., 2010. The Global Peatland $CO_2$ Picture: Peatland Status and Drainage Related Emissions in All Countries of the World. Wetlands International, Ede.

Joosten, H., Clarke, D., 2002. Wise Use of Mires and Peatlands. Saarijärvi, Finland: International Mire Conservation Group and International Peat Society, p. 304.

Kang, H., Kwon, M.J., Kim, S., Lee, S., Jones, T.G., Johncock, A.C., et al., 2018. Biologically driven DOC release from peatlands during recovery from acidification. Nature Communications 9, 3807.

Kiew, F., Hirata, R., Hirano, T., Wong, G.X., Aeries, E.B., Musin, K.K., et al., 2018. $CO_2$ balance of a secondary tropical peat swamp forest in Sarawak, Malaysia. Agricultural and Forest Meteorology 248, 494–501.

Klemme, A., Rixen, T., Müller-Dum, D., Müller, M., Notholt, J., Warneke, T., 2021. $CO_2$ emissions from peat-draining rivers regulated by water pH. Biogeosciences Discussions 2021, 1–20.

Kobayashi, S., 2016. Tropical peat swamp forest ecosystem and REDD+. In: Mizuno, K., Fujita, M.S., Kawai, S. (Eds.), Catastrophe & Regeneration in Indonesian's Peatlands: Ecology, Ecinomy & Society. Nus Press & Hyoto University Press, Singapore, pp. 211–237.

Köchy, M., Hiederer, R., Freibauer, A., 2015. Global distribution of soil organic carbon – Part 1: masses and frequency distributions of SOC stocks for the tropics, permafrost regions, wetlands, and the world. Soil 1, 351–365.

Kumagai, T.O., Saitoh, T.M., Sato, Y., Takahashi, H., Manfroi, O.J., Morooka, T., et al., 2005. Annual water balance and seasonality of evapotranspiration in a Bornean tropical rainforest. Agricultural and Forest Meteorology 128, 81–92.

Kuzyakov, Y., Friedel, J.K., Stahr, K., 2000. Review of mechanisms and quantification of priming effects. Soil Biology and Biochemistry 32, 1485–1498.

Langdon, C., Atkinson, M.J., 2005. Effect of elevated $pCO_2$ on photosynthesis and calcification of corals and interactions with seasonal change in temperature/irradiance and nutrient enrichment. Journal of Geophysical Research 110, C09S07.

Lapeyre, R., Pirard, R., Leimona, B., 2015. Payments for environmental services in Indonesia: what if economic signals were lost in translation? Land Use Policy 46, 283–291.

Lauerwald, R., Laruelle, G.G., Hartmann, J., Ciais, P., Regnier, P.A.G., 2015. Spatial patterns in $CO_2$ evasion from the global river network. Global Biogeochemical Cycles 29, 534–554.

Leifeld, J., Wüst-Galley, C., Page, S., 2019. Intact and managed peatland soils as a source and sink of GHGs from 1850 to 2100. Nature Climate Change 9, 945–947.

Lenssen, N.J.L., Schmidt, G.A., Hansen, J.E., Menne, M.J., Persin, A., Ruedy, R., et al., 2019. Improvements in the GISTEMP uncertainty model. Journal of Geophysical Research: Atmosphere 124, 6307–6326.

Lewis, E., Wallace, D., 1998. Program Develope Fro $CO_2$ System Calculations. Carbon Dioxide Information Analysis Center Oak Ridge National Laboratory, Oak Ridge, Tennessee, USA.

Lohberger, S., Stängel, M., Atwood, E.C., Siegert, F., 2018. Spatial evaluation of Indonesia's 2015 fire-affected area and estimated carbon emissions using Sentinel-1. Global Change Biology 24, 644–654.

Ludwig, W., Probst, J.-L., Kempe, S., 1996. Predicting the oceanic input of organic carbon by continental erosion. Global Biogeochemical Cycles 10, 23–41.

Madden, R.A., Julian, P.R., 1994. Observations of the 40–50-day tropical Oscillation—a review. Monthly Weather Review 122, 814–837.

Manoli, G., Meijide, A., Huth, N., Knohl, A., Kosugi, Y., Burlando, P., et al., 2018. Ecohydrological changes after tropical forest conversion to oil palm. Environmental Research Letters 13, 064035.

Mayer, B., Rixen, T., Pohlmann, T., 2018. The spatial and temporal variability of air-sea $CO_2$ fluxes and the effect of net coral reef calcification in the Indonesian seas: a numerical sensitivity study. Frontiers in Marine Science 5, 116.

McCulloch, M.T., D'Olivo, J.P., Falter, J., Holcomb, M., Trotter, J.A., 2017. Coral calcification in a changing World and the interactive dynamics of pH and DIC upregulation. Nature Communications 8, 15686.

McLeod, E., Chmura, G.L., Bouillon, S., Salm, R., Björk, M., Duarte, C.M., et al., 2011. A blueprint for blue carbon: toward an improved understanding of the role of vegetated coastal habitats in sequestering $CO_2$. Frontiers in Ecology and the Environment 9, 552–560.

Michel, J., Wiemers, K., Samhudi, H., Westphal, H., 2015. Molluscan assemblages under the influence of peat-draining rivers off East Sumatra, Indonesia. Molluscan Research 35, 81–94.

Middelburg, J.J., 2015. Escape by dilution. Science 348, 290–290.

Miettinen, J., Liew, S.C., 2010. Degradation and development of peatlands in Peninsular Malaysia and in the islands of Sumatra and Borneo since 1990. Land Degradation & Development 21, 285–296.

Miettinen, J., Shi, C., Liew, S.C., 2016. Land cover distribution in the peatlands of Peninsular Malaysia, Sumatra and Borneo in 2015 with changes since 1990. Global Ecology and Conservation 6, 67–78.

Miettinen, J., Hooijer, A., Vernimmen, R., Liew, S.C., Page, S.E., 2017. From carbon sink to carbon source: Extensive peat oxidation in insular southeast Asia since 1990. Environmental Research Letters 12, 024014.

MoEF, 2015. National Forest Reference Emission Level for REDD+ in the Context of Decision 1/CP.16 Paragraph 70, Directorate General of Climate Change. The Ministry of Environment and Forestry, Jakarta, Indonesia.

Molengraaff, G.A.F., 1921. Modern deep-sea research in the east Indian Archipelago. The Geographical Journal 57, 95–118.

Moore, S., Evans, C.D., Page, S.E., Garnett, M.H., Jones, T.G., Freeman, C., et al., 2013. Deep instability of deforested tropical peatlands revealed by fluvial organic carbon fluxes. Nature 493, 660–663.

Moosdorf, N., Oehler, T., 2017. Societal use of fresh submarine groundwater discharge: an overlooked water resource. Earth-Science Reviews 171, 338–348.

Morse, J.W., Arvidson, R.S., Lüttge, A., 2007. Calcium carbonate formation and dissolution. Chemical Reviews 107, 342–381.

Mucci, A., 1986. Growth kinetics and composition of magnesian calcite overgrowths precipitated from seawater: Quantitative influence of orthophosphate ions. Geochimica et Cosmochimica Acta 50, 2255–2265.

Müller, D., Warneke, T., Rixen, T., Müller, M., Jamahari, S., Denis, N., et al., 2015. Lateral carbon fluxes and $CO_2$ outgassing from a tropical peat-draining river. Biogeosciences 12, 5967–5979.

Müller, D., Warneke, T., Rixen, T., Müller, M., Mujahid, A., Bange, H.W., et al., 2016. Fate of terrestrial organic carbon and associated $CO_2$ and CO emissions from two Southeast Asian estuaries. Biogeosciences 13, 691–705.

Müller-Dum, D., Warneke, T., Rixen, T., Müller, M., Baum, A., Christodoulou, A., et al., 2019. Impact of peatlands on carbon dioxide ($CO_2$) emissions from the Rajang River and Estuary, Malaysia. Biogeosciences 16, 17–32.

Murdiyarso, D., Hergoualc'h, K., Verchot, L.V., 2010. Opportunities for reducing greenhouse gas emissions in tropical peatlands. Proceedings of the National Academy of Sciences 107, 19655.

Myhre, G., Shindell, D., Bréon, F.-M., Collins, W., Fuglestvedt, J., Huang, J., et al., 2013. Anthropogenic and natural radiative forcing. In: Stocker, T.F., Qin, D., Plattner, G.-K., Tignor, M., Allen, S.K., Boschung, J., et al. (Eds.), Climate Change 2013: The Physical Science Basis. Contribution of Working Group I to the Fifth Assessment Report of the Intergovernmental Panel on Climate Change. Cambridge University Press, Cambridge, United Kingdom and New York, NY, USA.

Nelsen, M.P., DiMichele, W.A., Peters, S.E., Boyce, C.K., 2016. Delayed fungal evolution did not cause the Paleozoic peak in coal production. Proceedings of the National Academy of Sciences 113, 2442.

Nightingale, P.D., Malin, G., Law, C.S., Watson, A.J., Liss, P.S., Liddicoat, M.I., et al., 2000. In situ evaluation of air-sea gas exchange parameterizations using novel conservative and volatile tracers. Global Biogeochemical Cycles 14, 373–387.

NOAA, Dlugokencky, E., 2016. $CO_2$ CCGG (Individual Flasks), Gobabeb Training and Research center. http://aftp.cmdl.noaa.gov/data/trace_gases/co2/flask/surface/co2_nmb_surface-flask_1_ccgg_month.txt.

Noleppa, S., Cartsburg, M., 2016. Auf der Ölspur – Berechnungen zu einer palmölfreieren Welt. WWF Germany, Berlin, Germany.

Osborn, T.J., Jones, P.D., 2014. The CRUTEM4 land-surface air temperature data set: construction, previous versions and dissemination via Google Earth. Earth System Science Data 6, 61–68.

Page, S.E., Siegert, F., Rieley, J.O., Boehm, H.-D.V., Jaya, A., Limin, S., 2002. The amount of carbon released from peat and forest fires in Indonesia during 1997. Nature 420, 61–65.

Page, S.E., Wüst, R.A.J., Weiss, D., Rieley, J.O., Shotyk, W., Limin, S.H., 2004. A record of Late Pleistocene and Holocene carbon accumulation and climate change from an equatorial peat bog (Kalimantan, Indonesia): implications for past, present and future carbon dynamics. Journal of Quaternary Science 19, 625–635.

Page, S.E., Rieley, J.O., Banks, C.J., 2011. Global and regional importance of the tropical peatland carbon pool. Global Change Biology 17, 798–818.

Parkhurst, D.L., Appelo, C.A.J., 2013. Description of Input and Examples for PHREEQC Version 3: A Computer Program for Speciation, Batch-Reaction, One-Dimensional Transport, and Inverse Geochemical Calculations (ebook).

Pind, A., Freeman, C., Lock, M.A., 1994. Enzymic degradation of phenolic materials in peatlands - measurement of phenol oxidase activity. Plant and Soil 159, 227–231.

Posa, M.R.C., Wijedasa, L.S., Corlett, R.T., 2011. Biodiversity and conservation of tropical peat swamp forests. BioScience 61, 49–57.

Raymond, P., Cole, J., 2001. Gas exchange in rivers and estuaries: choosing a gas transfer velocity. Estuaries and Coasts 24, 312–317.

Raymond, P.A., Hartmann, J., Lauerwald, R., Sobek, S., McDonald, C., Hoover, M., et al., 2013. Global carbon dioxide emissions from inland waters. Nature 503, 355–359.

Regnier, P., Friedlingstein, P., Ciais, P., Mackenzie, F.T., Gruber, N., Janssens, I.A., et al., 2013. Anthropogenic perturbation of the carbon fluxes from land to ocean. Nature Geoscience 6, 597–607.

Ricciardulli, L., Wentz, F.J., 2015. A scatterometer geophysical model function for climate-quality winds: QuikSCAT Ku-2011. Journal of Atmospheric and Oceanic Technology 32, 1829–1846.

Richey, J.E., Hedges, J.I., Devol, A.H., Quay, P.D., Victoria, R.L., Martinelli, L.A., et al., 1990. Biogeochemistry of carbon in the Amazon river. Limnology and Oceanography 35, 352–371.

Ridgwell, A.J., Watson, A.J., Maslin, M.A., Kaplan, J.O., 2003. Implications of coral reef buildup for the controls on atmospheric $CO_2$ since the Last Glacial Maximum. Paleoceanography 18.

Rixen, T., Baum, A., Pohlmann, T., Balzer, W., Samiaji, J., Jose, C., 2008. The Siak, a tropical black water river in central Sumatra on the verge of anoxia. Biogeochemistry 90, 129–140.

Rixen, T., Baum, A., Wit, F., Samiaji, J., 2016. Carbon leaching from tropical peat soils and consequences for carbon balances. Frontiers in Earth Science 4.

Rixen, T., Gaye, B., Emeis, K.C., Ramaswamy, V., 2019. The ballast effect of lithogenic matter and its influences on the carbon fluxes in the Indian Ocean. Biogeosciences 16, 485–503.

Rödenbeck, C., Houweling, S., Gloor, M., Heimann, M., 2003. $CO_2$ flux history 1982–2001 inferred from atmospheric data using a global inversion of atmospheric transport. Atmospheric Chemistry and Physics 3, 1919–1964.

Rogelj, J., Forster, P.M., Kriegler, E., Smith, C.J., Séférian, R., 2019. Estimating and tracking the remaining carbon budget for stringent climate targets. Nature 571, 335–342.

Saderne, V., Cusack, M., Almahasheer, H., Serrano, O., Masqué, P., Arias-Ortiz, A., et al., 2018. Accumulation of carbonates contributes to coastal vegetated ecosystems keeping Pace with sea level rise in an arid region (Arabian Peninsula). Journal of Geophysical Research: Biogeosciences 123, 1498–1510.

Saji, N.H., Goswami, B.N., Vinayachandran, P.N., Yamagata, T., 1999. A dipole mode in the tropical Inian Ocean. Nature 401, 360–363.

Sargeant, H.J., 2001. Report, forest Fire Prevention and Control Project; Oil palm Agriculture in the Wetlands of Sumatra: Destruction or Development?. Government of Indonesia Ministry of Forestry & European Union.

Saunois, M., Stavert, A.R., Poulter, B., Bousquet, P., Canadell, J.G., Jackson, R.B., et al., 2020. The global methane budget 2000–2017. Earth System Science Data 12, 1561–1623.

Schneider, U., Becker, A., Finger, P., Meyer-Christoffer, A., Rudolf, B., Ziese, M., 2011. GPCC Monitoring Product: Near Real-Time Monthly Land-Surface Precipitation from Rain-Gauges Based on SYNOP and CLIMAT Data. https://doi.org/10.5676/DWD_GPCC/MP_M_V5_100.

Siegel, H., Stottmeister, I., Reißmann, J., Gerth, M., Jose, C., Samiaji, J., 2009. Siak river system – east-Sumatra: characterisation of sources, estuarine processes, and discharge into the Malacca Strait. Journal of Marine Systems 77, 148–159.

Siegel, H., Gerth, M., Stottmeister, I., Baum, A., Samiaji, J., 2019. Remote sensing of coastal discharge of SE Sumatra (Indonesia). In: Barale, V., Gade, M. (Eds.), Remote Sensing of the Asian Seas. Springer International Publishing, Cham, pp. 359–376.

Smith, D.G., Bryson, C., Thompson, E.M., Young, E.G., 1958. Chemical composition of the peat bogs of the maritime provinces. Canadian Journal of Soil Science 38, 122–129.

Smith, T.M., Reynolds, R.W., Peterson, T.C., Lawrimore, J., 2008. Improvements to NOAA's historical merged land-ocean surface temperature analysis (1880–2006). Journal of Climate 21, 2283–2296.

Sorensen, K.W., 1993. Indonesian peat swamp forests and their role as a carbon sink. Chemosphere 27, 1065–1082.

Spalding, M.D., Ravilious, C., Green, E.P., 2001. World Atlas of Coral Reefs. University of California Press, Berkeley.

Suzuki, S., Ishida, T., Nagano, T., Waijaroen, S., 1999. Influences of deforestation on carbon balance in a natural tropical peat swamp forest in Thailand. Environment Control in Biology 37, 115–128.

Syvitski, J.P.M., Vörösmarty, C.J., Kettner, A.J., Green, P., 2005. Impact of humans on the flux of terrestrial sediment to the global ocean. Science 308, 376–380.

Syvitski, J.P.M., Kettner, A.J., Overeem, I., Hutton, E.W.H., Hannon, M.T., Brakenridge, G.R., et al., 2009. Sinking deltas due to human activities. Nature Geoscience 2, 681–686.

Takahashi, T., Sutherland, S.C., Chipman, D.W., Goddard, J.G., Ho, C., Newberger, T., et al., 2014. Climatological distributions of pH, $pCO_2$, total $CO_2$, alkalinity, and $CaCO_3$ saturation in the global surface ocean, and temporal changes at selected locations. Marine Chemistry 164, 95–125.

Tang Che Ing, A., 2017. Land-atmosphere exchange of carbon and energy at a tropical peat swmap forest in Sarawak, Malaysia. In: Doctor of Philosophy. Montana State University, Bozeman, Montana, USA, p. 110 vol. PhD.

Tata, H.L., van Noordwijk, M., Ruysschaert, D., Mulia, R., Rahayu, S., Mulyoutami, E., et al., 2014. Will funding to Reduce Emissions from Deforestation and (forest) Degradation (REDD+) stop conversion of peat swamps to oil palm in orangutan habitat in Tripa in Aceh, Indonesia? Mitigation and Adaptation Strategies for Global Change 19, 693–713.

Turetsky, M.R., Benscoter, B., Page, S., Rein, G., van der Werf, G.R., Watts, A., 2014. Global vulnerability of peatlands to fire and carbon loss. Nature Geoscience 8, 11–14.

UNEP, 2019. Emissions Gap Report 2019. United Nations Environment Programme (UNEP), Nairobi, Kenya.

UNFCCC, 2011. Compilation of Information on Nationally Appropriate Mitigation Actions to Be Implemented by Parties Not Included in Annex I to the Convention.

USDA, 1999. Soil Taxonomy A Basic System of Soil Classification for Making and Interpreting Soil Surveys. United States Department of Agriculture, Washington, DC, USA.

van der Werf, G.R., Dempewolf, J., Trigg, S.N., Randerson, J.T., Kasibhatla, P.S., Giglio, L., et al., 2008. Climate regulation of fire emissions and deforestation in equatorial Asia. Proceedings of the National Academy of Sciences 105, 20350–20355.

van der Werf, G.R., Randerson, J.T., Giglio, L., Collatz, G.J., Mu, M., Kasibhatla, P.S., et al., 2010. Global fire emissions and the contribution of deforestation, savanna, forest, agricultural, and peat fires (1997–2009). Atmospheric Chemistry and Physics 10, 11707–11735.

van der Werf, G.R., Randerson, J.T., Giglio, L., van Leeuwen, T.T., Chen, Y., Rogers, B.M., et al., 2017. Global fire emissions estimates during 1997–2016. Earth System Science Data 9, 697–720.

Vermeer, M., Rahmstorf, S., 2009. Global sea level linked to global temperature. Proceedings of the National Academy of Sciences 106, 21527.

Wang, P., 2009. Global monsoon in a geological perspective. Chinese Science Bulletin 54, 1113–1136.

Wang, Y., Roulet, N.T., Frolking, S., Mysak, L.A., 2009. The importance of Northern Peatlands in global carbon systems during the Holocene. Climate of the Past 5, 683–693.

Wanninkhof, R., 1992. Relationship between gas exchange and wind speed over the ocean. Journal of Geophysical Research 97, 7373–7381.

Wanninkhof, R., 2014. Relationship between wind speed and gas exchange over the ocean revisited. Limnology and Oceanography: Methods 12, 351–362.

Ward, N.D., Bianchi, T.S., Medeiros, P.M., Seidel, M., Richey, J.E., Keil, R.G., et al., 2017. Where carbon goes when water flows: carbon cycling across the aquatic continuum. Frontiers in Marine Science 4, 7.

Warren, M.W., Kauffman, J.B., Murdiyarso, D., Anshari, G., Hergoualc'h, K., Kurnianto, S., et al., 2012. A cost-efficient method to assess carbon stocks in tropical peat soil. Biogeosciences 9, 4477–4485.

WCED, 1987. World Commission on Environment and Development – Our Common Future. Oxford, UK.

Webster, P.J., Moore, A.M., Loschnigg, J.P., Leben, R.R., 1999. Coupled ocean-atmosphere dynamics in the Indian Ocean during 1997–98. Nature 401, 356–360.

Weyhenmeyer, G.A., Fröberg, M., Karltun, E., Khalili, M., Kothawala, D., Temnerud, J., et al., 2012. Selective decay of terrestrial organic carbon during transport from land to sea. Global Change Biology 18, 349–355.

Whittle, A., Gallego-Sala, A.V., 2016. Vulnerability of the peatland carbon sink to sea-level rise. Scientific Reports 6, 28758.

Wiggins, E.B., Czimczik, C.I., Santos, G.M., Chen, Y., Xu, X., Holden, S.R., et al., 2018. Smoke radiocarbon measurements from Indonesian fires provide evidence for burning of millennia-aged peat. Proceedings of the National Academy of Sciences 115, 12419.

Wit, F., Muller, D., Baum, A., Warneke, T., Pranowo, W.S., Muller, M., et al., 2015. The impact of disturbed peatlands on river outgassing in Southeast Asia. Nature Communications 6.

Wit, F., Rixen, T., Baum, A., Pranowo, W.S., Hutahaean, A.A., 2018. The Invisible Carbon Footprint as a hidden impact of peatland degradation inducing marine carbonate dissolution in Sumatra, Indonesia. Scientific Reports 8, 17403.

Woodroffe, C.D., Thom, B.G., Chappell, J., 1985. Development of widespread mangrove swamps in mid-Holocene times in northern Australia. Nature 317, 711–713.

Wosten, J.H.M., Ismail, A.B., van Wijk, A.L.M., 1997. Peat subsidence and its practical implications: a case study in Malaysia. Geoderma 78, 25–36.

Xu, J., Morris, P.J., Liu, J., Holden, J., 2018. PEATMAP: refining estimates of global peatland distribution based on a meta-analysis. Catena 160, 134–140.

Yu, Z.C., 2012. Northern peatland carbon stocks and dynamics: a review. Biogeosciences 9, 4071–4085.

Yu, Z., Loisel, J., Brosseau, D.P., Beilman, D.W., Hunt, S.J., 2010. Global peatland dynamics since the last glacial maximum. Geophysical Research Letters 37.

Yule, C., Gomez, L., 2009. Leaf litter decomposition in a tropical peat swamp forest in Peninsular Malaysia. Wetlands Ecology and Management 17, 231–241.

Yule, C.M., Lim, Y.Y., Lim, T.Y., 2018. Recycling of phenolic compounds in Borneo's tropical peat swamp forests. Carbon Balance and Management 13, 3.

Zigah, P.K., McNichol, A.P., Xu, L., Johnson, C., Santinelli, C., Karl, D.M., et al., 2017. Allochthonous sources and dynamic cycling of ocean dissolved organic carbon revealed by carbon isotopes. Geophysical Research Letters 44, 2407–2415.

# 5

# Coral reef social–ecological systems under pressure in Southern Sulawesi

Hauke Reuter[1,2], Annette Breckwoldt[1], Tina Dohna[3], Sebastian Ferse[1], Astrid Gärdes[1], Marion Glaser[1,17], Filip Huyghe[4], Hauke Kegler[1], Leyla Knittweis[5], Marc Kochzius[4], Wiebke Elsbeth Kraemer[4], Johannes Leins[1,6], Muhammad Lukman[7], Hawis Madduppa[8], Agus Nuryanto[9], Min Hui[10], Sara Miñarro[1,11], Gabriela Navarrete Forero[1,12], Sainab Husain Paragay[13], Jeremiah Plass-Johnson[1], Hajaniaina Andrianavalonarivo Ratsimbazafy[14], Claudio Richter[15], Yvonne Sawall[16], Kathleen Schwerdtner Máñez[1], Mirta Teichberg[1], Janne Timm[2], Rosa van der Ven[4], Jamaluddin Jompa[7]

[1]LEIBNIZ CENTRE FOR TROPICAL MARINE RESEARCH (ZMT), BREMEN, GERMANY; [2]FACULTY FOR BIOLOGY AND CHEMISTRY, UNIVERSITY OF BREMEN, BREMEN, GERMANY; [3]MARUM - CENTER FOR MARINE ENVIRONMENTAL SCIENCES, UNIVERSITY OF BREMEN, GERMANY; [4]ECOLOGY & BIODIVERSITY - MARINE BIOLOGY, VRIJE UNIVERSITEIT BRUSSEL (VUB), BRUSSELS, BELGIUM; [5]DEPARTMENT OF BIOLOGY, UNIVERSITY OF MALTA, MSIDA, MALTA; [6]HELMHOLTZ CENTRE FOR ENVIRONMENTAL RESEARCH (UFZ), LEIPZIG, GERMANY; [7]DEPARTMENT OF MARINE SCIENCE, HASANUDDIN UNIVERSITY (UNHAS), MAKASSAR, INDONESIA; [8]AGRICULTURAL UNIVERSITY BOGOR (IPB), BOGOR, INDONESIA; [9]JENDERAL SOEDIRMAN UNIVERSITY (UNSOED), PURWOKERTO, INDONESIA; [10]INSTITUTE OF OCEANOLOGY, CHINESE ACADEMY OF SCIENCES, QINGDAO, CHINA; [11]INSTITUTE OF ENVIRONMENTAL SCIENCE AND TECHNOLOGY (ICTA), UNIVERSITAT AUTÒNOMA DE BARCELONA, BELLATERRA, SPAIN; [12]AVENIDA RÍO TOACHI Y CALLE BAMBÚES, SANTO DOMINGO DE LOS COLORADOS, ECUADOR; [13]ENLIGHTENING INDONESIA, MAKASSAR, SOUTH SULAWESI, INDONESIA; [14]SYSTEMS ECOLOGY & RESOURCE MANAGEMENT UNIT, UNIVERSITÉ LIBRE DE BRUXELLES (ULB), BRUSSELS, BELGIUM; [15]DIVISION BIOSCIENCES/BENTHO-PELAGIC PROCESSES, ALFRED WEGENER INSTITUTE, BREMERHAVEN, GERMANY; [16]BERMUDA INSTITUTE OF OCEAN SCIENCES (BIOS), ST. GEORGE'S, BERMUDA; [17]INSTITUTE OF GEOGRAPHY, UNIVERSITY OF BREMEN, BREMEN, GERMANY

**Science for the Protection of Indonesian Coastal Ecosystems (SPICE). https://doi.org/10.1016/B978-0-12-815050-4.00005-5**

## Abstract

*Ecological and social processes of the Spermonde Archipelago, South Sulawesi, Indonesia, have been intensively studied during the Science for the Protection of Indonesian Coastal Ecosystems (SPICE) program. The archipelago is of specific interest to better understand how intensive exploitation of marine resources results in the degradation of reef systems. The projects specifically targeted (1) ecological processes in coral reefs, (2) genetic structure of populations, and (3) social–ecological dynamics relating to resource use, social networks, and governance structures. A modeling component emphasized (4) the integration of different ecological, social, and environmental components. Results indicated that reef resources in the Spermonde Archipelago are intensively exploited and further stressed by pollution effluents from hinterland processes. The lack of alternative livelihoods perpetuates dependencies within the patron–client system of the artisanal fisheries and supports high exploitation and also destructive resource uses. Greater inclusion of local stakeholders in the governance may result in better conservation practices, sustainable resource use, and improved livelihoods for the people.*

## Abstrak

*Selama proyek SPICE, aspek ekologis dan sosial di perairan Kepulauan Spermonde, Sulawesi Selatan, Indonesia, dipelajari secara intensif. Kepulauan dan perairan ini mengundang banyak perhatian untuk lebih memahami bagaimana tingkat pemanfaatan sumber daya laut yang intensif berakibat pada degradasi sistem terumbu karang. Kerangka kerja proyek penelitian SPICE, secara khusus, menargetkan kajian-kajian mengenai (a) proses ekologis dalam sistem terumbu karang, (b) struktur genetik populasi untuk menentukan konektifitasnya, dan (c) dinamika sosial-ekologi yang berkaitan dengan penggunaan sumber daya, jejaring sosial dan struktur tata kelola. Komponen pemodelan juga dimasukkan kedalam kajian mengenai (d) integrasi berbagai komponen ekologi, sosial dan lingkungan dalam karakterisasi sistem sosial-ekologis. Hasil-hasil penelitian menunjukkan bahwa sumber daya terumbu karang di Kepulauan Spermonde sangat tereksploitasi dan terancam juga oleh limbah polusi dari proses di daratan. Kurangnya mata pencaharian alternatif memperkuat ketergantungan para nelayan tradisional pada sistem patron-klien dan meningkatkan penggunaan sumber daya yang tidak berkelanjutan dan merusak. Keikutsertaan pemangku kepentingan lokal yang lebih erat dalam tata kelola sumber daya alam Kepulauan Spermonde sangat memungkinkan hasil praktik konservasi yang lebih baik, penggunaan sumber daya berkelanjutan, dan peningkatan mata pencaharian bagi masyarakat.*

## Chapter outline

**5.1 Introduction—coral reefs in Indonesia and the Spermonde Archipelago ... 145**

**5.2 Functioning of coral reefs ... 148**

5.2.1 Water quality and biogeochemical processes ... 149

5.2.2 Benthic coral reef community dynamics of Spermonde Archipelago ... 150

5.2.3 Bacterial communities and biofilms ... 151

5.2.4 Coral reef recruitment processes ... 152

5.2.5 Coral physiology ... 152

5.2.6 Relationships between benthic and fish communities ... 152

5.2.7 Consequences of disturbances for coral reef functioning ... 153

**5.3 Genetic connectivity of reefs in the Coral Triangle region** ...... **154**
5.3.1 Large-scale connectivity across the Coral Triangle region ...... 157
5.3.2 Small-scale connectivity in the Spermonde Archipelago ...... 162
5.3.3 Self-recruitment at the islands of Barrang Lompo and Samalona ...... 163
5.3.4 Application of connectivity data in marine-protected area network design ...... 164
**5.4 Social systems associated with the use of coral-based resources and reef-specific challenges** ...... **165**
5.4.1 Participatory assessment of Spermonde's coral reef fisheries ...... 165
5.4.2 Investigating marine social–ecological feedbacks and dynamics ...... 167
5.4.3 Reef-related livelihoods and implications for the present and future health of fishers and reefs ...... 168
5.4.4 Changing target species, perceptions of reef resources, and implications for food security ...... 169
5.4.5 Conclusions for the management of coral reef resources in the Spermonde Archipelago ...... 171
**5.5 Modeling to support the management of reef systems** ...... **171**
5.5.1 Simulating the impact of fisheries on coral reef dynamics ...... 172
5.5.2 A model on gear choices of fishermen ...... 175
5.5.3 Spatial patterns of fishing ground distribution ...... 176
**5.6 Summary and outlook** ...... **178**
**Acknowledgments** ...... **180**
**References** ...... **180**
**Appendix A5** ...... **189**

## 5.1 Introduction—coral reefs in Indonesia and the Spermonde Archipelago

With its roughly 17,500 islands, Indonesia is located in the Coral Triangle along with Malaysia, Papua New Guinea, the Philippines, the Solomon Islands, and East Timor. Indonesia is renowned for a very high biodiversity in coastal ecosystems, especially in coral reefs. Of 590 known coral species, it is believed that more than 75% have been identified in Indonesian waters (Burke et al., 2012; Veron et al., 2009). It is estimated that more than 50% of Indo-Pacific and more than 37% of world reef fish species are represented in Indonesian waters (Asian Development Bank, 2014a; UNEP-WCMC, 2014). This extremely high biodiversity is also found in other coastal habitats such as mangroves (41 of 54 true mangrove species (Asian Development Bank, 2014a; Hogarth, 1999)) and seagrass meadows (13 species of tropical seagrasses (Asian Development Bank, 2014a)).

Coral reefs cover around 39,500 km$^2$ of Indonesian coastal waters amounting to approximately 16% of global coral reefs (Burke et al., 2012) and providing many crucial ecosystem functions and services. The length of Indonesia's coastline is around 81,000 km, with 270,000 $km^2$ of shelf areas (depth up to 200 m). More than 900 of the Indonesian islands are permanently inhabited with much of the population living within 10 km of the coast. Thus, these communities are highly dependent on goods and services from coral reefs for livelihoods and other resources for local (coastal) economies. In 2007, fisheries contributed 2.4% to the national GDP and 1.9% to the total export value of all products, with more than 2.1 million people being employed in this sector (Asian Development Bank, 2014b). From 2000 until 2010, Indonesia provided 70% of the corals for the global ornamental and aquaria trade (Wood et al., 2012), while the whole of Southeast Asia accounted for 85% (Asian Development Bank, 2014b). Burke et al. (2002) specify the annual economic net benefit of a healthy Southeast Asian coral reef in a range between \$23,100 to \$270,000 per $km^2$ when tourism and esthetic values are added to fisheries and coastal protection. Overall, the marine resources of the Coral Triangle support ~130 million people locally, with tens of millions more through exports (Burke et al., 2011).

Currently, local human activities connected to intensive resource use, e.g., "overfishing," including destructive techniques, threaten nearly 85% of Indonesian coral reefs (Burke et al., 2012). According to Burke et al. (2011), Indonesia is one of the nine countries on earth, most vulnerable to the effects of coral reef degradation. Furthermore, the effects of coastal development and watershed pollution are major contributors to the decline of coral reefs in the region over the past three decades (Allen, 2008) and continue to constitute imminent threats. Since 1998, resource use has increased on more than 50% of the Indonesian reefs, mainly driven by unsustainable fishing practices and destructive resource extraction and partly due to growing coastal populations. As a result, 35% of the reefs are classified in a high or very high threat category (Burke et al., 2012), with only slight improvements since the situation in 2002 (Burke et al., 2002). Due to complex relationships in reef systems, even small changes in reef cover have the potential for trophic cascades and changing cycles of organic matter, which may subsequently affect ecosystem services and the people that rely on them (Arias-Gonzalez et al., 2011).

Besides local anthropogenic threats, regional natural disturbances, as well as global climate change play an important role. Large bleaching events have increased in frequency and are no longer limited to El Niño years. Reports from the Coral Bleaching Network's rapid survey indicate that bleaching affected 25%–75% of reefs in 21 of 22 Indonesian provinces in 2016. West Papua was the only province without a bleaching report for 2016 (Agung et al., 2018). Localized, human-derived pressure on coral reefs within Indonesia is considered to be of greater immediate concern than large-scale global pressures, and it cannot be ignored that local or regional disturbances severely aggravate the capacity for coral reefs to recover from global disturbances, such as bleaching events (Carilli et al., 2009).

The Spermonde Archipelago is an area of particular interest because of its history of intensive coral reef resource exploitation and massive reef degradation. It was therefore chosen as a research focus area of projects in all three Science for the Protection of Indonesian Coastal Ecosystems (SPICE) phases. The Spermonde Archipelago is located in the southwestern coastline of Sulawesi between the city of Makassar, the capital city of South Sulawesi, and the Makassar Strait. Makassar has a population of roughly 1.5 million people, and a further 1 million people inhabit the greater metropolitan area. This large population creates a high demand for the marine resources extracted from the archipelago (Schwerdtner-Máñez and Ferse, 2010; Ferse et al., 2014). Makassar is the largest city in eastern Indonesia, and its large maritime port (second largest in Indonesia) and airport serve as gateways to eastern Indonesia, thereby turning the city into a central export hub.

The Spermonde Archipelago consists of around 150 coral cays and reef islands of which approximately 50 are inhabited. The many submerged patch reefs, on a shallow limestone shelf, extend about 60 km offshore and 50 km along the coast and exhibit a clear nearshore–offshore gradient of biophysical conditions and anthropogenic impacts (Sawall et al., 2011; Seemann et al., 2013; Teichberg et al., 2018).

Depending on the availability of drinking water (Schwerdtner Máñez et al., 2012), the islands can be densely populated, with the population relying extensively on marine resources and offering residents very limited employment alternatives to fishing (Ferrol-Schulte et al., 2013). The local fishers employ various fishing methods, targeting a wide range of marine species and leading to a temporally and spatially variable exploitation pattern (Ferse et al., 2014). To a large extent, the Spermonde fishery functions in a tightly knit web of patron–client relationships. In these, patrons link artisanal fishers (clients) with national and international buyers via connections they have to traders in Makassar (Ferse et al., 2014; Miñarro et al., 2016), thereby transmitting changing demands and providing fishing techniques and equipment (see Section 5.4.2).

In recent decades, the coral reefs in the Spermonde Archipelago have undergone dramatic ecological changes, making the region an example for the contribution of very intensive fishing to the gradual destruction of a coastal ecosystem. The reefs are subject to a wide range of disturbances, derived both regionally from the main land and locally from the island's populations. The archipelago is home to the largest coral reef fishery in Indonesia (Pet-Soede and Erdmann, 1998), where fishers traditionally use long lines and fish traps. Over the past four decades, the use of more efficient and destructive fishing techniques (e.g., explosives, cyanide, compressors for diving) has increased the pressure on coastal ecosystems (Pet-Soede and Erdmann, 1998; Schwerdtner Máñez and Ferse, 2010). Glaser et al. (2015) report that blast fishing still continues, turning the reef into rubble unsuitable for coral recruitment. For the Pangkep Regency, the northern part of the Spermonde Archipelago, Yasir Haya and Fujii (2017) calculated that the proportion of live coral cover diminished from 7716 ha in 1994 to 4236 ha in 2014. For the island of Ballang Lompo, satellite images revealed an 80% loss of live coral cover between 1972 and 2016 and a more than 55% loss for seagrass habitat in the same timeframe (Nurdin et al., 2016).

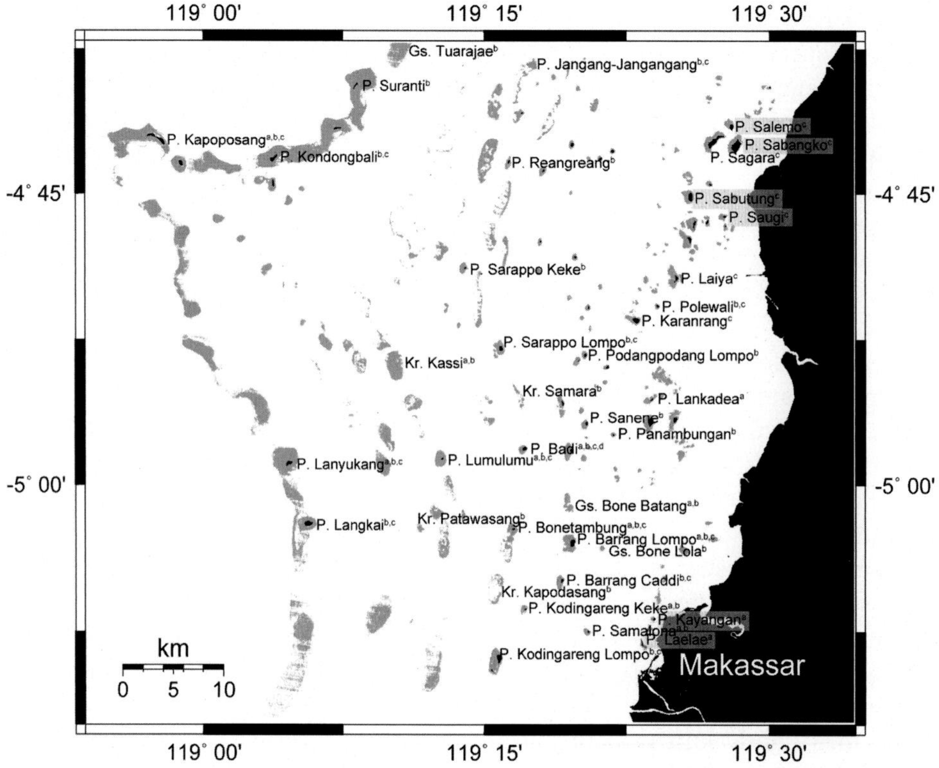


**FIGURE 5.1** Map of the Spermonde Archipelago indicating the islands investigated during the SPICE program. Superscripts indicate in which project islands were explicitly investigated (a: ecology, b: genetics, c: social science, d: modeling).

In the following sections, we provide details of the work completed during the three phases of the SPICE program (2003–16) in the Spermonde Archipelago. This chapter covers the extensive ecological studies: from studies of the individual coral to studies of changes at the level of the reef community (Section 5.2), of the genetic population structure and population connectivity for a diversity of organisms (Section 5.3), and of social dependencies and governance (Section 5.4, see also Chapter 11), as well as analytical approaches, such as modeling (Section 5.5). The studies during the SPICE program covered many islands of the Spermonde Archipelago (Fig. 5.1).

## 5.2 Functioning of coral reefs

The coral reefs of the Spermonde Archipelago have been subject to a wide range of disturbances. Observed patterns within the coral reef communities of Spermonde are closely linked to anthropogenic disturbances originating from regional and/or local sources, which are superimposed on the effects of global climate change. Spatially and temporally varying disturbances from local sources determine persistent ecological states of the coral reef systems, where differences in community composition and biodiversity are formed along a gradient of distance from shore. Indeed, studies have identified a positive relationship between species richness and distance from Makassar in coral (Edinger et al., 2000; Hoeksema, 2012), sponge (de Voogd et al., 2006), and foraminifera communities (Cleary and Renema, 2007). Beta diversity, or the difference in diversity between samples, of sea urchins, sponges, and corals displays a negative relationship with distance, suggesting communities closer to shore are more similar in their species composition (Becking et al., 2006) likely due in part to a loss of species diversity caused by disturbance. Among the many physical disturbances sustained by the reefs, some of the most impactful include destructive harvesting techniques of natural resources, such as blast fishing (Pet-Soede and Erdmann, 1998a,b; Pet-Soede et al., 1999; Pet-Soede et al., 2001), cyanide fishing (Erdman and Pet-Soede, 1997), and coral mining. Additionally, sewage and river discharge create a clear nearshore to offshore gradient of water quality (Nasir et al., 2016), thus contributing to coral degradation in nearshore reefs.

Through the three phases of the SPICE project, the coral reef ecosystems of the Spermonde Archipelago were intensely studied by joint teams from the Leibniz Centre for Tropical Marine Research, the University of Bremen, and the Hasanuddin University in Makassar (Research and Development Center for Marine, Coast, & Small Islands). Research comprised the organismal (Borell et al., 2008; Borell and Bischof, 2008; Sawall et al., 2011; Seemann et al., 2012; Sawall et al., 2014, population (Knittweis et al., 2009a,b; Knittweis and Wolff, 2010), community, and ecosystem levels (Sawall et al., 2012; Sawall et al., 2013; Plass-Johnson et al., 2015a,b; Plass-Johnson et al., 2016a,b; Plass-Johnson et al., 2018; Kegler et al., 2017a,b; Kegler et al., 2017; Kegler et al., 2018a; Teichberg et al., 2018). Importantly, the varying states of habitat degradation and the closely linked differences in coral reef ecosystem functioning are representative of many other coral reef ecosystems in the greater Southeast Asian region. Indicating a high dependency on natural resources derived from coral reefs and high levels of local and global disturbances, scientific output relating to the Spermonde Archipelago and the SPICE project provides critical ecological data to assist environmental management decisions at the local and regional levels. Given that these works have disentangled ecological, biological, and biophysical processes relating to disturbed coral reefs, the SPICE project has significantly enhanced fundamental understanding of coral reef system processes under high anthropogenic and natural stress—a situation likely to face reefs around the globe in the not-so-distant future (Hughes et al., 2017).

### 5.2.1 Water quality and biogeochemical processes

Nutrient input from various sources, including wastewater and fertilizer, is a major anthropogenic driver impacting coral reef functioning in the Spermonde Archipelago. Freshwater, riverine inputs from Tallo, Pangkep, and Maros estuaries have increased the supply of nutrients to the coastal waters of Spermonde Archipelago (Nasir et al., 2015), resulting in a water quality gradient from near- to offshore reefs. Nutrient concentrations in the estuaries are higher in nitrogen (N), primarily ammonium and phosphorus (P), increasing in the rainy season. These nutrient inputs drive seasonal increases in phytoplankton concentrations within the estuaries and nearshore sites, ranging from 1 to 8 mg $m^{-3}$. Riverine water is quickly mixed with marine waters, diluting nutrient concentrations at nearshore reef sites. However, long-term trends of poor water quality have been found, including high concentrations of chlorophyll *a* and suspended particulate matter (SPM), increased light attenuation, and occurrence of pathogenic bacteria (Kegler et al., 2017b; Sawall et al., 2011; Teichberg et al., 2018) during the many years of the SPICE program. These water quality indices generally improve at a short distance from shore (Sawall et al., 2011; Kegler et al., 2017b; Teichberg et al., 2018). Local inorganic and organic nutrient inputs from the islands, however, have been shown to influence primary production and bacterial concentrations on a smaller spatial scale, i.e., an individual coral reef system (Kegler et al., 2018a). Inorganic N, P, dissolved organic carbon (DOC), and transparent exopolymer particles (TEPs) were all elevated in back reef waters around an inhabited island, compared with an uninhabited island. This indicated that a lack of sewage treatment among inhabited islands can lead to detrimental water quality in reef systems (Kegler et al., 2018a). Mesozooplankton showed a remarkable difference in taxa composition between coast, shelf, and offshore areas in the Spermonde Archipelago, where the coastal zone was characterized by a high abundance of meroplankton and neritic copepod species, in contrast to an offshore community dominated by holoplanktonic organisms and oceanic copepod species (Cornils et al., 2010).

### 5.2.2 Benthic coral reef community dynamics of Spermonde Archipelago

The reefs of Spermonde are under intense pressure from more than 40,000 people inhabiting the Archipelago's islands, as well as from the coastal city of Makassar and agricultural activities on the coastal plains of South Sulawesi. Many of these activities cause indirect (yet strong) impacts on the reefs, e.g., via sedimentation and eutrophication related to terrestrial runoff and sewage seepage from the islands. In contrast, the harvesting of species with important ecological functions such as herbivory results in direct observable impacts on benthic communities. Blast fishing, which has persisted in the area for decades (Pet-Soede and Erdmann, 1998), has resulted in 20%–60% of the benthic live coral cover being reduced to rubble at 40%–60% of sites studied (Sawall et al., 2013; Teichberg et al., 2018). Prior to the 2018 ban on ornamental coral exports, some species, such as *Heliofungia actiniformis*, were harvested at sizes well below reproductive maturity, resulting in a change to their demographics at targeted reefs (Knittweis and Wolff, 2010). Partly, as a result of missing fisheries data, harvest quotas did not match population levels, and it was estimated that for some species, over 90% of the population was permitted to be removed within a year (Bruckner and Borneman, 2006).

The most notable observed change of the benthic community was a shift from live coral cover toward turf algae, macroalgae, and invertebrates (e.g., sponges and cnidarians).

However, this change was greater at nearshore than offshore reefs (Sawall et al., 2013; Teichberg et al., 2018; Plass-Johnson et al., 2018b). Furthermore, when there was more rubble, there were also more turf algae (Teichberg et al., 2018). When rubble is covered with turf algae rather than crustose coralline algae, reef recovery is impeded because coral recruits are attracted to the latter (Vermeij et al., 2011). Also, crustose coralline algae consolidate loose rubble, thereby increasing structural security of the rubble field (Smith et al., 2010). The increase in rubble may be caused by a number of factors including destructive fishing practices and coral predators, such as COTS, which consume live coral tissue, causing the carbonate skeleton to breakdown.

## 5.2.3 Bacterial communities and biofilms

Inorganic and organic nutrient inputs from different sources (i.e., terrestrial inputs from the mainland or sewage effluent from local populated islands) have an effect on bacterial abundance and community composition in the water, in sediments, on hard substrate and coral-associated agglomerations (Kegler et al., 2017b; Kegler et al., 2081a; Sawall et al., 2012). Bacterial communities were distinct across habitat types, including sediment, water-column (free-living and particle-attached), and coral mucus (Kegler et al., 2017b). Across the near- to offshore gradient, there were strong changes in bacterial community composition in the water column and sediment samples. Alarmingly, there was generally a high prevalence of potentially pathogenic bacteria in the water column across the entire gradient, e.g., high abundances of *Vibrio* spp. associated to diseases in nearshore sites (Kegler et al., 2017b). Additionally, although limited compared with large-scale impacts from the mainland, local island populations also influence water quality and bacterial community composition in the vicinity of populated islands (Kegler et al., 2018a). Several key water quality parameters, such as nitrate, phosphate, chlorophyll *a*, and TEP, were significantly higher at an inhabited than at an uninhabited island. Bacterial communities in sediments and particle-attached communities were significantly different between the two island types, with bacterial taxa commonly associated with nutrient and organic matter-rich conditions occurring in higher proportions at the inhabited island (Kegler et al., 2018a). Studies on biofilms on settlement tiles placed along the near- to offshore gradient showed a higher number of microbial operational taxonomic units (OTU) at nearshore sites, indicating a higher microbial diversity and a higher abundance of microbes, where the community was dominated by filamentous and turf algae (Sawall et al., 2012). It was also shown that nearshore pollution overrides the seasonal dynamics of microbial community structures, which may play a role in larvae settlement of sessile reef invertebrates near- versus offshore (Sawall et al., 2012). In another study by Kegler et al. (2017a), the most abundant bacteria on natural substrate and artificial tiles were Gammaproteobacteria, Alphaproteobacteria, and Cyanobacteria. Bacterial community composition (BCC) was strongly correlated with water quality, and significant differences in BCC between the inshore site and nearshore/midshelf were found. On artificial substrates, there was a significant difference in BCC in line with exposure time in the reef.

### 5.2.4 Coral reef recruitment processes

Scleractinian coral recruitment is high in Spermonde with up to 700 recruits $m^{-2}$ $yr^{-1}$ occurring predominantly during the dry season (July–October). Some recruitment also occurred during the wet (November–February) and transitional season (March–June) in particular at midshelf and offshore reefs (Sawall et al., 2013). Despite a strong cross-shelf gradient in environmental conditions, coral recruitment displayed little variation among sites, suggesting adequate source populations. However, there was a reduction in diversity at the most nearshore sites (Lae Lae; Sawall et al., 2013). Lower diversity in recruitment at nearshore sites is most likely due to increased turf algae, which were found to be important space competitors of coral recruits (Plass-Johnson et al., 2016a). Thus, new recruits encounter an increased probability of interacting with turf algae with every site closer to shore. Recruitment may depend on species-specific interactions among corals and algae, but further testing is needed to determine whether some corals are able to outcompete turf algae.

### 5.2.5 Coral physiology

An important factor that may impact coral development is the physiological adaptability to strong variation in abiotic conditions. Some coral species compete better under changing environmental conditions (Sawall et al., 2011, 2014). In studies on *Stylophora pistillata*, starvation led to reduced photosynthetic yield compared with fed corals when exposed to elevated water temperatures, indicating a link between heterotrophic feeding and photoinhibition under thermal stress (Borell and Bischof, 2008). For *S. pistillata* and *Galaxea fascicularis*, feeding with zooplankton resulted in sustained photosynthetic activity under elevated water temperatures, while starvation resulted in decreased zooxanthellae densities and photosynthesis. Both corals displayed reduced protein concentrations when starved, and lipid levels decreased in starved *S. pistillata* (Borell et al., 2008). The scleractinian coral *Stylophora subseriata* showed high physiological plasticity across the cross-shelf gradient: Colonies in more eutrophic nearshore waters showed higher photosynthesis and calcification, photosynthetic efficiency, zooxanthellae density, chlorophyll *a* concentrations as well as protein and lipid content (Sawall et al., 2011; Seemann et al., 2012). *Porites lutea* also showed higher photosynthetic rates and zooxanthellae densities in eutrophic nearshore reefs compared with oligotrophic midshore reefs but did not show enhanced rates of calcification as found in *S. subseriata*, indicating that additional energy is allocated toward stress mitigation (e.g., mucus production) rather than growth in nearshore reefs (Sawall et al., 2014). Despite overall decreases in live coral cover, the potential for acclimation of a few coral species to varying environmental conditions indicates some level of resilience to local disturbances.

### 5.2.6 Relationships between benthic and fish communities

In agreement with indicators of coral reef habitat health (e.g., hard coral cover), indices of fish community health, such as species richness, abundances, and biomass, also increased with distance from shore (Plass-Johnson et al., 2015a, 2016b, 2018b). In the Spermonde Archipelago, fish surveys revealed that communities at nearshore sites had greater variability in their trait-based functional composition than offshore sites (Plass-Johnson et al., 2016a,b). This is likely driven by the patchiness of hard coral habitat at disturbed sites in comparison with the offshore sites. Thus, nearshore sites were composed of some fish species that exploit hard coral resources and others that used rubble, turf algae, or sand. While variation in trait-based diversity across levels of habitat disturbance begins to reveal mechanisms underlying the relationship between habitat and fish, understanding the trophic plasticity of fish species may reveal strategies for dealing with changing environments. For instance, the parrotfish *Chlorurus bleekeri* and the farming damselfish *Dischistodus prosopotaenia* used differing feeding strategies, to either expand (parrotfish) or maintain (damselfish) (respectively) their nutritional resources in response to a changing habitat (Plass-Johnson et al., 2018a).

Further work conducted in the Spermonde Archipelago shows that the herbivorous fish functional group may display functional compensation based on the high species diversity within the region. Although there was high species turnover among all investigated sites, the important function of herbivory was continuously maintained at most sites by different species compositions (Plass-Johnson et al., 2015a,b), with often high grazing rates on macroalgae. The apparent high regional-level functional redundancy for macroalgal browsers contrasts with observations from less species-rich assemblages, notably reefs in the Caribbean (Micheli et al., 2014; Roff and Mumby, 2012). Coral reef herbivores play an important ecological role because they clear the substratum of biota that compete with scleractinian recruits, often repelling the larvae before settlement. While the results of this study display the potential for significant algal removal by herbivorous fish species, herbivory was not observed at the site nearest to shore, indicating that high levels of degradation also strongly affect ecologically important fish community functions and possibly resulting in the formation of negative feedback loops (Fig. 5.2).

## 5.2.7 Consequences of disturbances for coral reef functioning

Recent research has noted that the coral reefs of Southeast Asia are currently more threatened by localized human impacts, such as terrestrial effluents and resource use than by global stressors such as warming water temperatures and acidification (Burke et al., 2011). The various reef states found in the Spermonde Archipelago along a cross-shelf gradient suggest that phase shifts from coral-dominated states to algal-dominated states are driven by both bottom-up and top-down processes and might be influenced by negative social–ecological feedbacks reinforcing these negative states (Glaser et al., 2018). The long-term effects of water quality and overfishing on coral reef benthic composition can also be overshadowed by localized intense and acute disturbances (Baum et al., 2015). This is more evident further offshore, where water quality and herbivore abundances are relatively high, but other disturbances, including outbreaks of

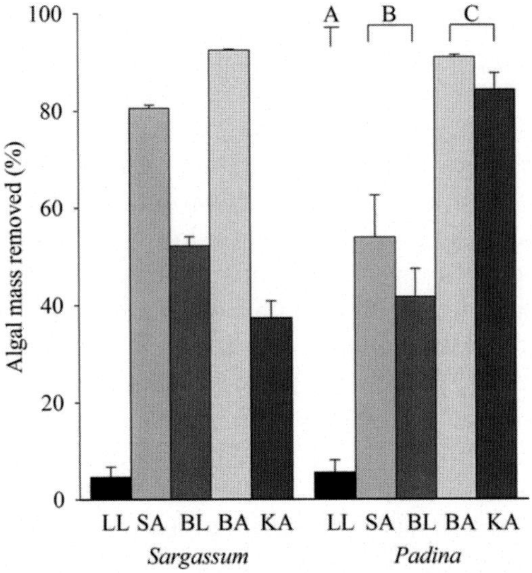


**FIGURE 5.2** Results from herbivory assays. Average (±SE) amount of *Sargassum* and *Padina* removed per 24-h period. Islands are arranged from closest to furthest from Makassar (*BA*, Badi; *BL*, Barrang Lompo; *LL*, Lae Lae; *KA*, Kapoposang; *SA*, Samalona; see Fig. 5.1). Data represent total mass loss after the correction for controls. Lettering above Padina indicates islands that were not significantly different in the PERMANOVA pairwise post hoc test. Sargassum treatments were all significantly different from each other. *From Plass-Johnson et al. (2015a), Figure 2.*

crown-of-thorn starfish and destructive fishing practices, such as blast fishing, lead to declines in reef health (Teichberg et al., 2018).

When integrating all water quality, benthic, and pelagic indices in the Spermonde Archipelago, it is possible to see differing levels of the state of health of the reefs with increasing distance from the mainland (Plass-Johnson et al., 2018b). Additionally, the benthic condition index, calculated as the percent live coral cover divided by the sum of the percent cover of other important benthic groups (macroalgae, turf algae, sponge, and cyanobacteria), showed a general decreasing trend closer to shore (Teichberg et al., 2018, Fig. 5.3). The evident trend is related to distance from the mainland; however, further midreef and outer-reef islands also showed evidence of how localized disturbances, such as outbreaks of crown-of-thorns starfish (Plass-Johnson et al., 2015b, Fig. 5.4), or overgrowth of cyanobacteria and sponges in areas with intensive bomb fishing (Teichberg et al., 2018), which can compete with large-scale gradients. This index also showed signs of quick recovery from acute local disturbances, indicating some resilience to disturbances with time.

## 5.3 Genetic connectivity of reefs in the Coral Triangle region

Connectivity is the exchange of individuals among populations, which is the driving force for colonization of new areas, replenishment of depleted populations,

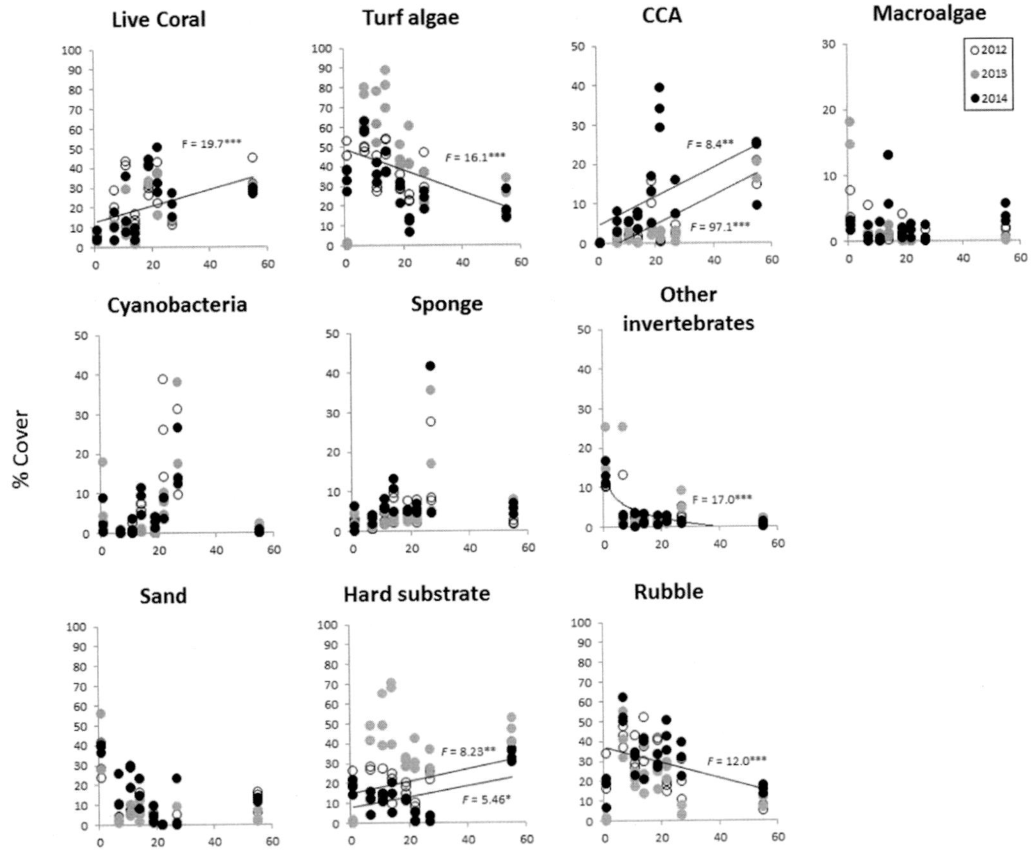


**FIGURE 5.3** Percentage of benthic cover of dominant groups of coral reef organisms and substrate type measured at different islands in 2012 (white), 2013 (gray), and 2014 (black). Sites are represented by distance from shore (km) on the x-axis. Regressions are indicated for variables that showed a significant increase or decrease with distance. Asterisks indicate level of significance (*** indicates $P < .001$; ** indicates $P > .001$ and $< 0.1$, * indicates $P > .01$ and $< 0.5$). *From Teichberg et al. (2018), Figure 5.*

recolonization, and maintenance of genetic diversity (Cowen and Sponaugle, 2009). Virtually all coral reef species disperse in their early life history stages as eggs and/or larvae, utilizing ocean currents as vectors. However, the extent and direction of dispersal, i.e., the actual distances and directions traveled with ocean currents, remain unclear (Jones et al., 2009). Do they return to their parental reef (self-recruitment), do they disperse only short distances to neighboring reefs, or do they undergo large distance dispersal of hundreds or even thousands of kilometers? Are these populations open or closed, i.e., show an influx of recruits from other populations and vice versa (open) or predominantly retain their own offspring to sustain their population (closed)? In which direction do eggs and larvae disperse? Genetic analysis can help unravel some of these questions by detecting the genetic "footprint" of dispersed larvae or migrating adults in the local population's gene pool. In addition to using this method to track the exchange

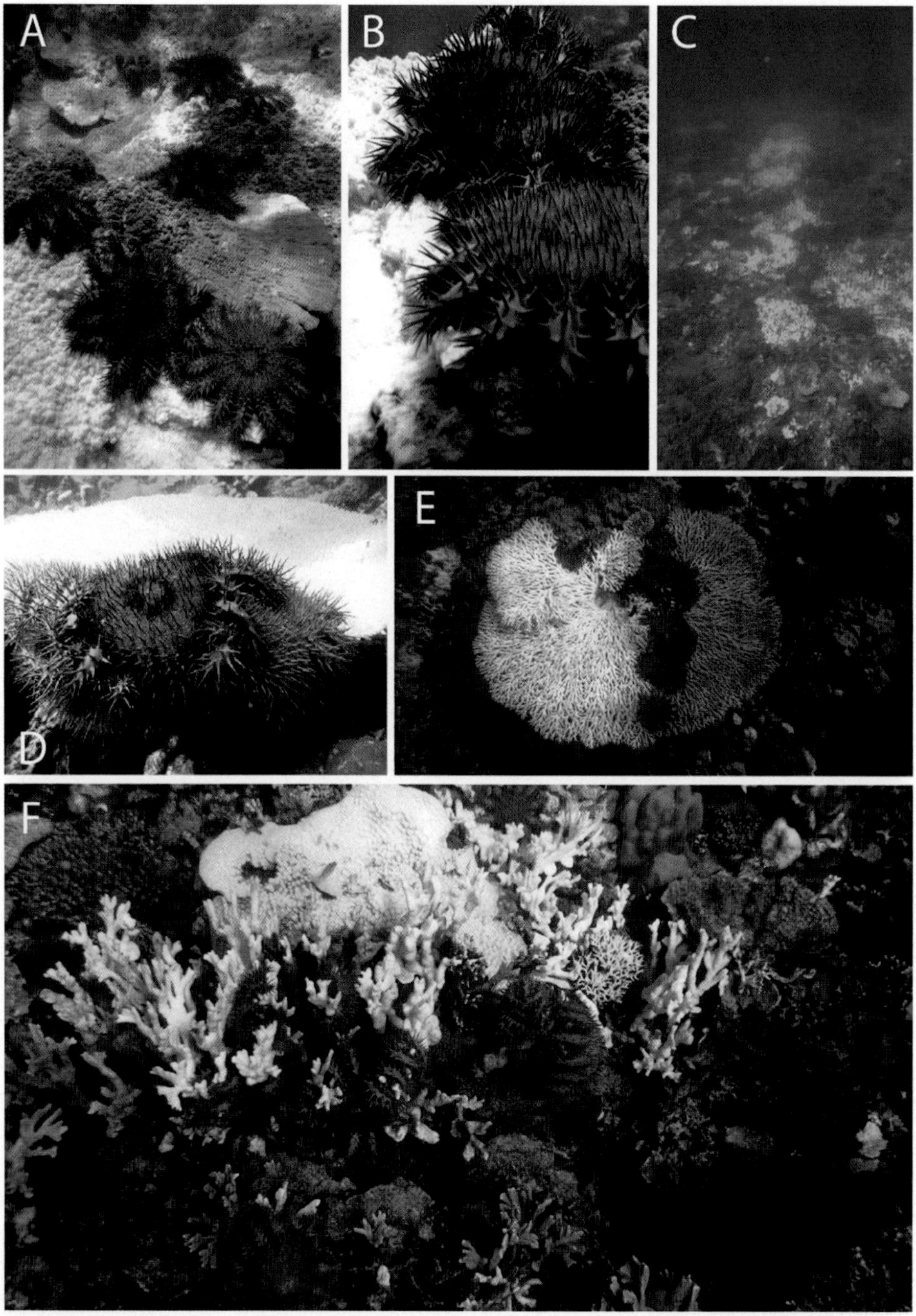


**FIGURE 5.4** Crown-of-thorns starfish (*Acanthaster planci*, COTS) activity at Barrang Lompo (A, D and F), Bonetambung (B and E), and Lumulumu (C) recorded during September 2012. At Barrang Lompo, COTS were observed feeding on multiple genera of coral in very high densities. Eighteen COTS were observed feeding in roughly 2 $m^2$ at Barrang Lompo (F). (C) A track, approximately 20 m long, of fresh feeding scars left by COTS in September 2012. *From Plass-Johnson et al. (2015b).*

of individuals between populations, genetic diversity indices can also serve as a proxy for population size and thereby help to identify very reduced, isolated, or vulnerable populations.

Knowledge about dispersal distances and patterns is a very important baseline information for conservation efforts and the spatial design of marine-protected area (MPA) networks. As explained in the introduction, the Coral Triangle is the global center of marine coastal biodiversity. However, its coral reefs are threatened by anthropogenic impacts, such as overexploitation, pollution, and climate change, as has been shown in the previous sections of this chapter. To provide vital knowledge for conservation efforts, this subproject of SPICE aimed to investigate connectivity of coral reefs at different spatial scales, zooming in from the Coral Triangle to Spermonde and single reefs within this archipelago.

There are different approaches to investigate connectivity, ranging from in situ methods, such as chemically tagging larval fish otoliths or analyzing their natural isotopic composition, to in silico biophysical modeling, which integrates knowledge on the ecology of early life history stages with oceanography. However, by far, the most commonly applied approach investigates the genetic population structure to estimate gene flow, i.e., the exchange of genetic material among populations, as a proxy for connectivity. The advantage of genetic methods is their universal applicability to all kinds of organisms, while the aforementioned analysis of otoliths is restricted to fish and small spatial scales.

In the framework of the SPICE project, the genetic population structure of 12 species of coral reef organisms was investigated at different spatial scales. This included different life history traits and dispersal potentials, ranging from ecosystem engineers, such as stony corals and giant clams, to sea stars, ascidians, and anemonefish (Fig. 5.5; Table 5.1). The mode of spawning (i.e., pelagic or demersal eggs; brooders) as well as the pelagic larval duration (PLD) can have a pronounced effect on the dispersal potential. Species that spawn eggs into the water column and/or have a long PLD are expected to disperse over long distances and therefore frequently show a weak genetic differentiation among populations. On the contrary, species that release larvae into the water column and/or have a short PLD are expected to show limited connectivity and thus exhibit populations that are genetically strongly differentiated.

To investigate the genetic population structure of these species, different population genetic markers were applied, ranging from single-locus mitochondrial and nuclear DNA sequences to multilocus nuclear microsatellite analyses. The latter was also applied for parentage analysis to estimate self-recruitment. Thousands of tissue samples were collected in a minimal invasive way underwater while SCUBA-diving at 63 sites in the Coral Triangle region (Fig. 5.6).

## 5.3.1 Large-scale connectivity across the Coral Triangle region

A general population genetic structure that can be observed in many marine taxa on this large scale is the genetic differentiation between Indian and Pacific Ocean populations (Crandall et al., 2019). Sea level low stands of up to 120 m during the Pliocene and

**FIGURE 5.5** Species investigated for the connectivity study: (A) clown anemonefish, *Amphiprion ocellaris*; (B) pink anemonefish, *Amphiprion perideraion*; (C) skunk clownfish, *Amphiprion akallopisos*; (D) boring giant clam, *Tridacna crocea*; (E) small giant clam, *Tridacna maxima*; (F) fluted giant clam, *Tridacna squamosa*; (G) mushroom coral, *Heliofungia actiniformis*; (H) bird's nest coral, *Seriatopora hystrix*; (I) branching coral, *Acropora millepora*; (J) blue sea star, *Linckia laevigata*; (K) ectoparasitic gastropod, *Thyca crystallina*; (L) sea squirt, *Polycarpa aurata*. *All photographs by M. Kochzius, except (G) by L. Knittweis.*

**Table 5.1** Early life history traits of the study species and their dispersal potential.

| Species | Mode of spawning | PLD (days) | Dispersal potential | Genetic marker | Tested barriers | Detected barrier | Strength of the barrier | Population genetic study |
|---|---|---|---|---|---|---|---|---|
| **Stony coral** | | | | | | | | |
| *Acropora millepora* | PE, PL | 60 | High | Microsats | ii | ii | Weak | van der Ven et al. (2021) |
| *Heliofungia actiniformis* | BL, PL | 3 | Low | ITS | ii, iii | ii, iii | Strong | Knittweis et al. (2009b) |
| *Seriatopora hystrix* | BL, PL | <0.5–7 | Low | Microsats | ii | ii | Strong | van der Ven et al. (2021) |
| **Giant clam** | | | | | | | | |
| *Tridacna crocea* | PE, PL | 9–12 | Medium | COI, Microsats | i, ii, iii, iv | i, ii, iv | Strong | Kochzius and Nuryanto (2008); Hui et al. (2012), Hui et al. (2016, 2017) |
| *Tridacna maxima* | PE, PL | 9–12 | Medium | COI | i, ii, iv | ii, iv | Strong | Nuryanto and Kochzius (2009); Hui et al. (2016) |
| *Tridacna squamosa* | PE, PL | 9–12 | Medium | COI | ii, iii, iv | iv | Strong | Hui et al. (2016) |
| *Tridacna noae* | PE, PL | 9–12 | Medium | COI | iii, iv | iv | Strong | Keyse et al. (2018) |
| **Sea star** | | | | | | | | |
| *Linckia laevigata* | PE, PL | 22 | High | COI, microsats | i, ii, iii, iv | i, iv | Weak | Kochzius et al. (2009); Alcazar and Kochzius (2016); Ratzimbazafy (2019) |
| **Parasitic gastropod** | | | | | | | | |
| *Thyca crystallina* | ? | ? | High | COI | ii, iii, iv | none | n/a | Kochzius et al. (2009) |
| **Sea squirt** | | | | | | | | |
| *Polycarpa aurata* | PE, PL | 0.5–7 | Low | Microsats | n/a | n/a | n/a | Timm et al. (2017) |
| **Anemonefish** | | | | | | | | |
| *Amphiprion ocellaris* | DE, PL | 8–12 | Low | CR, microsats | i, ii, iii | ii, iii | Strong, weak | Timm and Kochzius (2008); Timm et al. (2012) |
| *Amphiprion perideraion* | DE, PL | 18 | Medium | CR, microsats | ii, iii, iv | ii, iii or iv | Strong, weak | Dohna et al. (2015) |
| *Amphiprion akallopisos* | DE. PL | 15 | Medium | CR, microsats | i | none | n/a | Huyghe and Kochzius (2017, 2018) |

Utilized genetic markers: mitochondrial COI (cytochrome oxidase I) and CR (control region); nuclear ITS (internal transcribed spacer) and microsats (microsatellites). Detected barriers for gene flow: for codes, see Fig. 5.5. *BL*, brooded larvae; *DE*, demersal eggs; *PE*, pelagic eggs; *PL*, pelagic larvae; *PLD*, pelagic larval duration.

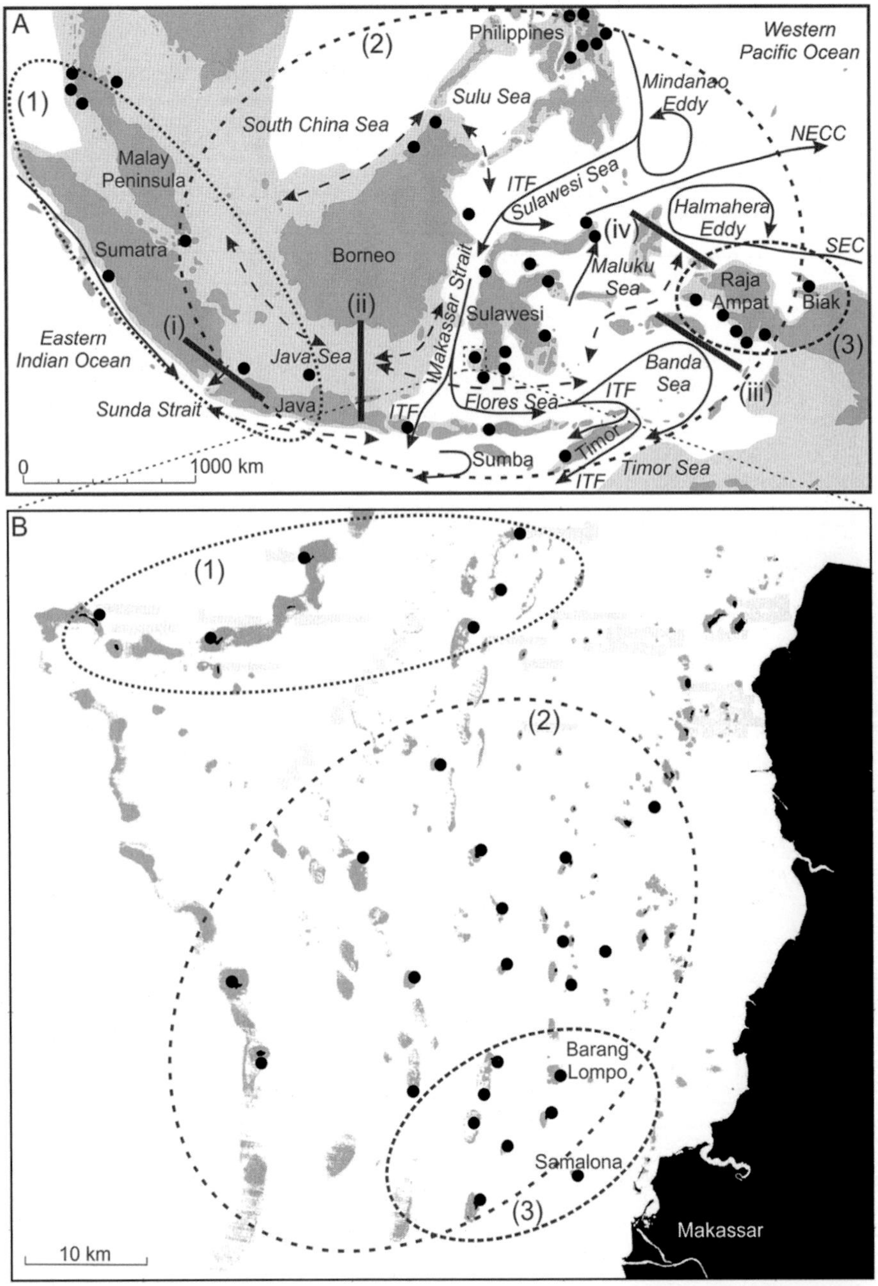


**FIGURE 5.6** (A) Map of the Coral Triangle region with sample sites for the species listed in Table 5.1 (not all species were sampled at all sites), as well as oceanographic patterns with dominant (*solid lines*) and seasonally changing (*dashed lines*) currents (Gordon, 2005; Gordon and Fine, 1996; Wyrtki, 1961). *ITF*, Indonesian throughflow; *NECC*, Northern Equatorial Counter Current; *SEC*, Southern Equatorial Current. Pleistocene maximum sea level low stand of 120 m is indicated by the light gray area, land by the dark gray area (Voris, 2000). Groups of genetically similar populations: (1) Eastern Indian Ocean and Java Sea, (2) Philippines, South China Sea, Sulawesi Sea, Makassar Strait, Flores Sea, Maluku Sea, and Savu Sea (3) Radja Ampat and Western Pacific. (i–iv) Potential barriers for dispersal. (B) Map of Spermonde Archipelago with samples sites (not all species were sampled at all sites). Groups of genetically similar populations: (1) outer-shelf reefs, (2) midshelf reefs, and (3) inner-shelf reefs. The light gray area indicates coral reefs, and the black area indicates land.

Pleistocene glacials—exposed shelves in Southeast Asia and Australia formed the Indo-Pacific Barrier, which separated populations of the two ocean basins (Hoeksema, 2007, Fig. 5.6A). In these periods of separation, the fauna diverged by allopatric or parapatric speciation but subsequently came in contact again after sea level rise during interglacials (Bowen et al., 2013). This separation left a unique signature in the genomes of many species, which can be used to trace dispersal after the rise of the sea level at the end of the last glacial.

Population genetic analyses show genetic differentiation between populations originating from the Indian and the Pacific Ocean in all taxa studied on this large scale (Table 5.1), except the gastropod *Thyca crystallina*, which is an ectoparasite of the sea star *Linckia laevigata*. This parasite does not show any significant genetic differentiation, which indicates high connectivity at the studied scales. The larvae of *T. crystallina* need to disperse widely to encounter their high-dispersed host *L. laevigata*, which shows a shallow but significant genetic population structure. The absence of population genetic patterns in *T. crystallina*, driven by its exceptional ecological demands, is not considered representative for the majority of organisms that were studied and will therefore not be included in the further discussion.

Across the Coral Triangle, a general pattern of genetic differentiation among the western, central, and eastern parts emerges from the population structures of the most analyzed species: (1) Eastern Indian Ocean and Java Sea, (2) Philippines, South China Sea, Sulawesi Sea, Makassar Strait, Flores Sea, Maluku Sea, and Savu Sea, and (3) Raja Ampat and Western Pacific. Additionally, two gene flow barriers (i, ii) were identified in the western part of the Coral Triangle in the Sunda Strait and the Java Sea, while two others appeared around Raja Ampat (iii, iv) in the eastern part of the Coral Triangle (Fig. 5.6A, Table 5.1).

A clear separation of the Eastern Indian Ocean populations can be observed in all taxa, which includes the Java Sea in some species, such as in the giant clam *Tridacna maxima* (Nuryanto and Kochzius, 2009; Hui et al., 2017) and the anemonefish *Amphiprion ocellaris* (Timm and Kochzius 2008; Timm et al., 2012). This separation of the Eastern Indian Ocean is also observed in connectivity studies based on biophysical modeling (Kool et al., 2011; Treml et al., 2015). Other species also show a clear genetic differentiation of the Java Sea populations from the central part of the Cora Triangle, including the stony corals *H. actiniformis* (Knittweis et al., 2009b) and *Seriatopora hystrix* (van der Ven et al., 2021), as well as the anemonefish *Amphiprion perideraion* (Dohna et al., 2015). However, in some species, this differentiation is very weak, for instance, in the broadcast spawning stony coral *A. millepora* (van der Ven et al., 2021). Conversely, in the giant clams *Tridacna crocea* (Kochzius and Nuryanto, 2008; Hui et al., 2016, 2017) and *Tridacna squamosa* (Hui et al., 2016), as well as in the star fish *L. laevigata* (Alcazar and Kochzius, 2016; Kochzius et al., 2009), populations from the Java Sea belong genetically to the central part of the Coral Triangle region.

Populations of all studied species in the central part of the Coral Triangle region seem to be well connected along the Indonesian Throughflow (ITF), all the way from the

Philippines through the Strait of Makassar down to the Flores Sea. This pattern of connectivity is also supported by biophysical modeling (Treml et al., 2015). Populations in the South China Sea along the coast of northern Borneo also belong to this genetic cluster, indicating high exchange. This pattern of high connectivity was also revealed by biophysical modeling, showing larval dispersal from the South China Sea off northern Borneo across the Sulu Sea into the Sulawesi Sea, following the ITF to the south (Kool et al., 2011).

Different genetic marker systems provide slightly different results in the anemonefish *A. ocellaris* (Timm et al., 2012). Based on nuclear microsatellites, the population close to the southern tip of the Malay Peninsula also belongs to the genetic cluster of the central Coral Triangle region, which is also the case for the giant clam *T. squamosa* (Hui et al., 2016). However, the mitochondrial control region in the anemonefish *A. ocellaris* shows a unique genetic signature, separating this population from all other sites, which is supported by biophysical modeling (Kool et al., 2011).

Raja Ampat and Biak in the East also show a strong genetic differentiation in all species, indicating limited connectivity of the eastern Coral Triangle region to the central part. Limited gene flow from the Western Pacific (Biak) is probably caused by the Halmahera Eddy, which redirects the westward flow of the South Equatorial Current (SEC) into the eastward Northern Equatorial Counter Current (NECC). Biophysical modeling also shows that this causes a barrier for dispersal (Treml et al., 2015). However, again the mitochondrial and nuclear markers show a different picture in the anemonefish *A. ocellaris* (Timm et al., 2012) and also in the giant clam *Tridacna crocea* (Hui et al., 2017), with the nuclear microsatellites indicating strong connectivity of Raja Ampat with the central Coral Triangle region.

## 5.3.2 Small-scale connectivity in the Spermonde Archipelago

The Spermonde Archipelago (Fig. 5.6B) shows an ecological zonation of the zoobenthos along a continuous inshore-to-offshore and shallow-to-deep environmental gradient, which makes it possible to divide reefs into inner-shelf, midshelf, and outer-shelf zones (Cleary et al., 2005, Section 5.2). The aim of this study was to test if this environmental gradient is reflected in the connectivity among the different reefs by analyzing the genetic population structure of several taxa, such as the stony corals *H. actiniformis* (Knittweis et al., 2009b), *S. hystrix*, and *A. millepora* (van der Ven et al., 2021), as well as the sea squirt *Polycarpa aurata* and the anemonefish *A. ocellaris* (Timm et al., 2017). An overall pattern of north–south differentiation into three genetic groups can be seen that roughly follows the zonation of inner-shelf, midshelf, and outer-shelf reefs (Fig. 5.6). Populations of the northern outer-shelf show more exchange among each other than inner-shelf populations do, where connectivity seems to be much more restricted (Timm et al., 2017). While the populations of the northern outer-shelf reefs are clearly genetically separated from mid- and inner-shelf reefs in all species, the geographic location of the barrier differs among species. The genetic differentiation within Spermonde might be due to different oceanographic conditions. Especially the northern outer-shelf reefs are strongly under the influence of the ITF, which probably imports larvae from upstream and facilitates dispersal among reefs. This input of new genetic material leads to a higher genetic diversity in the anemonefish *A. ocellaris,* which was significant in the mitochondrial marker. However, this pattern was not observed in the nuclear microsatellite analysis (Timm et al., 2017). Further inshore, the currents might gradually become weaker, and therefore, connectivity among populations is reduced, leading to a strong genetic differentiation, especially in the anemonefish *A. ocellaris* and the sea squirt *P. aurata* (Timm et al., 2017).

### 5.3.3 Self-recruitment at the islands of Barrang Lompo and Samalona

The antagonist of dispersal is self-recruitment, which determines to which extent a population is self-containing or depends on the influx of recruits from other populations by dispersal. It is another important aspect in the context of connectivity for the spatial design of MPAs. To study self-recruitment, the anemonefishes *A. ocellaris* and *A. perideraion* were chosen as model species and were investigated at the islands of Barrang Lompo and Samalona in the Spermonde Archipelago (Madduppa et al., 2014a). Fin clips of almost the whole adult population and juvenile individuals were collected in the reefs around these two islands. Nuclear microsatellite markers were used in a parentage analysis approach to match offspring with their parents. This was done for *A. ocellaris* at Barrang Lompo in 2008 and 2009, as well as at Samalona in 2009. *A. perideraion* was sampled in 2009 at Barrang Lompo.

The analysis shows interspecific, spatial, and temporal variability of self-recruitment in the two anemonefish species (Table 5.2). *A. ocellaris* shows a very high self-recruitment of 65.2% in Samalona, while both species had a lower but very similar

**Table 5.2** Comparison of percent self-recruitment (% SR) in different anemonefishes (*Amphiprion* spp.).

| Species | PLD | Location | Country | Reef type | % SR | References |
|---|---|---|---|---|---|---|
| *A. ocellaris* | 8–12 | Barrang Lompo (2008) | Indonesia | IR | 44 | Madduppa et al. (2014a) |
| *A. ocellaris* | 8–12 | Barrang Lompo (2009) | Indonesia | IR | 52 | Madduppa et al. (2014a) |
| *A. ocellaris* | 8–12 | Samalona | Indonesia | IR | 65.2 | Madduppa et al. (2014a) |
| *A. perideraion* | 18 | Barrang Lompo | Indonesia | IR | 46.9 | Madduppa et al. (2014a) |
| *A. percula* | 10–13 | Kimbe Island (2007) | Papua New Guinea | IR | 64 | Berumen et al. (2012) |
| *A. percula* | 10–13 | Kimbe Island (2004) | Papua New Guinea | IR | 42 | Planes et al. (2009) |
| *A. percula* | 10–13 | Kimbe Bay (2009) | Papua New Guinea | IR, CR | 12.9 | Almany et al. (2017) |
| *A. percula* | 10–13 | Kimbe Bay (2011) | Papua New Guinea | IR, CR | 20.2 | Almany et al. (2017) |
| *A. polymnus* | 9–12 | Schumann Island | Papua New Guinea | CR | 31.5 | Jones et al. (2005) |
| *A. polymnus* | 9–12 | Bootless Bay (2005) | Papua New Guinea | CR | 25 | Saenz-Agudelo et al. (2009) |
| *A. polymnus* | 9–12 | Bootless Bay (2008) | Papua New Guinea | CR | 7.1 | Saenz-Agudelo et al. (2011) |
| *A. akallopisos* | | Zanzibar | Tanzania | IR | 21.3 | Huyghe (2018) |
| *A. bicinctus* | 11 | Qita al Girsh | Saudi Arabia (Red Sea) | IR | 0.6 | Nanninga et al. (2015) |

*CR*, coastal reef; *IR*, isolated reef; *PLD*, pelagic larval duration.

self-recruitment of 47.4% (*A. ocellaris*) and 46.9% (*A. perideraion*) in Barrang Lompo. An interannual variation in self-recruitment is observed in *A. ocellaris* from the coral reef in Barrang Lompo, with 44% in 2008 and 52% in 2009. A temporal variation in self-recruitment was also observed in the anemonefishes *Amphiprion percula* and *Amphiprion polymnus* (Table 5.2). Site fidelity, i.e., returning to the same part of the reef where the larva had hatched, ranged from 0% to 44% in *A. ocellaris* and from 0% to 19% in *A. perideraion* at Barrang Lompo. At Samalona, site fidelity ranged from 8% to 11% in *A. ocellaris*. Exchange among the populations of *A. ocellaris* between the two islands was also investigated, showing that individuals of the larger adult population from Samalona are identified as parents of 21% of the juveniles from Barrang Lompo, while adults from the latter are parents of only 4% of the juveniles from Samalona. Percentage of self-recruitment observed in *A. ocellaris* and *A. perideraion* is higher than in all other investigated anemonefish species so far but are similar to Kimbe Island in Papua New Guinea (Table 5.2). This high amount of self-recruitment at Samalona and Barrang Lompo is concordant with the strong genetic differentiation of the *A. ocellaris* populations at these islands, which is further supported by the limited connectivity in the sea squirt *P. aurata* (Timm et al., 2017).

### 5.3.4 Application of connectivity data in marine-protected area network design

The majority of the studied species, such as stony corals, giant clams, blue starfish, and anemonefish, are important in the marine ornamental trade, in which Indonesia is one of the main exporters (UNEP-WCMC, 2014; Wabnitz et al., 2003). In the Spermonde Archipelago, marine ornamental fishery started in the late 1980s, and by the late 1990s, a wholesaler company from Makassar received a permit for the collection of marine ornamentals in Spermonde, which are shipped from Makassar to Jakarta or Bali by airplane. Even though marine ornamental fishery is not the main source of income in Spermonde, it covers 13%–43% of the expenses of island households and is thus economically important. Indeed, on one studied island, marine ornamental fishery even covered 84% of the household expenses (Madduppa et al., 2014b).

By far, the most collected marine ornamental species is the anemonefish *A. ocellaris*, which is very popular due to the animation film "Finding Nemo." Its host anemones are also collected for the international aquarium trade. It is estimated that each year, about 140,000 specimens of the anemonefish *A. ocellaris* and 31,000 host anemones (*Heteractis magnifica*, *Stichodactyla gigantea*, and *Stichodactyla mertensii*) are collected in Spermonde. The resulting high fishing pressure has a very strong impact on these target species, and underwater surveys show significantly lower densities on reefs with high exploitation (HE) in comparison with reefs with low exploitation (LE) rates. Also, body length and group size of the anemonefish *A. ocellaris* were significantly smaller at HE sites than at LE sites (Madduppa et al., 2014b). High fishing pressure also has negative implications for the genetic diversity of the anemonefish *A. ocellaris*. In Barrang Lompo

(HE), where the population density was threefold smaller than in Samalona (LE), genetic diversity (the number of alleles, private alleles, and allelic richness in nuclear microsatellites) was also significantly reduced (Madduppa et al., 2018).

Based on the results of the aforementioned studies on connectivity in the Coral Triangle region, the following management recommendations can be made. On a large scale, at least five regions should be considered as separate management units in the Coral Triangle region, taking into account the four detected barriers to gene flow (Fig. 5.6A): (1) Eastern Indian Ocean, (2) Java Sea, (3) South China Sea, Sulu Sea, Philippines, Sulawesi Sea, Makassar Strait, Flores Sea, and Banda Sea, (4) Raja Ampat, and (5) Western Pacific off Papua New Guinea. In the third large-scale management unit along the ITF, especially upstream populations (South China, Sulu Sea, Philippines, and Sulawesi Sea) need special conservation attention, because they are source populations for all other populations downstream along the ITF (Makassar Strait, Flores Sea, and the Banda Sea).

In the Spermonde Archipelago from the population genetic view, at least three different management units should be considered: (1) northern outer-shelf reefs, (2) central outer- and midshelf reefs, and (3) southern inner-shelf reefs (Fig. 5.6B). Since reefs of the northern outer-shelf are well connected with each other, MPAs do not need to be closely spaced. However, in the southern mid- and inner-shelf reefs, the situation is different, because populations show restricted connectivity even among closely located islands, with high self-recruitment shown in the anemonefish *A. ocellaris*. Therefore, a network of MPAs on each reef should be considered as the best management option.

## 5.4 Social systems associated with the use of coral-based resources and reef-specific challenges

The social–ecological system (SES) conceptualizations (Glaser et al., 2012) that relate to Spermonde Archipelago differ in accordance with the issue or problem in focus. The social actors and systems associated with Spermonde's ornamental and live reef fish fisheries are mainly international (foreign traders and consumers), while those associated with blast fishing and pollution are predominantly local and regional; therefore, there are also at least two reef-related, issue-specific SES definitions.

### 5.4.1 Participatory assessment of Spermonde's coral reef fisheries

During the second and third phases of SPICE, social scientists visited and worked on more than a third of the roughly 50 inhabited islands of the Spermonde Archipelago (Fig. 5.1). Drawing on natural science work from the first project phase, in particular on research on the ornamental coral trade, an extensive participatory social science research agenda was developed. For this, social scientists from the Leibniz Centre for Tropical Marine Research and the University of Bremen worked together with their counterparts at the Coral Reef Research Centre (now Research and Development Center for Marine, Coast & Small Islands [sic]) and the Anthropology Department of Hasanuddin University in Makassar. The overarching interest was to assess the livelihoods of fisherfolk using the coral reefs of Spermonde and the governance institutions covering those waters. Key themes were fishery and trade in ornamental coral, life reef fish and other reef products, reef governance and management, including the emergence of associated rules and institutions (see Chapter 11), and fishing techniques.

A number of preparatory visits to several of the islands between 2004 and 2007 were followed by the funded social science research on the reef-based SES of the Spermonde Archipelago. This began in 2007, in the second phase of the SPICE project, with a joint workshop in Makassar. A focus of the research was to be on several key coral reef fisheries (ornamental fish and corals, live reef fish, sea cucumbers) and on destructive fishing methods. Islands identified as locations of particular relevance or hubs of activity for these fisheries were therefore selected as focal locations for the social science research (Ferse et al., 2014; Glaser et al., 2015). The backbone of the approach were five ship-based excursions conducted in 2009, 2010, 2012, and 2013, each of which brought bilateral teams of a total of about 20 social scientists and students per excursion to multiple islands. The first excursion combined icebreaking activities (film screenings, town hall meetings, and photo sessions) to familiarize project teams and village communities with each other and with main shared themes, with focus group discussions, individual key-informant interviews, and an interview-based survey. This was followed by theme-focused interviews, with extensive daily team discussions and complementary data collection during the second excursion. During a third ship-based excursion, results were presented back to and discussed with communities, making use of a short documentary film produced on the research work during the first excursion (https://vimeo.com/5246067; Glaser et al., 2010c).

Two additional excursions, including similar multinational research teams conducting focus group discussions, surveys, and interviews, were conducted in the final phase of SPICE. These ship-based excursions were complemented by a range of additional research activities conducted mostly by students and doctoral candidates from several Indonesian and German partner institutions. A main underlying objective was to obtain local perspectives on contested issues, such as the involvement in and organization of destructive fishing practices. A participatory and transdisciplinary research approach was used, which allowed for the articulation of different local perceptions, opinions, and aspirations. The importance of the social science team's inclusive and enabling approach to generating new and relevant knowledge together with local stakeholders was acknowledged by several of the local respondents. These had previously voiced disappointment with government representatives visiting islands in the frame of a government-initiated program for the 'participatory' development of community-based MPAs. Actual participation in the official program planning process was found to be low (as was the resulting awareness of and compliance with the thus generated protected areas, see Glaser et al., 2010a). As one reason, a local respondent stated about the government representatives: "They always tell us what we cannot do, but never what we can do".

### 5.4.2 Investigating marine social–ecological feedbacks and dynamics

Similar to coral reef fisheries elsewhere in the Coral Triangle and beyond, the reefs of Spermonde are used in many different ways and covered by a range of multiple, sometimes overlapping, governance institutions. These include rules-in-use surrounding particular fishing gears, informal regulations pertaining to the reefs surrounding individual islands, village-level MPAs, and district-level regulations (Glaser et al., 2010a; Deswandi, 2012, p. 207; Gorris, 2016; see Chapter 11 for a more detailed description of the governance aspects). At least 24 distinct fishing methods (see Table A5.1, Fig. A5.1) are used in the reefs and surrounding waters, comprising a variety of net, hook-and-line and trap fishing techniques, gleaning, and spear fishing, as well as highly destructive methods (e.g., blast and cyanide fishing) that additionally pose significant health risks to the fishers (Ferse et al., 2014).

The different islands of the Spermonde Archipelago form a kaleidoscope of multiple fishing methods and cultures, with individual islands often concentrating on particular fishing methods. For example, there are islands with a particularly large number of bamboo trap fishers (Bonetambung), while others are hubs in the live reef fish fishery (e.g., Barrang Caddi) or the ornamental coral fishery (e.g., Barrang Lompo) with specific techniques and gears. In assessing the fishery, a SES analysis approach was taken to understand in particular the links between multiple ecological and social components of the focal system. In defining the SES(s) of interest, a problem-focused definition was used, comprising three elements: (1) a biogeophysical system (i.e., the Spermonde Archipelago), (2) its associated social agents with their institutions, which are not necessarily situated within the biogeophysical system, and (3) a specific problem context, such as the overuse of a particular resource (e.g., sea cucumbers) (Glaser et al., 2010b). Several of the reef resources in Spermonde are harvested for international, rather than local or national, markets. Thus, developments on international markets, such as changes in aquarium technology, demand for particular species or changes in the regulation of other important sources for target species and drive the dynamics of these fisheries (Ferse et al., 2014). Furthermore, many of the fisheries, in particular those that employ destructive and illegal gears and target overseas markets, are informally organized in a patron–client system, locally known as *punggawa/sawi* (Radjawali, 2011; Ferse et al., 2012). In this system, fishers form dyadic relationships with patrons that comprise financial loans and gear provided to fishers on credit by patrons. In return, fishers are expected to sell their catch to their particular patron at submarket prices in return for their loans. Patrons furthermore provide links to traders on other islands or in Makassar via their personal connections and, in the case of illegal fishing gear, provide a form of security from prosecution by relying on personal contacts to law enforcement agencies in Makassar (Radjawali, 2011). As fishers usually lack the connection to traders in Makassar and beyond, the link to a patron is a key mechanism for them to access markets beyond the local island, and in some cases to pursue illegal but lucrative fisheries. The patron–client institution shapes those fisheries, influencing target species and gears and modifying fishers' freedom of choice and room for maneuver (Ferse et al., 2014).

### 5.4.3 Reef-related livelihoods and implications for the present and future health of fishers and reefs

In contrast to the mostly specialized fisheries in temperate regions, fishers exploiting the coral reefs of Spermonde are often utilizing a range of fishing techniques, which they switch and combine based on, for example, demand, seasons, or weather patterns. While the vast majority of households on the islands depend in some way on marine resources, fishing is not necessarily the main, and certainly not the only, livelihood of most islanders (Ferse et al., 2014). For example, although the Archipelago is a main source region for ornamental corals, none of the ornamental coral collectors interviewed depended exclusively on the collection of corals, and most collected corals opportunistically, along with other species they deemed valuable (Ferse et al., 2012). Yet, marine resources constitute the mainstay of the majority of islanders' livelihoods, either directly via fishery, or indirectly via processing and trade. As many of the target species with the highest market value (e.g., sea cucumbers, ornamental corals, and groupers) are reef associated, the livelihoods of islanders depend directly on the future health and productivity of coral-dominated reef ecosystems. The poor soil and scarcity in available land on most of the inhabited islands leave little scope for alternative, non-reef–dependent livelihoods such as farming, with the exception of small garden plots on some of the islands. Freshwater resources are also increasingly scarce on many islands. This may be related to larger climate change–related drivers such as sea level rise but is also connected to increasing demand from rising numbers of inhabitants so that many Spermonde islands are approaching the limits of their carrying capacity (Schwerdtner Máñez et al., 2012). A further impairment to expanding livelihood options beyond marine resource use is the fact that many boys on the islands join their fathers or elder brothers in fishing at an early age, quitting school and thus foregoing options for formal employment, for example, in the nearby provincial capital Makassar (Glaser et al., 2015).

Prior to a blanket ban on all exports of ornamental corals in May 2018, over 70% of all corals in the marine aquarium trade originated from Indonesia. The Spermonde Archipelago was one of the four main collection areas for marine ornamentals in Indonesia (Ferse et al., 2012). According to a statement by the global trade group Ornamental Fish International, this ban has resulted in the loss of 10,000 jobs in the Indonesian aquarium coral fishery and trade within a year of being issued. In 2015, the use of compressors and scuba gear for diving was prohibited throughout the Archipelago (Hafez Muhammad and Sainab Husain Paragay, pers. comm. 2015). These bans underline the precariousness of reef resource-related livelihoods in Spermonde. While the versatility in gear use and access to credit and gear via patrons is likely to increase the adaptive capacity of fishers faced with such challenges, there is also the possibility of unintended consequences, such as increased indebtedness or the shift to high-yield destructive techniques. For example, Jaiteh et al. (2017) found that a ban on shark finning in Indonesia has resulted in an increase in high-risk activities such as blast fishing, illegal transboundary fishing, and people smuggling.

### 5.4.4 Changing target species, perceptions of reef resources, and implications for food security

The reef fisheries of Spermonde are highly dynamic, and opportunity and innovation appear to have been an important feature of these fisheries since the first fishers began settling the sandy islands of the Archipelago, which in some islands was over 200 years ago (Schwerdtner Máñez et al., 2012). Some species, such as sea cucumbers, have been targeted and exported for centuries, before recent changes in fishing technology (in this case, the introduction of compressor diving) allowed for the exploitation of previously untapped deeper parts of the reefs and for longer collection periods underwater. This resulted in a rapid increase in harvested amounts and an eventual overexploitation and decrease of the fishery as a whole (Schwerdtner Máñez and Ferse, 2010). The introduction of compressor diving and the overseas trade links established in export-oriented fisheries of earlier periods enabled the successive introduction of other fisheries using similar gears and trade routes, such as live reef fish and ornamental corals (Ferse et al., 2014). In general, the reef fisheries are highly versatile, and new techniques and target species are readily taken up by fishers. A key role in this is played by the fishing patrons, who enable connections to traders on other islands and to exporters in Makassar, introduce knowledge of new target species, and facilitate the uptake of new fishing methods. This is achieved by, for example, providing credit, gear, or information on the use of gears, gained by tapping into their extensive social networks including connections to traders in Makassar.

As a result of the versatility of the fishery and the rapid response to new market opportunities, a number of distinct peaks in fishing activity and production could be observed over time for a range of target species, such as sea cucumbers, live reef fish, ornamental corals, and moray eels (Schwerdtner Máñez and Husain Paragay, 2013; Ferse et al., 2014). Patrons are also key knowledge brokers in the reef fishery, acting as nodes in the flow of resources and information between international markets and exporters on one end of the trade chain, and individual fishers at the other. As a result, there is a disconnect in terms of information on the state of the reef ecosystem, which means that little information on ecological dynamics and feedbacks is available to actors further down the trade chain, such as traders and other patrons who are influential opinion leaders and decision-makers. Surveyors and employees of line agencies responsible for the management of fisheries and marine environments often lack resources for fishery monitoring in the field and instead rely on visits to fish markets or information from points of export, thus receiving distorted or incomplete information. The impairment or even lack of essential information flow on ecosystem changes from those who directly work with the ecosystem to main decision-makers is further aggravated by patrons who actively promote questionable views such as "it is impossible to overfish the sea" or "there will always be fish in the sea as long as there are leaves on trees," often invoking religious texts in doing so. The diversity of patrons trading different kinds of marine resources additionally instills a sense in fishers that there is a patron to sell each and every (marine) resource that might be fished or extracted to. This further exacerbates the overharvesting of reef resources (Ferse et al., 2014). The increased commodification of reef resources also has direct consequences for islanders' food security (Fig. 5.7): fish with a high nutritional value and high market value are frequently sold to Makassar rather than consumed locally, and financial returns are then used to buy processed food with lower nutritional value.

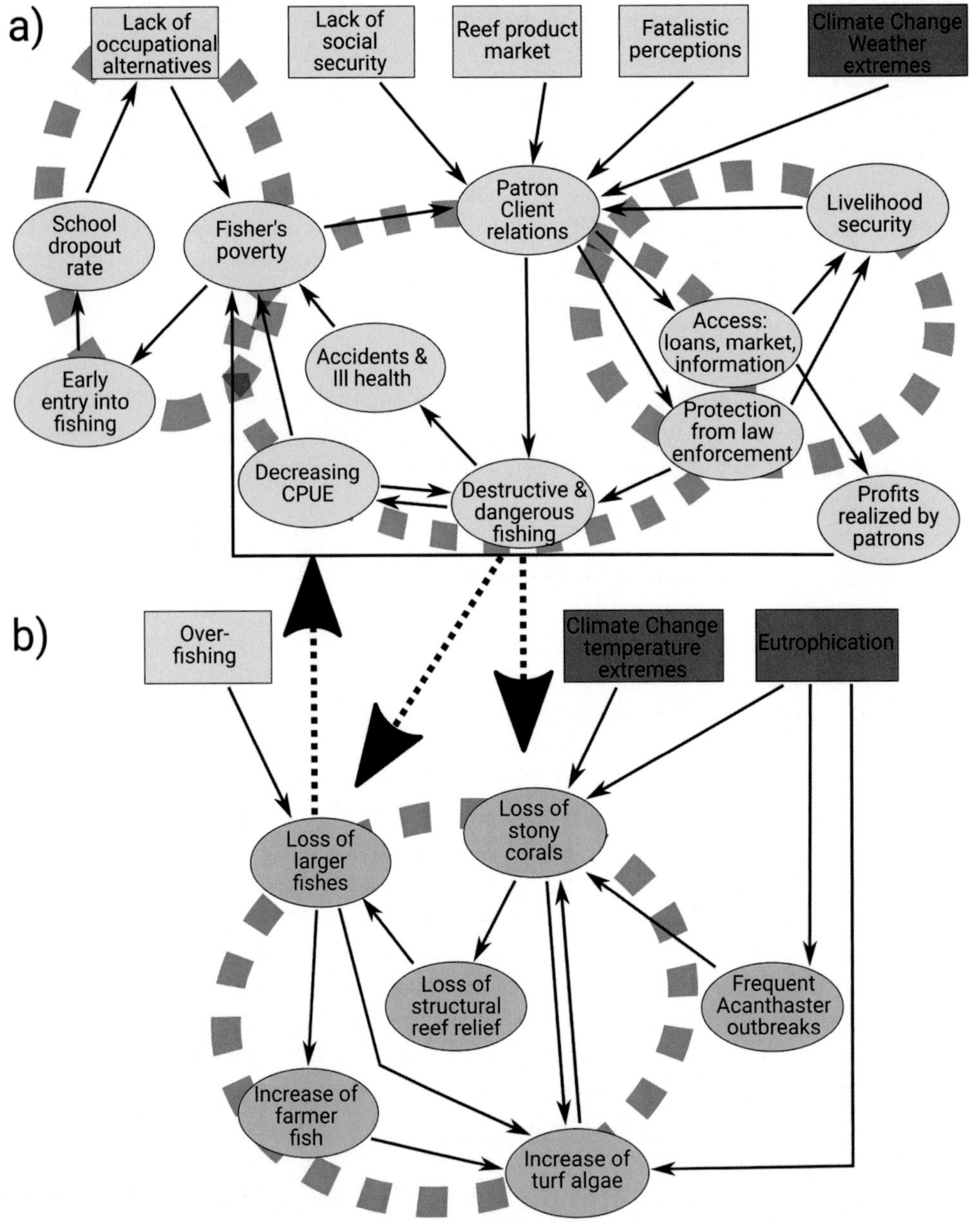


**FIGURE 5.7** In the Spermonde Archipelago, an interconnected set of predominantly social vicious cycles is locking the social–ecological system in an undesirable and increasingly resilient state. *From Glaser et al. (2018), Fig. 1.*

### 5.4.5 Conclusions for the management of coral reef resources in the Spermonde Archipelago

The social–ecological research conducted by the interdisciplinary team within the SPICE project generated three main conclusions. These all concern inclusiveness and are that (1) knowledge of the coral reef ecosystems of the Archipelago, (2) transparency of conservation-related benefits and their distribution, and (3) integration of coastal management planning and implementation need to be improved among and for all stakeholders (Glaser et al., 2015, see also Chapter 11 in this volume). A more socially and ecologically sustainable development and adaptive coastal governance is possible when these critical issues are addressed over time horizons that span beyond terms of political office. Local hierarchies and elites that form and maintain their own networks remain challenging on the road toward developing sustainable development options. In particular, the key roles played by fishing patrons need to be recognized and taken into account in the development of governance frameworks and management interventions as well as in the design of environmental and social policies. Patrons are the core knowledge brokers, enablers of unsustainable fishing, and importantly, the main providers of social security and adaptive capacity for vulnerable fishers. While many discussions have taken place on how to achieve meaningful participation in decision-making processes, a substantial and effective support in law enforcement (e.g., to reduce illegal, unreported, and unregulated fishing) seems to be the preferred and also more effective solution, not only for Spermonde but also for other regions in Indonesia. Finally, any governance and management interventions need to ensure a strengthening of local adaptive capacity and distribution of conservation benefits, which is regarded as legitimate and equitable by concerned stakeholders, if the danger of undermining the sustainability of the coral reef social–ecological system is to be avoided.

## 5.5 Modeling to support the management of reef systems

The SPICE project in the Spermonde Archipelago developed applications to improve the management of marine resources use in an integrative interdisciplinary modeling approach of reef processes and associated social drivers. The rationale for emphasizing model development lies in the importance of scientifically informed decisions, and the approach employed involves the extrapolation of dynamics and an application of possible scenarios, for instance, for environmental change or pollution trajectories. Together with the involvement of human actors and the consideration of institutional rule settings, the resulting simulations may provide valid projections to guide decisions in resource management. For the Spermonde Archipelago, these modeling efforts focused on managing the resilience of reef systems to human impact in the context of analyzing the effects of global climate change on reef dynamics by identifying preconditions for phase shifts in the context of local environmental conditions. Through the identification of driving factors, it is possible to distinguish between global and local factors, which can then be addressed through management (e.g., Kubicek et al., 2012). Thereby, the integration of information from different available sources proves immensely important not only for representing spatial heterogeneity but also for considering ecological processes on specific integration levels to facilitate the analysis of cross-level processes (Reuter et al., 2010). Ultimately, this should help to estimate the impact of human resource use under different management schemes and evaluate options for future pathways of change (mitigation, adaptation and transformation) under a set of potential environmental trajectories.

### 5.5.1 Simulating the impact of fisheries on coral reef dynamics

The spatially explicit simulation model **SEAMANCORE** (Spatially Explicit simulation model for Assisting the local MANagement of COral REefs, Miñarro et al., 2018) was developed to serve as a decision support tool for local resource management of Indo-Pacific coral reefs. A number of models address the effects of human impacts on ecological dynamics to inform management approaches. However, these models frequently ignore benthic or fish dynamics reaction to external anthropogenic stressors, and virtually none offers a user-friendly platform for nonscientist managers to access easily (Miñarro et al., 2018).

The model developed in this SPICE project may be used to explore the likely outcomes of different combinations of resource management and environmental scenarios by simulating the population dynamics of a coral reef community in reaction to environmental factors and different fishing approaches. It represents the spatial dynamics of a coral reef under the influence of local and global stressors, focusing on selected fish and benthic functional groups, the ecological relationships among them, and how they are affected by external stressors. The model integrates the available knowledge (for instance, social drivers and management strategies) derived in the SPICE projects in Spermonde and in past studies on Indonesian reef systems.

Modeling studies on reef benthos dynamics have targeted different stressors on reef dynamics (e.g., Mumby et al., 2007; Sandin and McNamara, 2012), its resilience (Bozec and Mumby, 2015), and dynamics of coral diseases (Brandt and McManus, 2009) but often exclude the feedback processes with upper trophic levels, e.g., fish functional groups. Models linking fish and benthic dynamics (e.g., Ainsworth and Mumby, 2015; Rogers et al., 2014) are frequently not spatially explicit, thus missing specific impacts on fishing grounds. The small-scale approach of SEAMANCORE is aimed at local management. Many models used for managing marine resources are more spatially aggregated to cover entire jurisdictional areas (e.g., Buddemeier et al., 2008; Weijerman et al., 2015). However, in the case of tropical islands, community-based management is important, and in Spermonde as elsewhere, it is often determined locally (McClanahan et al., 2006; Mumby and Steneck 2008; Glaser et al., 2010a).

For Pacific coral reef models, parameterization is challenging due to lack of data until the 1980s and regionally varying historical baseline data (Bruno and Selig, 2007).

Exceptions for model studies for the region include an ECOSIM model by Ainsworth et al. (2008) targeting management research priorities for Raja Ampat and a spatially explicit mean-field model (Gurney et al., 2013) assessing bleaching scenarios relating to the effect of fisheries on coral reef dynamics. Building upon earlier models developed to aid local coral reef management (e.g., Buddemeier et al., 2008; Holmes and Johnstone, 2010; Weijerman et al., 2015), we propose a spatially explicit simulation model as a new user-friendly ecological tool to assess the effects of simultaneous global and local stressors on coral reef communities. As an extension to the small-scale processes of coral–algae interaction developed in Kubicek et al. (2012), Kubicek et al. (2019), components, such as selected fish and benthic functional groups, are added as well as how they are affected by external stressors. The model may be used to explore likely outcomes of different resource management and environmental scenarios by simulating the population dynamics of a coral reef and the effects of different fishing methods. Results thus include assessing the trade-offs involved in the different options for the reef fishery regulations.

SEAMANCORE was designed to assess the influence of local processes at the coral reef scale, such as spatial competition outcomes or fishery exploitation, coupled with the effects of management strategies. It is implemented as a two-dimensional continuous grid representing the benthos (coral and algae interaction) and three interacting functional fish groups (Browsers & Grazers, Scrapers, and Carnivores). Currently, the maximum grid size is $1000 \times 1000$ m (1 $km^2$). The model represents interactions within the benthos (e.g., corals, algae) as well as trophic interactions within the fish groups and with the benthos. Different environmental scenarios and scenarios of fisheries (i.e., fishing methods and intensities) may be specified to represent different gear choices, including both legal and illegal fishing methods. Fishing is explicitly modeled with parameters for intensity, zonation, and selectivity, while the other stressors, bleaching and eutrophication, are more broadly specified through scenario settings. A gray-scale bathymetric map of the field site may be set up to define depth categories influencing the probability of occurrence of organisms and ecological processes and to determine whether the grid cells will be affected by stressors.

The model is parameterized to represent coral reefs of the Indo-Pacific, and input parameters and initial conditions can be changed according to the case-specific abundance of functional groups, fishing pressure, and biophysical conditions, such as the depth profile, bleaching frequency, and nutrient input. The flexibility in its input parameters facilitates scenario testing by local stakeholders. The current exemplary study is parameterized using field data collected from islands of the Spermonde Archipelago exposed to different stressor combinations and intensities (Miñarro et al., 2018). The model is designed with a user-friendly menu-based interface that does not require programming experience for simulating numerous scenarios that can be customized with depth profile maps and initial coral reef conditions of fish and benthos functional group abundance. All detailed settings of the model are accessible in the advanced mode. The model can be used to explore the likely trajectories of a specific coral reef ecosystem under different scenarios of depth profile, nutrient regimes, bleaching frequency, and fishery management strategies.

In a first application, the model was used to explore various scenarios of stressors, fishing, and bleaching regimes in a virtual coral reef, using approaches defined by Kubicek et al. (2015) to validate the model. Being spatially explicit, reef zonation patterns emerged as a result of model rules. The model was tested using an extensive sensitivity analysis for evaluating the influence of uncertainty due to parameter variability to ensure its usage under different scenario conditions (Fig. 5.8). Outputs include the relative abundances of key functional groups over time, which are commonly accepted indicators of ecosystem health and comparable against target conservation values or acceptable published levels for such abundances.

Specifically, the simulations show the expected complex behavior of the reef and some of its associated fish species with a number of positive and negative feedback

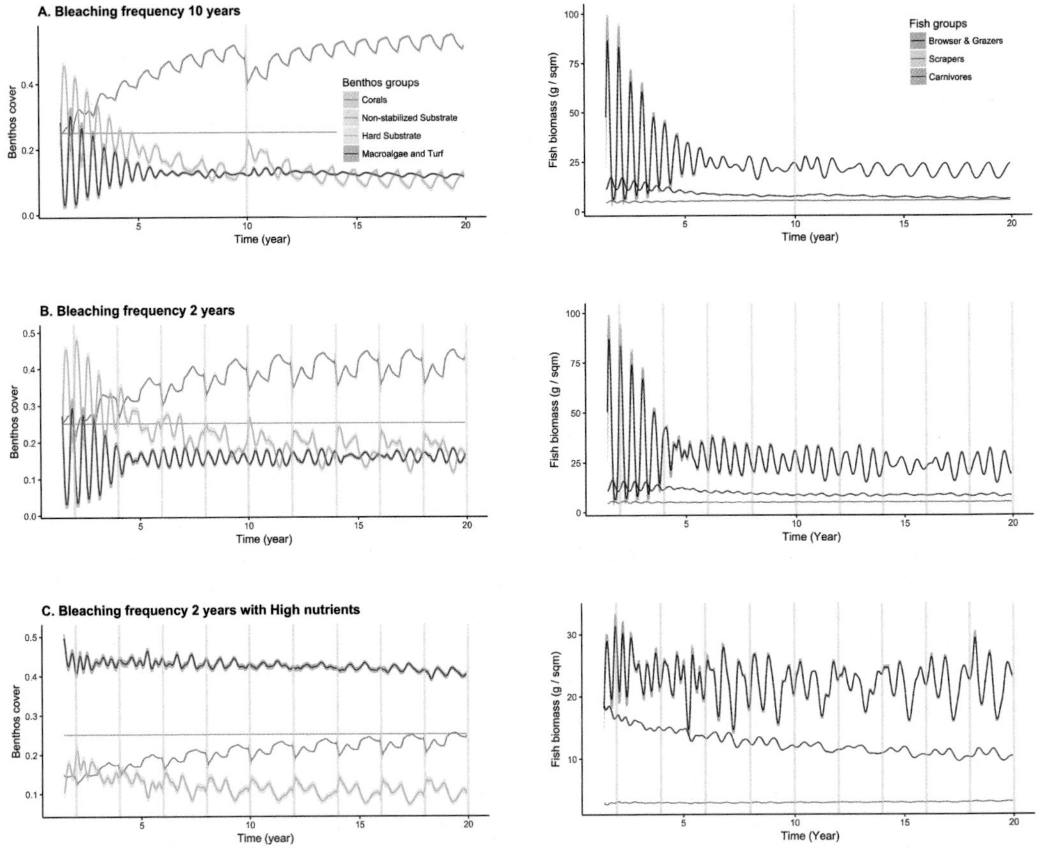


**FIGURE 5.8** Results from SEAMANCORE model runs for different bleaching frequencies and nutrient levels. Three scenarios of bleaching frequency: every 10 (A), 2 (B), and 2 years with high nutrients (C). The time series considered 20 years, benthos cover is given as a proportion, fish biomass in g $m^{-2}$. All data are averaged across the three depth categories of the depth map. *From Miñarro et al. (2018), Fig. 5.*

relationships that shape the trajectories of benthos dynamics and fish functional group in difficult to foresee ways. In particular, the feedback processes between coral cover and fish biomass (Graham et al., 2006; Sandin et al., 2008) are shaping the dynamics, but also the interactions involving the benthos. Simulations show the reef capacity to sustain moderate levels of disturbance below the threshold of natural variation, but a severely reduced capacity to cope if scenarios combine high nutrient levels with high bleaching frequencies (Miñarro et al., 2018). For future applications, we expect that linking fish demographics with changing habitat quality derived from simulating feedback processes on and within benthos will prove insightful for fisheries management.

### 5.5.2 A model on gear choices of fishermen

For the Spermonde Archipelago, the immense importance of the patron–client system for organizing fisheries with respect to gear choice and target organisms is well documented (e.g., Ferse et al., 2012; Ferrol-Schulte et al., 2014; Miñarro et al., 2016, see also Chapters 5.4 and 11). Artisanal fishermen (clients) are obliged to sell their catch to fish traders (patrons) who act as middlemen for the regional market. Clients are often highly indebted to the patrons and trade social security in times of low catches for targeting specific fish groups demanded by patrons, often with concessions to patrons in terms of product price received. This system is thought to potentially put high pressure on stocks, thereby risking unsustainable exploitation strategies (Nurdin and Grydehøj, 2014).

The modeling study by Leins (2017) supplements the simulation model by Miñarro et al. (2018) with a detailed social–ecological component and investigates the implications of the patron–client system on the sustainability of fishing in Spermonde. The model represents the social system of fisheries in the Spermonde Archipelago with respect to patron–client dependencies as well as independent fishers and depicts different fishing methods, their economic implications for the fishers, and their effects on spatially distributed resources.

Fishers are represented as autonomously acting agents with a distinct behavioral repertoire (activity rules), which depends on their social status, available assets and equipment, as well as acquired knowledge and external conditions. The actions of a fisher may again change the model's status and environmental settings. The knowledge of a fisher agent is distinguished into three categories: (1) *general knowledge* that has fixed values and equally applies for all fishers; (2) *individual knowledge* that changes depending on the individual fisher's actions and experience, e.g., catch history, own economic situation, and (3) *perception* of global and local conditions from the (partial) point of view of an individual fisher. Two types of fish resources (high and low value) are spatially distributed on the map, thus characterizing the fishing grounds. Dynamics of a stock depend on its specific growth function as well as on exploitation intensity and choice of gear, thus determining whether a stock thrives or decreases with time.

In the model, each fisher decides every day upon their activities depending on their knowledge. This involves, for instance, the decision to go fishing, which fishing ground to target, how long to stay there, and when to leave. The experiences from these activities will then complement the fisher's knowledge and directly impact his household's economic situation, for instance, depending on catch values and costs of the fishing trip. Ultimately, fishers may also change their social status (dependent or independent fisher), depending on their economic trajectory. Model rules and parameterization were very much based on the knowledge derived in the SPICE project by many project partners (e.g., Deswandi, 2012, p. p207; Ferse et al., 2012; Miñarro, 2017; Miñarro et al., 2016; Navarrete-Forero et al., 2017; Radjawali, 2012).

We applied the model for a set of different scenarios relating to potential management strategies, such as increasing awareness to reduce illegal fishing activities, economic setting affecting fish prices, or strict law enforcement (e.g., intensive control of illegal activities and different values for fines). Results indicated that for none of the scenarios featuring a majority of dependent fishers, it was possible to concurrently achieve recovering reef quality and healthy fish stocks. Interestingly, with the implemented algorithms, an increase in law enforcement also increased the economic stress of fishers, leading to intensified exploitation (Leins, 2017). Here, and also for details of the financial agreements/transaction between patrons and clients, limited knowledge restricts model implementation. Further studies should also include alternative livelihood options and the behavior of the patrons who have strong personal interest to keep the system intact. Thus, an extended model could facilitate more detailed simulations with a wider set of potential management strategies toward a sustainable and economically and socially beneficial artisanal coral reef fishery in the Spermonde Archipelago.

## 5.5.3 Spatial patterns of fishing ground distribution

Artisanal fisheries in the Spermonde Archipelago nearly exclusively target reef species and exhibit a highly diverse approach with localized impacts on reefs (see Chapter 5.4). A fundamental issue for impact assessment is the explicit extent and location of fishing activities, which can then be related to the status of specific coral reefs and allow for the establishment of a direct relation between ecosystem state and the spatial extent and effect of specific fishing methods. This information is needed to understand diverging spatial dynamics in reefs and derive targeted management recommendations such as spatial regulations, for instance, for MPAs. Such studies demand regular fisheries monitoring and the identification of the missing spatial information on resource use impact to relate information on habitat quality to specific anthropogenic activities.

In comparison with industrial fishing enterprises, artisanal fishers and fisheries managers only rarely use technical approaches to determine fishing locations. Their methods often include anecdotal information derived from other fishers (e.g., Gorris, 2016) or conducting a census using patrols or research vessels (Pet-Soede et al., 2001; Turner et al., 2015). These approaches either lack precision or are very cost and time intensive. Small and easy-to-use GPS trackers, however, provide a precise tool to quantify and locate fishing at a relatively low cost involving participatory sampling (Metcalfe et al., 2016).

In the SPICE project, we used a combination of measuring catch landings (biomass and species composition), interviews of hook and line fishermen about the use of the different fishing grounds, and the distribution of GPS data loggers among fishers to identify fishing grounds by gear-dependent patterns of boat movement (Navarrete Forero et al., 2017). The study was conducted during the wet season 2014/15 on Badi Island of the Spermonde Archipelago. The tracks of fishermen were used to generate a map of fishing grounds, thereby also indicating how frequently and for what time period each of these fishing grounds was visited. As typical boat speeds can be related to specific fishing gears, the movement information of the GPS tracks allowed the derivation of a spatial pattern of gear used and target species. This was then evaluated using catch composition assessment of landings by participating fishermen.

The results of this study show that patterns of fishing grounds' distribution can be related to target species and catch composition, and results may be interpreted in the context of patron–client relationships that structure the artisanal fisheries in Spermonde (Miñarro et al., 2016). Fishers traveled up to 100 km per day in small engine-powered canoes. Most of the fishing activities by fishers of the Island of Badi involved two gears (octopus bait and trolling line for live groupers) and three major fishing grounds (Fig. 5.9). These areas were located northwest of Badi, the home port, and on the way to Kondongbali Island. Typically, octopus dominated the catch landings in terms of biomass (CPUE = 10.1 kg boat day$^{-1}$), whereas groupers (Serranidae, CPUE = 1.7 kg boat

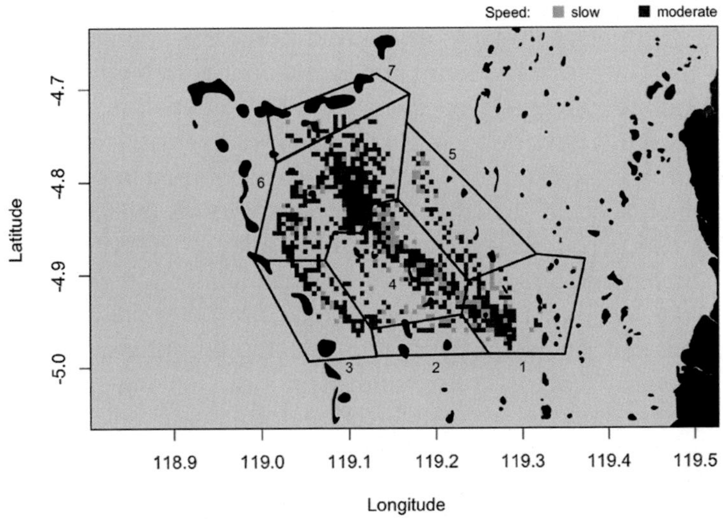


**FIGURE 5.9** Map of fishing activity of hook and line fishermen from Badi Island, Spermonde Archpelago. Cell color represents the prevalent fishing speed (slow: 0–3 km h$^{-1}$, moderate 4–8 km h$^{-1}$) associated with different gears. Polygon numbers represent names of fishing grounds: (1) Badi, (2) Lumu-lumu, (3) Lanyukan, (4) Tintingan, (5) Sarappo, (6) Malla'bang, (7) Kondongbali. *From Navarrete-Forero et al. (2017), Fig. 3.*

day$^{-1}$), targeted for their high market value (around 42 $ per kg vs. 2.3 $ per kg for Octopus), gave the highest revenues. The Serranidae fish family also constituted the most diverse group in the landed catch with the leopard coral grouper (*Plectropomus leopardus*) being the species with the highest catch statistics (52.5%).

The study by Navarrete Forero et al. (2017) shows that boat tracking with cheap and easy-to-use GPS tracker is a powerful tool that provides information on spatial resource use when combined with interviews and catch surveys. In this case, the study was benefited from a high level of participation by fishermen.

## 5.6 Summary and outlook

Located at the heart of the Coral Triangle, the Spermonde Archipelago was a treasure trove of marine biodiversity. But anthropogenic impacts from a growing coastal population, along with global warming, have led to a widespread demise of coral reefs and the related ecosystem services. The complex geography, great variety of stakeholders, and a number of threats including the unsustainable and destructive exploitation of resources and pollution pose serious challenges to the management of the archipelago. Results from the research of the SPICE project show the following: (1) The ecology of the archipelago is characterized by widespread habitat degradation and several documented environmental gradients that drive differences in species distribution. (2) Missing livelihood alternatives and a system of middlemen (patrons) and dependent clients stabilize unsustainable approaches to resource use by the islanders of the archipelago. (3) The analysis of genetic population structure and connectivity among populations in the Spermonde Archipelago provides evidence for different connectivity patterns in the outer-, middle-, and inner-shelf systems and separation of the southern inner-shelf reefs. Beyond the Spermonde Archipelago, especially the outer-shelf reefs are well connected to reefs upstream the ITF. (4) Mechanistic models allow for the integration of specific characteristics of the system, the most relevant components to sustainable reef management, their specific dynamics or life histories, and interactions to test the ecosystem-wide effects of management schemes under different biophysical scenarios.

While the implementation of effective management strategies may provide improvements at the local scale, small-scale efforts are potentially undermined by the impact of regional and global stressors, such as pollution inflows from the hinterland, outbreaks of crown-of-thorns starfish, and heat waves so that a return from algae-dominated (or other benthos groups) to coral-dominated reefs will be increasingly difficult in the future. However, research in the SPICE program has also shown that some resilient corals are able to survive in marginal environments and polluted coastal waters. This resilience might provide some frontloading to other stressors such as warming and acidification, so that despite the observed loss in coral cover and diversity, it could be possible that the species-rich coral community may be replaced by an impoverished, but resilient coral community in the near future.

Despite the improved understanding of the social–ecological system of Spermonde Archipelago generated by the SPICE program and many other actors, more investigations are needed to monitor the state of the reefs, changes in the coral community, and the processes of acclimation and adaptation to a changing environment. A better understanding of social networks and the governance system is also required, and resilient but undesirable social–ecological dynamics need to be addressed.

Our research also highlights that more information, in particular on spatial differences in resource use intensity and feedback processes with benthos, is needed to increase the scope of the developed model. Last but not least, stakeholder involvement and participation of local actors in decisions should be more extensively employed to both increase knowledge exchange between science, practitioners, and decision-makers and support capacity development on many different levels. This would facilitate the co-development of context-appropriate rules for reef use and fishing behavior and of innovative alternative livelihoods to provide better conditions for sustainable resource use.

Knowledge gaps and directions of future research

The SPICE project in the Spermonde Archipelago pursued a holistic interdisciplinary investigation of coral reef dynamics under a range of environmental constraints and scenarios of anthropogenic exploitation and pollution. All the actors in the SPICE program and many more generated an improved understanding of the social–ecological system of the Spermonde Archipelago through their research.

However, more and detailed investigations are needed to

- monitor the state of the reefs, changes in the coral community and the processes of acclimation and adaptation to a changing environment;
- specifically generate an improved understanding of social networks and the governance system to facilitate a better knowledge exchange between science, practitioners, and decision-makers for developing codevelopment of context-appropriate rules for reef use and fishing behavior and of innovative alternative livelihoods; and
- increase the projection accuracy of the developed models by including more detailed information and processes, in particular on spatial differences in resource use intensity and the feedback processes with benthos communities.

We are certain that all these further research activities together would provide valuable support for a sustainable resource use in the Spermonde Archipelago.

Implications/recommendations for policy and society

Investigations during SPICE showed that the Spermonde Archipelago is characterized by a severe degradation of its marine habitats. Additionally, we identified environmental gradients from the coast to the outer-shelf islands which shape benthic as well as fish communities. Missing livelihood alternatives and depleting marine resources result in island populations increasing efforts to sustain livelihood and the necessity of utilizing new resources.

*Continued*

Implications/recommendations for policy and society—cont'd

Key management recommendations include the following critical issues:

- Recognition of the multifacetted key role played by fishing patrons. This needs to be taken in account in developing concepts for environmental and social policies and in the development of governance frameworks and management interventions.
- Extended knowledge of effects of intensive exploitation on ecology of reef systems is required. However, small-scale efforts for sustainable management are potentially undermined by regional and global stressors. These will have to be addressed on the according spatial scales and political levels.
- Any development of a management framework should ensure a strengthening of local adaptive capacity and distribution of conservation benefits. The considered time horizons should span beyond terms of political office to ensure a more socially and ecologically sustainable development and adaptive coastal governance.

As one of the most important points, stakeholder involvement and participation of local actors in decisions should be more extensively employed. The increased knowledge exchange between science, practitioners, and decision-makers should aim at facilitating the co-development of context-appropriate rules for sustainable resource use in the Archipelago.

# Acknowledgments

The bilateral Indonesian–German research program SPICE was conducted under a governmental agreement between the German Federal Ministry of Education and Research (BMBF), the Indonesian Ministry of Maritime Affairs and Fisheries (KKP), and the Indonesian Ministry for Research and Technology (RISTEK) with involvement of many more governmental bodies. We thank all of them for their support, providing permissions and the funding (BMBF grant nos. 03F0390A-B, 03F0472A-C, 03F0474A, 03F0643A-B). We also thank the Projektträger Jülich for their extremely helpful support with the project organization. Numerous PhD, master, and bachelor students, as well as student helpers and lab technicians, have decisively supported research in this project. We thank them very much for this contribution. For continuing interest, support, kindness, and hospitality for over more than a decade of collaboration, we cordially thank the Spermonde Islanders and their leaders. We also thank our two reviewers for stimulating and encouraging comments.

# References

Agung, M.H., Prabuning, D., Yudiarso, P., Dewanto, H.Y., Sari, S.K., Kimura, T., Tun, K., Chou, L.M. (Eds.), 2018. Status of Coral Reefs in East Asian Seas Region. Ministry of the Environment of Japan and Japan Wildlife Research Center, Tokyo, Japan, p. 58 (Chapter 2.2).

Ainsworth, C.H., Mumby, P., 2015. Coral-algal phase shifts alter fish communities and reduce fisheries production. Global Change Biology 21 (1), 165–172.

Ainsworth, C.H., Varkey, D.A., Pitcher, T.J., 2008. Ecosystem simulations supporting ecosystem-based fisheries management in the Coral triangle, Indonesia. Ecological Modelling 214 (2–4), 361–374.

Alcazar, D.S.R., Kochzius, M., 2016. Genetic population structure of the blue sea star *Linckia laevigata* in the Visayas (Philippines). Journal of the Marine Biological Association of the United Kingdom 96, 707–713.

Allen, G.R., 2008. Conservation hotspots of biodiversity and endemism for Indo-Pacific coral reef fishes. Aquatic Conservation: Marine and Freshwater Ecosystems 18, 541–556.

Almany, G.R., Planes, S., Thorrold, S.R., Berumen, M.L., Bode, M., Saenz-Agudelo, P., et al., 2017. Larval fish dispersal in a coral-reef seascape. Nature Ecology and Evolution 1, 0148.

Arias-Gonzalez, J.E., Johnson, C., Seymour, R.M., Perez, P., Alino, P., 2011. Scaling up models of the dynamics of coral reef ecosystems: an approach for science-based management of global change. In: Dubinsky, Z., Stambler, N. (Eds.), Coral Reefs: An Ecosystem in Transition. Springer, pp. 373–390.

Asian Development Bank, 2014a. *Regional State of the Coral Triangle—Coral TRIANGLE Marine Resources: Their Status, Economies, and* Management. Mandaluyong City, Philippines.

Asian Development Bank, 2014b. State of the Coral Triangle: Indonesia. Mandaluyong City, Philippines.

Baum, G., Januar, H.I., Ferse, S.C.A., Kunzmann, A., 2015. Local and regional impacts of pollution on coral reefs along the thousand islands north of the Megacity Jakarta, Indonesia. PLoS One 10, e0138271.

Becking, L., Cleary, D., de Voogd, N., Renema, W., de Beer, M., van Soest, R.W.M., Hoeksema, B.W., 2006. Beta diversity of tropical marine benthic assemblages in the Spermonde Archipelago, Indonesia. Marine Ecology 27, 67–88. https://doi.org/10.1111/j.1439-0485.2005.00051.x.

Berumen, M.L., Almany, G.R., Planes, S., Jones, G.P., Saenz-Agudelo, P., Thorrold, S.R., 2012. Persistence of self-recruitment and patterns of larval connectivity in a marine protected area network. Ecology and Evolution 2, 444–452.

Borell, E., Bischof, K., 2008. Feeding sustains photosynthetic quantum yield of a scleractinian coral during thermal stress. Oecologia 157 (4), 593–601.

Borell, E.M., Yuliantri, A.R., Bischof, K., Richter, C., 2008. The effect of heterotrophy on photosynthesis and tissue composition of two scleractinian corals under elevated temperature. Journal of Experimental Marine Biology and Ecology 364 (2), 116–123.

Bowen, B.W., Rocha, L.A., Toonen, R.J., Karl, S.A., 2013. The origins of tropical marine biodiversity. Trends in Ecology and Evolution 28, 359–366.

Bozec, Y.M., Mumby, P.J., 2015. Synergistic impacts of global warming on the resilience of coral reefs. Philosophical Transactions of the Royal Society B: Biological Science 370 (1659), 20130267.

Brandt, M.E., McManus, J.W., 2009. Dynamics and impact of the coral disease white plague: insights from a simulation model. Diseases of Aquatic Organisms 87 (1–2), 117–133.

Bruckner, A.W., Borneman, E.H., 2006. Developing a sustainable harvest regime for Indonesia's stony coral fishery with application to other coral exporting countries. In: Proceedings of the 10th International Coral Reef Symposium, pp. 1692–1697.

Bruno, J.F., Selig, E.R., 2007. Regional decline of coral cover in the Indo-Pacific: timing, extent, and subregional comparisons. PLoS One 2 (8), e711.

Buddemeier, R.W., Jokiel, P.L., Zimmerman, K.M., Lane, D.R., Carey, J.M., Bohling, G.C., et al., 2008. A modeling tool to evaluate regional coral reef responses to changes in climate and ocean chemistry. Limnology and Oceanography: Methods 6 (9), 395–411.

Burke, L., Reytar, K., Spalding, M., Perry, A., 2011. Reefs at Risk Revisited. World Resource Institute, Washington, DC.

Burke, L., Reytar, K., Spalding, M., Perry, A., 2012. Reefs at Risk Revisited in the Coral Triangle. World Resource Institute, Washington, DC.

Burke, L., Selig, E., Spalding, M., 2002. Reefs at Risk in South-East Asia. World Resource Institute, Washington, DC.

Carilli, J.E., Norris, R.D., Black, B.A., Walsh, S.M., McField, M., 2009. Local stressors reduce coral resilience to bleaching. PLoS One 4 (7), e6324.

Cleary, D.F.R., Becking, L.E., de Voogd, N.J., Renema, W., de Beer, M., van Soest, R.W.M., et al., 2005. Variation in the diversity and composition of benthic taxa as a function of distance offshore, depth and exposure to the Spermonde Archipelago, Indonesia. Estuarine, Coastal and Shelf Science 65, 557–570.

Cleary, D.F.R., Renema, W., 2007. Relating species traits of foraminifera to environmental variables in the Sper monde Archipelago, Indonesia. Marine Ecology Progress Series 334, 73–82. https://doi.org/10.3354/meps334073.

Cornils, A., Schulz, J., Schmitt, P., Lanuru, M., Richter, C., Schnack-Schiel, S.B., 2010. Mesozooplankton distribution in the Spermonde archipelago (Indonesia, Sulawesi) with special reference to the Calanoida (Copepoda). Deep Sea Research Part II: Topical Studies in Oceanography 57, 2076–2088.

Cowen, R.K., Sponaugle, S., 2009. Larval dispersal and marine population connectivity. Annual Review of Marine Science 1, 443–466.

Crandall, E.D., Riginos, C., Bird, C., Liggins, L., Treml, E., Beger, M., et al., 2019. The molecular biogeography of the Indo-Pacific: testing hypotheses with multispecies genetic patterns. Global Ecology and Biogeography 28, 943–960.

de Voogd, N., Cleary, D., Hoeksema, B., Noor, A., van Soes, R., 2006. Sponge beta diversity in the Spermonde Archipelago, SW Sulawesi, Indonesia. Marine Ecology Progress Series 309, 131–142. https://doi.org/10.3354/meps309131.

Deswandi, R., 2012. Understanding Institutional Dynamics: The Emergence, Persistence, and Change of Institutions in Capture Fisheries in Makassar, Spermonde Archipelago, South Sulawesi, Indonesia (Ph.D. thesis). University of Bremen, p. p207.

Dohna, T.A., Timm, J., Hamid, L., Kochzius, M., 2015. Limited connectivity and a phylogeographic break characterize populations of the pink anemonefish, *Amphiprion perideraion*, in the Indo-Malay archipelago: inferences from a mitochondrial and microsatellite loci. Ecology and Evolution 5, 1717–1733.

Edinger, E.N., Kolasa, J., Risk, M.J., 2000. Biogeographic variation in coral species diversity on coral reefs in three regions of Indonesia. Divers Distribution 6, 113–127. https://doi.org/10.1046/j.1472-4642.2000.00076.x.

Erdman, L., Pet-Soede, M., 1997. How fresh is too fresh? The live reef food fish trade in eastern Indonesia. Live Reef Fish Information Bulletin 3, 41–45.

Ferrol-Schulte, D., Ferse, S.C.A., Glaser, M., 2014. Patron-client relationships, livelihoods and natural resource management in tropical coastal communities. Ocean and Coastal Management 100, 63–73.

Ferrol-Schulte, D., Wolff, M., Ferse, S., Glaser, M., 2013. Sustainable livelihoods approach in tropical coastal and marine social–ecological systems: a review. Marine Policy 42, 253–258. https://doi.org/10.1016/j.marpol.2013.03.007.

Ferse, S.C.A., Glaser, M., Neil, M., Schwerdtner Máñez, K., 2014. To cope or to sustain? Eroding long-term sustainability in an Indonesian coral reef fishery. Regional Environmental Change 14, 2053–2065.

Ferse, S.C.A., Knittweis, L., Krause, G., Maddusila, A., Glaser, M., 2012. Livelihoods of ornamental coral fishermen in South Sulawesi/Indonesia: implications for management. Coastal Management 40, 525–555.

Glaser, M., Baitoningsih, W., Ferse, S.C.A., Neil, M., Deswandi, R., 2010a. Whose sustainability? Top-down participation and emergent rules in marine protected area management in Indonesia. Marine Policy 34, 1215–1225.

Glaser, M., Breckwoldt, A., Deswandi, R., Radjawali, I., Baitoningsih, W., Ferse, S.C.A., 2015. Of exploited reefs and Fishers – a holistic view on participatory coastal and marine management in an Indonesian archipelago. Ocean and Coastal Management 116, 193–213.

Glaser, M., Ferse, S., Neil, M., Plass-Johnson, J., Satari, D.Y., Teichberg, M., Reuter, H., 2018. Breaking resilience for a sustainable future: Thoughts for the Anthropocene. Frontiers in Marine Science 5, 34.

Glaser, M., Krause, G., Halliday, A., Glaeser, B., 2012. Towards global sustainability analysis in the Anthropocene. In: Glaser, M., Krause, G., Ratter, B.M.W., Welp, M. (Eds.), Human-nature Interaction in the Anthropocene: Potentials of Social-Ecological Systems Analysis. Routledge, pp. 193–222 (Chapter 10).

Glaser, M., Krause, G., Oliveira, R.S., Fontalvo-Herazo, M., 2010b. Mangroves and people: a social-ecological system. In: Saint-Paul, U., Schneider, H. (Eds.), Mangrove Dynamics and Management in north Brazil. Springer, Heidelberg, Germany, pp. 307–351.

Glaser, M., Radjawali, I., Ferse, S.C.A., Glaeser, B., 2010c. Nested' participation in hierarchical societies? Lessons for social-ecological research and management. International Journal of Society Systems Science 2, 390–414.

Gordon, A.L., 2005. Oceanography of the Indonesian seas and their throughflow. Oceanography 18, 14–27.

Gordon, A.L., Fine, R.A., 1996. Pathways of water between the Pacific and Indian oceans in the Indonesian seas. Nature 379, 146–149.

Gorris, P., 2016. Deconstructing the reality of community-based management of marine resources in a small island context in Indonesia. Frontiers in Marine Science 3. https://doi.org/10.3389/fmars.2016.00120.

Graham, N.A., Wilson, S.K., Jennings, S., Polunin, N.V., Bijoux, J.P., Robinson, J., 2006. Dynamic fragility of oceanic coral reef ecosystems. Proceedings of the National Academy of Sciences of the United States of America 103 (22), 8425–8429.

Gurney, G.G., Melbourne-Thomas, J., Geronimo, R.C., Aliño, P.M., Johnson, C.R., 2013. Modelling coral reef futures to inform management: can reducing local-scale stressors conserve reefs under climate change? PLoS One 8 (11), e80137.

Hoeksema, B., 2007. Delineation of the Indo-Malayan centre of maximum marine biodiversity: the coral triangle. In: Renema, W. (Ed.), Biogeography Time, and Place: Distributions, Barriers, and Islands, pp. 117–178.

Hoeksema, B., 2012. Distribution patterns of mushroom corals (Scleractinia: Fungiidae) across the Spermonde shelf, South Sulawesi. The Raffles Bulletin of Zoology 60, 183–212.

Hogarth, P.J., 1999. The Biology of Mangroves. Oxford University Press, Oxford, England.

Holmes, G., Johnstone, R.W., 2010. Modelling coral reef ecosystems with limited observational data. Ecological Modelling 221 (8), 1173–1183.

Hughes, T.P., Barnes, M.L., Bellwood, D.R., Cinner, J.E., Cumming, G.S., Jackson, J.B.C., et al., 2017. Coral reefs in the anthropocene. Nature 546, 82–90.

Hui, M., Kochzius, M., Leese, F., 2012. Isolation and characterisation of nine microsatellite markers in the boring giant clam (*Tridacna crocea*) and cross-amplification in five other tridacnid species. Marine Biodiversity 42, 285–287.

Hui, M., Kraemer, W.E., Seidel, C., Nuryanto, A., Joshi, A., Kochzius, M., 2016. Comparative genetic population structure of three endangered giant clams (Tridacnidae) throughout the Indo-west Pacific: implications for divergence, connectivity, and conservation. Journal of Molluscan Studies 82, 403–414.

Hui, M., Nuryanto, A., Kochzius, M., 2017. Concordance of microsatellite and mitochondrial DNA markers in detecting genetic population structure in the boring giant clam, *Tridacna crocea*, across the Indo-Malay Archipelago. Marine Ecology - An Evolutionary Perspective 38, e12389.

Huyghe, F., 2018. Evolutionary and Ecological Connectivity of the Skunk Clown Fish in the Indian Ocean and Their Importance for the Design of marine Protected Areas (Ph.D. Thesis). Vrije Universiteit Brussel (VUB), Belgium.

Huyghe, F., Kochzius, M., 2017. Highly restricted gene flow between disjunct populations of the skunk clownfish (*Amphiprion akallopisos*) in the Indian Ocean. Marine Ecology - An Evolutionary Perspective 38, e12357.

Huyghe, F., Kochzius, M., 2018. Sea surface currents and geographic isolation shape the genetic population structure of a coral reef fish in the Indian Ocean. PLoS One 13 (3), e0193825.

Jaiteh, V.F., Loneragan, N.R., Warren, C., 2017. The end of shark finning? Impacts of declining catches and fin demand on coastal community livelihoods. Marine Policy 82, 224–233.

Jones, G.P., Almany, G.R., Russ, G.R., Sale, P.F., Steneck, R.S., van Oppen, M.J.H., et al., 2009. Larval retention and connectivity among populations of corals and reef fishes: history, advances and challenges. Coral Reefs 28, 307–325.

Jones, G.P., Planes, S., Thorrold, S.R., 2005. Coral reef fish larvae settle close to home. Current Biology 15, 1314–1318.

Kegler, P., Kegler, H.F., Gärdes, A., Ferse, S.C.A., Lukman, M., Alfiansah, Y.R., et al., 2017a. Bacterial biofilm communities and coral larvae settlement at different levels of anthropogenic impact in the Spermonde Archipelago, Indonesia. Frontiers in Marine Science 4, 270. https://doi.org/10.3389/fmars.2017.00270.

Kegler, H.F., Kegler, P., Luckmann, A., Jennerjahn, T.C., Gärdes, A., 2018a. Small scale shifts in microbial communities due to local island populations in the Spermonde Archipelago, Indonesia. PeerJ 6, e4555.

Kegler, H.F., Lukman, M., Teichberg, M., Plass-Johnson, J., Hassenrück, C., Wild, C., et al., 2017b. Bacterial community composition and potential driving factors in different reef habitats of the Spermonde archipelago, Indonesia. Frontiers in Microbiology 8 (662). https://doi.org/10.3389/fmicb.2017.00662.

Keyse, J., Treml, E.A., Huelsken, T., Barber, P.H., DeBoer, T., Kochzius, M., Rigino, C., 2018b. Historical divergences associated with intermittent land bridges overshadow isolation by larval dispersal in co-distributed species of *Tridacna* giant clams. Journal of Biogeography 45, 848–858. https://doi.org/10.1111/jbi.13163.

Knittweis, L., Jompa, J., Richter, C., Wolff, M., 2009a. Population dynamics of the mushroom coral *Heliofungia actiniformis* in the Spermonde archipelago, South Sulawesi, Indonesia. Coral Reefs 28 (3), 793–804.

Knittweis, L., Kraemer, W., Timm, J., Kochzius, M., 2009b. Genetic structure of *Heliofungia actiniformis* (Scleractinia: Fungiidae) populations in the Indo-Malay Archipelago: implications for live coral trade management efforts. Conservation Genetics 10, 241–249. https://doi.org/10.1007/s10592-008-9566-5.

Knittweis, L., Wolff, M., 2010. Live coral trade impacts on the mushroom coral *Heliofungia actiniformis* in Indonesia: potential future management approaches. Biological Conservation 143 (11), 2722–2729.

Kochzius, M., Nuryanto, A., 2008. Strong genetic population structure in the boring giant clam *Tridacna crocea* across the Indo-Malay archipelago: implications related to evolutionary processes and connectivity. Molecular Ecology 17, 3775–3787.

Kochzius, M., Seidel, C., Hauschild, J., Kirchhoff, S., Mester, P., Meyer-Wachsmuth, I., et al., 2009. Genetic population structures of the blue starfish *Linckia laevigata* and its gastropod ectoparasite *Thyca crystallina*. Marine Ecology Progress Series 396, 211–219.

Kool, J.T., Paris, C.B., Barber, P.H., Cowen, R.K., 2011. Connectivity and the development of population genetic structure in Indo-West Pacific coral reef communities. Global Ecology and Biogeography 20, 695–706.

Kubicek, A., Breckling, B., Hoegh-Guldberg, O., Reuter, H., 2019. Climate change drives trait-shifts in coral reef communities. Scientific Reports 9. Article number: 3721.

Kubicek, A., Jopp, F., Breckling, B., Lange, C., Reuter, H., 2015. Context-oriented model validation of individual-based models in ecology: a hierarchically structured approach to validate qualitative, compositional and quantitative characteristics. Ecological Complexity 22, 178–191.

Kubicek, A., Muhando, C., Reuter, H., 2012. Simulations of long-term community dynamics in coral reefs - how perturbations shape trajectories. PLoS Computational Biology 8 (11), e1002791. https://doi.org/10.1371/journal.pcbi.1002791.

Leins, J., 2017. Social Dependencies in Artisanal Reef Fisheries: Drawbacks and Improvement Strategies —Application of an Agent-Based Model (M.Sc. thesis). University of Oldenburg.

Madduppa, H.H., Timm, J., Kochzius, M., 2014a. Interspecific, spatial and temporal variability of self-recruitment in anemonefishes. PLoS One 9 (2), e90648.

Madduppa, H.H., Timm, J., Kochzius, M., 2018. Reduced genetic diversity in clown anemonefish (*Amphiprion ocellaris*) in exploited reefs of the Spermonde Archipelago, Indonesia. Frontiers in Marine Science 5, 80.

Madduppa, H.H., von Juterzenka, K., Syakir, M., Kochzius, M., 2014b. Socio-economy of marine ornamental fishery and its impact on the population structure of the clown anemonefish *Amphiprion ocellaris* and its host anemones in Spermonde Archipelago, Indonesia. Ocean and Coastal Management 100, 41–50.

McClanahan, T.R., Marnane, M.J., Cinner, J.E., Kiene, W.E., 2006. A comparison of marine protected areas and alternative approaches to coral-reef management. Current Biology 16 (14), 1408–1413.

Metcalfe, K., Collins, T., Abernethy, K.E., Boumba, R., Dengui, J.C., Miyalou, R., et al., 2016. Addressing uncertainty in marine resource management; combining community engagement and tracking technology to characterize human behavior. Conservation Letters 10, 460–469.

Micheli, F., Mumby, P.J., Brumbaugh, D.R., Broad, K., Dahlgren, C.P., Harborne, A.R., et al., 2014. High vulnerability of ecosystem function and services to diversity loss in Caribbean coral reefs. Biological Conservation 171, 186–194.

Miñarro, S., Leins, J., Acevedo-Trejos, E., Fulton, E.A., Reuter, H., 2018. SEAMANCORE: a spatially explicit simulation model for assisting the local MANagement of COral REefs. Ecological Modelling 384, 296–307.

Miñarro, S., Navarrete Forero, G., Reuter, H., van Putten, I.E., 2016. The role of patron-client relations on the fishing behaviour of artisanal fishermen in the Spermonde Archipelago (Indonesia). Marine Policy 69, 73–83.

Miñarro, S., 2017. Modelling Coral Reefs to Support Their Local Management: A Case Study in the Spermonde Archipelago, Indonesia (Dissertation). Universität Bremen, Bremen.

Mumby, P.J., Hastings, A., Edwards, H.J., 2007. Thresholds and the resilience of Caribbean coral reefs. Nature 450 (7166), 98.

Mumby, P.J., Steneck, R.S., 2008. Coral reef management and conservation in light of rapidly evolving ecological paradigms. Trends in Ecology and Evolution 23 (10), 555–563.

Nanninga, G.B., Saenz-Agudelo, P., Zhan, P., Hoteit, I., Berumen, M.L., 2015. Not finding Nemo: limited reef-scale retention in a coral reef fish. Coral Reefs 34, 383–392.

Nasir, A., Lukman, M., Tuwo, A., Hatta, M., Tambaru, R., Nurfadilah, 2016. The use of C/N ratio in assessing the influence of land-based material in coastal water of South Sulawesi and Spermonde Archipelago, Indonesia. Frontiers in Marine Science 3, 266. https://doi.org/10.3389/fmars.2016.00266.

Nasir, A., Tuwo, A., Lukman, M., Usman, H., 2015. Impact of increased nutrient on the variability of chlorophyll-a in the west coast of South Sulawesi, Indonesia. International Journal of Scientific Engineering and Research 6, 821–826.

Navarrete-Forero, G., Miñarro, S., Mildenberger, T.K., Breckwoldt, A., Sudirman, S., Reuter, H., 2017. Participatory boat tracking reveals spatial fishing patterns in an Indonesian artisanal fishery. Frontiers in Marine Science 4 (409), 1–12. https://doi.org/10.3389/fmars.2017.00409.

Nurdin, N., Grydehøj, A., 2014. Informal governance through patron–client relationships and destructive fishing in Spermonde Archipelago, Indonesia. J. Mar. Isl. Cult. 3, 54–59.

Nurdin, N., Komatsu, T., Rani, C., Fakhriyyah, S., 2016. Coral reef destruction of Small island in 44 years and destructive fishing in Spermonde Archipelago, Indonesia. IOP Conference Series: Earth and Environmental Science 47, 012011.

Nuryanto, A., Kochzius, M., 2009. Highly restricted gene flow and deep evolutionary lineages in the giant clam *Tridacna maxima*. Coral Reefs 28, 607–619.

Pet-Soede, L., Cesar, H.S.J., Pet, J.S., 1999. An economic analysis of blast fishing on Indonesian coral reefs. Environmental Conservation 26, 83–93.

Pet-Soede, L., Erdmann, M.V., 1998a. Blast fishing in southwest Sulawesi, Indonesia. Naga, the ICLARM Quarterly 21 (2), 4–9.

Pet-Soede, C., Van Densen, W.L.T., Hiddink, J.G., Kuyl, S., Machiels, M.A.M., 2001. Can fishermen allocate their fishing effort in space and time on the basis of their catch rates? An example from Spermonde archipelago, SW Sulawesi, Indonesia. Fisheries Management and Ecology 8, 15–36.

Pet-Soede, L., Erdmann, M., 1998b. An overview and com-parison of destructive fishing practices in Indonesia. SPC Live Reef Fish Information Bulletin 4, 28–36.

Planes, S., Jones, G.P., Thorrold, S.R., 2009. Larval dispersal connects fish populations in a network of marine protected areas. Proceedings of the National Academy of Sciences of the United States of America 106, 5693–5697.

Plass-Johnson, J.G., Bednarz, V.N., Hill, J.M., Jompa, J., Ferse, S.C.A., Teichberg, M., 2018a. Contrasting responses in the niches of two coral reef herbivores along a gradient of habitat disturbance in the Spermonde Archipelago, Indonesia. Frontiers in Marine Science 5, 32. https://doi.org/10.3389/fmars.2018.00032.

Plass-Johnson, J.G., Ferse, S.C.A., Jompa, J., Wild, C., Teichberg, M., 2015a. Fish herbivory as key ecological function in a heavily degraded coral reef system. Limnology and Oceanography 60, 1382–1391.

Plass-Johnson, J.G., Heiden, J., Abu, N., Lukman, M., Teichberg, M., 2016a. Experimental analysis of the effects of consumer exclusion on recruitment and succession of a coral reef system along a water quality gradient in the Spermonde Archipelago, Indonesia. Coral Reefs 35, 229–243.

Plass-Johnson, J.G., Schwieder, H., Heiden, J., Weiand, L., Wild, C., Jompa, J., et al., 2015b. *Acanthaster planci* outbreak in the Spermonde Archipelago, Indonesia. Regional Environmental Change 15, 1157–1162.

Plass-Johnson, J.G., Taylor, M.H., Husain, A.A.A., Teichberg, M., Ferse, S.C.A., 2016b. Non-random variability in functional composition of coral reef fish communities along an environmental gradient. PLoS One 11, e0154014.

Plass-Johnson, J.G., Teichberg, M., Bednarz, V.N., Gärdes, A., Heiden, J., Lukman, M., et al., 2018b. Spatio-temporal patterns in coral reef communities of the Spermonde Archipelago, 2012–2014, II: fish assemblages display structured variation related to benthic condition. Frontiers in Marine Science 5, 36. https://doi.org/10.3389/fmars.2018.00036.

Radjawali, I., 2011. Social networks and the live reef food fish trade: examining sustainability. Journal of Indonesian Social Sciences and Humanities 4, 65–100.

Radjawali, I., 2012. Examining local conservation and development: live reef foodfishing in the Spermonde Archipelago, Indonesia. Journal of Integrated Coastal Zone Management 12.

Ratsimbazafy, H.A., 2019. A Spatial Arrangement and Governance System for the Future Malagasy Network of marine Protected Areas (Ph.D. Thesis). Belgium. Vrije Universiteit Brussel (VUB).

Reuter, H., Jopp, F., Blanco-Moreno, J.-M., Damgaard, C., Matsinos, Y., DeAngelis, D., 2010. Ecological hierarchies and self-organisation– pattern analysis, modelling and process integration across scales. Basic and Applied Ecology 11, 572–581. https://doi.org/10.1016/j.baae.2010.08.002.

Roff, G., Mumby, P.J., 2012. Global disparity in the resilience of coral reefs. Trends in Ecology and Evolution 27, 404–413.

Rogers, A., Blanchard, J.L., Mumby, P.J., 2014. Vulnerability of coral reef fisheries to a loss of structural complexity. Current Biology 24 (9), 1000–1005.

Saenz-Agudelo, P., Jones, G.P., Thorrold, S.R., Planes, S., 2009. Estimating connectivity in marine populations: an empirical evaluation of assignment tests and parentage analysis under different gene flow scenarios. Molecular Ecology 18, 1765–1776.

Saenz-Agudelo, P., Jones, G.P., Thorrold, S.R., Planes, S., 2011. Connectivity dominates larval replenishment in a coastal reef fish metapopulation. Proceedings of the Royal Society B: Biological Sciences 278, 2954–2961.

Sandin, S.A., McNamara, D.E., 2012. Spatial dynamics of benthic competition on coral reefs. Oecologia 168, 1079–1090.

Sandin, S.A., Smith, J.E., DeMartini, E.E., Dinsdale, E.A., Donner, S.D., Friedlander, A.M., et al., 2008. Baselines and degradation of coral reefs in the northern line islands. PLoS One 3 (2), e1548.

Sawall, Y., Jompa, J., Litaay, M., Maddusila, A., Richter, C., 2013. Coral recruitment and potential recovery of eutrophied and blast fishing impacted reefs in Spermonde Archipelago, Indonesia. Marine Pollution Bulletin 74 (1), 374–382.

Sawall, Y., Richter, C., Ramette, A., 2012. Effects of eutrophication, seasonality and macrofouling on the diversity of bacterial biofilms associated with coral reefs of the Spermonde Archipelago, Indonesia. PLoS One 7 (7), e39951.

Sawall, Y., Teichberg, M., Seemann, J., Litaay, M., Jompa, J., Richter, C., 2011. Nutritional status and metabolism of the coral *Stylophora subseriata* along a eutrophication gradient in Spermonde Archipelago (Indonesia). Coral Reefs 30, 841–853.

Sawall, Y., Ttiwong, S.K., Jompa, J., Richter, C., 2014. Calcification, photosynthesis and nutritional status of the hermatypic coral *Porites lutea*: contrasting case studies from Indonesia and Thailand. Galaxea, Journal of Coral Reef Studies 16, 1–10.

Schwerdtner Máñez, K., Ferse, S.C.A., 2010. The history of Makassan Trepang fishing and trade. PLoS One 5, e11346.

Schwerdtner Máñez, K., Husain, P.S., 2013. First evidence of targeted moray eel fishing in the Spermonde Archipelago, South Sulawesi, Indonesia. Traffic Bulletin 25, 4–7.

Schwerdtner Máñez, K., Husain, S., Ferse, S.C.A., Máñez Costa, M., 2012. Water scarcity in the Spermonde Archipelago, Sulawesi, Indonesia: past, present and future. Environmental Science and Policy 23, 74–84.

Seeman, J., Sawall, Y., Auel, H., Richter, C., 2013. The use of lipids and fatty acids to measure the trophic plasticity of the coral *Stylophora subseriata*. Lipids 48 (3), 275–286.

Seemann, J., Carballo-Bolanos, R., Berry, K.L., González, C.T., Richter, C., Leinfelder, R.R., 2012. Importance of heterotrophic adaptations of corals to maintain energy reserves. In: Proceedings of the 12th International CoralReef Symposium, Cairns, Australi, p. 19A.

Smith, J.E., Hunter, C.L., Smith, C.M., 2010. The effects of top-down versus bottom-up control on benthic coralreef community structure. Oecologia 163, 497–507. https://doi.org/10.1007/s00442-009-1546-z.

Teichberg, M., Wild, C., Bednarz, V.N., Kegler, H., Lukman, M., Gärdes, A., et al., 2018. Spatio-temporal patterns in coral reef communities of the Spermonde Archipelago, 2012-2014, I: Comprehensive reef monitoring of water and benthic indicators reflect changes in reef health. Frontiers in Marine Science 5, 33. https://doi.org/10.3389/fmars.2018.00033.

Timm, J., Kochzius, M., 2008. Geological history and oceanography of the Indo-Malay Archipelago shape the genetic population structure in the False Clown Anemonefish (*Amphiprion ocellaris*). Molecular Ecology 17, 3999–4014.

Timm, J., Kochzius, M., Madduppa, H.H., Neuhaus, A.I., Dohna, T., 2017. Small-scale genetic population structure of coral reef organisms in Spermonde Archipelago, Indonesia. Frontiers in Marine Science 4, 294.

Timm, J., Planes, S., Kochzius, M., 2012. High similarity of genetic population structure in the False Clown Anemonefish (*Amphiprion ocellaris*) found in microsatellite and mitochondrial control region analysis. Conservation Genetics 13, 693–706.

Treml, E.A., Roberts, J., Halpin, P.N., Possingham, H., Riginos, C., 2015. The emergent geography of biophysical dispersal barriers across the Indo-West Pacific. Diversity and Distributions 21, 465–476.

Turner, R.A., Polunin, N.V.C., Stead, S.M., 2015. Mapping inshore fisheries: comparing observed and perceived distributions of pot fishing activity in Northumberland. Marine Policy 51, 173–181.

UNEP-WCMC, 2014. Review of Corals from Indonesia (Coral Species Subject to EU Decisions where Identification to Genus Level Is Acceptable for Trade Purposes). UNEP-WCMC, Cambridge.

van der Ven, M.R., Heynderickx, H., Kochzius, M., 2021. Differences in genetic diversity and divergence between brooding and broadcast spawning corals across two spatial scales in the Coral Triangle region. Marine Biology 168. https://doi.org/10.1007/s00227-020-03813-8.

Vermeij, Dailer, Smith, 2011. Crustose coralline algae can suppressmacroalgal growth and recruitment on Hawaiian coral reefs. Marine Ecology Progress Series 422, 1–7.

Veron, J.E.N., Devantier, L.M., Turak, E., Green, A.L., Kininmonth, S., Stafford-Smith, M., et al., 2009. Delineating the coral triangle. *Galaxea*. Journal of Coral Reef Studies 11, 91–100.

Voris, H.K., 2000. Maps of Pleistocene sea levels in Southeast Asia: shorelines, river systems and time duration. Journal of Biogeography 27, 1153–1167.

Wabnitz, C., Taylor, M., Green, E., Razak, T., 2003. From Ocean to Aquarium: The Global Trade in marine Ornamental Species. UNEP-WCMC, Cambridge, UK.

Weijerman, M., Fulton, E.A., Kaplan, I.C., Gorton, R., Leemans, R., Mooij, W.M., et al., 2015. An integrated coral reef ecosystem model to support resource management under a changing climate. PLoS One 10 (12), e0144165.

Wood, E., Malsch, K., Miller, J., 2012. International trade in hard corals: review of management, sustainability and trends. In: Proceedings of the 12th International Coral Reef Symposium. Cairns, Australia, pp. 9–13, 9–13 July 2012.

Wyrtki, K., 1961. Physical Oceanography of the Southeast Asian Waters. University of California, La Jolla, CA.

Yasir Haya, L.O.M., Fujii, M., 2017. Mapping the change of coral reefs using remote sensing and in situ measurements: a case study in Pangkajene and Kepulauan regency, Spermonde Archipelago, Indonesia. Journal of Oceanography 73, 623–645.

# Appendix A5

**Table A5.1** List of different gear types used in the Spermonde Archipelago and the targeted species.

| Category | Gear type | Local name | Target species | Main market | Legal status |
|---|---|---|---|---|---|
| Hook and line | Horizontal long line | Rawe[a] | Lutjanidae, Carangidae, Serranidae, Nemipteridae | Local | Legal |
| | Vertical long line | Rinta | Clupaeidae, pelagic bait fish | Local | Legal |
| | Trolling for piscivores | Kedo-kedo | Serranidae, Scombridae | Local/ export | Legal |
| | Trolling for squid | Doang-doang | Teuthida | Local | Legal |
| | Octopus bait | Pocong-pocong | Octopoda | Local/ export | Legal |
| | Shark bait | Tomba | Carcharhiniformes, Batoidea | Local/ export | Legal |
| Net | Gill net | Lanra | Clupeidae, Carangidae | Local | Legal |
| | Crab and shrimp gill net | Lanra | Crustacea | Local | Legal |
| | Scoop net | Sero | Pomacentridae, other small ornamentals | Export | Legal |
| | Purse seine | Gae/Rengge | Clupeidae, Engraulidae, Carangidae | Local | Legal |
| | Danish seine | Gae/Rengge | Leiognatidae, Synodontidae | Local | Legal |
| | Mobile lift net | Bagang Lopi | Carangidae, Clupeidae, Engraulidae, Teuthida | Local | Legal |
| | Stationary lift net | Bagang Tancap | Clupeidae, Leiognatidae, Teuthida | Local | Legal |
| | Beach seine | Jaring (mairo) | Miscellaneous | Local | Legal |
| | Mini trawl | Rere/ Renreng | Miscellaneous | Local/ export | Illegal |
| Traps | Fish trap | Bubu | Lethrinidae, Lutjanidae, Serranidae | Local/ export | Legal |
| | Crab trap | Rakkang | Crustacea | Local/ export | Legal |
| | Flying fish trap | Buaro/Bale-bale | Exocoetidae | Export | Legal |
| Others | Compressor diving | Penyalam/ Hookah | Holothuridae, Gastropoda, Nephropidae, Anthozoa | Export | Legal |
| | Spear gun | Patte' | Lutjanidae, Scombridae, Serranidae, Scaridae, Siganidae, Acanthuridae | Local | Legal |
| | Reef gleaning | ? | Gastropoda, Bivalvia | Local | Legal |
| | Fish-attracting device | Rumpon | Various pelagic fishes | Local/ export | Legal |
| | Blast fishing | Pembom/ Panges | Various fishes | Local | Illegal |
| | Cyanide fishing | Pembius/ Paselang | Live ornamental and food reef fishes, Nephropidae | Export | Illegal |

See Fig. A5.1 for illustrations. Gear types in blue are not illustrated.

[a]Rawe Makassar denotes a particular kind of horizontal long line, which is combined with a gill net.

From Ferse et al. (2014).

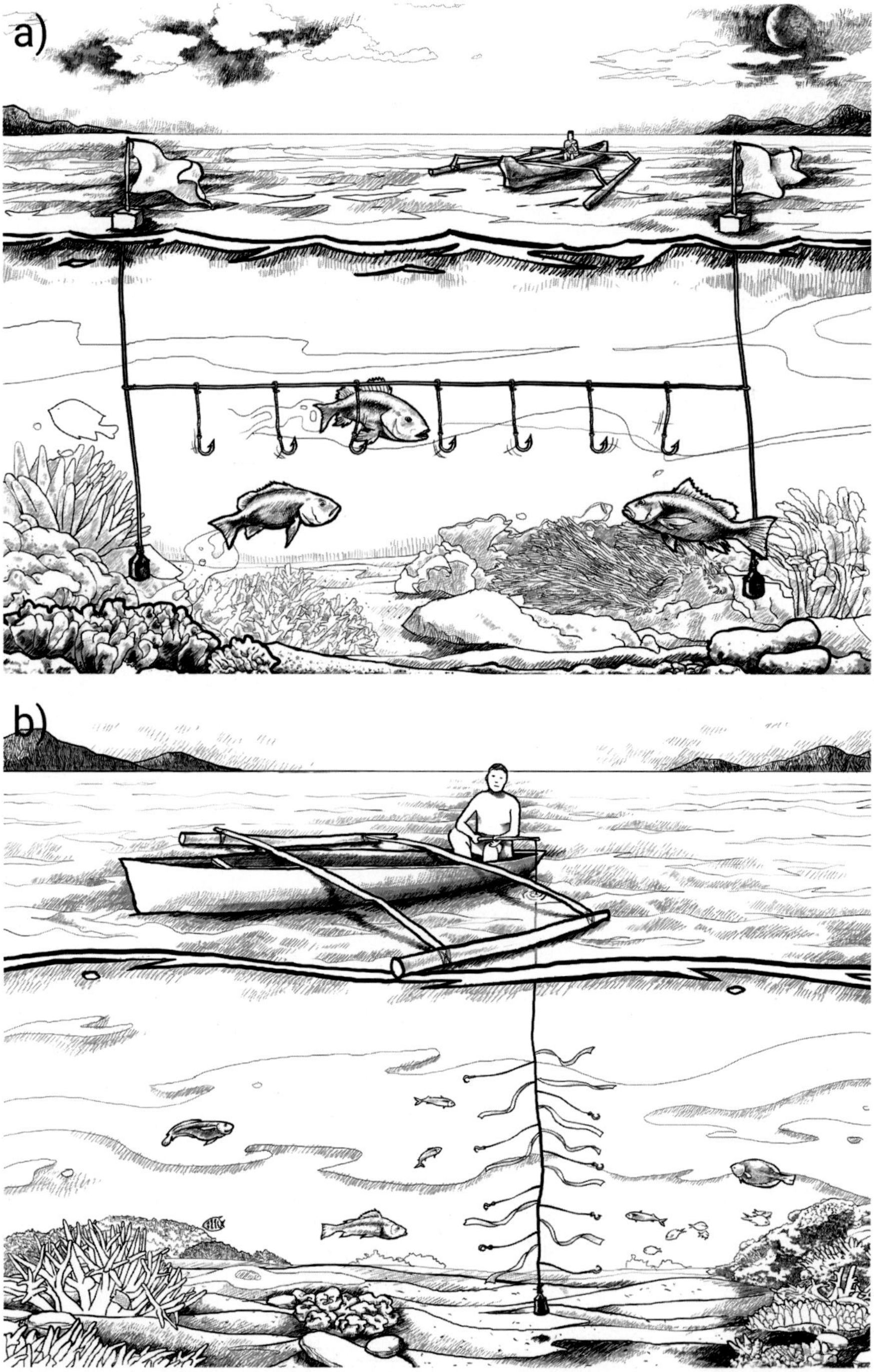


**FIGURE A5.1** Illustrations of different fishing methods. The drawings are by Saranat Tiemkeo. (A) Horizontal long line, (B) vertical long line, (C) trolling for piscivores, (D) trolling for squid, (E) octopus bait, (F) shark bait, (G) gill net, (H) purse seine, (I) mobile lift net, (J) stationary lift net, (K) mini trawl, (L) fish trap, (M) crab trap, (N) compressor diving, (O) fish-attracting device, (P) blast fishing, (Q) cyanide fishing.

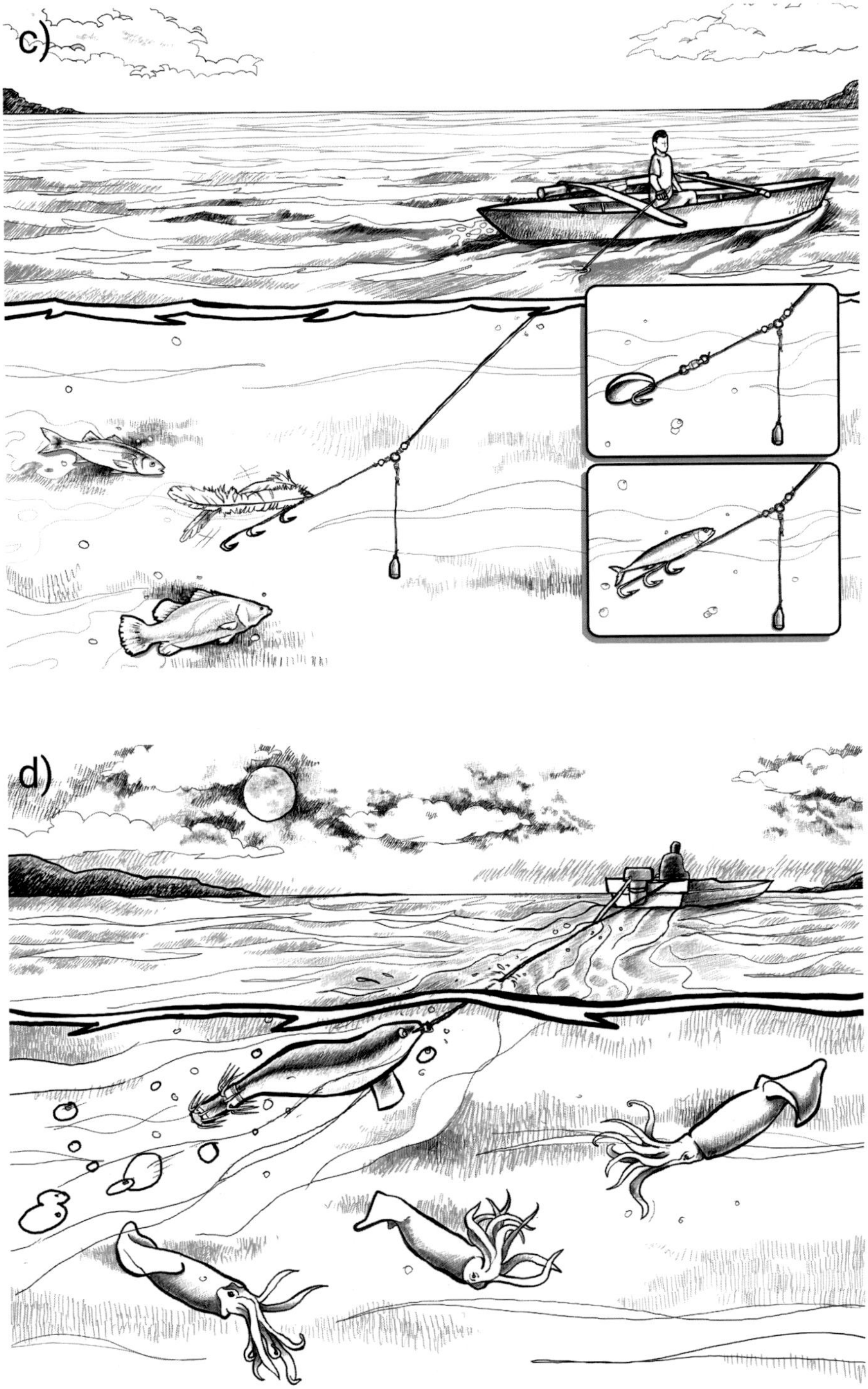


**FIGURE A5.1** Cont'd

**FIGURE A5.1** Cont'd

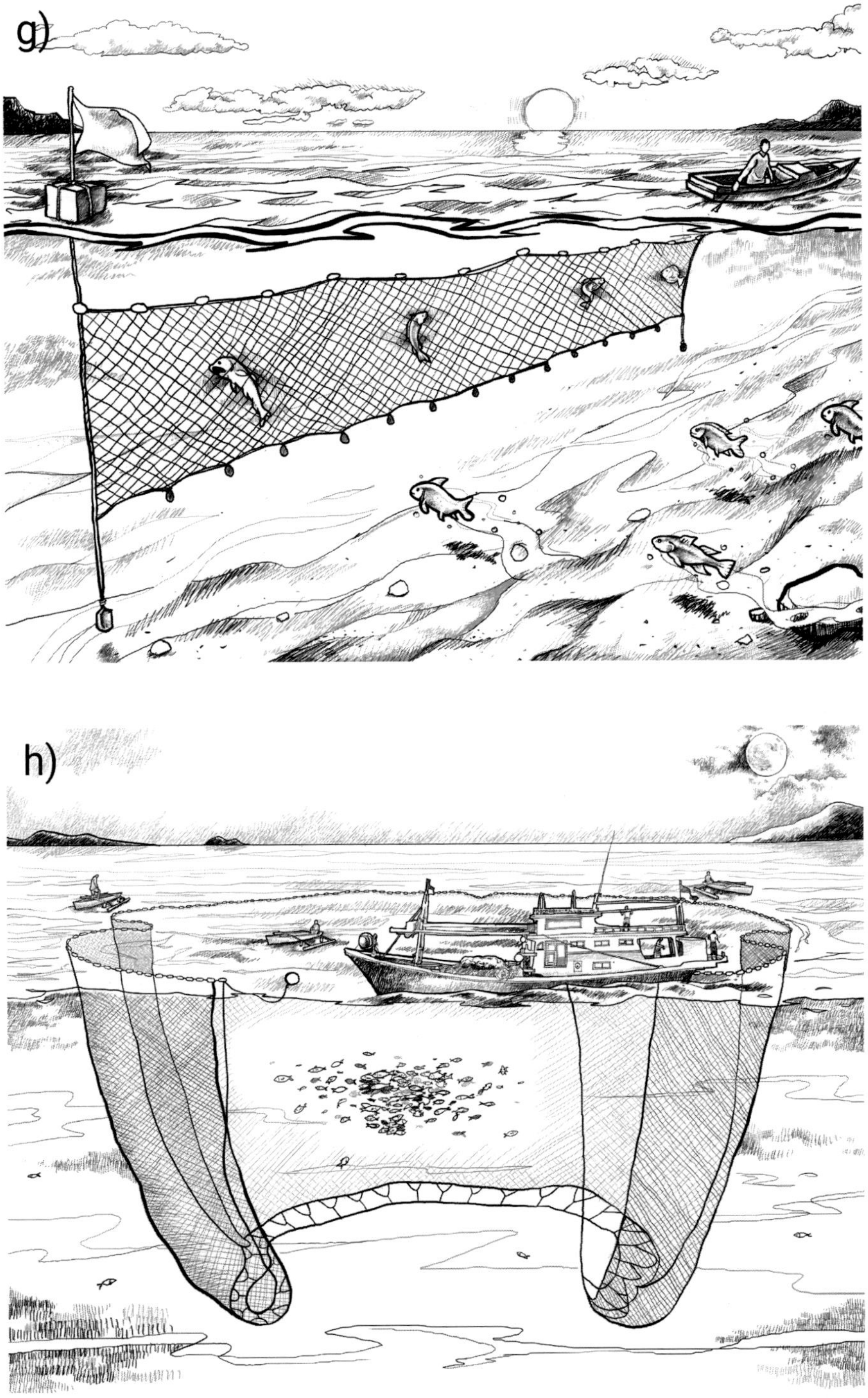


**FIGURE A5.1** Cont'd

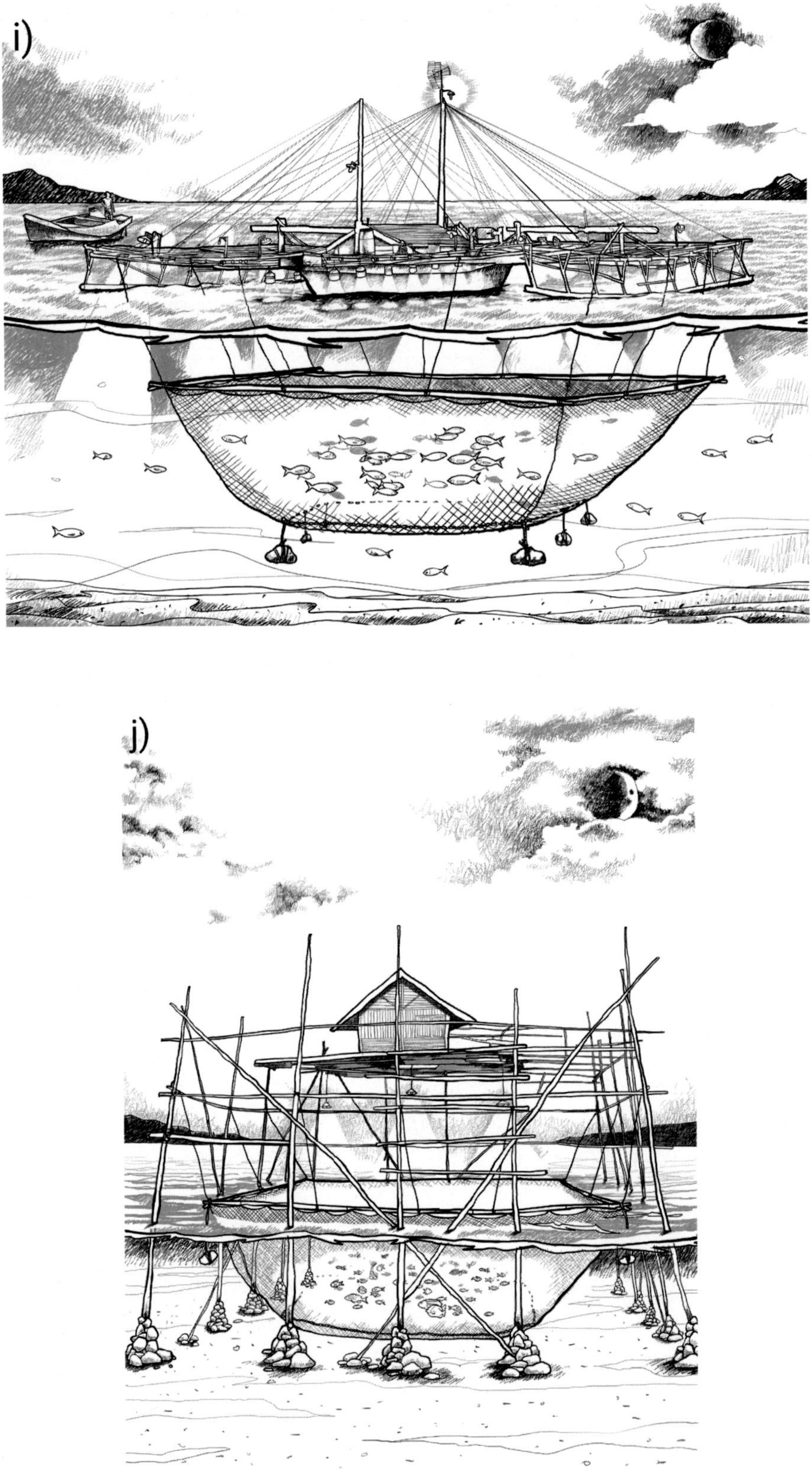


**FIGURE A5.1** Cont'd

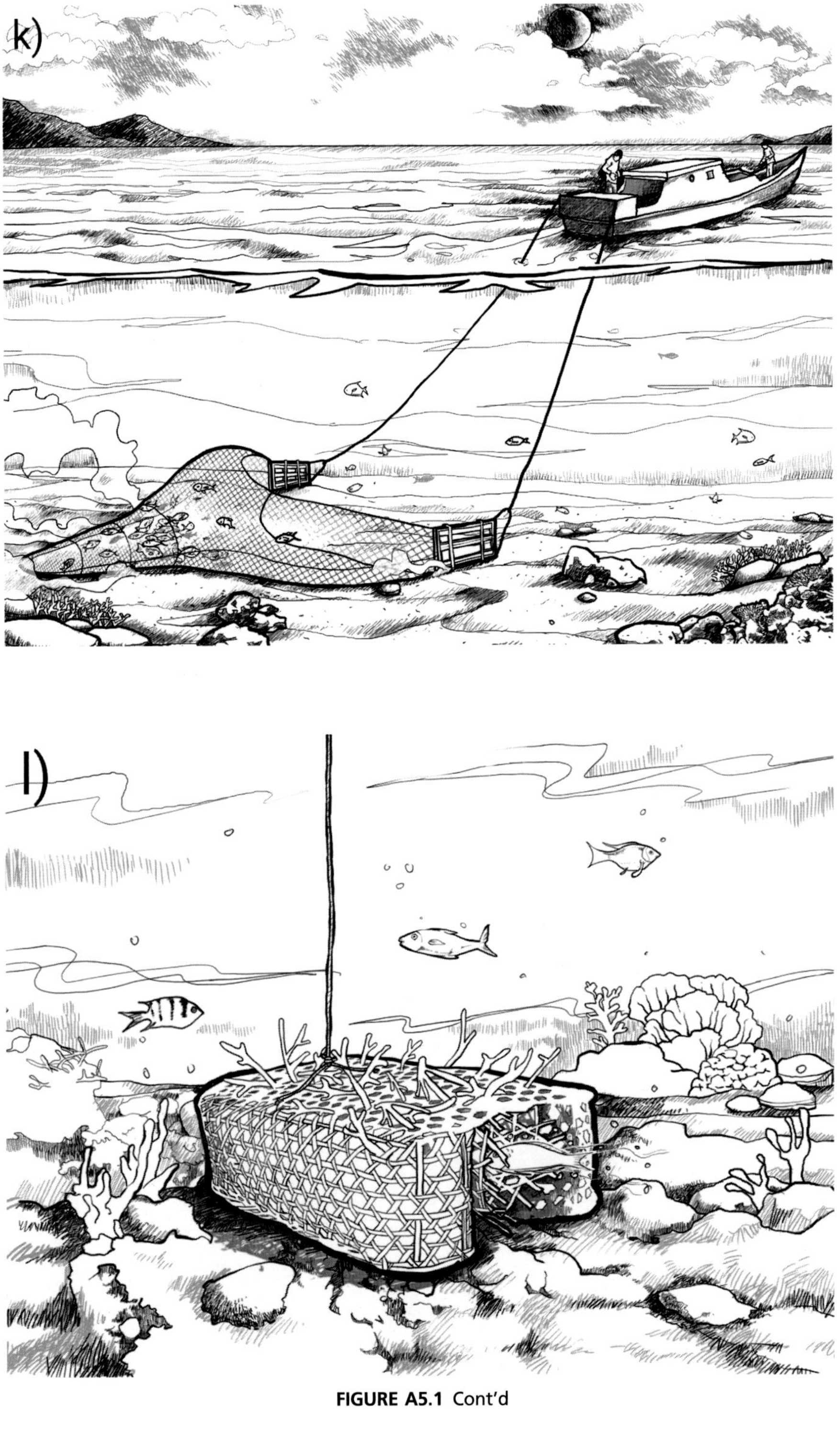


**FIGURE A5.1** Cont'd

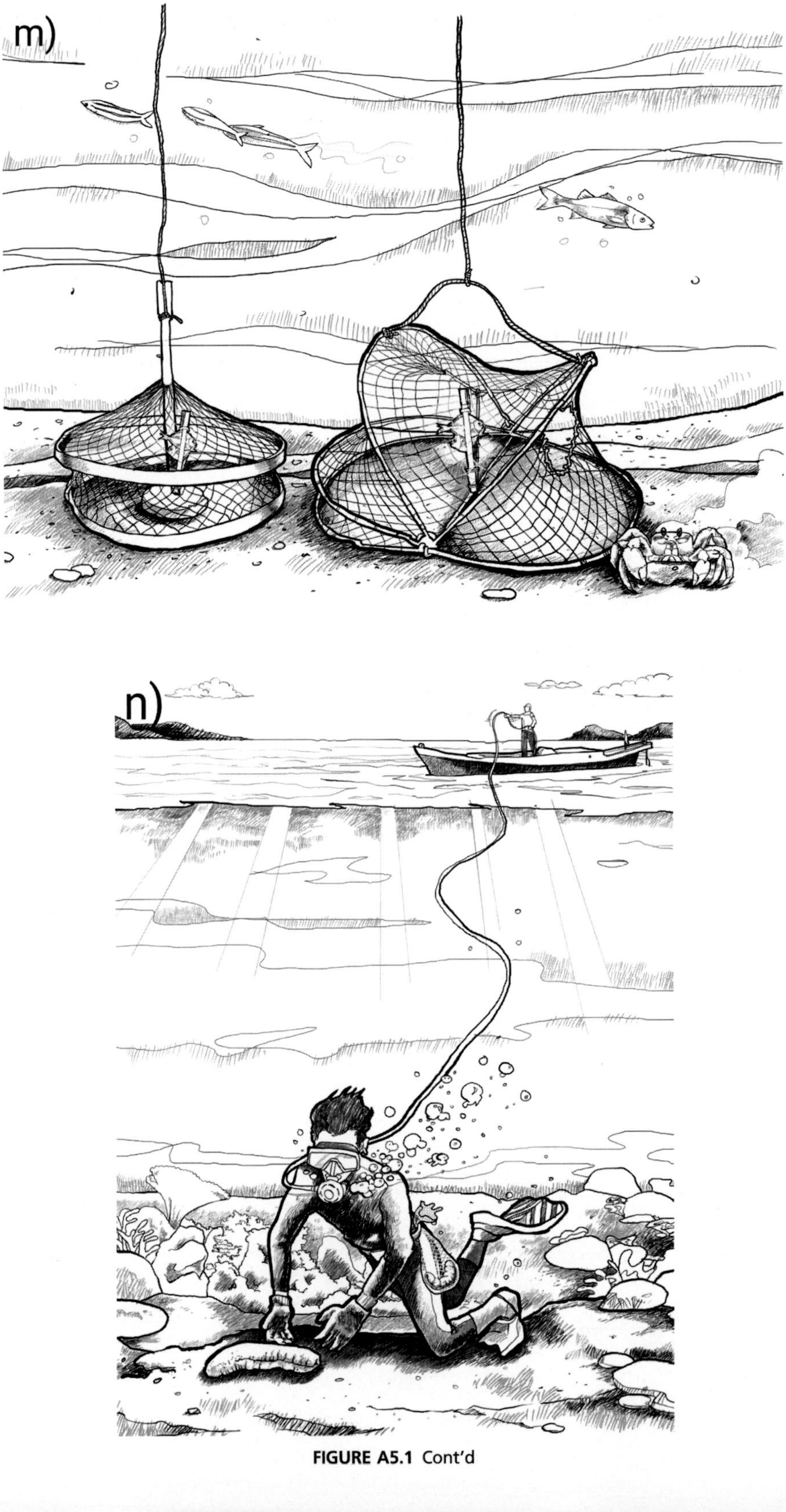


**FIGURE A5.1** Cont'd

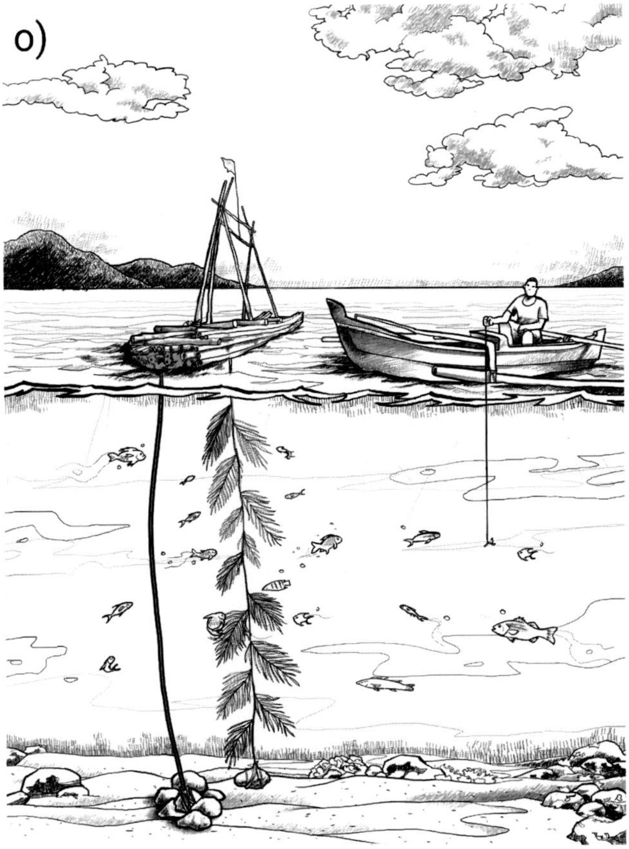


**FIGURE A5.1** Cont'd

**FIGURE A5.1** Cont'd

q)

**FIGURE A5.1** Cont'd

# 6

# Ecology of seagrass beds in Sulawesi—Multifunctional key habitats at the risk of destruction

Harald Asmus[1], Dominik Kneer[1], Claudia Pogoreutz[2], Sven Blankenhorn[1], Jamaluddin Jompa[3], Nadiarti Nurdin[3], Dody Priosambodo[4]

[1]ALFRED WEGENER INSTITUTE, HELMHOLTZ CENTRE FOR POLAR AND MARINE RESEARCH, LIST/SYLT, GERMANY; [2]RED SEA RESEARCH CENTER (RSRC), BIOLOGICAL, ENVIRONMENTAL SCIENCE AND ENGINEERING DIVISION (BESE), KING ABDULLAH UNIVERSITY OF SCIENCE AND TECHNOLOGY (KAUST), THUWAL, SAUDI ARABIA; [3]FACULTY OF MARINE SCIENCE AND FISHERIES, HASANUDDIN UNIVERSITY, MAKASSAR, SOUTH SULAWESI, INDONESIA; [4]DEPARTMENT OF BIOLOGY, FACULTY OF MATHEMATICS AND NATURAL SCIENCES, HASANUDDIN UNIVERSITY, MAKASSAR, SOUTH SULAWESI, INDONESIA

## Abstract

*Indonesian seagrass communities are among the most diverse compared with those of other tropical or temperate regions. In this chapter, we describe some of the results of our research on seagrass beds in Sulawesi during the German–Indonesian Research Project "Science for the Protection of Indonesian Coastal Ecosystems" (SPICE) from 2004 to 2016. We studied aspects of the distribution and characteristics of these ecosystems within the Spermonde Archipelago, the role of keystone species and eco-engineers, their function as a habitat for fishes, their impact on carbon flow and storage as well as the threat they face due to anthropogenic activities. Our results contributed to these topics either by confirming known data or by originating new ideas on the interactions of seagrasses with animals and physical drivers. The alarming loss of seagrass beds globally is a serious threat for the function of our oceans as carbon sink. To save the seagrass beds, we suggest immediate measures at a regional level for the Spermonde Archipelago. We further recommend detailed research on the role seagrass ecosystems play within the complex interactions between land use and coastal changes.*

## Abstrak

*Komunitas Lamun Indonesia termasuk salah satu yang paling beragam dibandingkan dengan wilayah tropis dan subtropis lainnya. Pada bab ini, kami menguraikan beberapa hasil penelitian yang dilakukan di padang lamun Sulawesi selama kegiatan riset bersama Jerman-Indonesia bertajuk "Sains untuk Perlindungan Ekosistem Pantai Indonesia" dari 2004 hingga 2016. Kami mengkaji aspek distribusi dan karakteristik dari ekosistem padang lamun yang ada di Kepulauan Spermonde, peran spesies kunci dan rekayasa ekologi, fungsi padang lamun sebagai habitat bagi ikan, dampak padang lamun terhadap aliran karbon dan penyimpanannya, termasuk ancaman yang dihadapi padang lamun akibat aktifitas manusia. Peringatan tentang berkurangnya luasan*

Science for the Protection of Indonesian Coastal Ecosystems (SPICE). https://doi.org/10.1016/B978-0-12-815050-4.00014-6 

*padang lamun di seluruh dunia adalah ancaman serius terhadap fungsi laut kita sebagai penyimpan karbon. Untuk menyelamatkan padang lamun ini, kami menyarankan untuk secepatnya melakukan penelitian dalam skala regional. Kami juga merekomendasikan dilakukannya riset yang lebih detail terkait peran ekosistem lamun dalam interaksi kompleks antara penggunaan lahan (land use) dan perubahan pantai.*

## Chapter outline

**6.1 General introduction to tropical Southeast Asian seagrass meadows ... 203**
6.1.1 High biodiversity of seagrasses in the coral triangle of the tropical Indo-West Pacific ... 203
6.1.2 Introduction to the Spermonde Archipelago and its seagrasses and mangroves ... 204
**6.2 The current distribution of seagrasses in the Spermonde Archipelago ... 206**
6.2.1 Area estimates and seagrass mapping ... 206
6.2.2 The structure of tropical seagrass bed systems ... 209
**6.3 Seagrass ecology ... 211**
6.3.1 The historic loss of megaherbivores and today's important role of burrowing shrimp ... 211
6.3.2 Macrobenthic communities ... 215
6.3.3 The food web and the trophic pyramid in tropical seagrass beds ... 217
6.3.4 The function of seagrass meadows as water filters and buffers for land runoff ... 221
6.3.5 Carbon storage ... 222
6.3.6 Seagrass beds as carbon sinks ... 222
6.3.7 Trophic transfers from seagrass meadows to nearby ecosystems ... 224
**6.4 Tropical seagrass beds as key habitat for fish species ... 225**
6.4.1 Tropical seagrasses and their associated fish communities ... 225
6.4.2 The seagrass canopy as a driver of fish communities ... 226
6.4.3 Differences in fish habitat utilization across seagrass meadows with distinct canopy structures ... 228
**6.5 Human—seagrass interactions ... 230**
6.5.1 Ecological value and ecosystem services ... 230
6.5.2 Fisheries on fish and invertebrates in seagrass beds ... 230
6.5.3 Seaweed farms ... 231
6.5.4 Human-made infrastructure ... 233
6.5.5 Current threats ... 234
**6.6 Conclusions and outlook ... 236**
**Acknowledgments ... 239**
**References ... 239**

# 6.1 General introduction to tropical Southeast Asian seagrass meadows

## 6.1.1 High biodiversity of seagrasses in the coral triangle of the tropical Indo-West Pacific

About 60 species of seagrasses occur worldwide where they form special habitats and attract an array of organisms from plants to sea mammals by their meadow-like stocks (den Hartog 1970, p. 275; Kuo and McComb, 1989; Brouns and Heijs, 1991). These associated organisms use the leaf or roots or the sediment as substrate (Borowitzka and Lethbridge, 1989; Howard et al., 1989), the shelter between the leaves as protection from predators or water movements, and the seagrass plants or their associated fauna and flora for food (Howard et al., 1989). Tropical seagrass beds exhibit the highest diversity of seagrasses and associated fauna on a global scale, and among them, seagrass beds of the Indonesian archipelago come first (Tomascik et al., 1997). The highest diversity of marine shallow water organisms is observed in a roughly triangular area that is situated between the east coast of Borneo in the west and the Solomon Islands in the east, extending to the north coast of New Guinea and Timor-Leste in the south and the Philippines in the north. Because this biodiversity hot spot has been defined firstly for corals, it is called the "coral triangle." The center of the "coral triangle" includes the shelf areas of Sulawesi and the lesser Sunda Islands. Seagrasses achieve a species number of 12 in this area, which is surprisingly low compared with other groups of organisms but is the highest for the different seagrass regions on earth (Kiswara, 1994; Kuriandewa et al., 2003; Short et al., 2007).

The reason for this high biodiversity in many organismic groups is not sufficiently explained; however, some contrasting hypotheses occur focusing either on the stability of environmental factors (McCoy and Heck, 1976) in the region that could be a precondition for a low extinction rate, or on the hydrographic separation of geographical units promoting local speciation (Jokiel and Martinelli, 1992). Concerning the recent distribution of corals, mangroves, and seagrasses, two opposing hypotheses have been proposed: McCoy and Heck (1976) supported the vicariance hypothesis and assumed a widely distributed biota that has been modified by tectonic events, climate change, and responding evolutionary mechanisms such as speciation and extinction. In contrast, den Hartog (1970, p. 275) explained the distribution patterns of seagrasses by their dispersal by radiation from a region of highest diversity, which is considered as the center of origin. For modern seagrasses, an area named "Malesia" (including the Indonesian archipelago, Malaysian Borneo, Papua New Guinea, and northern Australia) could be such a center of origin. This center-of-origin-hypothesis is supported by the fact that since the late Miocene the oceanic plates in the central Indo-West Pacific have been in the same position. As a further support for this hypothesis, Mukai (1993) showed that the higher the distance from the periphery of the coral triangle, the higher the species loss of seagrasses, and some adjacent areas such as Japan, Queensland, and Fiji show the lowest diversity of tropical seagrass species.

Few fossil records of seagrass species suggest that these plants originate from different families of freshwater and marsh plants returning to the sea in the late Cretaceous about 100 Mio years ago (den Hartog, 1970, p. 275), and representing until now the only angiosperm plants with a submerged marine life cycle. The area of origin for seagrasses is assumed to be the shallow coasts of the Tethys Sea, a large ocean bordered at that time by the continents of North and East Africa in the west, Central and South Asia in the north, and Madagascar, India (at that time), and Australia, including parts of Papua New Guinea, in the south. It was open to the North Atlantic in the northwest and the Pacific Ocean in the east. The Indonesian Archipelago was not developed at that time. The very complex and not very well-known geological history of the Indonesian Archipelago started to happen later and was the result of the elevation, migration, and collision of many small microcontinents, island arcs, volcanic islands, continental margins as well as the rise and fall of the sea level (Smith and Briden, 1977). Because the archipelago became stabilized more or less in its present shape about 10 Mio years ago (Pandolfi, 1993), the coral triangle could increase its biodiversity since that time. This also explains that coral reefs within the coral triangle are comparable young in geological timescales (Pandolfi, 1992). The birth of the archipelago was also accompanied by the appearance of extended but varying coastlines in a climatically stable and favorable area promoting biological invasions and survival of a wealth of coastal and shallow water plants and animals. Because the coastline altered its shape frequently with the rise and fall of sea level due to recurring glaciation periods, this may have led to speciation processes due to subsequent spatial separation and recombination of habitats. Therefore, the high biodiversity in the coral triangle was probably a result of both a center of origin and a kind of refuge, allowing high survival rates for invaders that may be extinct in other places.

### 6.1.2 Introduction to the Spermonde Archipelago and its seagrasses and mangroves

Most of the coastline of South Sulawesi is bordered by fringing reefs. Some of them grow directly from the shoreline, but in most cases, their seaward extension has created an intertidal to shallow subtidal sandy area between the beach and the inshore edge of coral growth, which measures anywhere between 100 and 1000 m across. The carbonate sand in those deposits and on the white beaches is a mix of broken coral, mollusk shells, urchin spines, foraminiferans, *Halimeda* chips, and other skeletal remains of once living reef organisms, which were pushed shoreward by waves breaking over the reef edge. On the west coast of South Sulawesi lies a shallow carbonate area, the Spermonde (or Sangkarang) Shelf (de Klerk, 1982). This is where the fringing reef veers off the coast and becomes a barrier reef, along a length of 150 km between Barru in the north and Takalar in the south and with a maximum width of 60 km from west to east along its central section (Fig. 6.1). Inside the lagoon defined by this barrier are more than 100 patch reefs, which are best described as coral outcrops rising from the lagoon floor. The largest of them measures several kilometers across, and in those that have reached the water surface, the same forces that form the sandy deposits and beaches behind fringing reefs have also filled their central areas with sand. Once enough of these unconsolidated sediments were pushed together on a focal point on those patch reef, low-lying islands

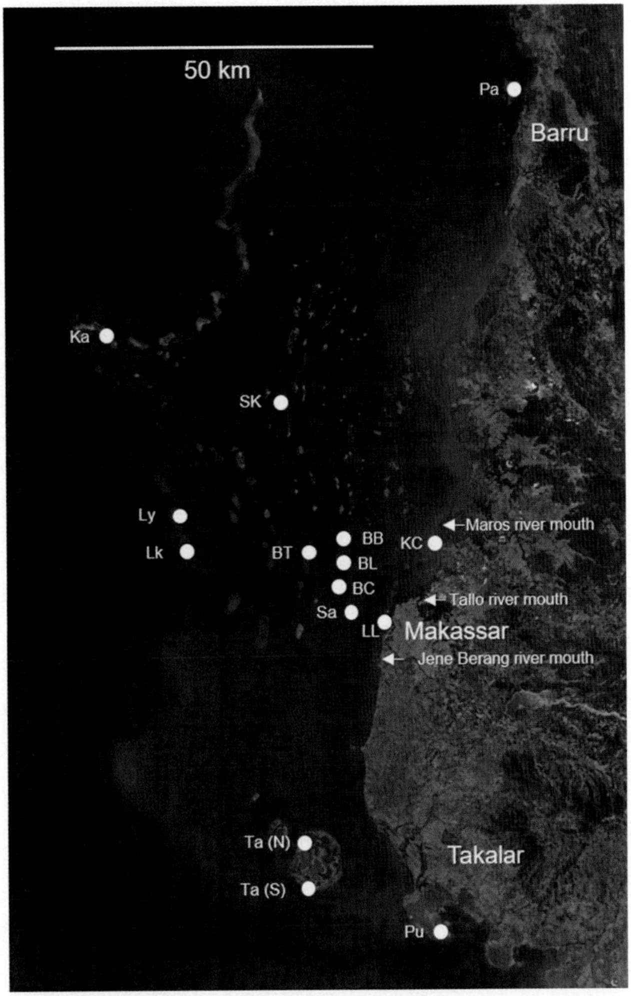


**FIGURE 6.1** Satellite map of the Spermonde Archipelago showing the barrier reef (left), patch reefs in the lagoon (center), and the South Sulawesi Mainland (right). Islands or locations that were studied during the SPICE project are indicated by the following abbreviations (in alphabetical order): **BB** (Bone Batang), **BC** (Barrang Caddi), **BL** (Barrang Lompo), **BT** (Bone Tambung), **Ka** (Kapoposang), **KC** (Kurri Caddi), **Lk** (Langkai), **LL** (Lae Lae), **Ly** (Lanyukan), **Pa** (Panikiang), **Pu** (Puntondo), **Sa** (Samalona), **SK** (Sarappo Keke), **Ta** (Tanakeke). Further indicated are the positions of the mouth of the rivers Maros, Tallo, and Jene Berang, the approximate locations of the Barru and Takalar regencies, and the city of Makassar. *Modified from Landsat/Copernicus. (December 31, 2014). www.google.com/earth/index.html.*

were formed (Hart and Kench, 2007). In the Spermonde Archipelago, there are several dozens of these so-called reef islands or coral cays, and today some of them are densely populated (e.g., Barrang Lompo Island with >200 people/ha or 20,000 people/$km^2$, a population density similar to the one found in the city states Macau and Monaco, the nations with the highest population densities on earth today).

While there are some seagrasses growing on the lagoon floor east of the barrier reef and around the patch reefs, the densest seagrass stands are concentrated on the shallow sandy deposits on fringing reefs and surrounding the reef islands away from the larger population centers. Some of the seagrass meadows that were once growing on muddy deposits near the mainland coast seem to have disappeared since they were first described. In some places, seagrass beds on muddy sediments survived, particularly where mangroves still occupy shallow muddy deposits between the seagrass meadows and the reef islands (e.g., Panikiang Island), or between the seagrass meadows and the beach (e.g., Tanakeke Island) (Fig. 6.1).

It is on these oceanic top reef habitats that most of the seagrass studies presented in this chapter were conducted. In the following sections, we summarized our results, and we try to give an overview over the distribution of seagrasses and seagrass beds in the Spermonde Archipelago (Section 6.2) and the ecology of these seagrass systems and their functioning (Section 6.3) with respect to their species interactions such as the important role of burrowing endofauna such as shrimps and worms (Section 6.3.1), the communities of associated macrobenthos (Section 6.3.2) as well as the interactions of seagrass species within the seagrass food web. We also summarize the role of the seagrass beds as water filters and buffers to land runoff (Section 6.3.4) and their role as carbon storage and carbon sinks (Sections 6.3.5 and 6.3.6) and try to sketch trophic transfers to nearby ecosystems. Special interest was laid in our project on the importance of seagrass beds for fish by conducting many experiments and investigations in this field, and therefore, we constructed a special section on this topic (Section 6.4). Seagrass beds are endangered in our days, and we tried to give an overview on the human interactions with this ecosystem (Section 6.5). Our conclusions, recommendations, and outlook and our evaluation for this special tropical area are given in Section 6.6.

## 6.2 The current distribution of seagrasses in the Spermonde Archipelago

### 6.2.1 Area estimates and seagrass mapping

Southeast Asian seagrasses systems are generally poorly studied; however, the Spermonde Archipelago is a notable exception (see Vonk, 2008, p. 10 and review by; Ooi et al., 2011). This is because the Spermonde barrier reef and lagoon systems have been the focal point of research expeditions for decades due to its relatively easy accessibility from a major population center (Makassar) (Hoeksema, 2015). In spite of that, no estimates for the total extent of area covered by seagrass meadows have been made to date.

So far, two studies have shown the feasibility of seagrass mapping using two different approaches (Kneer, 2013; Sawayama et al., 2015) (Fig. 6.2). However, both of these studies only covered a few islands, while a distribution map (as available for the Lesser

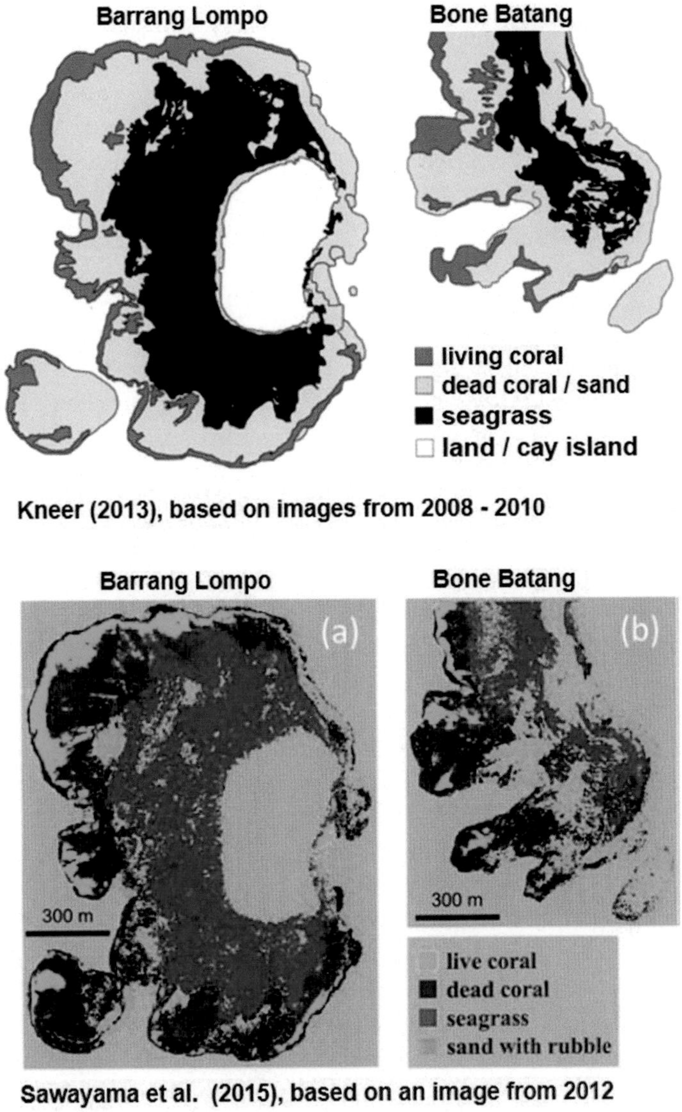


**FIGURE 6.2** Mapping approaches using ground truthing combined with aerial photographs (Kneer, 2013, images acquired 2008 to 2010) or satellite images (Sawayama et al., 2015, image acquired 2012) work well in the intertidal and shallow subtidal, as evidenced by the very similar results (note, however, that slightly different habitat classes were used in the two studies). Unfortunately, aerial and satellite imagery fail to detect meadows in deep water. *Reproduced from: Kneer (2013) and Sawayama et al.(2015).*

Sunda ecoregion, Torres-Pulliza et al., 2013) covering the whole Archipelago has yet to be created. It remains a challenge that the approaches used by the pioneering studies by Kneer (2013, aerial photography) and Sawayama et al. (2015, remote sensing) cannot detect sparse deep-water meadows (Erftemeijer and Stapel, 1999), which may cover extensive areas of the lagoon floor. *Halodule* sp. and *Halophila* sp. meadows have been detected in water depths of up to 43 m, but their total extent can only be determined by the extensive use of towed underwater video sleds or comparable novel technologies (McKenzie et al., 2020).

Based on the available data, the following estimate is made regarding the total seagrass area in the Spermonde Archipelago. There are ca. 120 reef islands (Janβen et al., 2017; Kench and Mann, 2017). Kneer (2013) studied five islands and found one to be without seagrasses, three with ca. 0.3 $km^2$, and one with 0.6 $km^2$ of seagrass area in the intertidal to shallow subtidal reef area. If these numbers are extrapolated to all 120 islands, there would be ca. 36 $km^2$ in the whole Archipelago. To put this number into perspective, 36 $km^2$ is three times the total area found all around Bali (12 $km^2$) and double the area found around Lombok or in the entire Komodo Marine National Park (18 $km^2$ each) (Torres-Pulliza et al., 2013). These 36 $km^2$ are the area which is in the intertidal and shallow subtidal and could be detected and confirmed by remote sensing. However, the discovery of sparse but extensive *Halodule* sp. and *Halophila* sp. meadows on the lagoon floor at water depths between 15 and 30 m (Erftemeijer and Stapel, 1999) (Fig. 6.3) indicates that there are large deep-water seagrass areas, which cannot be detected by remote sensing. Between 2014 and 2015, a dropcam was employed to take videos of the lagoon floor, and some of the deeper submerged reefs on 37 locations distributed over six transects from west to east between the Maros River mouth to the north and the mouth of the Jene Berang River to the south. Videos were taken over a wide range of different water depths, from less than 4 m at the three shallowest points to between 70 and 80 m at the deepest point. On five of these videos from water depths between 10 and 38 m, sparse stands of *Halophila* sp. were identified with certainty, and on four more videos from 4, 30, 41, and 43 m depth structures resembling *Halophila* leaves were seen, but the quality of the video did not allow a certain identification (Fig. 6.3). We are aware that not all areas of the shelf are suitable for seagrass growth; however, our sampling points were distributed evenly, and therefore, we make the rough estimation that at least 14% of the total shelf area (or at least 900 $km^2$ shelf area which is ca. 7000 $km^2$) could support deep-water meadows! These patches have a very low biomass compared with the dense top-reef meadows, so while the discovery potentially increases the total seagrass area 25-fold, their biomass per hectare is roughly two orders of magnitude less than that of the top-reef meadows (based on published values, e.g., 11 g $m^{-2}$ for deep-water *Halophila ovalis* compared with 900–1700 g $m^{-2}$ for mixed species shallow water meadows, Erftemeijer and Stapel, 1999). Therefore, in spite of their potentially very large area, their contribution to the total seagrass biomass in the Archipelago is unlikely to be more than one fraction of the total. However, despite their low biomass, these deep-water meadows have been shown to be highly productive

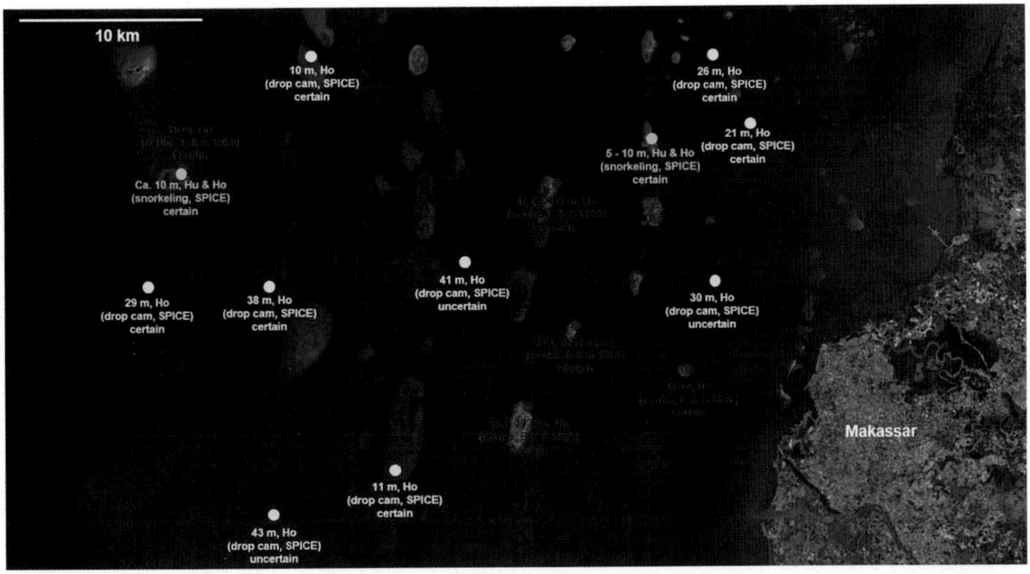


**FIGURE 6.3** Location of deep-water seagrass meadows (only > 4 m depicted) found during surveys conducted by Erftemeijer and Stapel (1999) (using SCUBA, dark green) and during the SPICE project (using snorkeling, light green; or using a dropcam, orange if certain, and red if doubtful). *White dots* show locations where the dropcam was deployed, but no (potential) seagrass was seen. While the majority of these communities were found at depths between 5 and 15 m, we identified *Halophila ovalis* with certainty at 38 m and with some uncertainty at 43 m depth. At the moment, these are only point records, and the actual extent of these stands remains unknown. Water depths in m and species codes (**Ho** = *Halophila ovalis*, **Hu** = *Halodule uninervis*) are also displayed on the map. *Modified from: Landsat/Copernicus. (December 31, 2016). www.google.com/earth/index.html.*

(Erftemeijer and Stapel, 1999) and could therefore play a significant role in the food webs and carbon budget of the lagoon floor habitat.

## 6.2.2 The structure of tropical seagrass bed systems

Seagrass systems represent communities or even ecosystems distinctly separated from adjacent benthic assemblages such as bare sand flats or coral reefs. They include the main structural elements of an ecosystem in one system, such as primary producers, consumers, decomposers, and abiotic parameters. Seagrass meadows form a special three-dimensional structure depending on the particular seagrass assemblage. This structure is developed poorly in pioneer assemblages dominated by small genera such as *Halophila* and *Halodule*. However, it can exert a distinct pattern as in those consisting of a mixture of tall *Enhalus acoroides* plants and *Thalassia hemprichii* as understory. Throughout the Indonesian Archipelago, multispecies or mixed seagrass beds consisting of up to eight species are of a relatively common occurrence (Ambo-Rappe, 2016). This is one of the key structural features clearly distinguishing Indonesian seagrass communities from those of the Caribbean, where monospecific beds are common.

In seagrass beds of the Spermonde Archipelago, Southwest Sulawesi, biomass of seagrasses ranged between 270 g and 7.5 g dw $m^{-2}$ depending on, to the particular community, the water depth, the exposure to currents, and the position of the seagrass bed at the shelf as well as the season (personal observations). In the SPICE project, biomass of seagrasses was recorded during rainy and dry season from two transects at each season and three islands showing a different position at the shelf area of the Spermonde Archipelago. At the most coastal location, the island of Barrang Lompo, four species, *E. acoroides, T. hemprichii, Cymodocea serrulata,* and *Cymodocea nodosa,* could be detected dominating the biomass at an average of 63 $\pm$ 30 g dw $m^{-2}$. In both transects, biomass was decreasing from the beach down to the reef edge, from 106 $\pm$ 47 g dw $m^{-2}$ to only 10 $\pm$ 16 g $m^{-2}$, respectively. Species diversity did not show a distinct trend along transects; however, species richness of seagrasses was higher in the upper stations close to the beach. In contrast to that, Erftemeier and Herman (1994) found a higher biomass range for *E. acoroides* (800 to 1441 g dw $m^{-2}$) and *T. hemprichii* (100 to 308 g dw $m^{-2}$) at the same island with maximum values in July and lower values in August to the end of the year. Vonk et al. (2008) found, at the island Bone Batang in the Spermonde Archipelago, a mean leaf and belowground biomass of 118 and 625 g dw $m^{-2}$ in dense, and 47 and 506 g dw $m^{-2}$ in sparse seagrass beds.

Further offshore to the central shelf, at the island Sarappokeke, the large seagrass *E. acoroides* is missing. In general, seagrass biomass there is in a range between 71 $\pm$ 69 to 8 $\pm$ 7 g dw $m^{-2}$ (i.e., 47 $\pm$ 35 on average); only in some very dense patches of *T. hemprichii,* it may reach a biomass of up to 124 $\pm$ 68 g $m^{-2}$. Although average biomass is lower, the species richness per transect is on average higher compared with the more coastal site Barrang Lompo. The highest biomass of seagrasses we found surprisingly in islands situated at the seaward shelf edge (outer rim) of the Spermonde Archipelago at the island of Kapoposang. The large seagrass *E. acoroides* was present at the nearshore stations of the transects, and thus, biomass was on average 111 $\pm$ 53 g dw $m^{-2}$, and on one transect, it reached up to 270 g dw $m^{-2}$.

Comparing the seagrass biomass with nutrient data available for that year, we suggest a congruence between the high seagrass biomass with higher nutrients at the coast of Sulawesi, lower nutrients and also biomasses of seagrasses in the central shelf area, and again higher nutrient levels at the shelf edge. The higher nutrient levels at the outer shelf edge were due to local upwelling from the adjacent deep-water regions of the strait of Makassar as a result of the currents parallel to the coast and the prevailing offshore winds. However, we need more evidence particularly from CN analyses of seagrass material to be sure whether a general nutrient gradient or factors at a minor or local scale are responsible for this pattern.

Because of the great number of leaves per unit area, seagrasses enlarge the surface area for settling of primary producers such as micro- and macroalgae or sessile zoobenthos from different taxa by a factor of 20 (Couchman, 1987, p. 4). In the Spermonde Archipelago, associated micro- and macroalgae augment the primary producer biomass by on average 17% related to dry mass. Calcareous algae and diatoms form the base for epiphyte growth, followed by red and green filamentous algae. The distribution of epiphyte biomass over the shelf area followed the same pattern as observed for seagrass biomass showing maximum values at the shelf edge ($14.7 \pm 8.9$ g dw $m^{-2}$) and minimum values at the central shelf ($5.2 \pm 8.4$ g dw $m^{-2}$) while increasing toward the coast ($10.3 \pm 10.1$ g dw $m^{-2}$).

The consumer assemblage of a tropical seagrass bed is highly diverse and consists of both very specialized residents and animals spending only parts of their life cycle within a seagrass bed. In the seagrass beds of the Spermonde Archipelago, we sampled 158 different taxa of macroinvertebrates, but only 86 of them could be identified to genus and even species level.

Macrobenthic animals are important components of tropical seagrass beds, and they settle either in the sediment of seagrass beds (infauna) or attached to the different parts of the plants (sessile epifauna). Other fauna components are mobile and live either at the sediment surface or move along the leaves or swim within the seagrass carpet. Most of the invertebrate faunal assemblage in the investigated sites of the Spermonde Archipelago is consisting of mollusks (52 species) and echinoderms (35 species), whereas hydrozoans (mainly corals and soft corals, 24 species), crustaceans (12 species), polychaetes, or other groups are less species rich. In our investigated sites also, sponges probably may reach high numbers of species (36), but we could not identify most of the species. The consumer assemblage represents not only many commercially important species, but also species that are endangered and already extinct in many places.

## 6.3 Seagrass ecology

### 6.3.1 The historic loss of megaherbivores and today's important role of burrowing shrimp

Historically, seagrass meadows in the tropical regions of the world were probably shaped by the sustained grazing activities of megaherbivores such as dugongs and green turtles, which in turn were controlled by large sharks (Heithaus et al., 2006). The interactions between large sharks and large herbivores might have had a substantial impact on the structure of the seagrass communities. While no information is available about the historical abundances of these animals in the Spermonde Archipelago, the fact that local communities living in the area have been part of large trading networks for centuries makes it likely that humans are the reason these animals are so rare today that they can be considered functionally extinct (Schwerdtner Máñez and Ferse 2010; Ferse et al., 2012; Moore et al., 2017). In theory, this should have led to an increase in the abundance of herbivorous fishes such as parrot fishes and rabbit fishes (Hemminga and Duarte, 2000; Kirsch et al., 2002; Ogden and Ziemann, 1977; Randall, 1965; Unsworth et al., 2007); however, a recent rapid human population growth resulted in an intense fishing pressure, and today about 50% of the seagrass primary leaf production enters the detrital pathway (Vonk, 2008). This may be the reason why today there is an abundant invertebrate community, comprising of the sea urchin *Tripneustes gratilla* and many species of burrowing alpheid and axiid shrimp. A comparative study found that invertebrate densities in seagrass meadows on Derawan Island (East Kalimantan), where turtles are still abundant and are significantly less than at Barrang Lompo and Bone Batang Island in the Spermonde Archipelago, supports this hypothesis. The high diversity of associated fauna in Indo-Pacific meadows might explain the resilience of the meadows in the Spermonde Archipelago. In contrast, Caribbean meadows have a lower seagrass and fauna diversity. There, the removal of large herbivores resulted in a buildup of detritus in the sediment, resulting in infections with slime molds, which caused seagrass die-offs (Jackson, 2001). Also, the seagrass *Halophila stipulacea*, native to the Red Sea, the Persian Gulf, and the Indian Ocean, was recorded in the Caribbean Sea and has meanwhile spread to most of the Caribbean islands (Winters et al., 2020).

Burrowing crustaceans from the decapod infraorders Caridea and Axiidea are very abundant in and around seagrass meadows in the Spermonde Archipelago and have been shown to play major ecological roles. Caridean shrimp in the family Alpheidea (the snapping shrimp) are known for their noncommensal mutualism with gobiid fishes of the genera *Amblygobius* and *Cryptocentrus*, where the fish serves as a watchdog while the shrimp (which has poor eyesight) is foraging aboveground. In turn, the fish are tolerated as commensals in the burrows (even though they do not contribute to burrow construction). Especially the species *Alpheus macellarius* (Fig. 6.4) has been shown to harvest living seagrass leaves by cutting them with their claws close to the sediment surface around their burrow openings and then translocating the leaf fragments belowground for storage in special chambers and later consumption (Stapel and Erftemeijer, 2000, as "*Alpheus edamensis*"; Nacorda, 2008; Vonk et al., 2008). Depending on shrimp density and activity, a substantial part of the leaf primary production can be harvested by snapping shrimp populations (Stapel and Erftemeijer, 2000, as "*A. edamensis*").

Axiidean shrimps (formerly grouped together with the infraorder Gebiidea in the paraphyletic "Thalassinidea") are also burrowers. Some representatives such as *Neaxius acanthus* (Fig. 6.4) and *Corallianassa coutierei* (Fig. 6.4) catch seagrass detritus, which drifts past their burrow entrance, but they never entirely leave their burrows (Kneer et al., 2008b; Vonk et al., 2008b). *N. acanthus* is also special because an entire community of commensal organisms has been found in its burrows. The most conspicuous are the bivalve *Barrimysia cumingii* and the enigmatic goby *Austrolethops wardi* (Kneer et al., 2008a,b), the only goby found to rely mainly on seagrass for its nutrition (Liu et al., 2008). An even wider variety of commensals were found in the burrow of *Axiopsis serratifrons* (Fig. 6.4) on Derawan Island in East Kalimantan (Borneo): *Ascidia subterranea*, a new species of tunicate, the small shrimp *Rostronia stylirostris* (inside the tunicate), four species of bivalves, three species of polychaetes, one gastropod, one polyplacophoran, and one sponge species (Kneer et al., 2013b). While individuals of *Axiopsis* were caught in the Spermonde Archipelago, no burrows of this species have yet been excavated here.

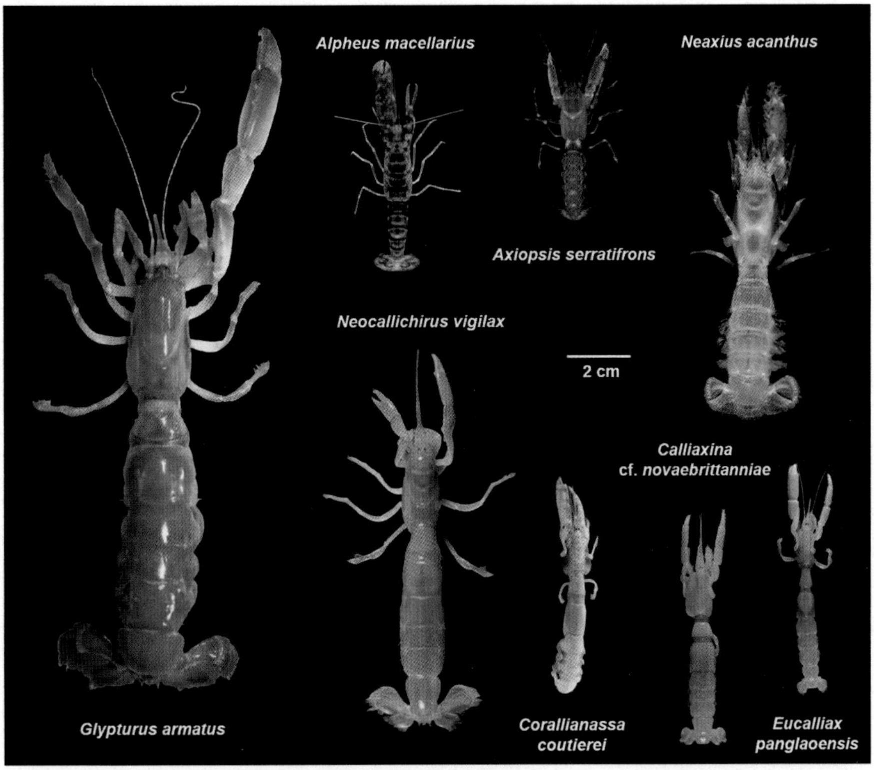


**FIGURE 6.4** Burrowing shrimp species with important ecological roles that were studied in the SPICE project: top from left to right: *Alpheus macellarius* (Caridea: Alpheidae), *Axiopsis serratifrons* (Axiidea: Axiidae), *Neaxius acanthus* (Axiidea: Strahlaxiidae); bottom from left to right: *Glypturus armatus* (Axiidea: Callianassidae), *Neocallichirus vigilax* (Axiidea: Callianassidae), *Corallianassa coutierei* (Axiidea: Callianassidae), *Calliaxina* cf. *novaebritanniae* (Axiidea: Callianassidae), *Eucalliax panglaoensis* (Axiidea: Callianassidae). *Reproduced from: Anker et al. (2015)* (A. macellarius, N. acanthus)*, Poore (2018)* (A. serratifrons)*, Dworschak (2011)* (N. vigilax)*, http://decapoda.free.fr* (C. coutierei)*, Dworschak (2018)* (C. *cf.* novaebritanniae, E. panglaoensis)*. Pictures have been resized to illustrate the relative size of fully grown specimens.*

Other axiidean species such as *Glypturus armatus* (Fig. 6.4) and *Neocallichirus vigilax* (Fig. 6.4) are mainly detritus feeders, which are almost never seen near the surface. However, they employ a feeding strategy called conveyer-belt deposit feeding, in which clean sediment is pumped out through an exhalant shaft and forms a conspicuous sediment mound. In the course of several days, the flanks of these mounds get overgrown with benthic microalgae and then fall back into funnels surrounding the mound where the shrimp then has access to surface-derived material without risking predation at the surface. In the intertidal, where *G. armatus* is most abundant, closely spaced sediment mounds coalesce into larger mounds. These composite structures are surprisingly long-lived, and the sediment turnover prevents them from being colonized by seagrasses (mainly *E. acoroides* and *T. hemprichii*), which are restricted to areas

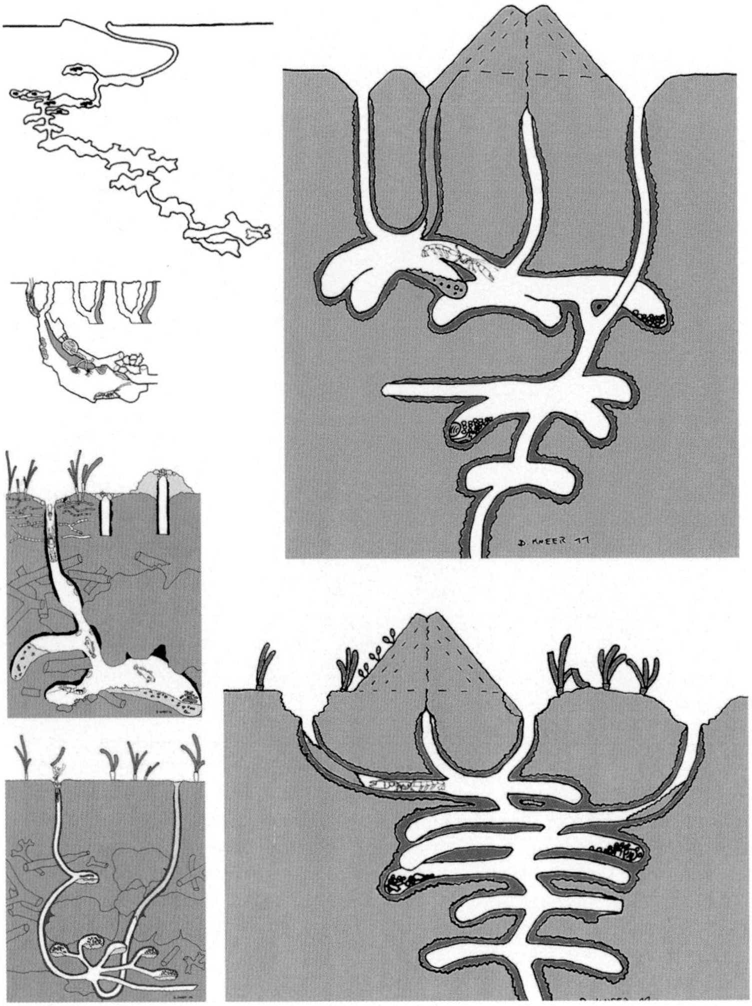


**FIGURE 6.5** Left from top to bottom: Reconstructed burrow shape of *Alpheus macellarius, Axiopsis serratifrons, Neaxius acanthus,* and *Corallianassa coutierei*; Right from top to bottom: Reconstructed burrow shape of *Neocallichirus vigilax* and *Glypturus armatus. Burrow of* A. macellarius *after Vonk et al. (2008c), Burrow of* Axiopsis serratifrons *after Kneer et al. (2013b).*

between mounds, and which become ponded at low tide (Kneer et al., 2013a; see also Curran and Martin, 2003 for *Glypturus acantochirus* vs. *Thalassia testudinum* and *Halodule wrightii*) (Figs. 6.5 and 6.6). In the subtidal, *G. armatus* gets replaced by *Neocallichirus vigilax*, but again seagrasses (mainly *Halophila ovalis*) are restricted to areas between mounds (Kneer et al., 2013a). This has been observed elsewhere (Suchanek, 1983 for *Neocallichirus maryae* and other species vs. *T. testudinum* and other species, Vaugelas, 1985 for *G. armatus* vs. *Halophila* sp.).

**FIGURE 6.6** Left: Women and children collecting invertebrates around large coalesced mounds of *Glypturus armatus* in the intertidal next to a mangrove fringe (*Rhizophora stylosa*) on Tanakeke Island; Right: Extensive intertidal seagrass meadow (*Enhalus acoroides* and *Thalassia hemprichii*) between coalesced mounds of *Glypturus armatus* on Tanakeke Island, the mounds are probably essential in retaining small tidal ponds that reduce desiccation stress for the seagrass plants.

Almost nothing is known about the lifestyle of the enigmatic *Calliaxina* cf. *novaebrittanniae* (Fig. 6.4) and *Eucalliax panglaoensis* (Fig. 6.4), except that they occur in large numbers in and around seagrass sediments and that they likely heavily depend on sulfide-metabolizing bacteria for their nutrition (as the isotopic signature of their muscle tissue resembles the signature of bivalve species for which such a dependence has been proven). While all other species could be kept and observed in Aquaria for many weeks or even months, *C.* cf. *novaebrittanniae* and *E. panglaoensis* always died within hours after the introduction into aquaria.

## 6.3.2 Macrobenthic communities

The presence of seagrass attracts many marine organisms for feeding, spawning, mating, nursing, and avoiding predators and settlement. Mutual interactions between fauna and seagrasses have created complex ecosystems. One of the important organism groups in seagrass beds is macrozoobenthic fauna. Investigating the macrofauna of seagrass beds of the Spermonde Archipelago, we focused on seagrass beds around Bone Batang Island. There, we found five genera and seven species of seagrass from two different families. These seagrass species mostly form mixed meadows. Macrozoobenthos is rich in species and most abundant compared with other groups of marine fauna in the Bone Batang seagrass bed. Based on its habitat, macrozoobenthos divides in two categories. Epifauna consists of animals that live and settle at the sediment surface and infauna lives buried or in burrows within the sediment.

Macrozoobenthos, especially infaunal species, are poorly known from tropical seagrass beds of Indonesia. Many species can settle in greater sediment depth as can be sampled by standard corers or grabs. Therefore, we used large plastic awnings of 12 $m^2$ to cover the sediment surface and to weigh this cover down with sand to reduce the oxygen within the sediment overnight. This was done to force the infaunal species to leave the sediment and to accumulate at the sediment surface where they could be easily collected. We found 273 species in seagrass beds. The number of benthic species from seagrass meadows was higher compared with bare sand with only 208 species. Macrozoobenthos consisted mainly of bivalves (24% of species), gastropods (22%), polychaetes (19%), crustaceans (29% i.e., crabs (16%), and shrimps (13%)).

In the seagrass bed, the highest density of macrozoobenthic fauna was found for the alpheid shrimp *A. macellarius* with an average density of 2.8 individuals/m$^2$, while in adjacent bare sands, the bivalves *Periglypta reticulata* and *Tellina virgata* dominated the assemblage with an average density of 1.8 individuals/m$^2$ and 1.2 individuals/m$^2$, respectively. Bray–Curtis similarity from all sampling stations varied between 22.5% and 49.8%. This showed that all stations have low similarity in macrozoobenthic composition. According to Hemminga and Duarte (2000), this relatively high variability may be caused by the age of the seagrass meadow in addition to environmental factors. Mature seagrass meadows have a higher macrozoobenthic species number compared with younger ones. Based on *Pearson Product Moment* Correlation Analysis, the correlation between seagrass density and macrozoobenthic density was relatively low (0.30231). Also increasing seagrass shoot density was not correlated with macrozoobenthic species density.

To investigate the influence of habitat characteristics and environmental factors on seagrass species composition and its spatial distribution, we took data of seagrass density from eight stations around the island reef flat. The results of Bray–Curtis similarity and n-MDS analysis showed that the seagrass composition among stations in Bone Batang Island consists of similar species with a similarity of more than 75% except for two sampling stations. The result of correspondence analysis showed that spatial distribution among stations was different and influenced by habitat characteristics.

Spatial distribution and species composition also showed that seagrass beds in Bone Batang were in a mature/climax stage. Associated fauna in Bone Batang seagrass beds are abundant and dominated by infaunal/burrower species. It can be concluded that mixed seagrass communities in Bone Batang revealed similar species composition. However, every station was dominated by different seagrass species and was influenced by other habitat characteristics and by a different associated fauna as well. How this fauna changed with the type of seagrass bed could be exemplified by the associated bivalve assemblage. Correlation analysis showed that the presence of bivalves in most types of seagrass meadows does depend neither on species number of seagrasses nor on seagrass shoot density, but it reflects on the phase of seagrass meadow succession.

Burrowing shrimps form an important infaunal group in tropical seagrass ecosystems due to their role in sediment aeration, water exchange, sediment mixing, and improving nutrient conditions within the sediment and the overlying water. In the Spermonde Archipelago, seagrass beds are abundant and covering every island reef flat. To throw more light on the species composition of this burrowing shrimp in the Spermonde Archipelago, we investigated species composition and diversity of this group at Bone

Batang Island. Species composition and diversity of burrowing shrimps were related to the different types of habitat and depended on life modes of the species, but hardly displayed a dependency on seagrass density or seagrass composition.

As mentioned earlier, burrowing shrimps are of high importance for the ecosystem functioning of tropical seagrass beds, as they are contributing mainly to the recycling of dead seagrass leaves in these systems. Using the aforementioned method of covering sediment areas with plastic awning, we collected and determined 22 species of burrowing shrimps from all stations. The highest density of burrowing shrimp was built by *A. macellarius* with 2.8 individuals/$m^2$. The diversity of burrowing shrimps tends to be higher in unvegetated sediments compared with the entire seagrass bed with a diversity index (Shannon Wiener) of 0.33–3.18 and 0.39–0.96, respectively. Species abundances of the different burrowing shrimp species were more uniformly distributed with higher values in adjacent bare sediments compared with the entire seagrass bed with 0.49–2.33 and 0.24–0.75, respectively.

### 6.3.3 The food web and the trophic pyramid in tropical seagrass beds

Primary producers are the base of most food webs exposed to light. Mainly plants represent this group by converting inorganic carbon by means of sunlight into organic carbon while releasing oxygen. Seagrass plants and their associated epiphytes represent the dominant primary producers in a seagrass meadow. Erftemeijer (1993, p. 173) also measured primary production by using light and dark bell jars and found net primary production rates generally below 500 mg C $m^{-2}$ $d^{-1}$ in the Spermonde Archipelago. He concluded that heterotrophic activity due to respiration and remineralization processes and also chemical oxygen uptake by dead organic material is responsible for the low net oxygen excess. Seagrasses on a reef flat, phytoplankton and microphytobenthos, such as epiphytes, are also a part of this community especially being present in the water column, in the sediment between the seagrass plants, above the canopy and at the seagrass leaves, respectively. All these primary producers require light, nutrients, and carbon dioxide for growth, and it is therefore inevitable that they compete for these resources. Epiphytes have been estimated to contribute with 36% to the primary production in Indonesian waters (Lindeboom and Sandee, 1989). Seagrass plants prefer habitats with oligotrophic waters, because they can obtain nutrients not only from leaves but also from the soil by their root system (Stapel et al., 1996), but they can also use organic substances and especially inorganic nitrogen in addition (Vonk et al., 2008c). Under conditions of increasing nutrient content in the water, they will be easily outcompeted by the accompanying epiphytic or filamentous algal community, and this can lead to heavy shading that hampers the development of seagrasses.

The material formed by plants of a community is transferred to the next trophic level in the food web, the primary consumers. Only few consumers directly graze on seagrass plants, because these plants contain phenolic compounds and tannins that are inedible for most of the herbivores (Jernakoff et al., 1996; Kikuchi and Perés, 1977; Klumpp et al., 1989). However, there are specialists for this group among crustaceans, echinoderms, fish, and sea mammals. In pristine tropical seagrass beds, macrograzers are very important keystone species structuring the plant community and sustaining a high diversity of their food plants (Sheppard et al., 2007). Especially dugongs (*Dugong dugong*) and green turtles (*Chelonia mydas*) graze upon small pioneer species in the succession of a seagrass bed (Bjorndal, 1980; Moore et al., 2017; Sheppard et al., 2007). These vertebrates form a mosaic of different developmental stages of seagrass beds in close spatial vicinity which gives a complex three-dimensional structure that is attractive also to other inhabitants such as crustaceans and fish. Thereby a single dugong consumes a remarkable amount of 30 kg fresh plant mass per day.

Green turtles (*C. mydas*) also eat as adults mainly seagrasses especially the species *T. hemprichii* grazing only its aboveground parts and let the roots untouched. Depending on the region, also other small seagrass species constitute the main part of the diet of green turtles, particularly *Halophila* species and the wiry and thin-leaved seagrasses *Syringodium isoetifolium*, *Halodule uninervis*, and *C. serrulata*. The food requirement of an adult green turtle of about 66 kg live weight is about 2 kg fresh mass (or approximately 218 g dry mass) per day. In some cases, green turtles can also induce overgrazing (Christianen et al., 2014).

Various tropical fish species feed on seagrasses, although they do not graze on whole leaves, but bite pieces out of the leaf so that characteristic bite marks remain. Especially rabbit fish (Siganidae) and parrot fish (Scaridae) provide for a rapid turnover of plant material (Unsworth et al., 2007b). Herbivorous fish of tropical seagrass beds not only utilize seagrass plants but also use those parts that contain a thick epiphytic and epizootic cover. Juvenile rabbit fish live in large flocks that graze only the epiphyte cover and thus play an important role in cleaning the seagrass leaves. To measure the impact of herbivorous fish on the seagrass system, we made experiments in which we included siganid fish in large cages and compared the epiphyte and seagrass growth. This study shows that epiphytes are an overutilized food resource for fish, while seagrass leaves are not substantially grazed in seagrass beds of South Sulawesi. Negative effects of epiphyte cover on seagrass vitality are therefore a rather long-term phenomenon. The importance of seagrass as direct food source is diminished by the high production of epiphytes. The present density of siganids depends on this sensitive resource. Abundance would suffer from reduced epiphyte cover directly and from seagrass decline indirectly. Siganid fish is economically and nutritionally important. The protection of its seagrass habitat is thus vital to the coastal population.

Among invertebrates, mainly sea urchins graze directly on seagrasses. In tropical seagrass meadows, sea urchins of the genus *Diadema* are abundant. These sea urchins are a boon for coral reefs cleaning them from vegetation cover, but a bane for seagrass beds where they can also demolish large parts of the vegetation. The most effective grazer on tropical seagrasses is the sea urchin *T. gratilla* that may consume a large amount of seagrasses (Vonk et al., 2008b). Investigations at the Spermonde Archipelago resulted in average densities of *T. gratilla* up to 1.55 $m^{-2}$. The species consumes up to 26% of the aboveground seagrass production and influences in this way the composition of species in a mixed seagrass bed. *T. gratilla* grazing also may play a role in preserving nitrogen for the meadow, because the heavy grazing stimulates growth of young seagrass leaves which accumule more nitrogen than older leaves, that passes into short circuits in N-cycling of leaf material (Vonk et al., 2008d).

Phytoplankton is the main food source for suspension feeders that are represented by bivalves. In seagrass beds of the Spermonde Archipelago, the species *Modiolus* sp. (named *micropterus*) is apparent in high densities, in some places up to 280 individuals per $m^{-2}$ (own unpublished data) and is firmly tacked by their byssus threats on the thick rhizomes of the large seagrass *E. acoroides*. The species is only sparsely covered by sediment and is utilizing phytoplankton by filtering the water layer at the sediment-water interface. These mussels reach high densities, but their impact and their filtration potential are still unknown, but may be expected as high as that of *Mytilus edulis* in temperate areas. Other important suspension feeders in South Sulawesi seagrass beds are the species of the family Pinnidae. Three species of this family occur in the Spermonde Archipelago, *Atrina vexillum*, *Pinna bicolor*, and *Pinna muricata*. All of them depend on phytoplankton as main food source. Phytoplankton is mainly present during conditions of higher nutrient contents in the water column, i.e. during the rainy season or in local upwelling areas. In tropical areas, suspension feeders have to survive often longer periods of phytoplankton deficiency, which has led in many species to special adaptations such as symbiosis with microalgae. This type of symbiosis is found especially in giant clams *Tridacna* spp. that are frequently found in seagrass beds, where they can use the high light availability. In tropical seagrass beds other invertebrate suspension feeders may also reach a high abundance, particularly sponges, tunicates, and polychaetes such as the family Serpulidae and Spionidae.

Most animals of a tropical seagrass bed do not use fresh and living plants, but dead organic material that is either produced within the seagrass bed directly (autochthonous detritus production) or is imported to the system by currents (allochthonous detritus production) and settles within the dense canopy. Due to the growth of bacteria that are decomposing this material slowly, the detritus receives organic nitrogen as a further nutrient component improving the quality of the food. As already mentioned in Section 6.3.1, especially burrowing shrimps (such as *A. macellarius* and *N. acanthus*) play an important role for the detritus decomposition by creating burrows containing special subsystems of the seagrass bed containing an own food web that depends solely on seagrass detritus (see Section 6.3.1 and Kneer et al., 2008a,b; Vonk et al., 2008b).

Other animals such as polychaetes and sipunculids use older detritus material as food. Other large detritus feeders live at the sediment surface such as the starfish *Protoreaster nodosus* and the sea cucumber *Opheodesoma* sp. The latter reaches a size of up to 1.50 m, and with its glutinous collar of tentacles, the animal wipes down seagrass leaves and bottom sediment and sucks food from its tentacles. Detritus feeders have developed many strategies to forage for its food from eating the sediment with rich organic compounds until creating complex burrow structures to preserve and to enrich the organic material to increase the food availability. In most cases detritus feeders also ingest bacteria as food, because these organisms are settling at the surface of the dead organic material. The percentage of bacteria within a diet spectrum of a detritus feeder is much smaller compared with the detrital material, because of the comparatively low biomass of the bacteria. However, because detritus is defined as a primary food source (contributing to trophic level 1), bacteria are the detritus feeder primary consumers. Considering bacteria as a part of the diet, detritus feeders become partly tertiary consumers. In addition to the adaptations for feeding on detritus and bacteria, some animals in tropical seagrass beds have developed a special symbiosis with bacteria to gain energy for growth and metabolic processes.

Other animals in the seagrass bed use very special pathways such as symbiosis with bacteria and are therefore difficult to integrate into a food web and energy flow scheme. Bivalves of the family Lucinidae, occurring frequently in tropical seagrass beds of the Spermonde Archipelago, developed a symbiosis with sulfur-oxidizing bacteria (Roeselers and Newton, 2012). These endobenthic bivalves live in sediments rich in sulfide (Seilacher, 1990). The bivalve draws sulfide-rich water into the inhalant siphon pouring it into its gills to make sulfur and oxygen available for its symbionts (Seilacher, 1990). The endosymbionts use these substrates to fix carbon into organic compounds, which can be directly used by the host as nutrients (König et al., 2015). During periods of starvation, lucinids may harvest and digest their symbionts as food, but a gut system is not developed in these bivalves.

The different animals forming the secondary and tertiary trophic level will be used as prey by predators. Most of the fish species in the tropical seagrass beds of the Spermonde Archipelago contribute to these higher trophic levels. Large amounts of zooplanktivorous fish such as *Atherinomorus lacunosus,* and some tropical clupeids forage over seagrass beds, where the availability of zooplankton is high due to the additional source of migrating benthic copepods that migrate from the bottom in seagrass beds to the overlying water, particularly at night (de Troch et al., 2001). Other zooplanktivorous fish such as the pipefish *Sygnathoides biaculeatus* feed the zooplankton within the canopy. The rich benthic resources are mainly exploited by benthivorous feeders such as the gerreid *Gerris oyena,* labrids *Cheilio inermis* and *Stethojulis strigiventer, the* lethrinid *Lethrinus obsoletus,* the nemipterid *Pentapodus trivittatus,* and the pomacentrid *Pomacentrus tripunctatus.* Both zooplankton and benthos and small fish can be eaten by omnivorous fish that are represented by the monocanthid *Acreichthys tomentosus* and the murray eel *Siderea picta.* The predator guild of fishes is highly diverse, and the species composition depends largely on the type and structure of the seagrass bed (see Section 6.4). The fish in the seagrass beds are mainly juvenile and are therefore of a small size, which is a suitable prey for piscivorus fish. These are mainly represented by belonids such as *Tylosurus crocodilus* and *Strongylura incisa,* but also small sharks such as *Carcharhinus limbatus* and barracudas such as *Sphyraena barracuda* and *Sphyraena obtusata* visit sometimes seagrass beds for foraging.

### 6.3.4 The function of seagrass meadows as water filters and buffers for land runoff

Due to the position between two important tropical coastal ecosystems, mangrove forests and coral reefs, tropical seagrass beds play an important role as buffers for the material export between both ecosystems and generally from the land to the open waters. However, the role of tropical seagrass beds in land–ocean interactions has rarely been a research objective. Investigations from Kenya emphasized the role seagrass beds play for the exchange of $CO_2$ with the atmosphere and for diminishing the outwelling of organic and inorganic carbon from mangrove areas to the seaward coastal ocean. Dissolved organic carbon is the dominant form of carbon in the water column over these seagrass beds, and the ratio of DOC/POC varies between 3 and 15 (Bouillon et al., 2004). Seagrass beds influence water movement by decreasing currents and waves and therefore increase sedimentation of particulate organic carbon suspended in the water column and reduce sediment resuspension. The material produced in seagrass beds has been suggested to contribute significantly to the export out of the system (Duarte and Cebrián, 1996; Stapel et al., 1997), but this was mainly observed for intertidal seagrass beds, and the exported material was trapped in the seaward subtidal beds where it contributes significantly to benthic remineralization (Boullion et al., 2004; Hemminga et al., 1994).

In the SPICE project, we could show that material losses, particularly leaf material, from the seagrass bed are prevented by burrowing shrimps, which collect drifting material and process it further within their burrows (see Section 6.3.1, Kneer et al., 2008b; Vonk et al., 2008b). This process stored the material within the seagrass bed where it was available for the seagrass plants as DIN and DIP after remineralization. This recycling is important in the frequently oligotrophic seagrass beds growing on coralligenous sediments. Biogeochemical processes in seagrass beds largely depend on the sediment type. In the Spermonde Archipelago, seagrass beds mainly grow on coralligenous carbonate sediments. These sediments are shown to be phosphorus limited because of their high carbonate binding capacity compared with seagrass beds on terrigenous sediments that are nitrogen limited (Short, 1987). However, in Indonesia, surprisingly high phosphate concentrations could be found in the upper layers of the sediment, suggesting a lower phosphate binding potential compared with the fine-grained sediments in the Caribbean (Erftemeyer and Middelburg, 1993; Short, 1987).

The exchange function of seagrass beds also changes with the location. The rich diversity of tropical seagrass beds has been characterized into four main categories such as "river estuaries," "coastal," "deep water," and "reef," and the main physical and chemical parameters that influence these systems are land and river runoff, physical disturbance, low light, and low nutrients, respectively (Carruthers et al., 2002).These different impacts form the exchange seagrass beds may have between bottom and water as well as between seagrass systems and adjacent communities and ecosystems within the coastal zone.

### 6.3.5 Carbon storage

There is a growing concern about anthropogenic climate change caused by the burning of fossil fuel, combined with land use change and cement production. As a result, research has recently been directed to quantifying the role seagrass meadows can play in mitigating global $CO_2$ emissions by taking up carbon and storing it in underground deposits that could last for millennia (Pergent-Martini et al., 2021). This interest was initially fueled by the discovery that *Posidonia oceanica* meadows in the Mediterranean grow on top of deposits of peat (termed "mattes") many meters thick, which are unparalleled in other seagrass systems around the world (Boudouresque et al., 1980). However, even though these deposits are mostly a lot thinner in the Southeast Asian region, the sheer extent of the seagrass meadows there still makes them valuable contributors to climate change mitigation. Fortes (2017) estimated that Southeast Asia has 30%–40% of the total seagrass and mangrove area of the world and that the amount of carbon sequestered annually by these two ecosystems together is equivalent to the amount emitted by all the 115 million cars in use in the region (see also Alongi et al., 2016).

Seagrass meadows have been known to provide a subsidy to nearby ecosystems through the export of particulate organic matter and plant and animal biomass (Heck et al., 2008). Therefore, the contribution of seagrass meadows to carbon sequestration is not limited to carbon burial within seagrass sediments. A significant proportion of the organic carbon in sediments outside seagrass meadows, as well as in the deep sea, can also be traced to seagrass sources (Duarte and Krause-Jensen, 2017; Kennedy et al., 2010). A series of measurements conducted during SPICE III showed this also for the Spermonde Archipelago. The isotopic signature of organic carbon in sediments from most locations on the lagoon floor closely resembled the signature of seagrass leaves. Even if the nearest meadow was several kilometers away, the isotopic signature of sediment organic carbon still resembled the signature of seagrass leaves and not the one of suspended matter/plankton. Only in sediment samples from those stations closest to the mainland and the big river mouths, the isotopic signature more closely resembled the signature of mangrove leaves. Thus, our results also support the notion that to fully account for the benefits of seagrass conservation in terms of climate change mitigation, the role of seagrass meadows in supporting carbon sequestration beyond their physical boundaries needs to be considered.

### 6.3.6 Seagrass beds as carbon sinks

Seagrasses are globally important; hence, an important component (10%–18%) of the total carbon burial in the ocean is assumed to take place in seagrass sediments (Kennedy et al., 2010).

Dense seagrass beds are sinks for carbon. The main carbon import is via organic particles, which settle among the dense seagrass canopy. Material loss by drifting feces of associated animals is too low to compensate for the sedimentation of particles (Asmus, 2011, pp. 1–209). Another mechanism is the carbon fixation by seagrass beds due to its photosynthetic activity. The fixed carbon will be stored in live tissue of the primary producers. Carbon stored in leaves has a shorter residence time than that of roots and rhizomes persisting longer in the sediment also as dead material buried in the sediment. DIC shows a net uptake by plant assimilation within the seagrass meadow. Respiration processes are distinctly lower (Asmus, 2011). Thus, a seagrass meadow is dependent on import of $CO_2$ from outside, and this is a distinct hint that seagrass systems may be limited by dissolved carbon components (Beer and Rehnberg, 1997; Zimmerman et al., 1997).

Organisms show a net export of carbon from the seagrass bed. An export of carbon by grazing animals that only visit seagrass beds for feeding can be strong and may lead to destruction of the total seagrass bed (Christianen et al., 2014). To sustain the living biomass within the system, productivity is the main regulator. Larval stages with low biomass enter the system; they grow within the seagrass bed by using the carbon resources (e.g., Moksnes, 2002), especially detritus and microphytobenthos, and emigrating after achieving larger biomasses or being eaten by fish or crabs. Production and subsequent export of organisms, but also of seagrass leaves (Peduzzi and Herndl, 1988), is therefore the main counteracting process in a seagrass bed to the prevailing particle accumulation and $CO_2$ assimilation. Information on exchange of DOC is poor. However, these components may show a low release, which only contributes a little to the total export of C from a dense seagrass bed. The surplus of assimilated C is stored within the seagrass bed mainly as refractory organic substance within the sediment.

Gullström et al. (2018) estimated the organic and inorganic carbon stocks in sediments of seagrass beds and compared it with that of unvegetated areas along a latitudinal gradient in the Western Indian Ocean from the tropics to the subtropics. They found a higher range of sedimentary organic carbon in seagrass sediments compared to unvegetated sediments and suggested that this community acts as an important carbon sink. On the other hand, the amount of inorganic carbon within the seagrass meadow sediments strongly correlated with the organic carbon. The authors interpreted the inorganic carbon as a potential carbon source that diminishes the function of seagrass beds as a carbon sink. However, seagrasses can also form $CaCO_3$, and this may also contribute to sediment inorganic carbon as has been shown for *T. testudinum* contributing to lime mud production (Enriquez and Schubert, 2014).

The dependence of the different carbon types on the sediment density, the landscape type, and the dominant seagrass species reveals that our understanding of plant-sediment interactions in seagrass beds is not yet sufficient to assess and evaluate the role of seagrass beds as carbon sinks. Furthermore, there is a requirement for more accurate knowledge of seagrass cover and better estimates for the total carbon sink capacity of seagrass meadows, and data from Southeast Asia are still scarce (Kennedy et al., 2010).

### 6.3.7 Trophic transfers from seagrass meadows to nearby ecosystems

Detached fresh seagrass leaves are positively buoyant, and on a global average, between 10% and 60% of the leaf primary production is exported from seagrass beds (Heck et al., 2008). Especially in tropical Southeast Asia, seagrasses often grow on fringing reefs, separated from open oceanic settings only by narrow expanses of reef slope. Previously, the halfbeak *Hemiramphus far* (Hemiramphidae) has been observed consuming floating leaves above seagrass meadows, which could contribute to nutrient retention (Vonk, 2008).

In the Wakatobi Marine National Park, Southeast Sulawesi, Indonesia, the black triggerfish *Melichthys niger* (Balistidae) dominates the fish community over reef slopes in terms of both abundance and biomass, as observed elsewhere (Kavanagh and Olney, 2006). This species is a broad omnivore, with sometimes "piranha-like" feeding behavior (Kavanagh and Olney, 2006). While snorkeling in the northwestern corner of Wangiwangi Island (Fig. 6.7A) on December 9 (5 degrees14′46″ S, 123 degrees31′47″ E)

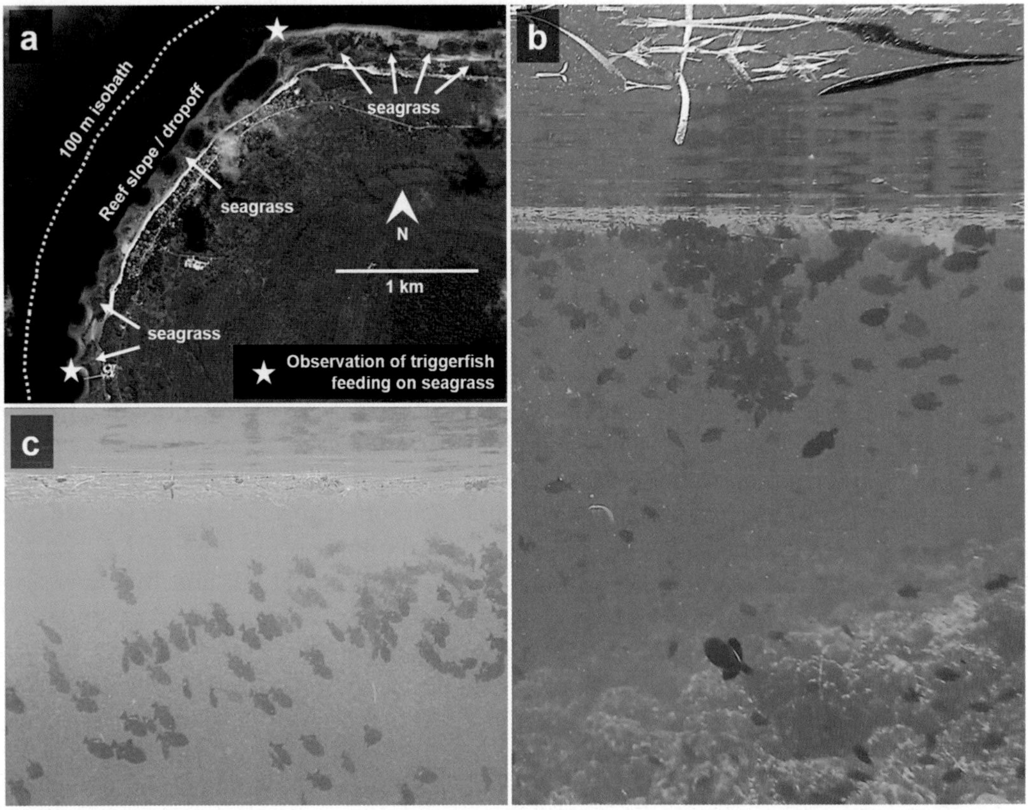


FIGURE 6.7 (A) The northwestern corner of Wangiwangi Island, bordered by a fringing reef covered by intertidal to shallow subtidal seagrass meadows and a steep drop off. Stars indicate where black triggerfish were observed feeding on drifting seagrass leaves. (B) As a raft of detrital seagrass leaf material passes over the reef slope, a school of black triggerfish rises to the surface to feed. (C) The raft has reached deep water and the fish school returns to the reef slope. *Based on: DigitalGlobe & TerraMetrics (Maxar Technologies). (June 1, 2013). www.google.com/earth/index.html.*

and 10 (5 degrees15′57″ S, 123 degrees31′03″ E), 2015, large schools were observed following rafts of floating seagrass *Thalassodendron ciliatum* (Cymodoceaceae) detritus and taking bites out of the material (Fig. 6.7B) until they had drifted over deep water, whereupon the fish either moved to the next raft or returned to near the reef slope (Fig. 6.7C) until another raft arrived. The fish engaged in this behavior almost continuously while observed, suggesting seagrass leaves or attached epibionts could constitute an important food source. This observation underlines the importance of tropical seagrass meadows for consumers in nearby ecosystems, and how their ongoing decline could have cascading effects beyond their boundaries.

## 6.4 Tropical seagrass beds as key habitat for fish species

### 6.4.1 Tropical seagrasses and their associated fish communities

The Coral Triangle in Southeast Asia harbors the highest diversity of marine fish globally (Fenner, 2007). Among the diversity of coastal ecosystems in the region, Indonesian seagrass beds boast high a diversity and biomass of fish and invertebrates (Cardinale et al., 2006; Gullström et al., 2002; MacArthur and Hyndes, 2007; Parrish, 1989). Of more than 80 fish species which can cooccur within a single seagrass meadow (Hutomo and Martosewojo, 1977; Pogoreutz et al., 2012; Unsworth et al., 2007), a significant proportion (>30 species) is commonly targeted by local fisheries (May, 2005). Thereby, seagrass ecosystems are a critical source of income for coastal populations, supporting the livelihoods of thousands of artisanal Indonesian fishermen (Duffy, 2006; Unsworth and Cullen, 2010).

Indeed, seagrass meadows are important key habitats for fish, which can be attributed to the seagrass plant canopy and the complex three-dimensional structure it creates. This three-dimensional structure provides a diversity of niches for associated flora and fauna (Duffy, 2006; Gullström et al., 2002; MacArthur and Hyndes, 2007; Parrish, 1989; Polte and Asmus, 2006). Thereby, tropical seagrass meadows constitute rich feeding grounds for a range of fish species drawing energy from various trophic levels, i.e., ranging from lower (detritivores and herbivores) to higher level consumers, i.e., different functional groups of predators (Connolly, 1994; de Troch et al., 2003; Du et al., 2016; Grenouillet et al., 2002; Heck Jr. and Orth, 1980, pp. 449–462; Jenkins and Hamer, 2001; Liu et al., 2008; Nakamura and Sano, 2004; Orth et al., 1984; Pogoreutz and Ahnelt, 2014; Vonk et al., 200a; Weinstein and Heck, 1977). In addition to food provision, the seagrass canopy provides critical refuge availability, i.e., shelter from predators, and therefore it can constitute critical nursery grounds for fish (Dorenbosch, Grol, Christianen, Nagelkerken and van der Velde, 2005; Hyndes et al., 1996; Lee and Lin, 2015; Lilley, 2014; Nadiarti et al., 2015; Nagelkerken et al., 2000; Nakamura et al., 2003; Allison, 2001; Horinouchi et al., 2009; Parrish, 1989; Pogoreutz et al., 2012; Young et al., 1997). Both factors, prey availability and shelter, ultimately affect fish distribution and density within seagrass meadows (Bell and Westoby, 1986; Grenouillet et al., 2002; Jenkins and Hamer, 2001; Nakamura and Sano, 2004).

In addition, the composition of fish assemblages in seagrass meadows can be subject to pronounced temporal fluctuations. Specifically, it may fluctuate depending on how frequently any given fish species utilizes the meadow, and during which stages of their life (along with patterns in recruitment or mortality; Bijoux et al., 2013). Fish utilizing seagrass meadows can be categorized into permanent residents (i.e., all life history stages present in the seagrass meadow), temporary residents (i.e., not present throughout all life history stages; e.g., exhibiting spawning or foraging migrations to and from other habitats), and occasional (rare) trespassers or visitors (Kuriandewa et al., 2003). Species belonging to the latter two categories may display distinct migration patterns subject to lunar, diel, or tidal cycles, which can result in predictable fluctuations in fish community composition (Bijoux et al., 2013; Helfman, 1986; Hobson, 1973; Kruse et al., 2015; Lee et al., 2014; Ogden and Quinn, 1994; Unsworth et al., 2007, 2008) and fish functional diversity (Berkström et al., 2013; Mumby, 2006; Nash et al., 2013). Thereby, fish migration ecologically and functionally links seagrass meadows with adjacent habitats, such as coral reefs (Appeldorn et al., 2009; Dorenbosch, Grol, Nagelkerken and van der Velde, 2006; Lee et al., 2014; Nagelkerken and van der Velde, 2002; Unsworth et al., 2008) via cross-habitat energy transfer and food web connectivity (Welsh and Bellwood, 2014).

## 6.4.2 The seagrass canopy as a driver of fish communities

The structural complexity (and thereby the three-dimensional structure, or niche space) created by the seagrass canopy is ultimately a function of its respective seagrass community. This can be attributed to a range of features, which, in combination, are unique for individual seagrass species. These features include the morphology of their emergent leaves, which vary in shape and dimensions (leaf length and height, leaf area index) (Bell and Westoby, 1986; Folkard, 2005; Gartner et al., 2013; Koch, 2001; Peterson et al., 2004), along with shoot stiffness and volume occupied by individual shoots (Bouma et al., 2005; Morris et al., 2008) (summarized in Gartner et al., 2013; Morris et al., 2008). On a meadow scale, the three-dimensional spatial arrangement of the emergent shoots and leaves can differ with vegetation density (Abdelrhman, 2003; Fonseca et al., 1982; Gambi et al., 1990; Peterson et al., 2004; Verduin and Backhaus, 2000) (summarized in Bell and Westoby, 1986; Peterson et al., 2004) and habitat patch size being perhaps the most critical variable (Fonseca et al., 1982). Consequently, multispecies meadows (i.e., comprised of several species of seagrass) will ultimately exhibit higher structural complexity and within-meadow patchiness compared with monospecific stands. For instance, seagrass meadows where large seagrasses with long leaves (e.g., *E. acoroides*) intermingle with shorter species of different leaf characteristics (e.g., leaf morphology, length, and dimensions, as observed for the common Indonesian seagrasses *Cymodocea* spp., *H. uninervis, Halophila* spp., *S. isoetifolium*, or *T. hemprichii*), can form particularly complex and diverse canopies. Seagrass meadows fringing Indonesian coasts and offshore habitats often sport multispecies stands, which may contain up to 10 species of seagrass (McKenzie et al., 2007). Commonly though, five or fewer species are observed, with one or two species dominating the seagrass community.

In meadows where the long-leaved *E. acoroides* is present, canopies consist of two stories, a lower story comprised by *E. acoroides* as well as one or multiple shorter species, and an upper story formed exclusively by the long *E. acoroides* leaves (Pogoreutz et al., 2012). In contrast, seagrass meadows consisting only or predominantly of one or several shorter species may form more homogenous canopies with a single story. Multispecies seagrass meadows dominated by shorter species, however, can still exhibit a certain degree of habitat complexity driven by features associated with different seagrass leaf characteristics and shoot density. While the combination of these features will ultimately affect hydrological properties of an individual seagrass meadow (Peterson et al., 2004), it may also affect the community composition of its associated fish fauna, as previously reported for temperate seagrass habitats (Hyndes et al., 2003; MacArthur and Hyndes, 2007). To validate this observation for small (i.e., tens of meters), but highly diverse tropical multispecies seagrass meadows off Indonesia, a case study on five seagrass meadows was conducted in the Indonesian Spermonde Archipelago.

The Spermonde Archipelago off Makassar (South Sulawesi) is home to diverse multispecies seagrass meadows (Erftemeijer and Allen, 1993; Vonk et al., 2008a,b). There, inter- and subtidal seagrass meadows off the densely populated island Barrang Lompo and the uninhabited coral cay Bone Batang exhibit distinctly different seagrass canopy structures due to differences in overall community composition, dominant species, and shoot density (Pogoreutz et al., 2012). Similar to other tropical seagrass meadows in the region (Hutomo and Martosewojo, 1977; Unsworth et al., 2007), associated fish faunas in the Spermonde region are diverse, hosting more than 120 taxa from 39 fish families, with up to 89 taxa observed within a single meadow. At both islands, the most speciose families of fish roaming these seagrass meadows are the Labridae (20 species), Pomacentridae (17 species), Nemipteridae (8 species), and Gobiidae (6 species). Of all fish taxa observed, only seven occur in population densities of more than 10 individuals 100 $m^{-2}$ at all or several sites, and about half of the total (61 taxa) constitute rare observations. For some taxa (*Cheilio inermis, Halichoeres argus, H. chloropterus, Pentapodus trivittatus, Apogon margaritiphorus*, Pomacentridae), both adults and juveniles are commonly observed in the area, while for others, only juvenile stages are found (e.g., Chaetodontidae, Haemulidae, Ephippidae). Notably, some of the most abundant species (Atherinidae: *Atherinomorus lacunosus*; Labridae: *C. inermis, H. argus, H. chloropterus*; Nemipteridae: *P. trivittatus*) can be commonly observed throughout the area, albeit abundances may vary considerably for some species. For instance, the abundances of *A. lacunosus* and *H. argus* may span an order of magnitude between individual seagrass meadows (about 2–97 and 1 to 22 individuals 100 $m^{-2}$, respectively). In contrast, the nemipterid *P. trivittatus* exhibits little variation between sites (ranging from one to nine individuals 100 $m^{-2}$). Thereby, changes in the abundance of dominant species drive differences in overall fish communities (Pogoreutz et al., 2012).

The observed differences in seagrass community structure (based on shoot density, leaf area index, biomass) align with water depth and are characterized by distinct community metrics of the associated fish fauna (species number, abundance, most common taxa) (Pogoreutz et al., 2012). While the abundance of the overall fish community as well as dominant species are both significantly affected by depth, overall abundances are in addition driven by seagrass shoot density, while abundances of dominant taxa were driven by seagrass leaf area index and biomass. Thereby, the composition and abundance of the fish communities and most abundant fish species differ between islands, inter- versus subtidal sites, and among individual sites. This finding suggests that fish diversity is a correlate of seagrass diversity (high diversity favoring habitat heterogeneity), while fish abundance is a correlate of seagrass shoot density (driving food availability and predation pressure; for a full presentation of results, please refer to Pogoreutz et al., 2012).

### 6.4.3 Differences in fish habitat utilization across seagrass meadows with distinct canopy structures

Differences in fish community composition and abundances across sites are ultimately attributable to species-specific habitat preferences of dominant fish species. Their distribution and abundance in seagrass meadows of the Spermonde Archipelago are first and foremost driven by seagrass leaf area index, biomass, and water depth. This is particularly interesting, as (with the exception of the planktivorous *A. lacunosus*) most dominant fish species are (macro)zoobenthivores. Differences in seagrass leaf blade area are a strong driver of zoobenthivorous fish populations, as it will considerably affect both refuge availability of prey and foraging efficiency of predators (Stoner, 1982). Seagrass biomass, a function of both leaf area index and seagrass shoot density, therefore constitutes a positive correlate of fish and macroinvertebrate abundance in seagrass meadows, due to high habitat complexity and thereby food provision in combination with reduced predation pressure (Atrill et al., 2000; Gullström et al., 2002; Hovel et al., 2002; Vonk et al., 2010; Schultz et al., 2009). Water depth, on the other hand, can generally support higher fish abundance and biomass in coastal marine environments, as it facilitates access to any given habitat for pelagic and schooling species (Pogoreutz et al., 2012; Thomas and Connolly, 2001).

The most abundant fish species recorded for multispecies seagrass meadows in the Spermonde region are either pelagic species or demersal habitat generalists, with the exception of the schooling herbivore *Siganus canaliculatus*, widely using the seagrass meadows as feeding grounds. Pelagic species such as *A. lacunosus*, *Tylosurus crocodilus*, or *Hemiramphus far* commonly venture into seagrass meadows for foraging rather than for shelter and may therefore not necessarily respond to differences in seagrass bed architecture, but rather to water depth. Similarly, demersal habitat generalists such as *H. argus*, *Pentapodus bifasciatus*, and *P. trivitattus* spend most of their lives within the seagrass meadow, but may also enter adjacent habitats or reef environments for foraging or reproductive purposes such as some Labridae and Nemipteridae (Kuriandewa et al., 2003). For some of the most common demersal species, both juvenile and adult stages can be regularly observed at the seagrass sites. In contrast, fish for which only juveniles are recorded in seagrass meadows (e.g., Chaetodontidae, Haemulidae, Ephippidae) are commonly coral reef–associated taxa, temporarily utilizing the seagrass habitat as nursery grounds (Nagelkerken et al., 2000; Pogoreutz et al., 2012; Polte et al., 2005). Thereby, while the first group can be considered permanent residents, the second group classifies as temporary residents (Pogoreutz et al., 2012).

Our study in the Spermonde Archipelago (Pogoreutz et al., 2012) demonstrates that seagrass community metrics (leaf area index, shoot density, biomass) strongly drive associated fish faunas on both assemblage and species levels, and differences in seagrass community metrics will ultimately support distinct fish communities, even at small spatial scales (meters to tens of meters). Both, diversity and population density of seagrasses are critical factors that differentially impact associated fish communities (Fig. 6.8). Thereby, to support functionally diverse, high biomass fish communities in tropical multispecies seagrass meadows, meaningful management actions need to consider the protection of even small and particularly patchy, heterogeneous seagrass stands.

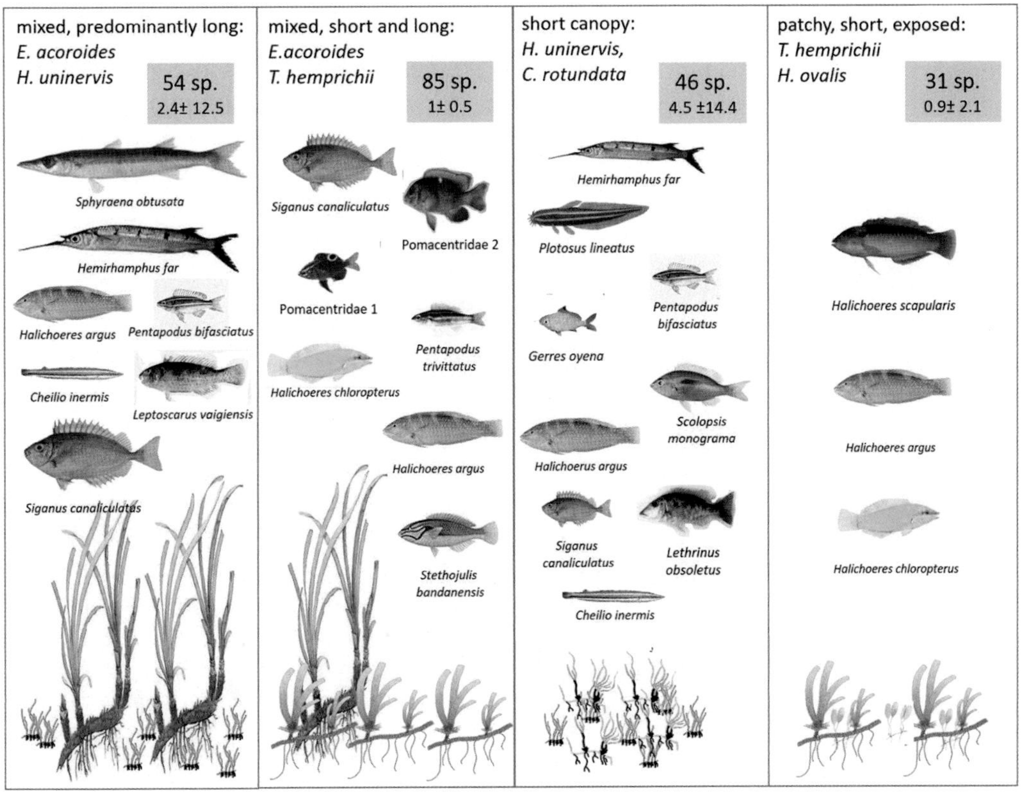


**FIGURE 6.8** Fish assemblages in four types of seagrass beds with a different type of canopy structure observed at the islands Barrang Lompo and Bone Batang, Spermonde Archipelago, SW Sulawesi. Initially, five seagrass beds were studied, but two of them turned out to be very similar in terms of seagrass characteristics and fish communities; we therefore group them together in this figure. Dominant fish species (more than one specimen per 100 $m^2$) are shown as icons with scientific name. Species number of fish and mean species abundance per 100 $m^2$ (±standard deviation) are shown in beige text fields. *After Pogoreutz et al. (2012).*

# 6.5 Human–seagrass interactions

## 6.5.1 Ecological value and ecosystem services

In Indonesia, seagrasses grow commonly in mixed beds, which are dominated by one or two seagrass species. The complexity of the seagrass bed structure supports the various ecological functions and ecosystem services they provide (Mtwana Nordlund et al., 2016).

The high primary productivity of seagrasses enables them to serve as significant food sources in marine ecosystems (Duarte, 1990). Seagrasses can be utilized directly and indirectly as food sources by various associated fauna. The life span of tropical seagrass leaves tends to be shorter than that of temperate seagrass leaves, resulting in a faster turnover and higher seagrass production (Hemminga et al., 1999). The comparatively lower fiber concentration compared with temperate seagrasses makes tropical seagrasses a preferred and nutritious food source for many grazers (Vonk et al., 2008a).

The different ecological functions of seagrasses have already been mentioned in Section 6.3. Besides their function as a food source, seagrass beds provide shelter and protection for associated fauna (Pogoreutz et al., 2012). Seagrass beds also provide essential nursery habitat to economically important vertebrate and invertebrate species. There is considerable evidence from many studies that seagrass beds can support sediment stabilization. Beds of the seagrass *Posidonia oceanica* can trap sediment particles from the water column (Dauby, 1995), while seagrass beds can also play a role in reducing or dissipating wave energy and regulating sediment surface elevation (Potouroglou et al., 2017).

## 6.5.2 Fisheries on fish and invertebrates in seagrass beds

In coastal areas in developing countries, artisanal fishery is an important source of income, often intermixed with agricultural activities (Cesar et al., 1997; Allison, 2001; van Oostenbrugge et al., 2004). Most fishermen depend on coastal fish stocks, because access to offshore fishing grounds is restricted by their equipment. Rapid population growth and increasing need for protein-rich food leads to increasing fishing pressure on local stocks (Johannes, 1998). Ecosystem health and services are prone to change with increasing fishing effort or techniques. Unsustainable methods directly damage the environment and, particularly in seagrass beds by altering physical conditions of the habitat (e.g., Jennings, 1998; Pet Soede, 2001; Valentine and Heck, 2005), interfere with biological factors and hence disturb the food web of the ecosystem (Sumaila et al., 2000; Arreguin-Sánchez et al., 2004; Campbell and Pardede, 2006). In Indonesia, decentralization processes (Satria and Matsuda, 2004) have led to increasing concern of local authorities about their marine resources. Evaluation of local fish stocks is much more common today, though interest of local fishermen is not necessarily fully addressed (Elliott et al., 2001).

In our study within the SPICE I project from an exemplified local fishing site including a seagrass bed at Puntondo, in South Sulawesi, we identified 208 fish species of

65 families. Ranked according to the number of species, the families Lutjanidae (13 species) Mullidae (12), Serranidae (12), Siganidae (12), Lethrinidae (11), Nemipteridae (11), Labridae (10), and Haemulidae (9) were most common. Fishery landings in the village Puntondo were dominated by a few fishing methods (nets and lines) and fishing grounds (coral reefs and beaches). Predatory fishes, such as Lethrinidae and Lutjanidae species, were most important for local economy. Catch abundance decreased especially over reefs and close to the beach. As these are also the preferred fishing grounds, fishing pressure in these areas should be decreased by either closing them seasonally or by gear restrictions. Fish from seagrass beds, rubble, and mangrove areas contributed only very little to overall fish landings in this particular village and consisted mainly of subadults. Seagrass beds and mangroves are nursery areas for reef fish (e.g., Dorenbosch et al., 2005; Nakamura and Sano, 2004), and species-, rather than size-, selective methods should be used there. Lift nets can be operated both species and size selective and are therefore recommended. In our study, fisheries for squids and crabs (*Portunus pelagicus*, Portunidae) as well as sea cucumbers were not evaluated though economically important. Especially crabs have sharply decreased in size and abundance and should also be considered when deciding on gear and fishing ground restrictions.

## 6.5.3 Seaweed farms

All over the world, beaches and shallow water ecosystems have been experiencing pressure from human activities for a very long time. Human activities are extremely diverse comprising direct and obvious impacts such as land reclamation or dredging, as well as more indirect damages caused by water traffic, as well as pollution and increased sediment loads. Among the shallow water ecosystems, seagrass beds have often been overlooked, and especially along tropical coasts, coral reefs and mangrove forests have received much more attention in the past.

Seagrass beds, in general, are very important spawning and nursery areas for fishes and crustaceans, and in tropical countries such as Indonesia, seagrass beds are also important fishing grounds. Seaweed farming, predominantly of the red algae *Eucheuma* spp. and *Kappaphycus* spp. is often promoted as an alternative livelihood for coastal areas in Indonesia where it is one of the very few economic goods that can be produced without land ownership and with limited financial resources. As a consequence, seaweed farming has been widely picked up by fishing communities, and often the most suitable areas for seaweed farming are situated within seagrass beds. Although seaweed farming at the current level is less detrimental than, for example, intensive shrimp farming, and therefore should be seen as a strong option for future aquaculture developments, intensive farming on seagrasses should be avoided or at least minimized by, for example, implementing other farming methods (Eklöf et al., 2006). The risk of ecosystem-level changes in large-scale and uncontrolled farm enterprises warrants a holistic and integrated coastal management approach, which considers all aspects of the tropical seascape including human societies and natural resource use (Eklöf et al., 2006).

However, seaweed farming has not replaced fishing, and seaweed farming directly competes with fishing in seagrass beds by restricting access and limiting gear selection and use. Indirect competition between those two sources of income was studied during the SPICE I project. We investigated the effects of the culturing activities, by installing a seaweed farm over a seagrass bed only for experimental work, controlled the different activities, and measured the effects. We specifically focus on the impacts of shading and trampling on seagrass (plant) communities, dependence of commercially important fishes on seagrass beds as habitat during their life cycles, as well as on socioeconomic dynamics between the two sectors.

The most common seaweed farming method in Indonesia—floating lines on the surface of the water—has significant impacts on seagrass plant communities although the individual lines are spaced widely enough to allow for enough light to penetrate to the sea floor (Fig. 6.9) but depends on the frequency of the working activity. According to our experiments, trampling on the plants when walking to and within the seaweed farms for maintenance and harvest seems to not have a significant impact. When the shading caused by the seaweed on the water surface was increased experimentally (i.e., to 72%, 52%, and 37% of the incoming light still available), the seagrass community changed. Especially at medium (52% of incoming light) and high (37% of incoming light) shading, biomass of smaller seagrass species (i.e., except *Enhalus*) decreased by up to 50%, which indicates that current farming practices are close to or at the limit of what the ecosystem can support. Often, seaweed farms are not located in the same area throughout the year, because farmers move them to the most suitable area (currents, waves, accessibility) allowing individual plots to recover. If more fishermen were to become seaweed farmers or existing farms would expand, we speculate that this rotation system would be restricted, and the stress on the seagrass ecosystem might be too high to sustain seaweed farming on the long term. Also, there are certain seaweed farming practices, which are

**FIGURE 6.9** Examples of seaweed farming on the bottom (left) and on floating lines (right) from Tanakeke Island. Farming on the bottom is mostly conducted in the intertidal and usually involves the digging of shallow ponds, which destroys the natural habitat, while farming on floating lines in the shallow subtidal can be sustainable as long as the lines are kept at a distance of ca. 1 m to avoid excessive shading.

obviously destructive to the natural habitat. Some farmers report that they remove seagrass under their farms, because they believe it harbors diseases (that has been proven wrong; if anything, seagrass meadows decrease bacterial loads in the water column; Lamb et al., 2017). In other places, farmers resort to farming on the bottom in the intertidal, probably due to lack of access to subtidal areas. There, they often dig shallow ponds and use the dug-up material to surround them with walls to avoid desiccation of the seaweed during low tide (Fig. 6.9). While this practice is still relatively rare in South Sulawesi, it has completely transformed most of the intertidal seagrass meadows around Nusa Lembongan and Nusa Ceningan near Bali (Campbell, 2009; pers. obs.). We do not expect that seagrass beds are able to reestablish in these areas when these activities may stop.

The most important commercial fishes in terms of revenue generation and protein supply in the study area in Takalar, South Sulawesi, do not directly depend on seagrass beds for habitat, shelter, or food, and changing seagrass communities due to seaweed farming impacts would likely not have a direct impact on those species (Blankenhorn 2007, p. 118). However, ecosystem services provided by the seagrass would eventually induce trickle-down effects to those commercially important species by changing their catch composition and reducing catch volumes. Before seaweed farming was introduced in South Sulawesi, coastal communities almost entirely relied on fishing, and seaweed farming is very much needed as additional income. Equally important, seaweed provides a more reliable income that can guarantee financial stability and enable fishers/farmers to send their children to better schools and tertiary education, which has previously been a struggle. The introduction of long-line surface seaweed farming in shallow water and seagrass areas could be kept sustainable environmentally; as long as this farming practice does not change into more extensive work, the effect to the benthic system is limited. Seaweed prices are volatile, and the plants are prone to disease outbreaks, and hence, the socioeconomic fishery seaweed system still needs diversification to provide communities with a stable future.

### 6.5.4 Human-made infrastructure

One case study showed an increase in seagrass area after an uninhabited island (Bone Batang) had been almost entirely removed by illegal sand mining activities, thereby creating more open space in the intertidal area suitable for seagrass growth (Kneer, 2013). Also, the construction of the Bili-Bili dam in 1997 (Agnes et al., 2009) reduced the sediment discharge of the Jene Berang River, the largest tributary to the Spermonde Archipelago, which may have allowed seagrasses to expand into subtidal areas on Barrang Lompo and Bone Batang Island, from which they were previously absent (see also Alongi et al., 2008 for an example of mangrove forests expanding following increased sedimentation). However, the vast majority of anthropogenic activities so far have been detrimental to the seagrass meadows, and the total area lost appears to be larger than the areas newly colonized. To determine the actual rate of seagrass loss in the future, baseline data are urgently needed.

## 6.5.5 Current threats

For the Spermonde Archipelago, only anecdotal information is available concerning areas that have been lost in the past. A series of studies conducted from 1991 to 1992 mention dense stands of *E. acoroides* in Gusung Tallang just north of the mouth of the Tallo River (Erftemeijer et al., 1993; Erftemeijer and Herman, 1994; Erftemeijer and Middelburg, 1993). These meadows appear to have disappeared before 2005 (Asmus pers. comm.; Vonk., pers. comm.), with sedimentation, dredging, and aquaculture activities all being potential causes. It is likely that more seagrass stands in comparable nearshore habitats have been lost due to land reclamation and coastal development. Currently increased turbidity caused by ongoing large-scale land reclamation activities, as well as the fortification of existing sea walls in front of Makassar harbor, poses a threat to remaining small patches of *E. acoroides* on Lae Lae Island (Fig. 6.10). As far as top-reef

**FIGURE 6.10** Left: The expansion of a breakwater covers a patch of *Enhalus acoroides* on Lae Lae Island. Top row middle: Aerial photographs of a mixed *Enhalus acoroides* and *Thalassia hemprichii* meadow on Langkai Island; the anchoring and movement of relatively few boats already leaves distinct scars. Top row right: Damage to seagrass and coral caused by the movement of boats across the reef as seen from the beach on Langkai Island. Bottom row middle: Using the reef flat as a garbage dump, Barrang Lompo Island. Note the stack of coral boulders mined for building purposes on the left. Bottom row right: Aerial photograph of rubbish on a beach in Barrang Lompo.

seagrass meadows on the more remote midshelf and offshore islands are concerned, there has been some loss due to increased boating activities in shallow water, land reclamation, and the construction of sea walls. These impacts are typically most severe on the most densely populated islands, as can be seen on satellite images. Drone pictures taken in 2015 suggest that even on relatively sparsely populated islands like Langkai, the effect of boats should not be underestimated (Fig. 6.10). Due to a lack of municipal waste management systems and a near total absence of environmental awareness in the general population, shorelines are often treated as communal waste dumps by villagers (Fig. 6.10).

In Sulawesi and also elsewhere in Indonesia (e.g., Bali), seagrass meadows as well as the fringing reefs around these habitats are often destroyed in the course of building infrastructure such as breakwaters, jetties, and harbors (Figs. 6.11–6.14).

### A word about seagrass restoration

Tropical seagrasses have a remarkable ability to (re)colonize suitable areas as soon as the cause for their initial disappearance has been removed, or new habitat becomes available. Quick recovery of seagrass beds has been documented for many cases, i.e., after the deliberate removal of shoots was stopped (Rollon et al., 1998), a stranded barge was removed (Olesen et al., 2004), and seaweed farming has ceased. The natural colonization of an area previously occupied by an island has been removed due to illegal sand mining (Kneer, 2013). This indicates that as long as there is a surviving meadow nearby which can serve as a natural source of propagules, the large-scale transplantation of shoots can be avoided. We believe that lessons can be learned from past attempts to restore mangroves: there, many costly attempts at planting have failed because species were put at the wrong tidal elevation, or into mudflats where mangroves have never existed. Because of these past experiences, a method called "ecological mangrove restoration" is now promoted. This method relies on the restoration of the original hydrology of places where mangroves formerly existed, followed by the natural colonization of propagules from nearby surviving stands (Brown et al., 2014; Lewis, 2009, 2014; Lewis et al., 2019).

During the SPICE project, we heard of an example where *Enhalus acoroides* was transplanted into a bare sand area, only for this area to be reclaimed for building purposes shortly thereafter (Kiswara, pers. comm.). We also witnessed the rather absurd situation where a group of researchers, having already obtained funds to conduct a restoration experiment, failed to find an unvegetated sandy plot near their research station; every suitable place they looked at was already covered with seagrasses! In the end, they resorted to transplant seagrass into a sandbank area which lacked seagrasses due to frequent disturbance by passing boats. This sandbank would probably experience natural recolonization if the boat traffic was redirected, but the restoration failed because boat traffic was not stopped. At another subtidal location, they sadly resorted to clearing an existing healthy patch to create a bare sand area for transplantation. While some lessons can still be learned from experiments such as this (Asriani et al., 2018; Williams et al., 2017), our point here is that the money and effort could in most cases have been more efficiently invested in investigating the underlying causes for the destruction of the meadows, and their potential for natural recolonization.

**FIGURE 6.11** Land reclaimed on top of a seagrass meadow and coral reef in the course of an illegal iron ore mining operation, Bangka Island, North Sulawesi. After a lengthy court battle with environmental organizations, the operator was forced to shut down the operation, but the damage is already done. *From: Maxar Technologies, Mai 24, 2013 and CNES/Airbus, Mai 5, 2017. www.google.com/earth/index.html.*

## 6.6 Conclusions and outlook

Research on Seagrass Ecology during the SPICE program over more than 10 years gave insight in many aspects of the seagrass beds especially in the Spermonde Archipelago. New insights could be gained in ecology of seagrass beds, particularly the associated macrofauna. Together with a research team from the Netherlands, we investigated the role of burrowing shrimps, which inhabit a diverse and specialized community within their burrows, which depends on seagrass and seagrass detritus as food. We used the role of these shrimps together with physical drivers of the environment, to get an impression on the natural limitation of seagrasses for settling and for building meadows. In small experiments, we looked on different compartments of the seagrass ecosystem, such as the role of herbivorous fish and the role of different types of seagrass beds as a habitat for different fish. We sampled data on most of the system compartments of tropical seagrass beds more than we could present in this book chapter, but we hope to give also some hints for later research, when showing many details that are not published yet. Besides the basic research, we always tried to do some advanced research, such as showing the influence of algal farms and artisanal fisheries on this ecosystem, showing the limits for damaging of the surrounding environment. We used remote sensing techniques to document changes in the extension of seagrass beds and to show destruction. We hope that our joint research efforts on seagrass bed ecology will be continued. Seagrass beds in

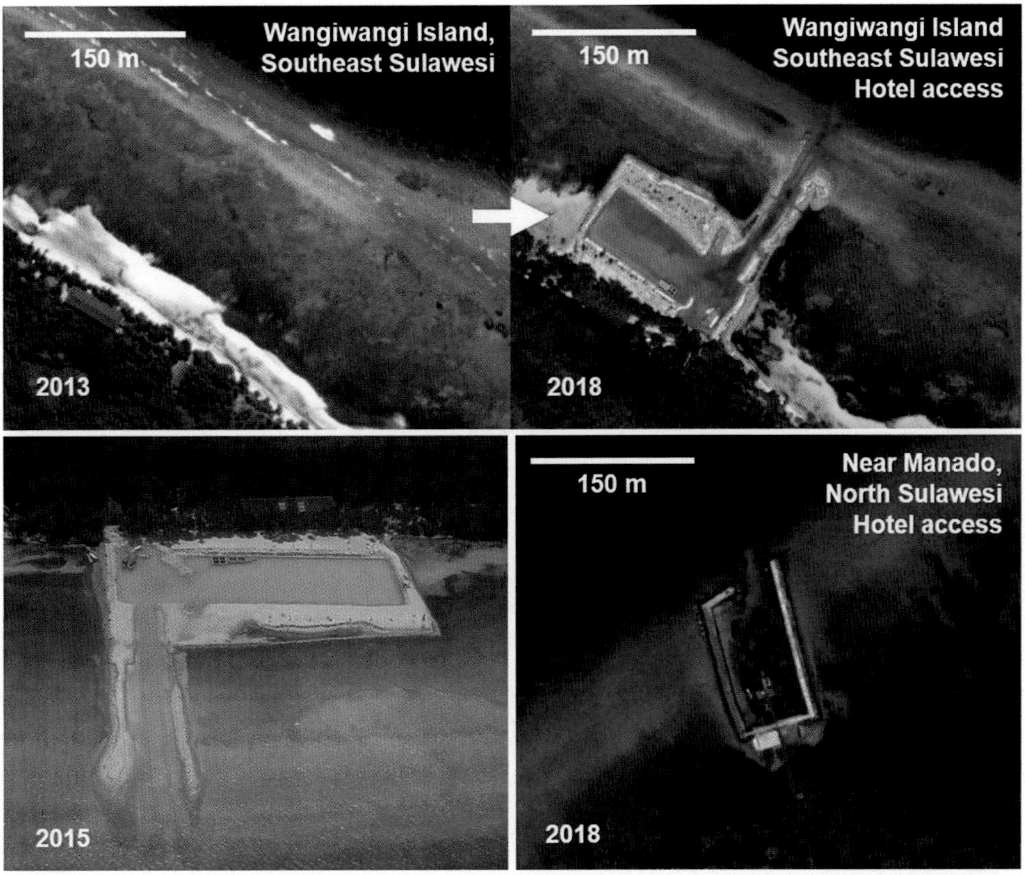


**FIGURE 6.12** Top: A small harbor dug into a fringing reef and seagrass meadow on Wangiwangi Island, Wakatobi, East Sulawesi, to create access to a hotel. Bottom left: The same structure seen from another angle. Bottom right: A similar hotel harbor in a fringing reef near Manado, North Sulawesi. *Modified from: Maxar Technologies, June 1, 2013 and October 21, 2018 (Wangiwangi), and Maxar Technologies, Mai 5, 2018 (Manado). www.google.com/earth/index.html.*

**FIGURE 6.13** Top left: Another harbor blasted into a fringing reef covered by seagrass, Bira, South Sulawesi. Top right: A breakwater built on top of a fringing reef and seagrass meadow in front of a new hotel, Nusa Dua, Bali. *Modified from: Maxar Technologies. (October 31, 2017). www.google.com/earth/index.html.*

**FIGURE 6.14** Top left: Pandawa Beach in Bali in 2011; the seagrass area was almost entirely occupied by seaweed farms dug into the intertidal reef flat. Top right: A closer look at seaweed farms on the bottom such as those in Pandawa Beach. Bottom row: Comparison of Pandawa beach in 2013 and 2018; after seaweed farming has ceased, the seagrass is naturally recovering. Note, however, how tourism infrastructure has taken over the beach next to the reef flat. *Picture from Nusa Lembongan, 2013, DigitalGlobe, May 5, 2013, and Maxar Technologies, November 3, 2018. www.google.com/earth/index.html.*

Indonesian waters are among the richest and most diverse on earth, and it is a great task to save this treasure for future generations.

Knowledge gaps and directions of future research

- A better understanding of the factors responsible for seagrass loss as well as of the factors promoting new habitat is required;
- There is a lack of areal extent data of seagrasses from Kalimantan, Central and Southeast Sulawesi, the Maluku Islands, and West Papua;
- We need to understand tropical seagrass ecosystems as a whole and that requires more quantitative data on all system compartments (i.e., seagrasses, associated plants, and associated fauna) which could be used as base for ecosystem and food web models. This would help to depict scenarios on the influence of certain factors on seagrass beds as a whole and would help to better recognize future threats due to human activities in coastal areas;
- We need more research on the impact of climate change on seagrass beds both experimentally and by modeling these systems.
- There are knowledge gaps in socio–cultural–economic themed research.

Implications/recommendations for policy and society

- The potential negative effects on seagrass beds have to be considered when decisions on land use in upstream regions are made;
- Measures to increase environmental awareness of decision-makers and the general public are recommended;
- With respect to climate change mitigation, a better protection of seagrass beds is recommended in order to maintain their carbon sink function as "Blue Carbon" ecosystem;
- In order to sustain the supply of major seagrass ecosystem services, a better control and limited use of resources in particular through fishing and algal farming is recommended;
- Measures to preserve seagrass ecosystems should be realistic, cost-effective, and understandable for the local population;
- Short-term measures required to protect coral reefs and seagrass meadows include the following: (i) to reduce plastic waste disposal in the sea, (ii) to abstain from all boating activities which destroy the benthic habitat (e.g., anchoring, crossing the reef flat at low tide, and building or repairing boats in the seagrass area), (iii) to stop illegal fishing in particular with destructive methods (bomb and cyanide fishing), (iv) to stop the killing of protected wildlife (dugongs, turtles, helmet shells, and giant clams), and (v) to stop all building activities near the shoreline (land reclamation, erection of sea walls, and mining of sand and coral).
- In the Spermonde Archipelago, a large part of the reef area on each island should be set aside as a strict no-take zone. Especially, seagrass beds on those islands that have a lower population density or are uninhabited should be considered to be totally protected.

# Acknowledgments

The authors would like to thank the Indonesian Ministry of Research and Technology (RisTek) for granting permission to carry out field work in Indonesia, and we thank the German Federal Ministry for Education and Research (BMBF; SPICE I: grant No. 03F0390C; SPICE II: grant No. 03F0472C, SPICE III: grant No. 03F0642A) and the Alfred-Wegener-Institute, Helmholtz Centre for Polar and Marine Research, for financial support.

# References

Abdelrahman, M.A., 2003. Effect of eelgrass *Zostera marina* canopies on flow and transport. Marine Ecology Progress Series 248, 67–83.

Agnes, R.D., Mochtar, S., Said, A., Fujikura, R., 2009. Effects of construction of the Bili-Bili Dam (Indonesia) on living conditions and their patterns of resettlement and return. International Journal of Water Resources Development 25 (3), 467–477 (11).

Allison Edward, H., Ellis, F., 2001. The livelihood approach and management of small-scale fisheries. Marine Policy 25, 377–388.

Alongi, D.M., Murdiyarso, D., Fourgurean, J.W., Kauffman, J.B., Hutahaenan, A., Crooks, S., et al., 2016. Indonesia's blue carbon: a globally significant and vulnerable sink for seagrass and mangrove carbon. Wetlands Ecology and Management 24, 3–13.

Alongi, D.M., Trott, L.A., Rachmansyah, R., Tirendi, F., McKinnon, A.D., Undu, M.C., 2008. Growth and development of mangrove forest overlying smothered coral reefs, Sulawesi and Sumatra, Indonesia. Marine Ecology Progress Series 370, 97–109.

Ambo-Rappe, R., 2016. Differences in richness and abundance of species assemblages in tropical seagrass beds of different structural complexity. Journal of Environmental Science and Technology 9, 246–256.

Anker, A., Pratama, I.S., Firdaus, M., Rahayu, D.L., 2015. On some interesting marine decapod crustaceans (Alpheidae, Laomediidae, Strahlaxiidae) from Lombok, Indonesia. Zootaxa 3911 (3), 301–342.

Appeldorn, R.S., Aguilar-Perera, A., Bouwmeester, B.L.K., Dennis, G.D., Hill, R.L., Merten, W., et al., 2009. Movement of fishes (Grunts: Haemulidae) across the coral reef seascape: a review of scales, patterns and processes. Caribbean Journal of Science 45, 304–316.

Arreguin-Sanchez, F., Hernandez Herrera, A., Ramırez-Rodrıguez, M., Pérez-España, H., 2004. Optimal management scenarios for the artisanal fisheries in the ecosystem of La Paz Bay, Baja California Sur, Mexico. Ecological Modelling 172 (2–4), 373–382. https://doi.org/10.1016/j.ecolmodel.2003.09.018.

Asmus, H., 2011. Material Exchange of the Sylt-Rømø Bight and its Relation to Habitat and Species Diversity Habilitation Thesis Part 1 Synopsis. Christian Albrechts Universität, Kiel.

Asriani, N., Ambo-Rappe, R., Lanuru, M., Williams, S.L., 2018. Species richness effects on the vegetative expansion of transplanted seagrass in Indonesia. Botanica Marina 61 (3), 205–211.

Atrill, M.J., Strong, J.A., Rowden, A.A., 2000. Are macroinvertebrate communities influenced by seagrass structural complexity? Ecography 23, 114–121.

Beer, S., Rehnberg, J., 1997. The acquisition of inorganic carbon by the seagrass *Zostera marina*. Aquatic Botany 56, 277–283.

Bell, J.D., Westoby, M., 1986. Importance of local changes in leaf height and density to fish and decapods associated with seagrasses. Journal of Experimental Biology 104, 249–274.

Berkström, C., Lindborg, R., Thuresson, M., Gullström, M., 2013. Assessing connectivity in a tropical embayment: fish migrations and seascape ecology. Biological Conservation 166, 43–53.

Bijoux, J.P., Dagorn, L., Gaertner, J.-C., Cowley, P.D., Robinson, J., 2013. The influence of natural cycles on coral reef fish movement: implications for underwater visual census (UVC) surveys. Coral Reefs 32, 1135–1140.

Bjorndal, K.A., 1980. Nutrition and grazing behavior of *Chelonia mydas*. Marine Biology 56, 147–154.

Blankenhorn, S., 2007. Seaweed Farming and Artisanal Fisheries in an Indonesian Seagrass Bed – Complementary or Competitive Usages Doctoral Thesis. University of Bremen.

Borowitzka, M.A., Lethbridge, R.C., 1989. Seagrass epiphytes. Chapter 14. In: Larkum, A.W.D., Mc Comb, A.J., Shepherd, S.A. (Eds.), Biology of Seagrasses – a Treatise on the Biology of Seagrasses with Special Reference to the Australian Region. Elsevier Science Publishers B. V., Amsterdam, pp. 458–499.

Boudouresque, C.F., Giraud, G., Thommeret, J., Thommeret, Y., 1980. First attempt at dating by $^{14}C$ the undersea beds of dead *Posidonia oceanica* in the bay of Port-Man (Port-Cros, Var, France). Travaux scientifiques du Parc national de Port-Cros 6, 239–242.

Bouillon, S., Moens, T., Overmeer, I., Koedam, N., Dehairs, F., 2004. Resource utilization patterns of epifauna from mangrove forests with contrasting inputs of local versus imported organic matter. Marine Ecology Progress Series 278, 77–88.

Bouma, T.J., de Vries, M.B., Low, E., Peralta, G., Tanczos, C., van de Koppel, J., et al., 2005. Trade-offs related to ecosystem engineering: a case study on stiffness of emerging macrophytes. Ecology 86, 2187–2199.

Brouns, J.J.W.M., Heijs, F.M.I., 1991. Seagrass ecosystems in the tropical West Pacific. In: Mathieson, A.C., Nienhuis, P.H. (Eds.), Ecosystems of the World 24, Intertidal and Littoral Ecosystems. Elsevier, Amsterdam, pp. 371–390.

Brown, B.M., Fadillah, R., Nurdin, Y., Soulsby, I., Ahmad, I., 2014. Community based ecological mangrove rehabilitation (CBEMR) in Indonesia. Surveys and Perspectives Integrating Environment and Society 7 (2), 1–12.

Campbell, S., 2009. Seaweed farms and seagrass – can they co-exist? Seagrass-Watch Magazine 39, 12–13.

Campbell, S.J., Pardede, S.T., Pardede, S.T., 2006. Reef fish structure and cascading effects in response to artisanal fishing pressure. Fisheries Research 79 (1–2), 75–83. https://doi.org/10.1016/j.fishres.2005.12.015.

Cardinale, B.J., Srivastava, D.S., Duffy, J.E., Wright, J.P., Downing, A.L., Sankaran, M., et al., 2006. Effects of biodiversity on the functioning of trophic groups and ecosystems. Nature 443 (7114), 989–992. https://doi.org/10.1038/nature05202.

Carruthers, T.J.B., Dennison, W.W., Longstaff, B.J., Waycott, M., Abal, E.G., McKenzie, L.J., et al., 2002. Seagrass habitats of Northeast Australia: models of key processes and controls. Bulletin of Marine Science 71 (3), 1153–1169.

Cesar, H., Lundin, C.G., Bettencourt, S., Dixon, J., 1997. Indonesian coral reefs- an economic analysis of a precious but threatened resource. Ambio 26 (6), 245–250. ISSN: 0044-7447.

Christianen, M.J.A., Herman, P.M.J., Bouma, T.J., Lamers, L.P.M., van Katwijk, M.M., van der Heide, T., et al., 2014. Habitat collapse due to overgrazing threatens turtle conservation in marine protected areas. Proceedings of the Royal Society 281. https://doi.org/10.1098/rspb.2013.2890, 20132890.

Connolly, R.M., 1994. Removal of seagrass canopy: effects on small fish and their prey. Journal of Experimental Marine Biology and Ecology 184, 99–110.

Couchman, D., 1987. Seagrasses: A Brief Look at Their Ecology and Biology. Queensland Departure of primary Industries, Brisbane Australia.

Curran, H.A., Martin, A.J., 2003. Complex decapod burrows and ecological relationships in modern and Pleistocene intertidal carbonate environments, San Salvador Island, Bahamas. Palaeogeography, Palaeoclimatology, Palaeoecology 192, 229–245.

Dauby, P., 1995. Particle fluxes over a Mediterranean seagrass bed: a one year case study. Marine Ecology Progress Series 126 (1–3), 233–246.

De Klerk, L.G., 1982. Sea Levels, Reefs and Coastal plains of South Sulawesi: A Morphogenetic Pedological Study PhD Thesis. The Netherlands [in Dutch]. University of Utrecht.

de Troch, M., Fiers, F., Vincx, M., 2001. Alpha and beta diversity of harpacticoid copepods in a tropical seagrass bed: the relation between diversity and species range size distribution. Marine Ecology Progress Series 215, 225–236.

de Troch, M., Fiers, F., Vincx, M., 2003. Niche segregation and habitat specialisation of harpacticoid copepods in a tropical Seagrass bed. Marine Biology 142, 345–355.

den Hartog, C., 1970. The Seagrasses of the World. North Holland Publishing Co., Amsterdam.

Dorenbosch, M., Grol, M.G.G., Christianen, M.J.A., Nagelkerken, I., van der Velde, G., 2005. Indo-Pacific seagrass beds and mangroves contribute to fish density and diversity on adjacent coral reefs. Marine Ecology Progress Series 302, 63–76.

Dorenbosch, M., Grol, M.G.G., Nagelkerken, I., van der Velde, G., 2006. Different surrounding landscapes may result in different fish assemblages in East African seagrass beds. Hydrobiologia 563, 45–60.

Duarte, C.M., 1990. Seagrass nutrient content. Marine Ecology Progress Series 67, 201–207. https://pdfs.semanticscholar.org/b7d9/6a6d500d4255f6da1a0a83aa41276aeb4c38.pdf.

Duarte, C.M., Cebrián, J., 1996. The fate of marine autotrophic production. Limnology and Oceanography 41, 1758–1766.

Duarte, C.M., Krause-Jensen, D., 2017. Export from seagrass meadows contributes to marine carbon sequestration. Frontiers in Marine Science 4 (13), 1–7.

Duffy, J.E., 2006. Biodiversity and the functioning of seagrass ecosystems. Marine Ecology Progress Series 311, 233–250.

Du, J., Zheng, X., Peristiwady, T., Liao, J., Makatipu, P.C., Yin, X., et al., 2016. Food sources and trophic structure of fishes and benthic macroinvertebrates in a tropical seagrass meadow revealed by stable isotope analysis. Marine Biology Research 12, 748–757.

Dworschak, P.C., 2011. Redescription of *Callianassa vigilax* de Man, 1916, a subjective senior synonym of *Neocallichirus denticulatus* Ngoc-Ho, 1994 (Crustacea: Decapoda: Callianassidae). Annalen des Naturhistorischen Museums in Wien, Series B 112, 137–151.

Dworschak, P.C., 2018. Axiidea of Panglao, the Philippines: families Callianideidae, Eucalliacidae and Callichiridae, with a redescription of Callianassa calmani Nobili, 1904. Annalen des Naturhistorischen Museums in Wien, Series B 120, 15–40.

Eklöf, J.S., Henriksson, R., Kautsky, N., 2006. Effects of tropical open-water seaweed farming on seagrass ecosystem structure and function Marine. Ecology Progress Series 325, 73–84. https://doi.org/10.3354/meps325073.

Elliott, G., Mitchell, B., Wiltshire, B., Manan, A.I., Wismer, S., 2001. Community participation in marine protected area management: Wakatobi National Park, Sulawesi, Indonesia. Coastal Management 29 (4), 295–316. https://doi.org/10.1080/089207501750475118.

Enriquez, S., Schubert, N., 2014. Direct contribution of the seagrass *Thalassia testudinum* to lime mud production. Nature Communications 5, 3835. https://doi.org/10.1038/ncomms4835.

Erftemeijer, P.L.A., 1993. Factors Limiting Growth and Production of Tropical Seagrasses: Nutrient dynamics in Indonesian Seagrass Beds PhD- Thesis. Nijmegen Catholic University, Nijmegen, The Netherlands.

Erftemeijer, P.L.A., Allen, G.R., 1993. Fish fauna of seagrass beds in south Sulawesi, Indonesia. Records of the Western Australian Museum 16, 269–277.

Erftemeijer, P.L.A., Drossaert, W.M.E., Smekens, M.J.E., 1993. Macrobenthos of two contrasting seagrass habitats in South Sulawesi, Indonesia. Wallaceana 70, 5–12.

Erftemeijer, P.L.A., Herman, P.M.J., 1994. Seasonal changes in environmental variables, biomass, production and nutrient contents in two contrasting tropical intertidal seagrass beds in South Sulawesi, Indonesia. Oecologia 99, 35–59.

Erftemeijer, P.L.A., Middelburg, J.J., 1993. Sediment-nutrient interactions in tropical seagrass beds: a comparison between terrigenous and a carbonate sedimentary environment in South Sulawesi (Indonesia). Marine Ecology Progress Series 102, 187–198.

Erftemeijer, P.L.A., Stapel, J., 1999. Primary production of deep-water *Halophila ovalis* meadows. Aquatic Botany 65, 71–82.

Fenner, D., 2007. The ecology of Papuan coral reefs. In: Marshall, A.J., Beehler, B.M. (Eds.), The Ecology of Papua. Part Two. Periplus Editions. Hk) Ltd, pp. 771–800.

Ferse, S.C.A., Glaser, M., Neil, M., Schwerdtner Máñez, K., 2012. To cope or to sustain? Eroding long-term sustainability in an Indonesian coral reef fishery. Regional Environmental Change. https://doi.org/10.1007/s10113-012-0342-1.

Folkard, A.M., 2005. Hydrodynamics of model *Posidonia oceanica* patches in shallow water. Limnology and Oceanography 50, 1592–1600.

Fonseca, M.S., Fisher, J.S., Zieman, J.C., Thayer, G.W., 1982. Influence of the seagrass, *Zostera marina*, on current flow. Estuarine, Coastal and Shelf Science 15, 351–364.

Fortes, M.D., 2017. Blue carbon stock in seagrasses and mangroves of SE Asia: implication to climate change mitigation. In: Abstract Book of the World Blue Carbon Conference, 7–9 September 2017, Jakarta, Indonesia, vol. 7.

Gambi, M.C., Nowell, A.R.M., Jumars, P.A., 1990. Flume observations on flow dynamics in *Zostera marina* (eelgrass) beds. Marine Ecology Progress Series 61, 159–169.

Gartner, A., Tuya, F., Lavery, P.S., McMahon, K., 2013. Habitat preferences of macroinvertebrate fauna among seagrasses with varying structural forms. Journal of Experimental Marine Biology and Ecology 439, 143–151.

Grenouillet, G.L., Pont, D., Seip, K.L., 2002. Abundance and species richness as a function of food resources and vegetation structure: juvenile fish assemblages in rivers. Ecography 25, 641–650.

Gullström, M., de la Torre Castro, M., Bandeira, S.O., Björk, M., Dahlberg, M., Kautsky, N., et al., 2002. Seagrass ecosystems in the western Indian Ocean. Ambio 31, 588–596.

Gullström, M., Lyimo, L.D., Dahl, M., Samuelsson, G.S., Eggertsen, M., Anderberg, E., et al., 2018. Blue carbon storage in tropical seagrass meadows relates to carbonate stock dynamics, plant–sediment processes, and landscape context: insights from the western Indian Ocean. Ecosystems 21, 551–566. https://doi.org/10.1007/s10021-017-0170-8.

Hart, D.E., Kench, P.S., 2007. Carbonate production on an emergent reef platform, Warraber Island, Torres Strait, Australia. Coral Reefs 26, 53–68.

Heck, K.L., Carruthers, T.J., Duarte, C.M., Hughes, A.R., Kendrick, G.A., Orth, R.J., et al., 2008. Trophic transfers from seagrass meadows subsidize diverse marine and terrestrial consumers. Ecosystems 11, 1198–1210.

Heck Jr., K.L., Orth, R.J., 1980. Seagrass Habitats: The Roles of Habitat Complexity, Competition and Predation in Structuring Associated Fish and mobile Macroinvertebrate Assemblages. Estuarine Perspectives.

Heithaus, M.R., Wirsing, A.J., Burkholder, D., Thomson, J., Dill, L.M., 2006. Validation of a randomization procedure to assess animal habitat preferences: microhabitat use of tiger sharks in a seagrass ecosystem. Journal of Animal Ecology 75, 666–676.

Helfman, G., 1986. Fish behaviour by day, night, and twilight. In: Pitcher, T. (Ed.), Behaviour of Teleost Fishes. Croom-Helm, London, pp. 366–387.

Hemminga, M.A., Duarte, C.M., 2000. Seagrass Ecology: An Introduction. Cambridge University Press, Cambridge.

Hemminga, M.A., Marbà, N., Stapel, J., 1999. Leaf nutrient resorption, leaf lifespan and the retention of nutrients in seagrass systems. Aquatic Botany 65 (1–4), 141–158.

Hemminga, M.A., Slim, F.J., Kazunug, J., Ganssen, G.M., Nieuwenhuize, J., Kruyt, N.M., 1994. Carbon outwelling from a mangrove forest with adjacent seagrass beds and coral reefs (Gazi Bay, Kenya). Marine Ecology Progress Series 106, 291–301.

Hobson, E.S., 1973. Diel feeding migrations in tropical reef fishes. Helgolaender Wissenschaftliche Meeresuntersuchungen 24, 361–370.

Hoeksema, B.W., 2015. The Spermonde archipelago: a model area for coral reef studies. In: Abstract Book of the International Conference on Small Islands Research in Tropical Regions (SIRTRE) 2015: The Spermonde Archipelago and Other Case Studies, 15–16 September 2015, Makassar, Indonesia, p. 11.

Horinouchi, M., Mizuno, N., Jo, Y., Fujita, M., Sano, M., Suzuki, Y., 2009. Seagrass habitat complexity does not always decrease foraging efficiencies of piscivorous fishes. Marine Ecology Progress Series 377, 43–49.

Hovel, K.A., Fonseca, M.S., Myer, D.L., Kenworthy, W.J., Whitfield, P.E., 2002. Effects of seagrass landscape structure, structural complexity and hydrodynamic regime on macrofaunal densities in North Carolina seagrass beds. Marine Ecology Progress Series 243, 11–24.

Howard, R.K., Edgar, G.J., Hutchings, P.A., 1989. Faunal assemblages of seagrass beds. Chapter 16. In: Larkum, A.W.D., Mc Comb, A.J., Shepherd, S.A. (Eds.), *B*iology of Seagrasses – A Treatise on the Biology of Seagrasses with Special Reference to the Australian Region. Elsevier Science Publishers B. V., Amsterdam, pp. 536–564.

Hutomo, M., Martosewojo, S., 1977. The fishes of seagrass community on the west side of Burung Island (Pari Islands, Seribu Islands) and their variations in abundances. Marine Research in Indonesia 17, 147–172.

Hyndes, G.A., Kendrick, A.J., MacArthur, L.D., Stewartt, E., 2003. Differences in the species- and size-composition of fish assemblages in three distinct seagrass habitats with differing plant and meadow structure. Marine Biology 142, 1195–1206.

Hyndes, G.A., Potter, I.C., Lenanton, R.C.J., 1996. Habitat partitioning by whiting species (Sillaginidae) in coastal waters. Environmental Biology of Fishes 45, 21–40.

Jackson, J.B.C., 2001. What was natural in the coastal oceans? Proceedings of the National Academy of Sciences (PNAS) 98 (10), 5411–5418.

Janßen, A., Wizeman, A., Klicpera, A., Satari, D.Y., Westphal, H., Mann, T., 2017. Sediment composition and facies of coral reef islands in the Spermonde Archipelago, Indonesia. Frontiers in Marine Science 4, 144.

Jenkins, G.P., Hamer, P.A., 2001. Spatial variation in the use of seagrass and unvegetated habitats by post-settlement King George whiting (Perioidei: Sillaginidae) in relation to meiofaunal distribution and macrophyte structure. Marine Ecology Progress Series 224, 219–229.

Jennings, S., Kaiser, M.J., 1998. The effect of fishing on marine ecosystems. Advances in Marine Biology 34, 201–212. https://doi.org/10.1016/S0065-2881(08)60212-6.

Jernakoff, P., Brearley, A., Nielsen, J., 1996. Factors affecting grazer-epiphyte interactions in temperate seagrass meadows. Oceanography and Marine Biology 34, 109–162.

Johannes, R.E., 1998. The case for data-less marine resource management: examples from tropical nearshore finfisheries. Trends in Ecology & Evolution 13 (6), 243–246. https://doi.org/10.1016/S0169-5347(98)01384-6.

Jokiel, P.I., Martinelli, F.J., 1992. The vortex model of coral reef biogeography. Journal of Biogeography 19, 449–458.

Kavanagh, K.D., Olney, J.E., 2006. Ecological correlates of population density and behavior in the circumtropical black triggerfish *Melichthys niger* (Balistidae). Environmental Biology of Fishes 76, 387–398.

Kench, P.S., Mann, T., 2017. Reef island evolution and dynamics: insights from the Indian and Pacific Oceans and perspectives for the Spermonde archipelago. Frontiers in Marine Science 4, 145.

Kennedy, H., Beggins, J., Duarte, C.M., Fourqurean, J.W., Holmer, M., Marbà, N., et al., 2010. Seagrass sediments as global carbon sinks: isotopic constraints. Global Biogeochemical Cycles 24, 1–8.

Kikuchi, T., Perés, J.M., 1977. Consumer ecology of seagrass beds. In: McRoy, C.P., Helfferich, C. (Eds.), Seagrass Ecosystems. Marcel Dekker,Inc, N.Y.

Kirsch, K.D., Valentine, J.F., Heck Jr., K.L., 2002. Parrotfish grazing on turtlegrass *Thalassia testudinum*: evidence for the importance of seagrass consumption in food web dynamics of the Florida Keys National Marine Sanctuary. Marine Ecology Progress Series 227, 71–85.

Kiswara, W., 1994. A review: seagrass ecosystem studies in Indonesian waters. In: Sudara, S., Wilkinson, C.R., Chou, L.M. (Eds.), Proceedings 3rd ASEAN-Australia Symposium on Living Coastal Resources. Chulalongkorn University, Bangkok I, pp. 259–281.

Klumpp, D.W., Howard, R.T., Pollard, D., 1989. Trophodynamics and nutritional ecology of seagrass communities. Chapter 13. In: Larkum, A.W.D., McComb, A.J., Sheperd, S.A. (Eds.), Biology of Seagrasses: A Treatise on the Biology of Seagrasses with Special Reference to the Australian Region. Elsevier, Amsterdam, pp. 394–457.

Kneer, D., 2013. Dynamics of Seagrasses in a Heterogeneous Tropical Reef Ecosystem PhD Thesis. Christian-Albrechts-University Kiel, Germany.

Kneer, D., Asmus, H., Ahnelt, H., Vonk, J.A., 2008a. Records of *Austrolethops wardi* Whitley (Teleostei: Gobiidae) as an inhabitant of burrows of the thalassinid shrimp *Neaxius acanthus* in tropical seagrass beds of the Spermonde Archipelago, Sulawesi, Indonesia. Journal of Fish Biology 72, 1095–1099.

Kneer, D., Asmus, H., Jompa, J., 2013a. Do burrowing callianassid shrimp control the lower boundary of tropical seagrass beds? Journal of Experimental Marine Biology and Ecology 446, 262–272.

Kneer, D., Asmus, H., Vonk, J.A., 2008b. Seagrass as the main food source of *Neaxius acanthus* (Thalassinidea: Strahlaxiidae), its burrow associates, and *Corallianassa coutierei* (Thalassinidea: Callianassidae). Estuarine, Coastal and Shelf Sciences 79, 620–630.

Kneer, D., Monniot, F., Stach, T., Christianen, M.J.A., 2013b. *Ascidia subterranea* sp. nov. (Phlebobranchia: Ascidiidae), a new tunicate belonging to the *A. sydneiensis* Stimpson, 1855, group, found as burrow associate of *Axiopsis serratifrons* A. Milne-Edwards, 1873 (Decapoda: Axiidae) on Derawan Island, Indonesia. Zootaxa 3616 (5), 485–494.

Koch, E.W., 2001. Beyond light: physical, geological, and geochemical parameters as possible submerged aquatic vegetation habitat requirements. Estuaries 24, 1–17.

König, S., Le Guyader, H., Gros, O., 2015. Thioautotrophic bacterial endosymbionts are degraded by enzymatic digestion during starvation: case study of two lucinids *Codakia orbicularis* and *C. orbiculata*. Microscopy Research and Technique 78 (2), 173–179. https://doi.org/10.1002/jemt.22458. ISSN 1097-0029.

Kruse, M., Taylor, M., Muhando, C.A., Reuter, H., 2015. Lunar, diel, and tidal changes in fish assemblages in an East African marine reserve. Regional Studies in Marine Science. https://doi.org/10.1016/j.rsma.2015.05.001.

Kuo, J., McComb, A.J., 1989. Seagrass taxonomy, structure and development. Chapter 2. In: Larkum, A.W.D., Mc Comb, A.J., Shepherd, S.A. (Eds.), Biology of Seagrasses – a Treatise on the Biology of Seagrasses with Special Reference to the Australian Region. Elsevier Science Publishers B. V., Amsterdam, pp. 6–73.

Kuriandewa, T.E., Kiswara, W., Hutomo, M., Soemodihardjo, S., 2003. The seagrasses of Indonesia. In: Green, E.P., Short, F.T. (Eds.), World Atlas of Seagrasses. University of California Press, pp. 171–182.

Lamb, J.B., van de Water, J.A.J.M., Bourne, D.G., Altier, C., Hein, M.Y., Fiorenza, E.A., et al., 2017. Seagrass ecosystems reduce exposure to bacterial pathogens of humans, fishes, and invertebrates. Science 355, 731–733.

Lee, C.-L., Huang, Y.-H., Chung, C.-Y., Lin, H.-J., 2014. Tidal variation in fish assemblages and trophic structures in tropical Indo-Pacific seagrass beds. Zoological Studies 53 (1), 56. https://doi.org/10.1186/s40555-014-0056-9.

Lee, C.L., Lin, H.J., 2015. Ontogenetic habitat utilization patterns of juvenile reef fish in low-predation habitats. Marine Biology 162 (9), 1799–1811. https://doi.org/10.1007/s00227-015-2712-y.

Lewis, R.R., 2009. Methods and criteria for successful mangrove restoration. Chapter 28. In: Perillo, G.M. E., Wolanski, E., Cahoon, D.R., Brinson, M.M. (Eds.), Coastal Wetlands: An Integrated Ecosystem Approach. Elsevier, p. 787.

Lewis, R.R., 2014. Mangrove forest restoration and the preservation of Mangrove biodiversity. Chapter 14.1. In: Bozzano, M., Jalonen, R., Thomas, E., Boshier, D., Gallo, L., Cavers, S., et al. (Eds.), Genetic Considerations in Ecosystem Restoration Using Native Tree Species. State of the World's forest Genetic Resources – Thematic Study. *FAO and Biodiversity International*, Rome.

Lewis, R.R., Brown, B.M., Flynn, L.L., 2019. Methods and Criteria for successful mangrove forest rehabilitation. Chapter 24. In: Perilo, G.M.E., Wolanski, E., Cahoon, D.R., Hopkinson, C.S. (Eds.), Coastal Wetlands: An Integrated Ecosystem Approach, second ed.

Lilley, R.J., October 2014. Atlantic Cod (*Gadus morhua*) benefits from the availability of seagrass (*Zostera marina*) nursery habitat. Global Ecology and Conservation 2, 367–377. https://doi.org/10.1016/j.gecco.2014.10.002.

Lindeboom, H.J., Sandee, A.J.J., 1989. Production and consumption of tropical seagrass fields in eastern Indonesia measured with bell jars and microelectrodes. Netherlands Journal of Sea Research 23, 181–190.

Liu, H.T.H., Kneer, D., Asmus, H., Ahnelt, H., 2008. The feeding habits of *Austrolethops wardi*, a gobiid fish inhabiting burrows of the thalassinidean shrimp *Neaxius acanthus*. Estuarine, Coastal and Shelf Science 79, 764–767.

MacArthur, L.D., Hyndes, G.A., 2007. Varying foraging strategies of Labridae in seagrass habitats: herbivory in temperate seagrass meadows? Journal of Experimental Marine Biology and Ecology 340, 247–258.

May, D., 2005. Folk taxonomy of reef fish and the value of participatory monitoring in Wakatobi National Park, southeast Sulawesi, Indonesia. South Pacific Commission Traditional Marine Resource Management and Knowledge Information Bulletin 18, 18–35.

McCoy, E.D., Heck Jr., K.L., 1976. Biogeography of corals, seagrasses and mangroves: an alternative to the center of origin concept. Systematic Zoology 25, 201–210.

McKenzie, L., Coles, R., Erftemeijer, P.L.A., 2007. Seagrass ecosystems of Papua. In: Marshall, A.J., Beehler, B.M. (Eds.), The Ecology of Papua (Pt. 2). Periplus Editions (Hk) Ltd, pp. 800–823.

McKenzie, L., Nordlund, L.M., Jones, B.L., Cullen-Unsworth, L.C., Roelfsema, C., Unsworth, R.K.F., 2020. The global distribution of seagrass meadows. Environmental Research Letters 15, 074041.

Moksnes, P.-O., 2002. The relative importance of habitat-specific settlement, predation and juvenile dispersal for distribution and abundance of young juvenile shore crabs *Carcinus maenas* L. Journal of Experimental Marine Biology and Ecology 271 (1), 41–73.

Moore, A.M., Ambo-Rappe, R., Ali, Y., 2017. "The lost Princess (putri duyung)" of the small islands: dugongs around Sulawesi in the Anthropocene. Frontiers in Marine Science 4, 284.

Morris, E.P., Peralta, G., Brun, F.G., van Duren, L., Bouma, T.J., Perez-Llorens, J.L., 2008. Interactions between hydrodynamics and seagrass canopy structure: spatially explicit effects on ammonium uptake rates. Limnology and Oceanography 53, 1531–1539.

Mtwana Nordlund, L., Koch, E.W., Barbier, E.B., Creed, J.C., 2016. Seagrass ecosystem services and their variability across genera and geographical regions. PLoS One 11 (10), 1–23.

Mukai, H., 1993. Biogeography of the tropical seagrasses in the western Pacific. Marine and Freshwater Research 44, 1–17.

Mumby, P.J., 2006. Connectivity of reef fish between mangroves and coral reefs: algorithms for the design of marine reserves at seascape scales. Biological Conservation 128, 215–222.

Nacorda, H.M., 2008. Burrowing Shrimps and Seagrass Dynamics in Shallow-Water Meadows off Bolinao (NW Philippines) PhD Thesis. Wageningen University, The Netherlands.

Nadiarti, J.J., Riani, E., Jamal, M., 2015. A comparison of fish distribution pattern in two different seagrass species-dominated beds in tropical waters. Journal of Engineering and Applied Sciences 10 (6), 147–153. http://medwelljournals.com/abstract/?doi=jeasci.2015.147.153.

Nagelkerken, I., van der Velde, G., 2002. Do non-estuarine mangroves harbour higher densities of juvenile fish than adjacent shallow-water and coral reef habitats in Curacao (Netherlands Antilles)? Marine Ecology Progress Series 245, 191–204.

Nagelkerken, I., van der Velde, G., Gorissen, M.W., Meijer, G.J., van't Hof, T., den Hartog, C., 2000. Importance of mangroves, seagrass beds, and the shallow coral reef as a nursery for important coral reef fishes, using a visual census technique. Estuarine, Coastal and Shelf Science 51, 31–44.

Nakamura, Y., Horinouchi, M., Nakai, T., Sano, M., 2003. Food habits of fishes in a Seagrass bed on a fringing reef at Iriomote Island, southern Japan. Ichthyological Research 50, 15–22.

Nakamura, Y., Sano, M., 2004. Comparison between community structures of fishes in *Enhalus acoroides*- and *Thalassia hemprichii*-dominated seagrass beds on fringing coral reefs in the Ryukyu Islands, Japan. Ichthyological Research 51, 38–45.

Nash, K.L., Graham, N.A.J., Bellwood, D.R., 2013. Fish foraging patterns, vulnerability to fishing, and implications for the management of ecosystem function across scales. Ecological Applications 23, 1632–1644.

Ogden, J.C., Quinn, T.P., 1994. Migration in coral reef fishes: ecological significance and orientation mechanisms. In: Migration in Fishes. Plenum Press, New York, pp. 293–398.

Ogden, J.C., Ziemann, J.C., 1977. Ecological aspects of coral reef-seagrass bed contacts in the Caribbean. In: Taylor, D.L. (Ed.), Proceedings, Third International Coral Reef Symposium. Biology, vol. 1. Rosenstiel School of Marine and Athmospheric Science, Miami, Florida, pp. 377–382.

Olesen, B., Marbà, N., Duarte, C.M., Savela, R.S., Fortes, M.D., 2004. Recolonization dynamics in a mixed seagrass meadow: the role of clonal versus sexual processes. Estuaries 27 (5), 770–780.

Ooi, J.L.S., Kendrick, G.A., Van Niel, K.P., Affendi, Y.A., 2011. Knowledge gaps in tropical Southeast Asian seagrass systems. Estuarine, Coastal and Shelf Science 92, 118–131.

Orth, R.J., Heck, K.L., van Montfrans, J., 1984. Faunal communities in seagrass beds: a review of the influence of plant structure and prey characteristics on predator-prey relationships. Estuaries 7, 339.

Pandolfi, M.G.L., 1992. Successive isolation rather than evolutionary centres for the origin of Indo-Pacific reef corals. Journal of Biogeography 92, 593–609.

Pandolfi, M.G.L., 1993. The review of the tectonic history of New Guinea and its significance for marine biodiversity. In: Proceedings of 7th International Coral Reef Symposium, Guam, vol. 2, pp. 718–728.

Parrish, J.D., 1989. Fish communities of interacting shallow-water habitats in tropical oceanic regions. Marine Ecology Progress Series 58, 143–160.

Peduzzi, P., Herndl, G.J., 1988. Decomposition and significance of seagrass leaf litter (*Cymodocea nodosa*) for the microbial food web in coastal waters (Gulf of Trieste, Northern Adriatic Sea). Marine Ecology Progress Series 71, 163–174.

Pergent-Martini, C., Pergent, G., Monnier, B., Boudouresque, M.C., Valette-Sansevin, A., 2021. Contribution of *Posidonia oceanica* meadows in the context of climate change mitigation in the Mediterranean Sea. Marine Environmental Research 165, 105236.

Peterson, C.H., Luettich Jr., R.A., Micheli, F., Skilleter, G.A., 2004. Attenuation of water flow inside seagrass canopies of differing structure. Marine Ecology Progress Series 268, 81–92.

Pet-Soede, L., 2001. Destructive fishing practices mini symposium. SPC Live Reef Fish Information Bulletin 8, 16–19.

Pogoreutz, C., Ahnelt, H., 2014. Gut morphology and relative gut content do not reliably reflect trophic level in gobiids: a comparison of four species from a tropical Indo-Pacific seagrass bed. Journal of Applied Ichthyology 30, 408–410.

Pogoreutz, C., Kneer, D., Litaay, M., Asmus, H., Ahnelt, H., July 2012. The influence of canopy structure and tidal level on fish assemblages in tropical Southeast Asian seagrass meadows. Estuarine, Coastal and Shelf Science 107, 58–68. https://doi.org/10.1016/j.ecss.2012.04.022.

Polte, P., Asmus, H., 2006. Intertidal seagrass beds (*Zostera noltii*) as spawning grounds for transient fishes in the Wadden Sea. Marine Ecology 312, 235–243.

Polte, P., Schanz, A., Asmus, H., 2005. The contribution of seagrass beds (*Zostera noltii*) to the function of tidal flats as a juvenile habitat for dominant, mobile epibenthos in the Wadden Sea. Marine Biology 147, 813–822.

Poore, G.C.B., 2018. Burrowing lobsters mostly from shallow coastal environments in Papua New Guinea (Crustacea: Axiidea: Axiidae, Micheleidae). Memoirs of Museum Victoria 77, 1–14.

Potouroglou, M., Bull, J.C., Krauss, K.W., Kennedy, H.A., Fusi, M., Daffonchio, D., et al., 2017. Measuring the role of seagrasses in regulating sediment surface elevation. Scientific Reports 7 (1), 11917. http://www.nature.com/articles/s41598-017-12354-y.

Randall, J.E., 1965. Grazing effect on sea grasses by herbivorous fish in the West Indies. Ecology 46, 255–260.

Roeselers, G., Newton, I.L.G., 2012. On the evolutionary ecology of symbioses between chemosynthetic bacteria and bivalves. Applied Microbiology and Biotechnology 94 (1), 1–10. https://doi.org/10.1007/s00253-011-3819-9. ISSN 0175-7598.

Rollon, R.R., De Ruyter van Steveninck, E.D., van Vierssen, W., Fortes, M.D., 1998. Contrasting recolonization strategies in multi-species seagrass meadows. Marine Pollution Bulletin 37 (8–12), 450–459.

Satria, A., Matsuda, Y., 2004. Decentralization of fisheries management in Indonesia. Marine Policy 28 (5), 237–250. https://doi.org/10.1016/j.marpol.2003.11.001.

Sawayama, S., Nurdin, N., Akbar, A.S.M., Sakamoto, S.X., Komatsu, T., 2015. Introduction of geospatial perspective to the ecology of fish-habitat relationships in Indonesian coral reefs: a remote sensing approach. Ocean Science Journal 50 (2), 343–352.

Schultz, S.T., Kruschel, C., Bakran-Petricioli, T., 2009. Influence of seagrass meadows on predator-prey habitat segregation in an Adriatic lagoon. Marine Ecology Progress Series 374, 85–99.

Schwerdtner Máñez, K., Ferse, S.C.A., 2010. The history of Makassan trepang fishing and trade. PLoS One 5 (6), 1–8.

Seilacher, A., 1990. Aberrations in bivalve evolution related to photo- and chemosymbiosis. Historical Biology 3 (4), 289–311. https://doi.org/10.1080/08912969009386528. ISSN 0891-2963.

Sheppard, J.K., Lawler, I.R., Marsh, H., 2007. Seagrass as pasture for seacows: landscape-level dugong habitat evaluation. Estuarine, Coastal and Shelf Science 71, 117–132.

Short, F., 1987. Effects of sediment nutrients on seagrasses: literature review and mesocosm experiment. Aquatic Botany 27, 41–57.

Short, F.T., Carruthers, T., Dennison, W., Waycott, M., 2007. Global seagrass distribution and diversity: a bioregional model. Journal of Experimental Marine Biology and Ecology 350, 3–20.

Smith, A.G., Briden, J.C., 1977. Mesozoic and Cenozoic Paleocontinental Maps. Cambridge University press, Cambridge.

Stapel, J., Aarts, T.L., van Duynhoven, B.H.M., de Groot, J.D., van den Hoogen, P.H.W., Hemminga, M.A., 1996. Nutrient uptake by leaves and roots of the seagrass *Thalassia hemprichii* in the Spermonde Archipelago, Indonesia. Marine Ecology Progress Series 134, 195–206.

Stapel, J., Erftemeijer, P.L.A., 2000. Leaf harvesting by burrowing alpheid shrimps in a *Thalassia hemprichii* meadow in South Sulawesi, Indonesia. Biologia Marina Mediterranea 7 (2), 282–285.

Stapel, J., Manuntun, R., Hemminga, M.A., 1997. Biomass loss and nutrient redistribution in an Indonesian *Thalassia hemprichii* seagrass bed following seasonal low tide exposure during daylight. Marine Ecology Progress Series 148, 251–262. https://doi.org/10.3354/meps148251. Bibcode: 1997MEPS.148.251S.

Stoner, A.W., 1982. The influence of benthic macrophytes on the foraging behavior of pinfish, *Lagodon rhomboides* (Linnaeus). Journal of Experimental Marine Biology and Ecology 58, 271–284.

Suchanek, T.H., 1983. Control of seagrass communities and sediment distribution by *Callianassa* (Crustacea, Thalassinidea) bioturbation. Journal of Marine Research 41, 281–298.

Sumaila, U.R., Guénette, S., Alder, J., Chuenpagdee, R., 2000. Addressing ecosystem effects of fishing using marine protected areas. ICES Journal of Marine Science 57, 752–760. https://doi.org/10.1006/jmsc.2000.0732.

Thomas, B.E., Connolly, R.M., 2001. Fish use of subtropical saltmarshes in Queensland, Australia: relationships with vegetation, water depth and distance onto the marsh. Marine Ecology Progress Series 209, 275–288.

Tomascik, T., Mar, A.J., Nontji, A., Moosa, M.K., 1997. The Ecology of the Indonesian Seas – Part Two. Periplus editions (HK) Ltd, ISBN 962-593-163-5, p. 1387.

Torres-Pulliza, D., Wilson, J.R., Darmawan, A., Campbell, S.J., Andréfouët, S., 2013. Ecoregional scale seagrass mapping: a tool to support resilient MPA network design in the coral triangle. Ocean and Coastal Management 80, 55–64.

Unsworth, R.K.F., Bell, J.J., Smith, D.J., 2007a. Tidal fish connectivity of reef and sea grass habitats in the Indo-Pacific. Journal of the Marine Biology Association of the UK 87, 1287–1296.

Unsworth, R.K.F., Cullen, L.C., 2010. Recognising the necessity for Indo-Pacific seagrass conservation. Conservation Letters 3, 63–73.

Unsworth, R.K.F., de Leon, P.S., Garrard, S.L., Jompa, J., Smith, D.J., Bell, J.J., 2008. High connectivity of Indo-Pacific seagrass fish assemblages with mangrove and coral reef habitats. Marine Ecology Progress Series 353, 213–224.

Unsworth, R.K.F., Taylor, J.D., Powell, A., Bell, J.J., Smith, D.J., 2007b. The contribution of scarid herbivory to seagrass ecosystem dynamics in the Indo-Pacific. Estuarine, Coastal and Shelf Science 74 (1–2), 53–62.

Unsworth, R.K.F., Wylie, E., Smith, D.J., Bell, J.L., 2007c. Diel trophic structuring of seagrass bed fish assemblages in the Wakatobi Marine National Park, Indonesia. Estuarine, Coastal and Shelf Science 72, 81–88.

Valentine, J.F., Heck, K.L., 2005. Perspective review of the impacts of overfishing on coral reef food web linkages. Coral Reefs 24, 209–213. https://doi.org/10.1007/s00338-004-0468-9. Download citation.

van Oostenbrügge, J.A.E., van Densen, W.L.T., Machiels, M.A.M., 2004. How the uncertain outcomes associated with aquatic and land resource use affect livelihood strategies in coastal communities in the Central Moluccas, Indonesia. Agricultural Systems 82 (1), 57–91.

Vaugelas, J de, 1985. On the presence of the mud-shrimp *Callichirus armatus* in the sediments of Mataiva Lagoon. In: Delesalle, B., Galzin, R., Salvat, B. (Eds.), Proceedings of the Fifth International Coral Reef congress, Tahiti, 25 May – 1 June 1985, vol. 1, pp. 314–316. Fig. 48–50.

Verduin, J.J., Backhaus, J.O., 2000. Dynamics of plant-flow interactions for the seagrass Amphibolus antarctica: field observations and model simulations. Estuarine, Coastal and Shelf Science 50, 185–204.

Vonk, J.A., 2008. Seagrass Nitrogen Dynamics – Growth Strategy and the Effects of Macrofaunal in Indonesian Mixed-Species Meadows PhD Thesis. Faculty of Science, Radboud University Nijmegen, the Netherlands.

Vonk, J.A., Christianen, M.J.A., Stapel, J., 2008a. Redefining the trophic importance of seagrasses for fauna in tropical Indo-Pacific meadows. Estuarine, Coastal and Shelf Science 79, 653–660.

Vonk, J.A., Christianen, M.J.A., Stapel, J., 2010. Abundance, edge effect, and seasonality of fauna in mixed-species seagrass meadows in southwest Sulawesi, Indonesia. Marine Biology Research 6, 282–291.

Vonk, J.A., Kneer, D., Stapel, J., Asmus, H., 2008b. Shrimp burrow in tropical seagrass meadows: an important sink for litter. Estuarine, Coastal and Shelf Science 79, 79–85.

Vonk, J.A., Middelburg, J.J., Stapel, J., Bouma, T.J., 2008c. Dissolved organic nitrogen uptake by seagrasses. Limnology and Oceanography 53 (2), 542–548.

Vonk, J.A., Pijnappels, M.H.J., Stapel, J., 2008d. In situ quantification of *Tripneustes gratilla* grazing and influence on three co-occurring tropical seagrasses. Marine Ecology Progress Series 360, 107–114. https://doi.org/10.3354/meps07362.

Weinstein, M.P., Heck, K.L., 1977. Ichthyofauna of seagrass meadows along the Caribbean coast of Panama and in the Gulf of Mexico: composition, structure and community ecology. Marine Biology 50, 97–107.

Welsh, J.Q., Bellwood, D.R., 2014. Herbivorous fishes, ecosystem function and mobile links on coral reefs. Coral Reefs 33, 303–311.

Williams, S.L., Ambo-Rappe, R., Sur, C., Abbott, M., Limbong, S.R., 2017. Species richness accelerates marine ecosystem restoration in the coral triangle. Proceedings of the National Academy of Sciences of the United States of America 114, 11986–11991.

Winters, G., Beer, S., Willette, D.A., Viana, I.G., Chiquillo, K.I., Beca-Carretero, P., et al., 2020. The tropical seagrass *Halophila stipulacea*: reviewing what we know from its native and invasive habitats, alongside identifying knowledge gaps. Frontiers in Marine Science 7, 300. https://doi.org/10.3380/fmars.2020.00300.

Young, G.C., Potter, I.C., Hyndes, G.A., Lestang, S., 1997. The ichthyofauna of an intermittently open estuary: implications of bar breaching and low salinities on faunal composition. Estuarine, Coastal and Shelf Science 45, 53–68.

Zimmerman, R.C., Kohrs, D.G., Steller, D.L., Alberte, R.S., 1997. Impacts of $CO_2$ enrichment on productivity and light requirements of eelgrass. Plant Physiology 115 (2), 599–607.

7

# Mangrove ecosystems under threat in Indonesia: the Segara Anakan Lagoon, Java, and other examples

Tim C. Jennerjahn[1,2], Erwin Riyanto Ardli[1,3], Jens Boy[4], Jill Heyde[5], Martin C. Lukas[1,5], Inga Nordhaus[1], Moh Husein Sastranegara[3], Kathleen Schwerdtner Máñez[1,6], Edy Yuwono[3,7]

[1]*LEIBNIZ CENTRE FOR TROPICAL MARINE RESEARCH (ZMT), BREMEN, GERMANY;* [2]*FACULTY OF GEOSCIENCE, UNIVERSITY OF BREMEN, BREMEN, GERMANY;* [3]*FAKULTAS BIOLOGI, UNIVERSITAS JENDERAL SOEDIRMAN, PURWOKERTO, JAVA, INDONESIA;* [4]*INSTITUTE OF SOIL SCIENCE, LEIBNIZ UNIVERSITÄT, HANNOVER, GERMANY;* [5]*SUSTAINABILITY RESEARCH CENTER (ARTEC), UNIVERSITY OF BREMEN, GERMANY;* [6]*PLACE NATURE CONSULTANCY, ASHAUSEN, GERMANY;* [7]*UNIVERSITAS WANITA INTERNASIONAL, BANDUNG, JAWA BARAT, INDONESIA*

## Abstract

*Indonesian mangrove forests are of major local and global importance for ecological and economic reasons. Indonesia has both the largest area of mangrove forests and the highest mangrove deforestation rate by country. Using the mangrove-fringed Segara Anakan Lagoon on Java as a prime example, this chapter explains the ecosystem services provided by mangrove-dominated coastal ecosystems, as well as the threats to it. Related governance approaches and interventions are discussed, while special emphasis is given to water quality, "Blue Carbon" storage, biodiversity, natural resource use, land use change, and the underlying political and societal dynamics. While ecosystem service supply is strongly impaired in the Segara Anakan Lagoon, mainly because of deforestation and high sediment deposition related to land use change, mangrove ecosystems in other areas appear to be in a better state. Finally, directions of future research and recommendations for policy and society are given.*

## Abstrak

*Hutan bakau Indonesia sangat penting secara lokal dan global karena memiliki nilai ekologis dan ekonomis. Indonesia merupakan negara dengan wilayah hutan bakau terluas, namun laju deforestasi hutan bakaunya tertinggi. Menggunakan Laguna Segara Anakan yang bertipe hutan bakau tepian pulau di Jawa sebagai contoh utama, bab ini membahas tentang jasa ekosistem yang diberikan oleh ekosistem pesisir yang didominasi hutan bakau maupun ancaman terhadap jasa ekosistem tersebut. Pembahasan meliputi pendekatan dan intervensi tata kelola yang terkait, sedangkan penekanan khususnya adalah pada kualitas air, cadangan karbon pada ekosistem pesisir dan laut ('Blue Carbon' storage), keanekaragaman hayati, pemanfaatan sumber daya alam, perubahan penggunaan lahan, dan dinamika sosial politik yang mendasarinya. Meskipun pasokan*

Science for the Protection of Indonesian Coastal Ecosystems (SPICE). https://doi.org/10.1016/B978-0-12-815050-4.00004-3

*jasa ekosistem di Laguna Segaran Anakan sangat buruk, terutama karena deforestasi dan tingginya sedimentasi terkait dengan perubahan penggunaan lahan, ekosistem bakau di daerah lain kondisinya lebih baik. Di bagian akhir bab ini disajikan arah penelitian kedepan dan rekomendasi untuk kebijakan dan masyarakat.*

**Chapter outline**

**7.1 Introduction** 252
**7.2 The study areas** 254
**7.3 Environmental setting and natural resource use** 257
7.3.1 The physical setting 257
7.3.2 Water quality, biogeochemistry, and pollution 258
7.3.3 Carbon sources and storage 260
7.3.4 Flora and fauna 266
7.3.5 Population and natural resource use in the Segara Anakan region 268
**7.4 Environmental change in the Segara Anakan Lagoon region: causes, drivers, and impacts** 269
7.4.1 Decline of marine species and fisheries 269
7.4.2 Sedimentation and its causes 270
7.4.3 Reclamation of land and conflicts over new land 273
**7.5 Threats to mangrove forests and their ecosystem services in Indonesia** 274
**7.6 Management programs** 276
**Acknowledgments** 279
**References** 279

## 7.1 Introduction

Mangrove forests are lining major parts of tropical and subtropical coasts and provide a wealth of functions and services that are important for healthy coastal ecosystems and human well-being. They are highly productive and support biodiversity, serve as nurseries for juvenile stages of coastal fish and other organisms, provide coastal protection against waves and storms, and help to maintain water quality. In addition, they are highly efficient natural sinks for carbon dioxide, provide nontimber forest products for the local population like, e.g., firewood and honey, and serve as sites for tourism (Huxham et al., 2017; Wells et al., 2006). While the intimately connected ecological and economic relevance of mangrove ecosystem functions and services increasingly comes into the focus of science and society, mangrove forests are under threat from a diverse array of human interventions and climate change effects (Chowdhury et al., 2017; Jennerjahn et al., 2017). This is of particular importance in Indonesia, which holds the largest mangrove area by country (3,112,989 ha; Giri et al., 2011), but also exhibits the highest area loss (−48,205 ha between 2000 and 2012; Richards and Friess, 2016). There, the major climate change effects, i.e., sea level rise, warming of air and surface waters, aridity, and increased storminess, are of minor importance compared with human interventions, mainly logging, conversion to other land uses, pollution, and overharvesting of wood and fishery resources (Alongi, 2015; Jennerjahn et al., 2017; Richards and Friess, 2016).

Maintaining the structures and functions of mangrove forests is not only of importance for the continued supply of all their ecosystem services but also for seagrass meadows and coral reefs, adjacent coastal ecosystems whose well-being is intimately linked to the existence of healthy mangrove forests. For example, coral reefs provide physical protection from waves and storms to seagrass meadows and mangrove forests, while the latter reduce the risk of eutrophication and siltation of seagrass meadows and coral reefs through accumulation of land-derived nutrients and sediments (e.g., Alongi, 2009; Guannel et al., 2016). Understanding the structure and functions of mangrove ecosystems under these multiple stressors is a major step toward developing measures for their conservation and sustainable use. The Indonesian–German research and education program "Science for the Protection of Indonesian Coastal Marine Ecosystems" (SPICE) provided the frame for this in its cluster "Mangrove Ecology and Sustainability." The structure, functions, and ecosystem services of mangroves as well as human impacts, their causes, and governance and management issues were investigated with multi- and interdisciplinary approaches in several regions of Indonesia, but with a geographic focus on the mangrove-fringed Segara Anakan Lagoon (SAL) in south central Java. Other regions that were less intensively studied are the Berau Regency in Kalimantan and the Togian Islands in east Sulawesi, but the results of which are also reported here.

The SAL is an area rich in natural resources and mainly nourished by the inputs from the Citanduy, the fifth largest river of Java, and the tidal exchange with the Indian Ocean. The lagoon ecosystem has been threatened by overfishing, logging of mangrove wood, high sediment and organic matter input from the catchment area, and pesticide and oil pollution (White et al., 1989). Management programs have been conducted for decades to overcome these problems (Asian Development Bank, 2006; Olive, 1997), but have had little success.

Therefore, more than a decade (2004–2016) of research was conducted in the SPICE cluster "Mangrove Ecology and Sustainability" to provide new insights on cause–effect relationships, to quantify these relationships, to assess ecosystem service supply, and to analyze governance and management regimes of Indonesia's mangrove ecosystems.

The long-term research program SPICE allowed us to advance the understanding of (1) the dynamics and drivers of social–ecological change, (2) the connectivity and adaptability of mangrove ecosystems and related social systems, and (3) governance and management challenges in times of global change. As the name says—Science for the Protection of Indonesian Coastal Marine Ecosystems—our research results provide the means to increase awareness among stakeholders on the ecological and economic relevance of mangrove forests, and they provide knowledge that can support the development of measures toward conservation and more sustainable use of these exceptional ecosystems. The research also directs attention to the critical roles of political structures, misfits between state-led management approaches and realities on the ground, and widespread conflicts over natural resources, which to date have limited the potential for more sustainable management of these resources.

## 7.2 The study areas

Mangrove forests occur in six environmental settings, which can be very different in their hydrodynamics, geomorphology, biogeochemistry, and flora and fauna composition. These are (1) river-dominated, (2) tide-dominated, (3) wave-dominated, (4) composite river- and wave-dominated, (5) drowned bed rock valley, and (6) carbonate settings (Woodroffe, 1992). The SAL is a combination of a river- and tide-dominated system, the areas under study in the Berau Regency are mainly tide-dominated with a partial river dominance, while the Togian Islands belong to the carbonate settings (Fig. 7.1).

The SAL is a 25-km-long, shallow estuarine lagoon located on the south coast of the densely populated island of Java (Figs. 7.2 and 7.3). It is separated from the Indian Ocean by the rocky mountainous Nusakambangan Island and connected to it by two inlets at its

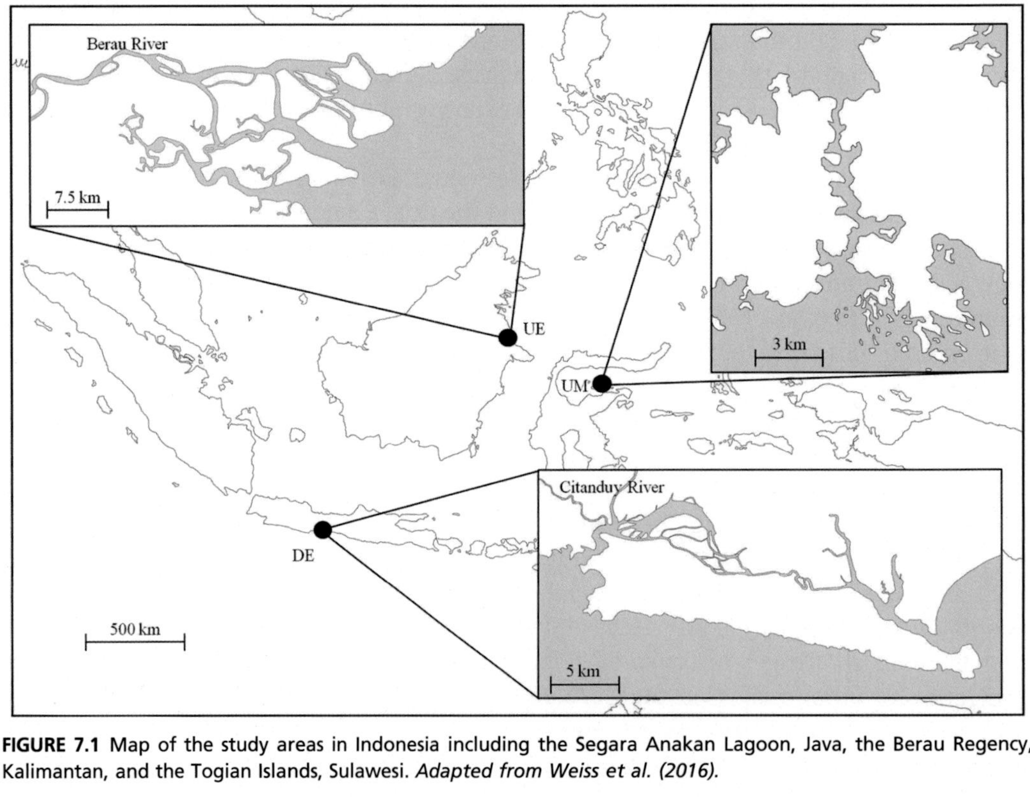


**FIGURE 7.1** Map of the study areas in Indonesia including the Segara Anakan Lagoon, Java, the Berau Regency, Kalimantan, and the Togian Islands, Sulawesi. *Adapted from Weiss et al. (2016).*

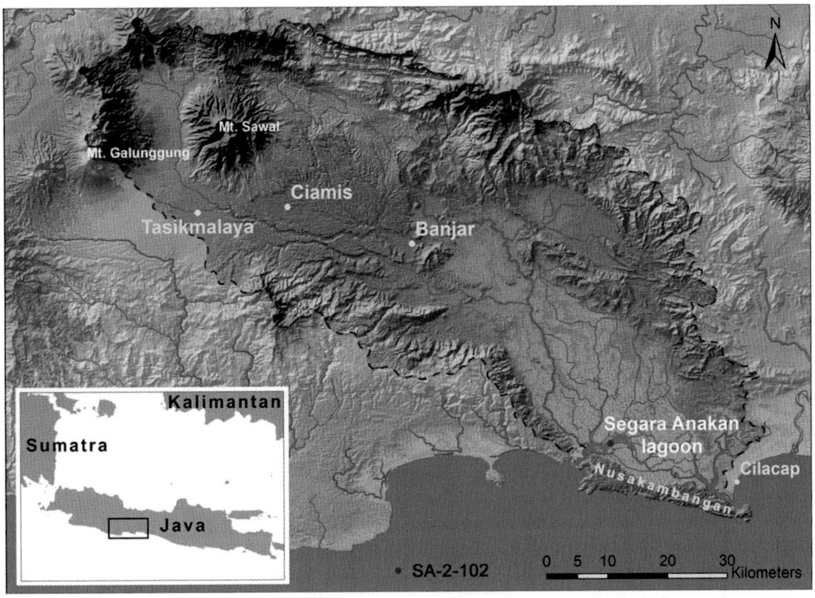


**FIGURE 7.2** The Segara Anakan Lagoon and its catchment area. *Adapted from Lukas (2014b).*

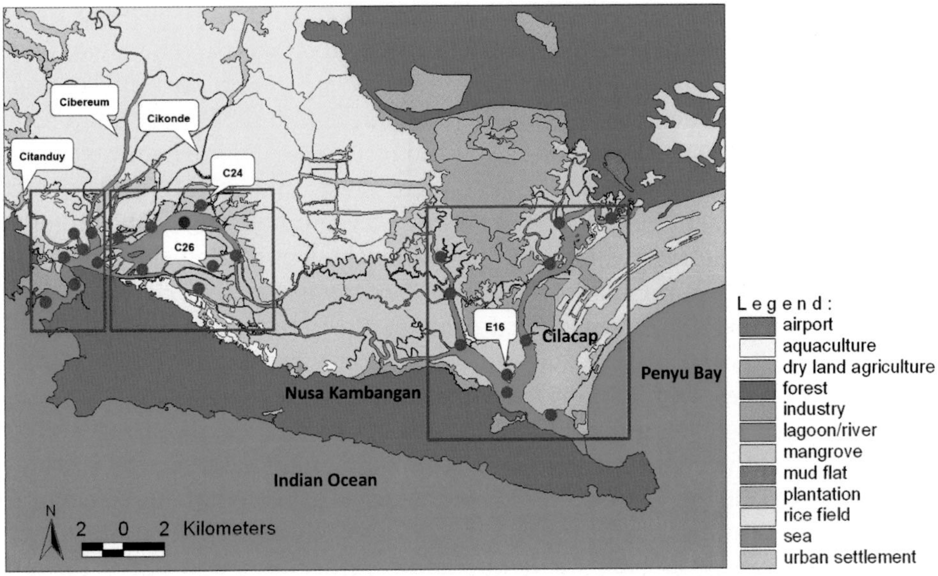


**FIGURE 7.3** Map of the Segara Anakan Lagoon in 2006 including land use, major rivers (Citanduy, Cibereum, Cikonde) and the city of Cilacap. The red rectangles denote major geographical areas of work (left to right: W = western, C = central, E = eastern lagoon) as later used. Red dots denote sampling stations in the lagoon and three mangrove stations C24, C26, and E16. The blue dot denotes the location of a sediment core taken in the central lagoon. *Adapted from Jennerjahn et al. (2009).*

western and eastern ends (Yuwono et al., 2007). It has a dominance of river input in its western part, where the Citanduy River discharges into the lagoon, and a dominance of tidal exchange in its eastern part, where freshwater input is very low (Holtermann et al., 2009). The lagoon is fringed by the largest remaining single mangrove forest on the south coast of Java with a total area of approximately 9000 ha (Ardli and Wolff, 2009), the flora of which consists of 15 mangrove tree species and 5 understory genera (Nordhaus et al., 2019). The density, diversity, and aboveground biomass of trees are lower in the central than in the eastern part of the lagoon, mainly as a consequence of environmental degradation due to logging and land use conversion (Hinrichs et al., 2009; Nordhaus et al., 2019). The lagoon has shrunk drastically and became shallower due to heavy sedimentation from rivers, mainly the Citanduy River, in the past 150 years (Yuwono et al., 2007; Lukas, 2014a,b). The land areas adjacent to the lagoon are dominated by settlements and agriculture, mainly cultivation of rice under irrigation, the area of which increased by 62% between 1987 and 2006 to a total of 19,274 ha, which makes up 21% of the Segara Anakan region (Fig. 7.2; Ardli and Wolff, 2009). The 450 $km^2$ large catchment area of the lagoon comprises a range of land use and land cover types, including irrigated rice agriculture, rainfed agriculture with a diversity of crops, settlements with house gardens, villagers' mixed forests, state forests (mainly comprising monocultures of teak and pine), and plantations (e.g., rubber and cocoa) (Lukas, 2015). The city of Cilacap with a population of about 234,000 (BPS Cilacap, 2018) is located at the eastern end of the lagoon and harbors the largest oil refinery of Indonesia in its port.

The Berau Regency is located in East Kalimantan (118°05′E, 2°25′N; Fig. 7.1). The hydrodynamics of the Berau coastal area are driven by the two rivers Berau and Tabalar in the west and strong currents in the east influenced by the Indonesian Throughflow (ITF) that connects the Pacific and Indian Oceans through the Makassar Strait (Gordon, 2005; Wiryawan et al., 2005). The coast is covered by dense mangrove forests, which are relatively undisturbed. Some areas near the Berau River estuary have been cleared for aquaculture, but these activities are still in an early development stage. The hinterland is covered by tropical rainforest, which has been partly converted into palm oil plantations (Weiss et al., 2016). With a population density of 10 persons per $km^2$, Berau is one of the least populated of the 300 regencies in Indonesia (BPS Kalimantan Timur, 2017). Its economy is characterized by a heavy dependence on the extraction of minerals and other natural resources, with a massive push of coal mining, logging, and timber plantation development following decentralization in the early 2000s (Keulartz and Zwart, 2004). About 37% of the original mangrove area of Kalimantan has been lost (Ilman et al., 2016). Many remaining areas have lost their ecological connections to adjacent freshwater or terrestrial systems due to agricultural and oil palm plantations that have replaced lowland forests. In 2005, the Berau marine protected area (1321 × 106 ha) was established, which has the second highest coral reef biodiversity in Indonesia and the most extensive remaining mangrove forest in Kalimantan covering an area of 56,000 ha (Siahainenia, 2016).

The Togian Islands are a ca. 120 km long chain of around 60 larger islands in the Bay of Tomini (Fig. 7.1). Since 2004, the area has belonged to the district of Ampana and has been designated a marine protection area in the same year. Although the Bay of Tomini is tucked away from greater flow-through systems and in general shallow, both of which are limiting the exchange of water, the mangrove forests and coral reefs are surprisingly rich in diversity (Wallace, 1999). Major income of the Togian population, which consists of a number of ethnic groups, is the trade of seafood, using the capital small town of Wakai as the hub for exportations, increasingly to China. There is, to a lesser extent, also dive tourism, which unfortunately suffered dramatically due to heavy dynamite and cyanide fishing practices, resulting in severely disturbed reefs with uncertain chances for regeneration. Unlike the underwater situation, the islands still harbor relatively undisturbed coastal mangrove forests, which receive freshwater input by karst runoff and precipitation. There are also mangrove patches found, which grow directly on intact coral reefs, sometimes called "clearwater mangroves" (Weiss et al., 2016).

## 7.3 Environmental setting and natural resource use

### 7.3.1 The physical setting

The physical setting is a major determinant of biogeochemical processes and fluxes and the flora and fauna composition in mangrove ecosystems. Precipitation in the SAL watershed is high due to the monsoon climate with maximum rainfall during the wet NW monsoon in austral summer (November–March). The long-term annual average precipitation in the coastal city of Cilacap amounts to 3340 mm $year^{-1}$ but can be as low as 1200 mm $year^{-1}$ during El Niño years (Jennerjahn et al., 2009; Weatherbase, 2019). In Sidareja in the center of the SAL, watershed precipitation amounts to 2770 mm $year^{-1}$ (Weatherbase, 2019). The SAL receives freshwater input from a number of rivers. The Citanduy river contributes >80% of the freshwater in the western part of the lagoon, with an annual average discharge of 227 $m^3\ s^{-1}$ (dry season 171 $m^3\ s^{-1}$, rainy season 283 $m^3$ $s^{-1}$), resulting in an annual total of 7.2 $km^3$ (Ludwig, 1985). Tidal exchange with the Indian Ocean occurs through two channels in the western and eastern parts of the lagoon. The mixed and predominantly semidiurnal tide ranges between 0.4 m during neap tide and 1.9 m during spring tide (Holtermann et al., 2009; White et al., 1989). The residence time of the water in the lagoon is generally short, which is on the order of 0–2 days near the western and eastern outlets and is at maximum 8 days (dry season) and 12 days (wet season) in the central lagoon (Holtermann et al., 2009). It is at the lower end of the range reported for lagoons in Brazil, Taiwan, and around the Mediterranean (Hung and Hung, 2003; Knoppers, 1991; Umgiesser et al., 2014). The sediment dynamics vary largely between the western and eastern parts of the lagoon. While total suspended matter concentration varied around 10 mg $L^{-1}$ in the latter due to the low freshwater input, it can be orders of magnitude higher (>100 to >1000 mg $L^{-1}$) in the west because of the high input of the Citanduy River during the wet season (Moll, 2011; Yuwono et al., 2007). There, it is much higher than the global average of 500 mg $L^{-1}$ because of the generally high physical and chemical weathering and high river discharge in the SE Asia region (Milliman and Farnsworth, 2011). The hydrodynamics and the sediment dynamics exert major control on the dispersal and accumulation of substances in the lagoon like, for example, sediments, carbon, nutrients and pollutants, and how they can affect flora and fauna and ultimately ecosystem health.

Kalimantan and its coastal areas including the Berau Regency are affected by the Asian-Australian monsoon. The wet season starts in October and extends until May. The dry season spans from July to September, and average annual rainfall in the Berau catchment varies from 2400 to 3350 mm year$^{-1}$. During the study period, monthly rainfall ranged between 194 mm in September 2015 and 484 mm in November 2015 (World Weather Online, 2019). Porewater salinity was 6.1 at the most northern mangrove site in the Berau Delta, whereas salinity ranged between 24.3 and 35.0 at all other sites (Tripathi, 2016).

The Togian Islands are located in the Bay of Tomini, between 121°37′ and 122°25′ E close to the equator. Roughly 60 larger islands span an area of 120 km in lateral extension. The precipitation is around 1600–1700 mm year$^{-1}$, which falls more or less equally distributed with a short drier period in August and September. Temperature ranges between 23 and 33°C without any observable seasonality. Climatically, the Togians belong to the "lands below the wind" famed in seafaring, as they lie just between the northern and southern monsoon systems dictating the climate of the rest of Sulawesi (Wallace, 1999).

## 7.3.2 Water quality, biogeochemistry, and pollution

Dissolved oxygen ranged between 5.4 and 6.2 mg L$^{-1}$ or 70%–90% of oxygen saturation in the SAL and was lower in the Citanduy River (3.9 mg L$^{-1}$, 50% saturation). Concentrations of the dissolved inorganic nutrients nitrate and phosphate, which to a large extent result from human activities in the hinterland, displayed large spatial and temporal variations. Concentrations were generally higher in the western (up to 35 μM nitrate) than in the eastern part of the lagoon (<10 μM nitrate), and they were higher during the wet (W: 5–35 μM nitrate, E: 3–10 μM nitrate) than during the dry season (W: 0–8 μM nitrate, E: 0–5 μM nitrate; Moll, 2011). The concentrations of chlorophyll *a* as an indicator of biomass of primary producers were also moderate (1–8 μg L$^{-1}$; Yuwono et al., 2007). It appears that the nutrient inventory of Segara Anakan results from a mixture of anthropogenic and natural sources and processes. Maximum inputs into the western lagoon were supplied by the Citanduy during the rainy season, while tidal exchange with the Indian Ocean dominated in the eastern lagoon, which generally has little freshwater input. Porewater profiles exhibited that losses from recycling processes in the mangrove forests are an additional source of nutrients to the lagoon during the dry season (Jennerjahn et al., 2009). Despite the high freshwater and nutrient input from the agriculture-dominated hinterland, the trophic status of SAL was mostly oligo- to mesotrophic and nutrient pollution on a low to moderate level compared with other lagoons around the globe (Jennerjahn et al., 2009).

In terms of organic pollution, the focus is often on the so-called "dirty dozen," the major persistent organic pollutants originally listed in the Stockholm Convention on

Persistent Organic Pollutants (2018), but it is known there are thousands of other organic contaminants. However, in most cases, little to nothing is known on their potential risk for the environment (Muir and Howard, 2006). With a nontarget screening approach, more than 50 organic contaminants were found in water, sediment, and macrobenthic invertebrates of the SAL in 2008. The level of contamination was low to moderate in all parts of the lagoon except for the eastern lagoon close to the oil refinery, where it was high even on a global scale (Fig. 7.4; Dsikowitzky et al., 2011; Syakti et al., 2013).

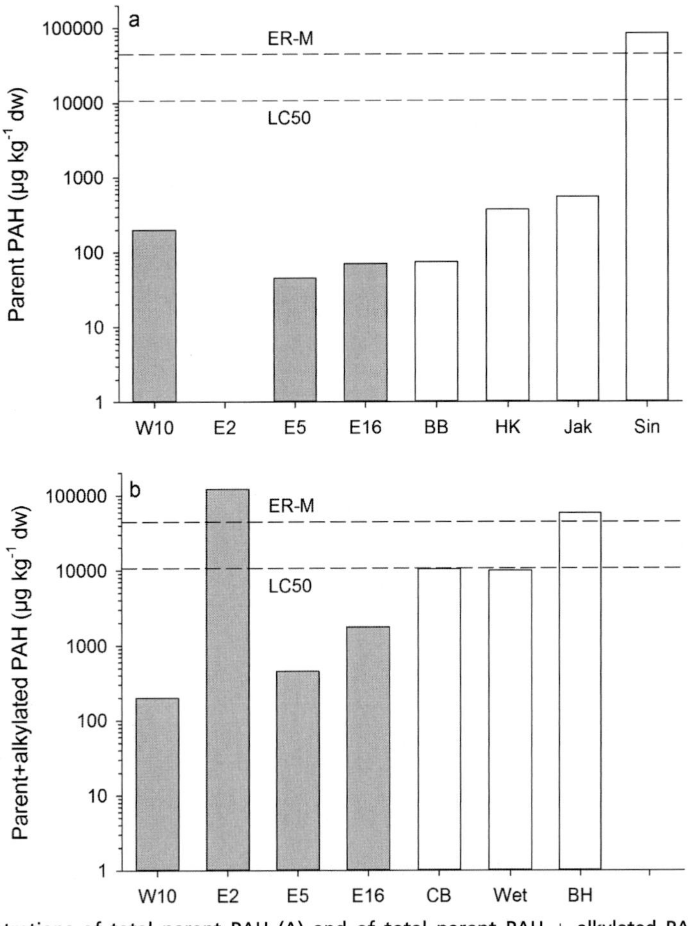


**FIGURE 7.4** Concentrations of total parent PAH (A) and of total parent PAH + alkylated PAH (B) in sediments from Segara Anakan (gray bars) and in sediments from other areas (white bars). Note the logarithmic scale for PAH concentrations. Sediment toxicity thresholds: LC50 of oil-polluted sediment as obtained by a standard amphipod bioassay using Rhepoxynius abronius, considering 20 parent PAHs and 19 alkylated PAHs (Page et al., 2002). LC50 is the concentration of a compound that is lethal for 50% of the exposed population. Effects range-median (ER-M) value for toxic effects of marine and estuarine sediments on aquatic organisms, considering 12 parent PAHs and methylnaphthalene (Long et al., 1995). The ER-M indicates the concentration of a compound above which toxic effects are generally observed. (A) BB, Banten Bay, Indonesia; HK, Nature reserve near Hongkong; Jak, Jakarta Bay, Indonesia; Sin, Singapore's coastal environments. (B) CB, Cienfugos Bay, Cuba; Wet, Wetlands, Alberta, Canada; BH, Boston harbor. *Adapted from and for data sources see Dsikowitzky et al. (2011).*

Substances resulting from municipal sewage input were found, but concentrations were low. Moreover, no contaminants were detected which can clearly be assigned to further pollution sources such as agriculture or other industries. Contamination close to the oil refinery mainly consisted of alkylated polycyclic aromatic hydrocarbons (PAHs), while concentrations of the "usual suspects" known from the Stockholm Convention, the parent PAHs were low. As yet not much attention has been paid to alkylated PAHs. However, the sum of parent PAHs together with alkylated PAHs in sediments close to the oil refinery exceeded published toxicity thresholds for aquatic invertebrates (Fig. 7.4). As these compounds are structurally similar, it is likely that alkylated PAHs have a health risk potential similar to that of parent PAHs. Macrobenthic invertebrates from the mangrove forest took up contaminants from different sources in different combinations suggesting a species-specific uptake related to habitat and feeding mode (Dsikowitzky et al., 2011). This, in turn, implies that the biotic responses to organic contaminants are not uniform and health risks for organisms can be manifold. Taking into account that organisms from the lagoon are an important food source for the local population and that some pollutants can become enriched along the food web, the consumption of benthic organisms like, for example, crabs, mussels, and snails may even impose a health risk on humans. However, assessing this risk would require to quantify possible bioaccumulation of contaminants in those organisms, which could not be done in the frame of the SPICE investigations.

The generally low to moderate concentrations of nutrients and organic contaminants are probably mainly related to the short residence time of the water in the SAL and indicate that a large portion of land-derived substances from agriculture, industry and households are rapidly exported to the Indian Ocean.

### 7.3.3 Carbon sources and storage

While mangrove forest is the dominant land cover in and around the lagoon area and therefore also a major source of carbon for storage and export, the tidal exchange with the Indian Ocean and the river inputs from the hinterland are other quantitatively important sources of carbon. The carbon and nitrogen stable isotope composition of plants and soils from the mangrove forest and from the rice-dominated hinterland as well as of lagoon sediments indicates a W–E gradient in organic matter composition that is mainly related to the land-derived inputs. While rice field soils have an average $\delta^{13}C_{org}$ of −26.1‰ and a $\delta^{15}N$ of 5.3‰, mangrove leaves of the dominant species form the other end of the spectrum with ranges of −30.9 to −27.7‰ for $\delta^{13}C_{org}$ and of −1.0 to 6.2‰ for $\delta^{15}N$ (Moll, 2011). The isotope signatures of sediments from the eastern lagoon are closer to those of the mangrove leaves, while those of sediments from the western and central lagoon are closer to those of rice soils and the Citanduy River (Moll, 2011; Weiss et al., 2016). Sediment cores obtained from an island in the central lagoon that only formed in the second half of the last century and is dominated by mangrove plants display an isotope signature close to that of rice soils and Citanduy sediment and suggest that a large portion of the sediment and carbon deposited there is not of mangrove but of hinterland origin (Fig. 7.5).

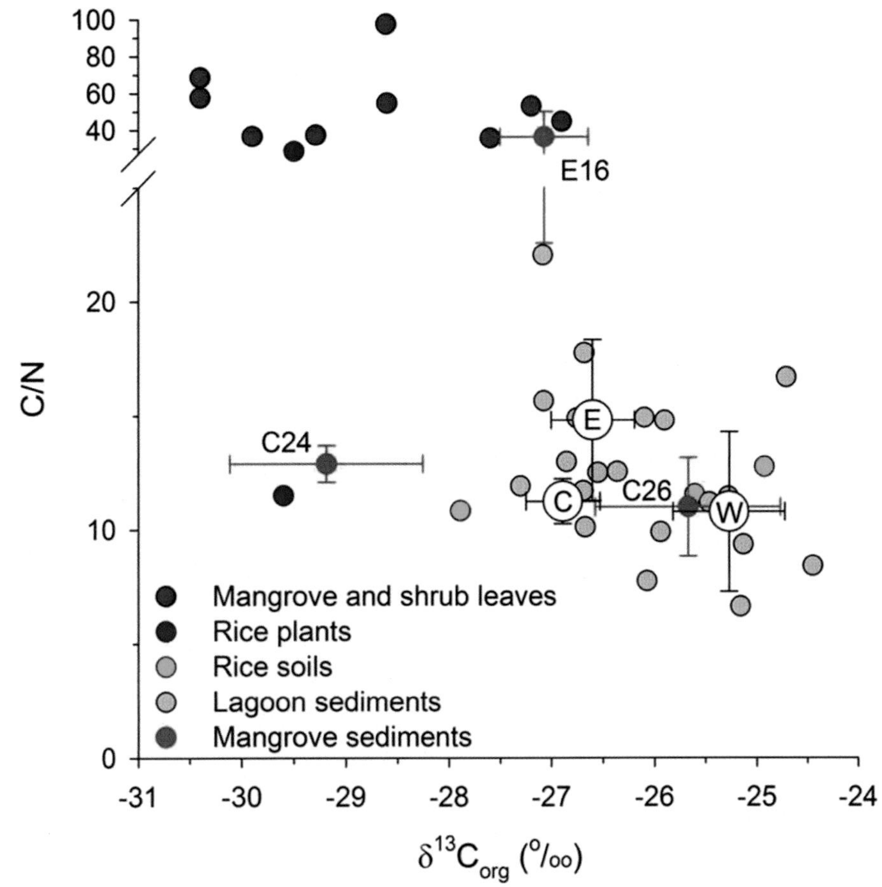


**FIGURE 7.5** Biogeochemical properties of Segara Anakan Lagoon (gray circles) and mangrove sediments (pink circles, E16 from eastern lagoon, C24 and C26 from central lagoon) and of potential sources (rice plants and soils, mangrove, and shrub leaves). *Data from Yuwono et al. (2007) and Jennerjahn (unpublished).*

The high relative carbon storage per unit area is an important ecosystem service of mangrove forests. Major part of that carbon is stored in sediments, hence, hydrodynamics and sediment dynamics play an important role. In accordance with the large differences in hydro- and sediment dynamics between the western and central versus the eastern lagoon, carbon stocks and accumulation rates are also very different. Carbon stocks in the upper meter of mangrove sediment including belowground biomass are on the order of 100–200 Mg C $ha^{-1}$ in the western and central lagoon and of 300–600 Mg C $ha^{-1}$ in the eastern lagoon (Kusumaningtyas et al., 2019; Weiss et al., 2016). The aboveground biomass of <20 Mg C $ha^{-1}$ is extremely low compared with other mangrove ecosystems in Indonesia and the global average (Alongi, 2014; Murdiyarso et al., 2015), which is probably mainly due to mangrove degradation caused by wood extraction (Hinrichs et al., 2009; Kusumaningtyas et al., 2019). Low phosphate concentrations and high N/P ratios of dissolved nutrients in combination with intensive soil nutrient recycling in logged areas with increasing pioneer vegetation (Weiss et al., 2016) indicate that P limitation may be an additional factor for the low carbon stocks in the western and central SAL.

In contrast, the carbon accumulation rate (CAR) with an average 658 g C $m^{-2}$ $year^{-1}$ is more than three times higher in the western and central than in the eastern lagoon and among the highest in the world (range ca. 50–1700 g C $m^{-2}$ $year^{-1}$; Kusumaningtyas et al., 2019). There, the autochthonous carbon supply of the mangrove plants is diluted by the high loads of mineral sediments from the hinterland, which results in the fairly low carbon stock. Moreover, the carbon isotope composition and C/N ratios indicate that part of the sedimentary carbon in the mangrove forest originates from the agricultural hinterland (Fig. 7.5; Kusumaningtyas et al., 2019). In the eastern SAL, the CAR is much lower, but still moderate on a global scale. Because of the low freshwater input, there is little dilution with allochthonous sediment input, leading to a high C stock but low CAR. Consequently, the carbon isotope composition and C/N ratios indicate the mainly autochthonous origin of organic matter deposited in the mangrove ecosystem.

The uniqueness of Segara Anakan is also important in terms of carbon storage in lagoon sediments. Mangrove productivity, in combination with the high sediment and carbon delivery from the hinterland, leads to the observed high carbon accumulation rates in mangrove sediments. However, the continuous tidal exchange between mangroves, lagoon, and the Indian Ocean is also responsible for a substantial export of carbon from the mangroves that are accumulating in lagoon sediments. A recent reconstruction of lagoon deposition history reveals that in the past 400 years, climate oscillations and human-induced land use change were responsible for variations in carbon accumulation. In the past century, the CAR in the central lagoon amounted to 153 g C $m^{-2}$ $year^{-1}$, was even higher during the past two decades, and almost half of it was of mangrove origin (Hapsari et al., 2020). In combination with the hinterland input, this indicates that the lagoon itself is also a quantitatively relevant carbon sink.

The formerly described influences on the organic carbon composition of the hinterland were less pronounced in Berau, as the conservational levels of both the Berau catchment and the mangrove forest itself are higher, and as the mangrove forest is larger, while the river catchment is smaller than in the case of the SAL. Organic carbon concentrations in the mangrove sediments of Berau (1.5%–8.5%) were approximately 50% higher than in the SAL (1.1%–4.6%). In terms of organic carbon stocks, a clear gradient from the main river to its oxbows existed, yielding up to twice the carbon stocks of the SAL in sites experiencing lower sedimentation rates (Fig. 7.6; Weiss et al., 2016). C/N ratios were similar to those in the SAL and much lower than in marine mangrove sediments (e.g., Togian Islands), indicating a better potential biodegradability of organic matter in these sediments. Unlike the sediments of marine mangrove forests, those of the estuarine mangroves showed no change of the C/N ratio with depth.

A suitable way to estimate the contribution of leaf and root litter to the organic carbon in sediment is the $\delta^{13}C$ value. It differs between roots and stem of woody mangroves and the invading halophytic herbs, which appear in case of mangrove disturbance.

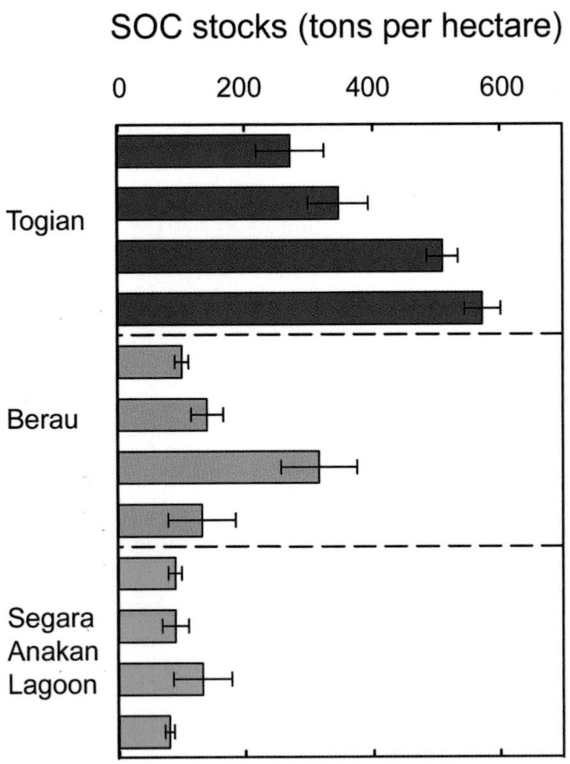


**FIGURE 7.6** Comparison of the sediment organic carbon stocks down to a depth of 1 m in the mangrove forests of the Segara Anakan Lagoon, Berau, and the Togians Islands. The bars represent individual cores taken in the three study areas. Error bars denote standard deviation. *Modified from Weiss et al. (2016).*

Consequently, mangrove sediments of Berau and the Togian Islands displayed a range between −30 and −28‰, while values in the SAL ranged between −27.5 and −25.5‰, indicating the higher disturbance level of the latter (Weiss et al., 2016). In terms of $\delta^{15}N$ of soil organic matter, values were similar in the two estuarine mangrove ecosystems, ranging between 3 and 8‰, and indicating a substantial influx of N via litter input to the ecosystem, lacking the clear nitrogen-fixing symbiont signature observed for the Togians (see the following).

Due to the lacking allochthonous sediment input and therefore the missing dilution effect of suspended mineral debris, the sediments of the Togian mangroves had the highest organic carbon concentrations among the three study regions (17.3%–26.2%; Weiss et al., 2016). Based on the comparison of the upper meter, the Togian mangrove sediments sequestered by far the most organic carbon of the three study regions (Fig. 7.6). This is most likely attributable to the very high C/N ratio of 25–60, impeding microbial turnover. The low N content in the marine mangrove sediments of the Togian Islands seems to be a result of the lacking nitrogen input estuarine mangroves receive from their agriculture-dominated hinterland. The low $\delta^{15}N$ value of 0–1‰ of the Togian mangrove sediments indicates fixation by microbial root symbionts as the main entrance gate of nitrogen to the ecosystem.

In the context of evaluating the factors controlling carbon sequestration in mangroves, a strong interplay between N and P supply of mangroves was found (Weiss et al., 2016). There is a strong gradient in P content of mangrove biomass toward the sea, especially of the root biomass, with the highest stocks close to the sea, which was also observed in other regions (Castañeda-Moya et al., 2013; Adame et al., 2014). Likewise, nitrogen supply is the lower the further away a mangrove is growing from the N-bearing waters of the river estuaries, resulting in litter of wide range of C/N ratios and hampering microbial degradation in case of the seaward mangroves. As a result, mangroves invest increasing fluxes of photoassimilates into nitrogen fixation by symbionts under these conditions. As an outcome, it appears that water-extractable P (the sum of P offered by exchange processes with the soil matrix and the microbial recycling activity) is a major control of carbon stocks across the observed mangrove types (Fig. 7.7). It indicates that carbon stocks generally are determined by the distance to the sea (the source of P) and to the next river (the source of nonsymbiotic N), i.e., the larger the distance the lower the carbon stock (Weiss et al., 2016). This illustrates that restoration of mangrove forests also contributes substantially to carbon sequestration, besides its obvious advantages in coastal protection, as seaward mangroves without direct access to easy available N

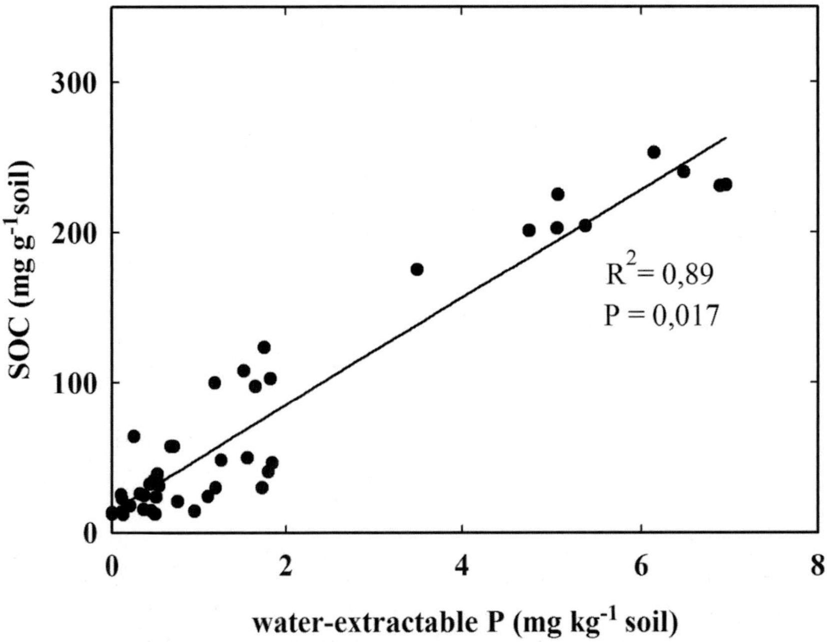


**FIGURE 7.7** A close correlation between water-extractable P and sediment organic carbon exists across the mangroves in the Segara Anakan Lagoon, the Berau River estuary, and the Togian Islands, thus holding true for estuarine and marine mangroves both. *From Weiss et al. (2016).*

sources appear to have a high potential for building up carbon-rich soils. A recent "Blue Carbon" study in the Perancak estuary on Bali found carbon sequestration in soils of a mangrove forest 10 years after restoration on abandoned aquaculture ponds almost as high as in soils of a nearby undisturbed mangrove forest (Sidik et al., 2019).

Carbon stocks and accumulation rates display a large variability, but they are generally high on a global scale even when considering the Segara Anakan mangrove forest a "degraded" one (Figs. 7.8 and 7.9). The large discrepancies between the carbon stock and the carbon accumulation rate in some areas (Kusumaningtyas et al., 2019) highlight the large differences in the underlying processes and the relevance of both measurements to assess the "Blue Carbon" storage potential and its relevance for PES and REDD schemes. Indonesian mangrove forests occur in all environmental settings, which display large differences in the input and accumulation of autochthonous versus

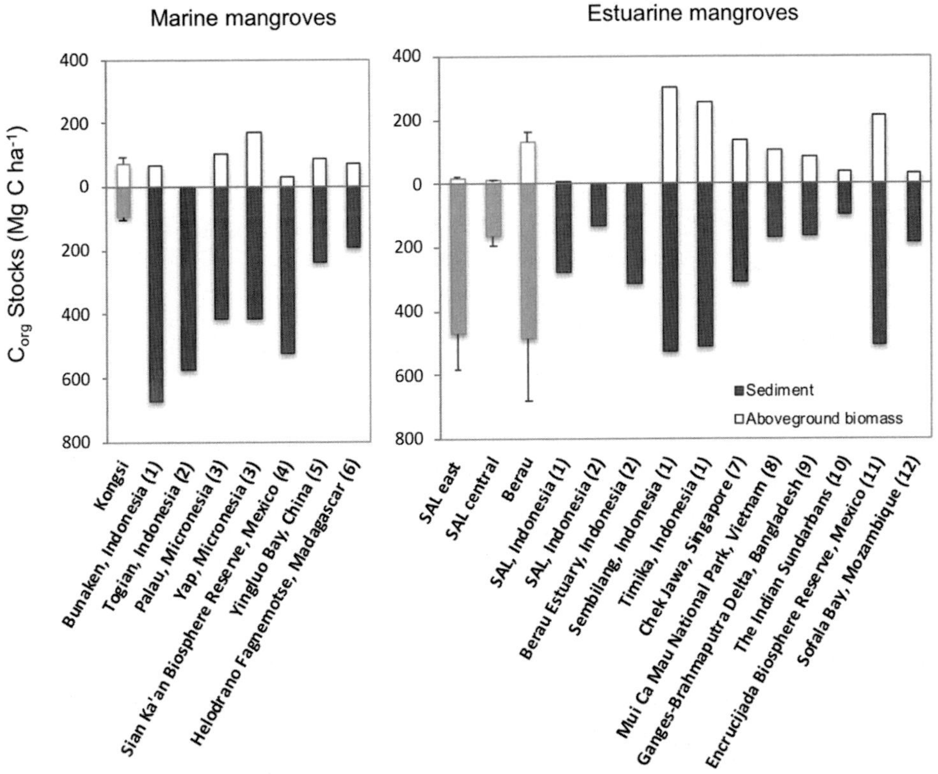


**FIGURE 7.8** Global-scale comparison of carbon stocks (aboveground biomass and sediment including belowground biomass to 1 m depth) in marine and estuarine mangroves from several sites. Data source: 1—Murdiyarso et al. (2015); 2—Weiss et al. (2016); 3—Kauffman et al. (2011); 4—Adame et al. (2013); 5—Wang et al. (2013); 6—Benson et al. (2017); 7—Phang et al. (2015); 8—Tue et al. (2014); 9—Donato et al. (2011); 10—Ray et al. (2011); 11—Adame et al. (2015); 12—Sitoe et al. (2014). Boxes in light gray denote sites in the study areas of Kusumaningtyas et al. (2019; Kongsi Island; Segara Ananan: SAL east, SAL central; Berau: Berau Delta). *From Kusumaningtyas et al. (2019).*

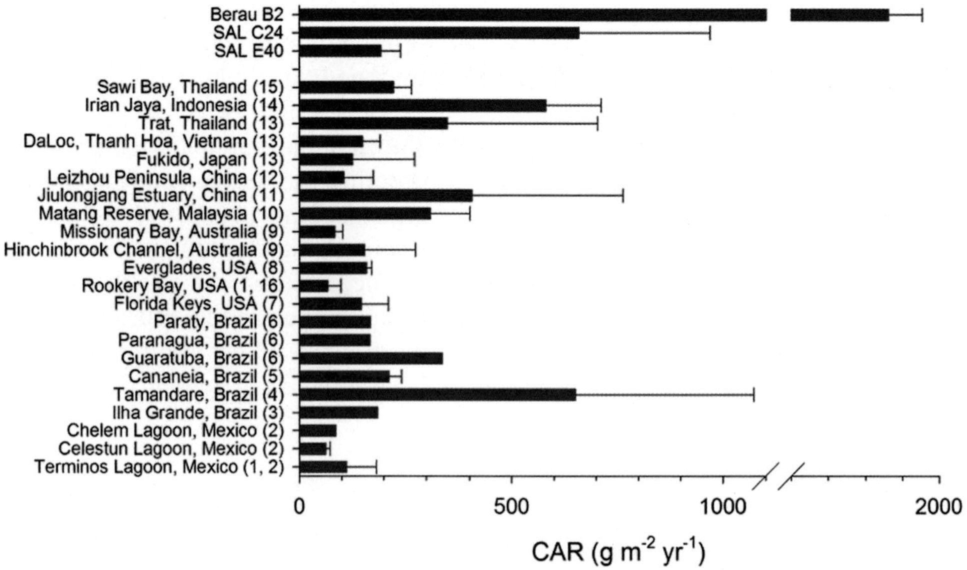


**FIGURE 7.9** Global-scale comparison of carbon accumulation rates in our study areas (Berau: B2; Segara Anakan: SAL C24, SAL C26) with other sites. Data sources: 1—Lynch (1989); 2—Gonneea et al. (2004); 3—Sanders et al. (2008); 4—Sanders et al. (2010a); 5—Sanders et al. (2010b); 6—Sanders et al. (2010c); 7—Callaway et al. (1997); 8—Smoak et al. (2013); 9—Brunskill et al. (2002); 10—Alongi et al. (2004); 11—Alongi et al. (2005); 12—Yang et al. (2014); 13—Tateda et al. (2005); 14—Brunskill et al. (2004); 15—Alongi et al. (2001); 16—Cahoon and Lynch, unpublished data in Chmura et al. (2003). *From Kusumaningtyas et al. (2019).*

allochthonous carbon and mineral sediments. In this context, carbon stocks allow to assess the potential of $CO_2$ being released upon degradation of the ecosystem (e.g., Murdiyarso et al., 2015; Weiss et al., 2016), but they do not provide information on the actual carbon sequestration rate of the ecosystem. In contrast, carbon accumulation rates allow to quantify the amount of $CO_2$ sequestered at present and also in the past. Therefore, those are the more relevant numbers for calculating the climate change mitigation potential (Kusumaningtyas et al., 2019). As yet, however, there is very little information on the CAR of Indonesian mangrove ecosystems available.

## 7.3.4 Flora and fauna

Mangrove vegetation and macrobenthic invertebrates were studied over a period of 10 years during which a considerable decline of biodiversity was recorded (Nordhaus et al., 2019). In 2005, 21 true mangrove tree species were identified, and the community was dominated by *Aegiceras corniculatum, Avicennia alba, Ceriops tagal, Rhizophora apiculata, Sonneratia caseolaris,* and the palm *Nypa fruticans* (Hinrichs et al., 2009). Species richness declined to 15 in 2015, and mean and maximum tree diameter and mean density declined for almost all species. Stand basal area and aboveground biomass ($6.3\ cm^2\ m^{-2}$ and $18.4\ t\ dm\ ha^{-1}$ in 2015, respectively) as well as the density of *Avicennia* spp., *Sonneratia* spp., *Ceriops* spp., *Aegiceras corniculatum,* and *Bruguiera* spp.

decreased significantly over the 10 years. By contrast, an increase of *Rhizophora apiculata* abundance and biomass occurred in the eastern area of the lagoon as a result of reforestation and natural regeneration (Nordhaus et al., 2019).

Large areas of the western and central parts of the lagoon are overgrown by the herbaceous plant *Acanthus* spp., which was favored by tree logging in combination with the high freshwater input through rivers (Hinrichs et al., 2009). In 2015, this area had a significantly lower tree density, stand basal area, species number, habitat complexity, and aboveground biomass than the eastern area of the lagoon. The forest in the central lagoon can be classified as severely degraded (Nordhaus et al., 2019).

The community of macrobenthic invertebrates is diverse and composed of 49 species of brachyuran crabs, 45 gastropod species, 10 bivalve species, and 19 polychaete species (Geist et al., 2012; Güldener, 2013; Nordhaus et al., 2009; Pamungkas, 2015). Species numbers of crabs and gastropods are lying at the higher end of the worldwide range. In contrast, biomass of benthic invertebrates is low compared with other Indo-West Pacific mangrove forests (Geist et al., 2012). Dominant species are *Perisesarma darwinense* (Sesarmidae), *Metaplax elegans* (Camptandriidae), *Ilyoplax strigicarpus* (Dotillidae), *Uca bellator*, and *Uca coarctata* (Ocypodidae). The community composition differed significantly between the central and the eastern parts of the lagoon due to environmental conditions, including differing pore water salinity, sediment grain size distribution, and vegetation composition (Geist et al., 2012; Nordhaus et al., 2009; Nordhaus et al., 2011). Concurrent with the decline in tree species richness, density, and aboveground biomass, species richness of crabs declined considerably between 2005 and 2015 (Nordhaus et al., 2019; Rose, 2015, p. 121). Several new invertebrate species were detected of which the meiobenthic species *Echinoderes applicitus* (Kinorhyncha) and the polychaetes species *Polymastigos javaensis* have already been described (Ostmann et al., 2012; Pamungkas, 2015).

The mangrove forest of the Berau Delta was composed of 13 tree species and one fern species (*Acrostichum speciosum*). The most frequent tree species were *Rhizophora apiculata* und *Bruguiera parviflora*. The forest was in a good condition at almost all sites with tree heights between 10 and 25 m. Two of the 10 sites were located near shrimp ponds for which large mangrove areas had been cut (Tripathi, 2016). A total of 42 crab and 37 gastropod species were identified. The crab communities were dominated by *Ilyoplax dentatus* (Dotillidae), followed by *Clistocoeloma merguiense* (Sesarmidae) and *Paracleistostoma laciniatum* (Camptandriidae). The most abundant gastropods were *Assiminea reticula*, *Melampus* spp., and *Laemodonta* spp. Mean crab and gastropod densities were $47.4 \pm 36.7\ m^{-2}$ and $8.6 \pm 6.0\ m^{-2}$, respectively (Tripathi, 2016). Some crab species were only recorded with one or two individuals, indicating them as rare species. The number of crab species was high, but that of gastropods was low compared with other mangrove forests in the Indo-West Pacific region.

The mangrove forest in the Togian Islands is dominated by *Rhizophora apiculata* and *Bruguiera parviflora*, similar to the sites in the Berau regency. *Rhizophora stylosa* and *R. mucronata* are mixed in, if broader coastal shelves are colonized (Weiss et al., 2016).

Toward the sea, *R. apiculata* forms almost pure stands directly limiting the seagrass ecosystem in front. Mangrove trees can reach up to 15 m height in the hinterland of the mangrove forest, but they are typically 6–8 m high under the influence of the wash of the waves. Crab and gastropod diversity appeared to be high but are not quantitatively evaluated so far.

### 7.3.5 Population and natural resource use in the Segara Anakan region

Segara Anakan is home to some 14,000 people living in three villages (*desa*) and a number of associated settlements (*dusun*). The villages are characterized by a nuclear structure and home to a mostly Javanese population, who were originally fisherfolk. Newer settlements often have a linear structure and developed together with the establishment of rice farming. Many of their inhabitants are migrants and often belong to the ethnic group of Sundanese from West Java, who profited from the formation of new land caused by sediment deposits. The settlement history of the lagoon is strongly linked to Java's colonial past. Written sources document the existence of 11 villages in 1706, but the spread of piracy throughout the entire Indonesian archipelago also reached Segara Anakan at the beginning of the 19th century. Its population suffered from terrible attacks by raiders acquiring slaves. In 1812, Segara Anakan's inhabitants had been either carried off or escaped inland. Resettlement started with the building of guardhouses to prevent the pirates from coming back. These guardhouses were built on stilts in the water, making them excellent places for fishing. Additional houses were soon built, and the guard stations eventually developed into fishing villages, the so-called *kampung laut* or sea villages (Schwerdtner Mañez, 2010). Segara Anakan's Javanese inhabitants still trace their origins back to a royal order assigning them to be the guardians of the coast (Reichel et al., 2009) and identify themselves as descendants of fisherfolk.

Sedimentation of the lagoon and overfishing have undermined the economic viability of fishing over the past decades. Today, fishing is largely a part-time activity, combined with the cultivation of irrigated rice, and the growing of other plants including soy beans and cassava. Fruits and vegetables are grown for subsistence needs and sold on a small-scale basis to local markets. In addition, a number of tree plantations established by lagoon residents deliver wood for paper production and other uses. In addition to agricultural expansion, aquaculture development has been a driver of mangrove conversion. Aquaculture ponds were established by external investors in the frame of the ADB-funded *Segara Anakan Conservation and Development Project*, implemented 1996–2005. These ponds were soon plundered by residents, who did not benefit from the development. The ponds have been abandoned or used mainly by immigrant Sundanese residents thereafter (Reichel et al., 2009). Another substantial mangrove area in the central part of the lagoon has gradually been converted to aquaculture by smaller-scale investors moving into the lagoon from the north coast of West Java.

Segara Anakan's mangroves have certainly always been used for different purposes. This included logging to obtain wood for construction and firewood. Charcoal production has also a long history, as evident from historical photographs. Between the 1870s and 1930s, fuelwood and charcoal production increased to an industrial level because of its use in sugar factories and for the operation of railways (De Haan, 1931). These industrial uses exceeded local residents' timber demands for houses and fishing stakes by far and contributed to extensive wood extraction, rendering first management attempts of the colonial forest administration ineffective (De Haan, 1931). Today, brickworks are an important buyer of charcoal, and increases in gas prices have contributed to a rising demand from households (Schwerdtner Mañez, Ring, Krause and Glaser, 2014). In addition to charcoal production and fuelwood collection, mangroves are used as a source of fodder. They are also important for the catching of crabs (*Scylla* spp.) and the collection of shellfish. In addition to local natural resource uses, off-farm employments and remittances have become an increasingly important source of livelihood for lagoon residents.

## 7.4 Environmental change in the Segara Anakan Lagoon region: causes, drivers, and impacts

The drivers of environmental change are manifold. Besides (1) climate change-related sea level rise, warming, and changes in atmospheric moisture transport, they include (2) human interventions, for example, through land use change and natural resource extraction and (3) extreme events, which can be a combination of both natural processes and human interventions, such as earthquakes, landslides, tsunamis, and a changing frequency and intensity of tropical storms (Jennerjahn and Mitchell, 2013). Most of these are relevant in Indonesia, except for the tropical storms, which are of minor relevance because of Indonesia's location in the equatorial calm zone. Indonesia has naturally high river fluxes of dissolved and particulate matter and sediments into the coastal zone; it harbors a large marine biodiversity; and in many parts, it has a dense coastal population that strongly relies on coastal natural resources. This, in turn, makes the coastal regions highly vulnerable to the outcomes of environmental change.

The island of Java is an extreme considering its magnitude of and its vulnerability to environmental change because of its exposure to natural events such as volcano eruptions and landslides, intensive land and natural resources use, and extremely high population densities of $>1000$ inhabitants per $km^2$. The environmental status and ecosystem services of the SAL are related to a large range of factors, shaping land, and natural resource use, including conflicts over resource access and control, political–economic structures, and development and management interventions in both the lagoon itself and its catchment area.

### 7.4.1 Decline of marine species and fisheries

Historical documents picture Segara Anakan as a diverse and species-rich environment. Dolphins, turtles, and crocodiles were common sights, and a "splendidly developed" mangrove forest grew at its shores (Beumee, 1929). Until the early 20th century, the Common Windowpane oyster (*Placuna placenta*) was an important resource collected for both consumption and its pearls (Schwerdtner Mañez, 2010). Local fisherfolk was also famous for *terasi* production, a paste made from fermented shrimp.

Impacts of environmental changes were documented nearly a century ago, when the decrease in lagoon size and depth began to affect fishing. Contemporary publications reported the disappearance of larger species and related income losses for fishers who were considered comparatively wealthy before (e.g., Schaafsma, 1927). In addition to fewer valuable species, family-owned fishing grounds got lost as former water area turned into mudflats. These fishing grounds had traditionally been used to set fixed nets, and losing them required the owners to change their gear, or to find alternative income sources. Decreasing depth was the change most mentioned by local fishermen. Decreasing salinity has also been found to be relevant, as it most likely reduces the quantity of marine fish entering the lagoon (Dudley, 2000). Another aspect is the shifting of currents resulting from changes in the water body, which has an influence on the setting of nets, such as *apong*. The tidal bag nets called *apong* are perhaps the most relevant gear change. They were introduced in the 1960s by fishermen who had been working on trawlers. Characterized by a small mesh size and set in tidal channels, *apong* nets are extremely effective and catch also juvenile species. Especially the catching of juvenile has caused conflicts between fishers operating in the lagoon and others operating in the ocean, who argue that the lagoon fishing activities decrease their catches (Dudley, 2000; Schwerdtner Mañez et al., 2014).

Livelihood diversification became the main strategy to adapt to the environmental changes. Fishers started rice farming on the new land, either part-time in combination with fishing or as a sole income strategy. Trading, sand or clay extraction, the construction of fish ponds, or employment for others became alternative sources of income. Still, fishing remained the primary income source of 90% of the population until the 1980s. After that, a vast transformation from fishing to farming took place (Olive, 1997). Research in SPICE revealed that only few full-time fishers remain, which explains the low number of interviewees. Out of 30 interviewed fishers, 19 reported species disappearances. Caught species are very small, and overall catches are low. Besides several finfish species, a number of shrimp species are caught, including *Metapenaeus elegans, Penaeus merguiensis, P. indicus* (Dudley, 2000), and crabs (*Scylla* spp.; I. Nordhaus, personal observation). In contrast to common narratives, fishing no longer is the most important livelihood strategy of Segara Anakan's inhabitants. It is, however, an important activity in terms of additional income, subsistence, and self-identification.

## 7.4.2 Sedimentation and its causes

Riverine sediment input has drastically reduced the size and depth of the SAL. However, despite predictions of its total infilling in the 1990s (White et al., 1989), the lagoon still exists. These predictions were based on unrealistically high sediment inputs of

5–10 × $10^6$ t $year^{-1}$ (PRC-ECI, 1987, p. 351) or even 17 × $10^6$ t $year^{-1}$ (Napitulu and Ramu, 1980). Our own estimates based on bathymetric and hydrographic measuring campaigns and modeling result in an annual sediment input from the hinterland of 1.09 × $10^6$ t $year^{-1}$, only 13% of which are deposited in the lagoon. The rest is directly exported into the Indian Ocean, 0.84 × $10^6$ t $year^{-1}$ through the western channel and 0.12 × $10^6$ t $year^{-1}$ through the eastern channel (Jennerjahn and Winter, unpublished data). Our investigations were conducted after a large-scale dredging program, which removed 9.3 million $m^3$ of sediments from an area of 512 ha mainly in the central SAL in 2002–05 (Asian Development Bank, 2006). The dredging had only a temporary effect on the sedimentation process, which accelerated in the late 19th and throughout much of the 20th century. The water surface area of the lagoon decreased from almost 9000 ha in 1857/60 to slightly more than 2000 ha in 2013 (Fig. 7.10; Lukas, 2014a, 2017a).

While seen as desirable in the context of state-led agricultural reclamation plans until the late 1970s, sedimentation has been regarded mainly as a threat by governmental representatives, consultants, scholars, and fishers since the 1980s (Lukas and Flitner, 2019). Substantial investments into watershed management in the frame of both internationally funded projects and national programs have since aimed at reducing river sediment loads and lagoon sedimentation, yet with limited success.

In line with simplistic assumptions about the causes of high river sediment loads in the whole of Java, these management interventions focused on reducing soil erosion on upland farmers' private agricultural plots through field terracing and tree planting. The knowledge of other sediment sources and their causes remained scarce. No watershed-wide analysis was undertaken for the area prior to SPICE. The assumption that farmers' private plots represent the single-most important sediment source provided a clear-cut narrative for political intervention and rendered inquiries into other causes of high river sediment loads seemingly unnecessary (Lukas, 2015).

Watershed-wide mapping of land use and sediment sources, combined with analyses of satellite images and historical maps, and social-scientific research methods in the frame of SPICE has shown that farmers' private agricultural plots are in fact only one of numerous sediment sources. In addition, there are a broad range of other historical and contemporary drivers of lagoon sedimentation that have been neglected to date (Lukas, 2015, Fig. 7.10). Among the most important of these drivers are conflicts over land and forest resources. Satellite image analysis and watershed-wide mapping identified some state forest and plantation lands as hot spots of land cover change and soil erosion (Lukas, 2015). Historically rooted conflicts over the access to and control of land and forest resources are a major cause of land cover change and soil erosion in these areas (Lukas, 2014b, 2015, 2017b). State forest management practices, involving large-scale clear cuts, also on steep slopes and in close proximity to water courses, are an additional cause of land cover change, erosion, and high river sediment loads (Lukas, 2015, 2017a).

Other important sediment sources include slope cuts to enlarge agricultural fields in valley floors (a practice called *ngaguguntur*), agriculture in riparian zones, and erosion in settlements and on roads, trails, and embankments (Lukas, 2017a). In addition, various

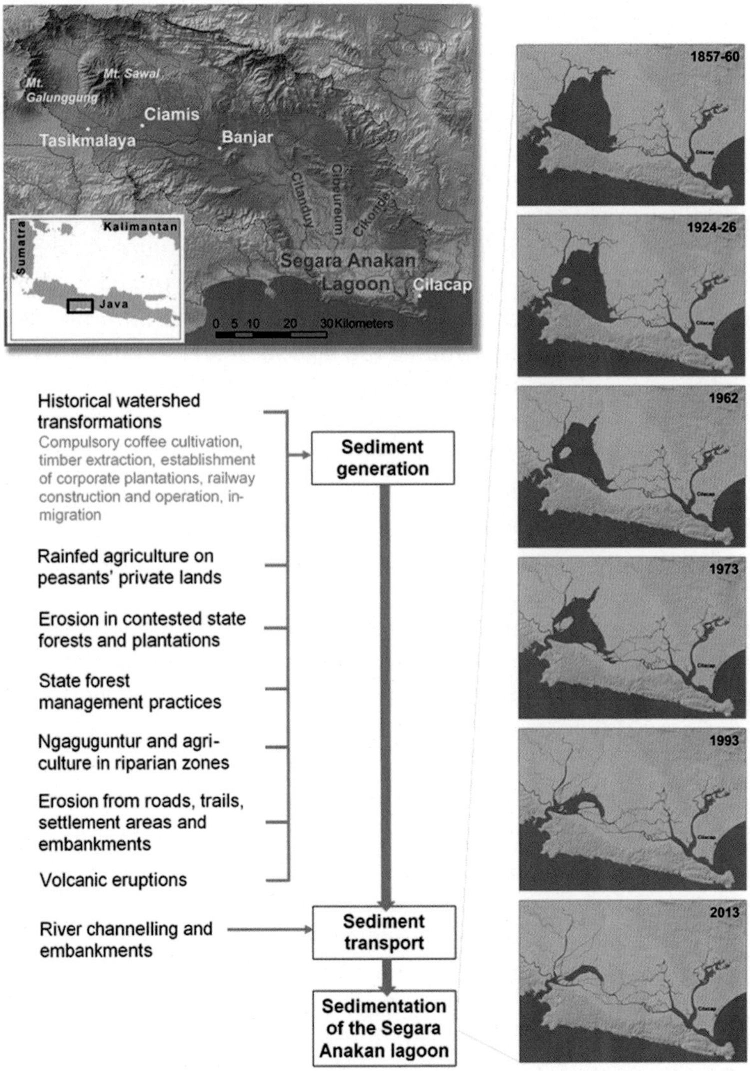


**FIGURE 7.10** The Segara Anakan Lagoon has rapidly shrunk due to sediment input from the Citanduy, Cibeureum, and Cikonde Rivers. This is the result of a broad range of drivers that have increased sediment generation and transport. *From Lukas (2017a).*

historical land use and watershed modifications, including coffee cultivation under colonial rule, timber extractions, establishment of corporate plantations, railway construction, and the rapid opening up of new agricultural land in the late 19th and early 20th century, appear to have contributed to lagoon sedimentation (Lukas, 2017a). Volcanic eruptions of Mount Galunggung have also temporarily contributed to river sediment loads (Lukas, 2017a). In addition, the straightening and embankment of the lagoon's tributaries and agricultural reclamation of the floodplains of the lower river basin in the frame of government and donor-funded interventions have accelerated lagoon sedimentation, particularly since the 1960/70s (Lukas, 2017a).

Sedimentation of the SAL is hence the result of a broad range of causes. The focus of debates and management interventions on only one of these factors, i.e., erosion on upland farmers' private agricultural plots, and the neglect of numerous other sediment sources, has further limited the effectiveness of watershed management (Lukas, 2015). Political entanglements of watershed management with politics of forest access and control have contributed to this narrow focus of debates and interventions (Lukas, 2015; Lukas and Flitner, 2019).

### 7.4.3 Reclamation of land and conflicts over new land

Sedimentation of the SAL has had profound effects on social–ecological conditions and resource uses, has triggered conflicts, and has been the target of government-led management interventions. As with watershed management, many interventions targeting the SAL have been based on narrow assumptions and a limited understanding of realities on the ground (Heyde, 2016; Heyde et al., 2017). Furthermore, like in the lagoon's catchment area, tenure contestation also limits the scope for sustainable environmental management. While sedimentation has been seen as a threat by government representatives and academics since the early 1980s, local residents and newcomers to the SAL have used various strategies to reclaim muddy emergent land and transform it for agricultural use, homes, and other village infrastructure (Heyde, 2016). As fishing livelihoods became increasingly insecure due to sedimentation and subsequent loss of fishing grounds, people started to turn to farming (Heyde, 2016; Olive, 1997). Their small-scale reclamation initiatives have hastened the transition from water to land. As described in more detail by Heyde (2016), over time people have used at least three different microreclamation approaches: (1) transporting soil to build up land for homes and village infrastructure, (2) constructing dikes to block saltwater intrusion into agricultural fields, and (3) intentional channeling of riverine sediment into low-lying fields to enhance agricultural potential.

Early attempts by residents to assert rights to emergent land in the lagoon were at times met with heavy-handed responses from state authorities, including eviction attempts. More recently, conflict has been less overt. For example, the unclear legal status of emergent land in Indonesia has resulted in claims based on different interpretations of the law, including whether emergent land should be designated (1) as part of the national forest estate, or (2) under the jurisdiction of local residents who had traditional rights (*hak ulayat*) to open water areas that were transformed to land (Heyde, 2016). Since the mid-2000s, the state forest corporation has twice tried in the lagoon to establish village-level institutions associated with the Joint Community Forest Management program. These attempts reflect the corporation's belief that they have authority over large areas of emergent land. In both cases, the institutions were rejected by local residents (Heyde, 2016). Rights to farm emergent land were initially granted by village governments and later recognized by the district, although not in the form of full ownership (certification) (Heyde, 2016). In the early 2000s, a cadastral survey was conducted in the western part of the lagoon, following which homes and their yards became eligible for certification. At the time of field research in early 2015, ownership of agricultural land was not allowed (Heyde, 2016). Additionally, over time, people's claims to emergent land have been strengthened as users have increasingly been incorporated into the local taxation system.

The failure to resolve tenure issues points to a persistent lack of trust and weak coordination between actors in the lagoon, including the state forest corporation, various district government agencies, and local residents. There is no forum for coordination of state agencies and little opportunity for engagement by nonstate actors, such as local residents (Heyde, 2016).

Large contradictions between the perspectives and interests of local residents on the one hand and state agencies on the other hand are also evident in the large-scale construction of ponds for shrimp production that started in 1996 in the frame of the ADB-funded Segara Anakan Conservation and Management Project. External town-based investors rather than local lagoon residents managed and benefited from the development. For local fishers and farmers, the shrimp pond development caused hardships due to the loss of mangrove trees and the degradation of soil quality in their rice fields. As a consequence, the ponds were plundered and abandoned in 2001 (Reichel et al., 2009). Unfortunately, the conversion of mangrove forests to shrimp ponds was irreversible. Restoration and manual replanting efforts would not readily regenerate the mangroves (Djohan, 2014). Although aquaculture was supposed to be an alternative source of income for the coastal community, the shrimp production in fact led to ecological and social problems in the Segara Anakan ecosystem and community.

## 7.5 Threats to mangrove forests and their ecosystem services in Indonesia

Recent estimates show that mangrove deforestation rates in Southeast Asia in the period 2000–2012 with an average loss rate of 0.18% per year were lower than previously thought (Richards and Friess, 2016). However, mangrove loss in Indonesia is substantial. Almost half of the lost mangrove habitat is converted to aquaculture, about one-sixth to oil palm plantations. Major recent losses occurred on the islands of Sumatra, Kalimantan, and Sulawesi with conversion to oil palm plantations being the major land use change on Sumatra, while aquaculture is the dominant conversion in other parts of Indonesia (Richards and Friess, 2016).

The conversion to other land uses for economic reasons does not only mean a loss of natural habitat in the coastal zone but also entails a loss of important ecosystem services. Although these have a "value" that is not directly marketable and therefore is hardly considered by political decisions in our economy-driven world, the loss of ecosystem services can have consequences that also result in economic losses. Therefore, attempts have been made to put a value on ecosystem services from coastal wetlands. A much-noticed study resulted in a value of 10,000 USD $ha^{-1}$ $year^{-1}$ for mangroves (Costanza et al., 1997). In another study, the annual economic values of mangroves, estimated by the cost of the products and services they provide, have been estimated to be 200,000–900,000 USD $ha^{-1}$ $year^{-1}$ (Wells et al., 2006). Mangroves can also be provided with an economic value based on the cost to replace the products and services that they provide, or the cost to restore or enhance mangroves that have been eliminated or degraded. The range of reported costs for mangrove restoration is 225 to 216,000 USD $ha^{-1}$, not including the cost of the land (Lewis, 2005). However, not all ecosystem services provided can be given a "price," there are others, in particular on a local scale, which have an important value in sustaining the livelihoods of people, for example, the collection of wood for use as fuel for cooking though not having a high direct economic value (Huxham et al., 2015).

The ecosystem services of Indonesia's mangrove forests will change during this century because of climate change and associated sea level rise, but probably much more by other human interventions (Huxham et al., 2017; Jennerjahn et al., 2017). Our studies show that the supply of many of those mangrove ecosystem services are at risk, in particular in the strongly altered SAL and its surroundings (Table 7.1). There, mangrove conversions and unsustainable natural resource uses threaten the supply of the "provisioning" and the "supporting" ecosystem services. As yet, this is less relevant in Berau and on the Togian Islands, but the largely expanding oil palm plantations and aquaculture ponds at the expense of mangrove forests on Kalimantan already endanger ecosystem service supply in Berau. The supply of the "regulating" ecosystem service is to some extent endangered in the SAL but is mainly uncritical, because of high freshwater supply, the generally high river fluxes of dissolved and particulate substances, and the short residence time of the water in these coastal areas (Holtermann et al., 2009; Jennerjahn et al., 2009; Milliman and Farnsworth, 2011).

The direct value of mangroves of Berau was calculated as 296 USD $ha^{-1}$ $year^{-1}$, the indirect value as 726 USD $ha^{-1}$ $year^{-1}$, and the option and existence value as 15 and 358 USD $ha^{-1}$ $year^{-1}$, respectively (Wiryawan and Mous, 2003). However, Kalimantan is the region of Indonesia for which the highest mangrove loss is forecasted for the next two decades mainly due to conversion into aquaculture ponds, followed by plantations (Ilman et al., 2016). The construction of aquaculture facilities is considered the main reason for the loss of mangrove areas on the east and west coast of Kalimantan (Karstens and Lukas, 2014; Richards and Friess, 2016); palm oil plantations replace mangroves on the west and south coast (Richards and Friess, 2016). Conservation efforts have remained minimal, and only a few mangrove areas are included in protected areas. For instance, the 4150 $km^2$ large Tanjung Puting National Park in south central Kalimantan has been recognized as a UNESCO biosphere reserve (Spalding et al., 2010). Yet, the example of large-scale conversions of legally protected mangrove forests to aquaculture in the Kapuas estuary in West Kalimantan by nonlocal investors backed by fisheries authorities

**Table 1** Mangrove ecosystem service supply at risk in Indonesia, in the Segara Anakan Lagoon (SAL), Java, in Berau, Kalimantan, and on the Togian Islands, Sulawesi. This table shows how resource exploitation and environmental change impair the future provision of mangrove ecosystem services. Ecosystem services of mangrove ecosystems defined by the UNEP World Conservation Monitoring Center (Wells et al., 2006; second column) are grouped in the four categories defined by the Millennium Ecosystem Assessment (2005; first column). Red – at high risk, yellow – at moderate risk, green – at no risk, grey – not applicable/not investigated.

| Category | Mangrove ecosystem service | SAL | Berau | Togian |
|---|---|---|---|---|
| Provisioning | Subsistence and commercial fisheries | | | |
| | Habitat | | | |
| | Honey | | | |
| | Fuelwood | | | |
| | Building materials | | | |
| | Traditional medicines | | | |
| Regulating | Protection of beaches and coastlines from storm surges, waves and floods | | | |
| | Reduction of beach and soil erosion | | | |
| | Stabilization of land by trapping sediments | | | |
| | Water quality maintenance (N and pollutant filter) | | | |
| | Climate regulation (C sequestration) | | | |
| Cultural | Tourism and recreation | | | |
| | Spiritual – sacred sites | | | |
| Supporting | Cycling of nutrients | | | |
| | Nursery habitats | | | |
| | Biodiversity | | | |

and state representatives demonstrates the weak enforcement of environmental law and the environmentally destructive power of corrupt political–economic networks (Karstens and Lukas, 2014a,b).

## 7.6 Management programs

For centuries, the natural resources and hence socioeconomic goods and services of the mangrove-fringed SAL and other mangrove ecosystems in Indonesia have been in use, but they have been increasingly threatened by conversion, degradation through unsustainable use, pollution from industry and households, and sedimentation.

The social–ecological changes in the rapidly shrinking SAL have been the subject of political and scholarly debates and of state-led management interventions for decades. Yet, rather than the ecological value of the lagoon and its mangrove forests, it was political instability and agricultural development plans that directed political attention to the lagoon and its adjacent river basin between the 1950 and 1970s (Lukas and Flitner, 2019). At that time, the lagoon and the adjacent mangrove and swamp forests were regarded as unproductive areas that were to be reclaimed for agriculture. State-initiated land swaps related to violent displacements of people in the uplands north of the lagoon and agricultural development projects pushed large-scale conversions of mangrove and swamp forests between the 1960s and 1980/90s (Lukas, 2014b; Lukas and Flitner, 2019). The declaration of the Citanduy River as national priority area for development and its selection as site for the implementation of the first US–Indonesian development project in 1969 set in motion decades of state-led interventions in the river basin, the SAL, and its watershed (Heyde, 2016; Lukas and Flitner, 2019).

In line with changing development paradigms, political interests, and material necessities created by the first projects, the focus of these interventions started to shift from agricultural reclamation to watershed and lagoon management in the late 1970s and early 1980s. Lagoon sedimentation, which was first seen as desirable and which the agricultural reclamation projects unintentionally accelerated, became a threat to the new agricultural irrigation schemes upstream, the in- and offshore fisheries, the lagoon ecosystem, and the livelihoods of lagoon residents (Lukas and Flitner, 2019). Substantial national and international funds have since been spent for upland conservation and lagoon management (Heyde, 2016; Lukas, 2015; Lukas and Flitner, 2019). Yet, the effects of these management interventions are regarded as limited. Lagoon sedimentation continues to be seen as threat by some residents, governmental representatives, and academics, while other residents hasten the transition from water to land through microreclamation strategies (Heyde, 2016). Limited knowledge of and misleading narratives about the causes of sedimentation and conflicts over land and forest resources have limited the effectiveness of sediment mitigation strategies (Lukas, 2014b, 2015, 2017a,b). River diversions to reduce sediment input and lagoon dredging only partly materialized, led to social conflicts and had only temporary effects (Heyde et al., 2017; Reichel et al., 2009). Conflicts over the control of emerging land and mangrove forests and unclear, competing jurisdictions basically thwart mangrove management and reforestation (Heyde, 2016). Furthermore, lacking trust and communication between actors, and mismatches between state-led planning and realities on the ground undermine natural resources management in both the lagoon and its watershed (Heyde, 2016; Lukas, 2017b). Based on an analysis of these issues, Heyde et al. (2017) concluded with six recommendations that should be considered to strengthen development and environmental policies in the Segara Anakan region (see Text Box 2).

More than a decade of SPICE research has resulted in robust scientific results that allowed (1) to identify cause–effect relationships in terms of environmental and socio-political issues and respective responses of the social–ecological system, (2) to quantify carbon storage and pollution (nutrients, organic contaminants) at least partly, (3) to identify knowledge gaps for future research directions, and (4) to give recommendations for policy and society (see Text Boxes 1 and 2). These recommendations as well as other results of SPICE research have been communicated to government representatives across political levels from villages to national ministries, as well as to civil society organizations. To share and discuss research results with policy-makers and civil society and to open a discussion about corresponding policy and management implications, J. Heyde and M. Lukas conducted a science-policy workshop in Cilacap in January 2016. This workshop brought together representatives of lagoon, river and watershed management authorities, forest and fisheries management authorities, agricultural and land registration agencies, district and village governments, community groups, and environmental and land rights organizations. Representatives of lagoon management authorities were also invited to the SPICE workshops held at Universitas Jenderal Soedirman, Purwokerto.

Knowledge gaps and directions of future research

- The filtering capacity of mangrove ecosystems for anthropogenic nutrients needs to be quantified to assess the risk of coastal eutrophication.
- Carbon accumulation in mangrove sediments needs to be quantified to assess Indonesia's natural carbon sinks and the potential for REDD+ schemes.
- To assess the sediment filling and the potential disappearance of the lagoon, sediment dynamics must be understood better and fluxes be quantified.
- Careful participatory mapping of competing land and resource claims and participatory design and implementation of conflict resolution mechanisms and community-based resource management approaches are the key to enhanced natural resource management in the SAL and its hinterland.

Implications/recommendations for policy and society

- Further education to raise stakeholder awareness of the value of mangrove ecosystem services and the vulnerability to overexploitation of resources is required.
- To keep the ecologically and economically important mangrove ecosystem services, logging and land conversion of mangrove forests need to be stopped.
- Because of the high competition for traditional livelihood designs heavily relying on natural resource extraction, alternative livelihood options need to be developed, for example, in the tourism, service, and finishing of goods sectors.
- Planning of initiatives to reduce river sediment loads needs to consider a broader range of sediment sources than has historically been the case.
- Conflicts between farmers and state agencies and plantation companies over land and forest resources in the watershed need to be addressed.
- Lagoon sedimentation should be regarded as a transformation of the ecosystem to which society can adapt, rather than as a threat that must be combated at any cost.
- The spatial plan for the lagoon needs to be updated to take into account claims to and usage of emergent land, and the changing environmental services of the lagoon.
- Coordination between state agencies involved in lagoon management needs to be strengthened.
- Payments for environmental services should not be prioritized as a means of addressing issues of soil erosion in the watershed and of lagoon sedimentation and degradation.

# Acknowledgments

We are strongly indebted to the huge number of enthusiastic students from Indonesia, Germany, and other parts of the world who invested a lot of time and effort in conducting thesis projects or internships or simply wanted to help. We are also grateful for the assistance provided by numerous technicians from the participating organizations. The research reported here was made possible by the continuing financial and administrative support of the German Federal Ministry of Education and Research (Grant Nos. 03F0391, 03F0471, 03F0644), the Indonesian Ministry for Research and Technology (RISTEK), the Indonesian Ministry for Maritime Affairs and Fisheries (KKP), and the German Academic Exchange Service (DAAD).

# References

Adame, M.F., Kauffman, J.B., Medina, I., Gamboa, J.N., Torres, O., Caamal, J.P., et al., 2013. Carbon stocks of tropical coastal wetlands within the karstic landscape of the Mexican Caribbean. PLoS One 8, e56569. https://doi.org/10.1371/journal.pone.0056569.

Adame, M.F., Teutli, C., Santini, S., Caarnal, J.P., Zaldivar-Jimenez, A., Hernandez, R., 2014. Root biomass and production of mangroves surrounding a karstic oligotrophic coastal lagoon. Wetlands 34, 479–488.

Adame, M.F., Santini, N.S., Tovilla, C., Vázquez-Lule, A., Castro, L., Guevara, M., 2015. Carbon stocks and soil sequestration rates of tropical riverine wetlands. Biogeosciences 12, 3805–3818.

Alongi, D.M., 2009. The Energetics of Mangrove Forests. Springer, Heidelberg, Germany.

Alongi, D.M., 2014. Carbon cycling and storage in mangrove forests. Annual Reviews in Marine Science 6, 195–219.

Alongi, D.M., 2015. The impact of climate change on mangrove forests. Current Climate Change Reports 1, 30–39.

Alongi, D.M., Wattayakorn, G., Pfitzner, J., Tirendi, F., Zagorskis, I., Brunskill, G., et al., 2001. Organic carbon accumulation and metabolic pathways in sediments of mangrove forests in southern Thailand. Marine Geology 179, 85–103.

Alongi, D.M., Sasekumar, A., Chong, V.C., Pfitzner, J., Trott, L.A., Tirendi, F., et al., 2004. Sediment accumulation and organic material flux in a managed mangrove ecosystem: estimates of land-ocean-atmosphere exchange in peninsular Malaysia. Marine Geology 208, 383–402.

Alongi, D.M., Pfitzner, J., Trott, L.A., Tirendi, F., Dixon, P., Klumpp, D.W., 2005. Rapid sediment accumulation and microbial mineralization in forests of the mangrove *Kandelia candel* in the Jiulongjiang Estuary, China. Estuarine, Coastal and Shelf Science 63, 605–618.

Ardli, E.R., Wolff, M., 2009. Land use and land cover change affecting habitat distribution in the Segara Anakan Lagoon, Java, Indonesia. Regional Environmental Change 9, 235–243.

Asian Development Bank, 2006. Indonesia: Segara Anakan Conservation and Development Project. Manila: Asian Development Bank, Completion Report.

Benson, L., Glass, L., Jones, T.G., Ravaoarinorotsihoarana, L., Rakotomahazo, C., 2017. Mangrove carbon stocks and ecosystem cover dynamics in southwest Madagascar and the implications for local management. Forests 8, 1–21.

Beumee, J.G.B., 1929. Bandoeng-Tjilatjap-Djoka. Java: Forth Pacific Science Congress.

BPS Kalimantan Timur, 2017. Tabel Luas Wilayah dan Kepadatan Penduduk Menurut Kabupaten/Kota Tahun 2015, Badan Pusat Statistik Provinsi Kalimantan Timur. Available at: https://kaltim.bps.go.id.

Brunskill, G.J., Zagorskis, I., Pfitzner, J., 2002. Carbon burial rates in sediments and a carbon mass balance for the Herbert River region of the Great Barrier Reef continental shelf, North Queensland, Australia. Estuarine, Coastal and Shelf Science 54, 677–700.

Brunskill, G.J., Zagorskis, I., Pfitzner, J., Ellison, J., 2004. Sediment and trace element depositional history from the Ajkwa River estuarine mangroves of Irian Jaya (West Papua), Indonesia. Continental Shelf Research 24, 2535–2551.

Callaway, J.C., DeLaune, R.D., Patrick Jr., W.H., 1997. Sediment accretion rates from four coastal wetlands along the Gulf of Mexico. Journal of Coastal Research 13, 181–191.

Castañeda-Moya, E., Twilley, R.R., Rivera-Monroy, V.H., 2013. Allocation of biomass and net primary productivity of mangrove forests along environmental gradients in the Florida Coastal Everglades, USA. Forest Ecology and Management 307, 226–241.

Chmura, G.L., Anisfeld, S.C., Cahoon, D.R., Lynch, J.C., 2003. Global carbon sequestration in tidal, saline wetland soils. Global Biogeochemical Cycles 17. https://doi.org/10.1029/2002GB001917.

Chowdhury, R.R., Uchida, E., Chen, L., Osorio, V., Yoder, L., 2017. Anthropogenic drivers of mangrove loss: geographic patterns and implications for livelihoods. In: Rivera-Monroy, V., Lee, S.Y., Kristensen, E., Twilley, R.R. (Eds.), Mangrove Ecosystems: A Global Biogeographic Perspective – Structure, Function and Services. Springer Publishing Company, New York, pp. 275–300.

Cilacap, B.P.S., 2018. Tabel Jumlah Penduduk, Kepadatan per KM persegi, dan Persentasenya menurut Kecamatan di Kabupaten Cilacap, 2010–2016, Badan Pusat Statistik Cilacap. Available at: https://cilacapkab.bps.go.id.

Costanza, R., d'Arge, R., de Groot, R., Farber, S., Grasso, M., Hannon, B., et al., 1997. The value of the world's ecosystem services and natural capital. Nature 387, 253–260.

De Haan, J.H., 1931. He teen en auder over Tjilatjapsche vloedbosschen. Tectona 24, 39–76.

Djohan, T.S., 2014. Colonization of mangrove forest at abandoned shrimp-pond of Segara Anakan-Cilacap. Journal Teknosains 4, 1–102.

Donato, D.C., Kauffman, J.B., Murdiyarso, D., Kurnianto, S., Stidham, M., Kanninen, M., 2011. Mangroves among the most carbon-rich forests in the tropics. Nature Geoscience 4, 293–297.

Dsikowitzky, L., Nordhaus, I., Jennerjahn, T., Khrycheva, P., Sivatharshan, Y., Yuwono, E., et al., 2011. Anthropogenic organic contaminants in water, sediments and benthic organisms of the mangrove-fringed Segara Anakan Lagoon, Java, Indonesia. Marine Pollution Bulletin 62, 851–862.

Dudley, R., 2000. Segara Anakan Conservation and Development Project Components B & C. Consultant's Report. Segara Anakan Fisheries Management Plan.

Geist, S., Nordhaus, I., Hinrichs, S., 2012. Occurrence of species-rich crab fauna in a human impacted mangrove forest questions the application of community analysis as an environmental assessment tool. Estuarine, Coastal and Shelf Science 96, 69–80.

Giri, C., Ochieng, E., Tieszen, L.L., Zhu, Z., Singh, A., Loveland, T., et al., 2011. Status and distribution of mangrove forests of the world using earth observation satellite data. Global Ecology and Biogeography 20, 154–159.

Gonneea, M.E., Paytan, A., Herrera-Silveira, J.A., 2004. Tracing organic matter sources and carbon burial in mangrove sediments over the past 160 years. Estuarine, Coastal and Shelf Science 61, 211–227.

Gordon, A.L., 2005. Oceanography of Indonesian seas. Oceanography 8 (14), 14–27.

Guannel, G., Arkema, K., Ruggiero, P., Verutes, G., 2016. The power of three: coral reefs, seagrasses and mangroves protect coastal regions and increase their resilience. PLoS One 11 (7), e0158094. https://doi.org/10.1371/journal.pone.015809.

Güldener, B., 2013. Impact of Biotic and Abiotic Parameters on the Diversity and Spatial Distribution of Mollusks in the Mangroves of Segara Anakan Lagoon. Bachelor thesis, University Bremen.

Hapsari, K.A., Jennerjahn, T.C., Lukas, M.C., Karius, V., Behling, H., 2020. Intertwined effects of climate and land use change on environmental dynamics and carbon accumulation in a mangrove-fringed coastal lagoon in Java, Indonesia. Global Change Biology 26, 1414–1431. https://doi.org/10.1111/gcb.14926.

Heyde, J., 2016. Environmental Governance and Resource Tenure in Times of Change: Experience from Indonesia. Doctoral thesis. University of Bremen, Germany. http://elib.suub.uni-bremen.de/edocs/00105588-1.pdf.

Heyde, J., Lukas, M.C., Flitner, M., 2017. Payments for Environmental Services: A New Instrument to Address Long-Standing Problems? Policy Paper. Artec-Paper 213, Sustainability Research Center (Artec). University of Bremen.

Hinrichs, S., Nordhaus, I., Geist, S.J., 2009. Status, diversity and distribution patterns of mangrove vegetation in the Segara Anakan Lagoon, Java, Indonesia. Regional Environmental Change 9, 275–289.

Holtermann, P., Burchard, H., Jennerjahn, T.C., 2009. Hydrodynamics of the Segara Anakan Lagoon. Regional Environmental Change 9, 259–274.

Hung, J.J., Hung, P.Y., 2003. Carbon and nutrient dynamics in a hypertrophic lagoon in southwestern Taiwan. Journal of Marine Systems 42, 97–114.

Huxham, M., Emerton, L., Kairo, J., Munyi, F., Abdirizak, H., Muriuki, T., et al., 2015. Applying climate compatible development and economic valuation to coastal management: a case study of Kenya's mangrove forests. Journal of Environmental Management 157, 168–181.

Huxham, M., Dencer-Brown, A., Diele, K., Kathiresan, K., Nagelkerken, I., Wanjiru, C., 2017. Mangroves and people: local ecosystem services in a changing climate. In: Rivera-Monroy, V., Lee, S.Y., Kristensen, E., Twilley, R.R. (Eds.), Mangrove Ecosystems: A Global Biogeographic Perspective – Structure, Function and Services. Springer Publishing Company, New York, pp. 245–274.

Ilman, M., Dargusch, P., Dart, P., Onrizal, O., 2016. A historical analysis of the drivers of loss and degradation of Indonesia's mangroves. Land Use Policy 54, 448–459.

Jennerjahn, T.C., Mitchell, S.B., 2013. Pressures, stresses, shocks and trends in estuarine ecosystems – an introduction and synthesis. Estuarine, Coastal and Shelf Science 130, 1–8.

Jennerjahn, T.C., Nasir, B., Pohlenga, I., 2009. Spatio-temporal variation of dissolved inorganic nutrients in the Segara Anakan Lagoon, Java, Indonesia. Regional Environmental Change 9, 259–274.

Jennerjahn, T.C., Gilman, E., Krauss, K.W., Lacerda, L.D., Nordhaus, I., Wolanski, E., 2017. Mangrove ecosystems under climate change. In: Rivera-Monroy, V., Lee, S.Y., Kristensen, E., Twilley, R.R. (Eds.), Mangrove Ecosystems: A Global Biogeographic Perspective – Structure, Function and Services. Springer Publishing Company, New York, pp. 211–244.

Karstens, S., Lukas, M.C., 2014. Contested aquaculture development in the protected mangrove forests of the Kapuas Estuary, West Kalimantan. Geo-oko . 35, 78–121.

Kauffman, J.B., Heider, C., Cole, T.G., Dwire, K.A., Donato, D.C., 2011. Ecosystem carbon stocks of micronesian mangrove forests. Wetlands 31, 343–352.

Keulartz, J., Zwart, H.A.E., 2004. Boundaries, Barriers, and Bridges. Philosophical Fieldwork in Derawan. From. http://www.filosofie.science.ru.nl/derawanreport.pdf.

Knoppers, B., Kjerfve, B., Carmouze, J.P., 1991. Trophic state and water turn-over time in six choked coastal lagoons in Brazil. Biogeochemistry 16, 149–166.

Kusumaningtyas, M.A., Hutahaean, A.A., Fischer, H.W., Pérez-Mayo, M., Pittauer, D., Jennerjahn, T.C., 2019. Variability in the organic carbon stocks, sources, and accumulation rates of Indonesian mangrove ecosystems. Estuarine, Coastal and Shelf Science 218, 310–323.

Lewis III, R.R., 2005. Ecological engineering for successful management and restoration of mangrove forests. Ecological Engineering 24, 403–418.

Long, E.R., Macdonald, D.D., Smith, S.L., Calder, F.D., 1995. Incidence of adverse biological effects within ranges of chemical concentrations in marine and estuarine sediments. Environmental Management 19, 81–97.

Ludwig, H.F., 1985. Segara Anakan Environmental Monitoring and Optimal Use Planning Project Final Report. Bandung, Indonesia. Institute of Hydraulic Engineering, Agency for Research and Development and Ministry of Public Works.

Lukas, M.C., 2014a. Cartographic reconstruction of historical environmental change. Cartographic Perspectives 78, 5–24.

Lukas, M.C., 2014b. Eroding battlefields: land degradation in Java reconsidered. Geoforum 56, 87–100.

Lukas, M.C., 2015. Reconstructing Contested Landscapes. Dynamics, Drivers and Political Framings of Land Use and Land Cover Change, Watershed Transformations and Coastal Sedimentation in Java. IndonesiaDoctoral Thesis Germany: University of Bremen. https://elib.suub.uni-bremen.de/edocs/00106383-1.pdf.

Lukas, M.C., 2017a. Widening the scope: linking coastal sedimentation with watershed dynamics in Java, Indonesia. Regional Environmental Change 17, 901–914.

Lukas, M.C., 2017b. Konservasi daerah aliran sungai di Pulau Jawa, Indonesia: Terjebak dalam konflik sumberdaya hutan. artec-Paper 212, Sustainability Research Center (artec). University of Bremen.

Lukas, M.C., Flitner, M., 2019. Fixing scales. Scalar politics of environmental management in Java, Indonesia. Environment and Planning E: Nature and Space 2 (3), 565–589.

Lynch, J.C., Meriwether, J.R., McKee, B.A., Vera-Herrera, F., Twilley, R.R., 1989. Recent accretion in mangrove ecosystems based on 137Cs and 210Pb. Estuaries 12, 284–299.

Millennium Ecosystem Assessment, 2005. Ecosystems and Human Well-Being: Synthesis. Island Press, Washington DC. www.unep.org/maweb.

Milliman, J.D., Farnsworth, K.L., 2011. River Discharge to the Coastal Ocean – A Global Synthesis. Cambridge University Press, Cambridge.

Moll, R., 2011. Impact of Mangroves and an Agriculture-Dominated Hinterland on the Carbon and Nutrient Biogeochemistry in the Segara Anakan Lagoon. PhD Thesis. University of Bremen, Java, Indonesia.

Muir, D.C.G., Howard, P.H., 2006. Are there other persistent organic pollutants? A challenge for environmental chemists. Environmental Science and Technology 40, 7157–7166.

Murdiyarso, D., Purbopuspito, J., Kauffman, J.B., Warren, M.W., Sasmito, S.D., Donato, D.C., et al., 2015. The potential of Indonesian mangrove forests for global climate change mitigation. Nature Climate Change 5, 1089–1092.

Napitupulu, M., Ramu, K.L.V., 1980. Development of the Segara Anakan area of central Java. In: Proceedings of the Workshop on Coastal Resources Management in the Cilacap Region, 20–24 August 1980. . Yogyakarta: Gadjah Mada University. Indonesian Institute of Sciences and United Nations University, Jakarta, Indonesia, pp. 66–82.

Nordhaus, I., Hadipudjana, F.A., Janssen, R., Pamungkas, J., 2009. Spatio-temporal variation of macrobenthic communities in the mangrove-fringed Segara Anakan Lagoon, Indonesia, affected by anthropogenic activities. In: Jennerjahn, T.C., Yuwono, E. (Eds.), Segara Anakan, Java, Indonesia, a Mangrove-Fringed Coastal Lagoon Affected by Human Activities, Special Issue, Regional Environmental Change, vol. 9, pp. 291–313.

Nordhaus, I., Salewski, T., Jennerjahn, T.C., 2011. Food preferences of mangrove crabs related to leaf nitrogen compounds in the Segara Anakan Lagoon, Java, Indonesia. Journal of Sea Research 65, 414–426.

Nordhaus, I., Toben, M., Fauziyah, A., 2019. Impact of deforestation on mangrove tree diversity, biomass and community dynamics in the Segara Anakan Lagoon, Java, Indonesia: a ten-year perspective. Estuarine, Coastal and Shelf Science 227, 106300.

Olive, C.A., 1997. Land Use Change and Sustainable Development in Segara Anakan, Java, Indonesia PhD Thesis. Interactions Among Society, Environment and Development. University of Waterloo.

Ostmann, A., Nordhaus, I., Sørensen, M.V., 2012. First recording of kinorhynchs from Java, with the description of a new brackish water species from a mangrove-fringed lagoon. Marine Biodiversity 42, 79–91.

Page, D.S., Boehm, P.D., Stubblefield, W.A., Parker, K.R., Gilfillan, E.S., Neff, J.M., et al., 2002. Hydrocarbon composition and toxicity of sediments following the Exxon Valdez oil spill in Prince William Sound, Alaska, USA. Environmental Toxicology and Chemistry 21, 1438–1450.

Pamungkas, J., 2015. The description of a new species *Polymastigos javaensis* n.sp. (Annelida: Capitellidae) from the Segara Anakan mangroves, central Java, Indonesia. Zootaxa 3980, 279–285.

Phang, V.X.H., Chou, L.M., Friess, D.A., 2015. Ecosystem carbon stocks across a tropical intertidal habitat mosaic of mangrove forest, seagrass meadow, mudflat and sandbar. Earth Surface Processes and Landforms 40, 1387–1400.

PRC-ECI, 1987. Segara Anakan Engineering Measures Study. Main Report Citanduy River basin Development Project. Banjar, Indonesia. PRC-Engincering Consultants, Inc.

Ray, R., Ganguly, D., Chowdhury, C., Dey, M., Das, S., Dutta, M.K., et al., 2011. Carbon sequestration and annual increase of carbon stock in a mangrove forest. Atmospheric Environment 45, 5016–5024.

Reichel, C., Frömming, U.U., Glaser, M., 2009. Conflicts between stakeholder groups affecting the ecology and economy of the Segara Anakan region. Regional Environmental Change 9, 335–343.

Richards, D.R., Friess, D.A., 2016. Rates and drivers of mangrove deforestation in Southeast Asia, 2000–2012. Proceedings of the National Academy of Sciences 113 (2), 344–349.

Rose, S., 2015. Crab Diversity and Community Structure in Relation to Environmental Change in the Segara Anakan Lagoon, Java, Indonesia. Master thesis, University Bremen.

Sanders, C.J., Smoak, J.M., Sathy Naidu, A., Patchineelam, S.R., 2008. Recent sediment accumulation in a mangrove forest and its relevance to local sea-level rise (Ilha Grande, Brazil). Journal of Coastal Research 24, 533–536.

Sanders, C.J., Smoak, J.M., Naidu, A.S., Sanders, L.M., Patchineelam, S.R., 2010a. Organic carbon burial in a mangrove forest, margin and intertidal mud flat. Estuarine, Coastal and Shelf Science 90, 168–172.

Sanders, C.J., Smoak, J.M., Sanders, L.M., Naidu, A.S., Patchineelam, S.R., 2010b. Organic carbon accumulation in Brazilian mangal sediments. Journal of South American Earth Sciences 30, 189–192.

Sanders, C.J., Smoak, J.M., Naidu, A.S., Araripe, D.R., Sanders, L.M., Patchineelam, S.R., 2010c. Mangrove forest sedimentation and its reference to sea level rise, Cananeia, Brazil. Environmental Earth Science 60, 1291–1301.

Schaafsma, J.M.G., 1927. Een en over de Segara Anakan (Kinderzee), 1926 Jaarsverslag van den Topographischen Dienst in Nederlandsch-Indie over 130–134.

Schwerdtner Mañez, K., 2010. Java's forgotten pearls: history and disappearance of pearl-fishing in the Segara Anakan Lagoon, South Java, Indonesia. Journal of Historical Geography 36, 367–376.

Schwerdtner Máñez, K., Ring, I., Krause, G., Glaser, M., 2014. The Gordian Knot of mangrove conservation: disentangling the role of scale, services and benefits. Global Environmental Change 28, 120–128.

Siahainenia, A.J., 2016. Food from the Sulawesi Sea – The Need for Integrated Sea Use Planning Unpublished. PhD Thesis. University of Wageningen, Netherlands.

Sidik, F., Adame, M.N., Lovelock, C.E., 2019. Carbon sequestration and fluxes of restored mangroves in abandoned aquaculture ponds. Journal of the Indian Ocean Region 15 (2), 177–192.

Sitoe, A.A., Mandlate, L.J.C., Guedes, B.S., 2014. Biomass and carbon stocks of Sofala Bay mangrove forests. Forests 5, 1967–1981.

Smoak, J.M., Breithaupt, J.L., Smith III, T.J., Sanders, C.J., 2013. Sediment accretion and organic carbon burial relative to sea-level rise and storm events in two mangrove forests in Everglades National Park. Catena 104, 58–66.

Spalding, M., Kainuma, M., Collins, L., 2010. World Atlas of Mangroves. Earthscan, London, Washington DC.

Stockholm Convention on Persistent Organic Pollutants. http://chm.pops.int/TheConvention/. (Accessed 26 June 2018).

Syakti, A.D., Hidayati, N.V., Hilmi, E., Piram, A., Doumenq, P., 2013. Source apportionment of sedimentary hydrocarbons in the Segara Anakan nature reserve, Indonesia. Marine Pollution Bulletin 74, 141–148.

Tateda, Y., Nhan, D.D., Wattayakorn, G., Toriumi, H., 2005. Preliminary evaluation of organic carbon sedimentation rates in Asian mangrove coastal ecosystems estimated by 210Pb chronology. Radioprotection 40, S527–S532.

Tripathi, S., 2016. Diversity and Community Composition of Crabs and Gastropods in Mangrove Forests of Berau, East Kalimantan. Master thesis. University Bremen.

Tue, N.T., Dung, L.V., Nhuan, M.T., Omori, K., 2014. Carbon storage of a tropical mangrove forest in Mui Ca Mau National Park, Vietnam. Catena 121, 119–126.

Umgiesser, G., Ferrarin, C., Cucco, A., De Pascalis, F., Bellafiore, D., Ghezzo, M., et al., 2014. Comparative hydrodynamics of ten Mediterranean lagoons by means of numerical modeling. Journal of Geophysical Research Oceans 119, 2212–2226.

Wallace, C.C., 1999. The Togian Islands: coral reefs with a unique coral fauna and an hypothesized Tethy Sea signature. Corals 18, 162.

Wang, G., Guan, D., Peart, M.R., Chen, Y., Peng, Y., 2013. Ecosystem carbon stocks of mangrove forest in Yingluo Bay, Guangdong Province of south China. Forest Ecology and Management 310, 539–546.

Weatherbase, 2019. www.weatherbase.de.

Weiss, C., Weiss, J., Boy, J., Iskandar, I., Mikutta, R., Guggenberger, G., 2016. Soil organic carbon stocks in estuarine and marine mangrove ecosystems are driven by nutrient colimitation of P and N. Ecology and Evolution 6, 5043–5056.

Wells, S., Ravilous, C., Corcoran, E., 2006. In the Front Line: Shoreline protection and Other Ecosystem Services from Mangroves and Coral Reefs. United Nations Environment Programme World Conservation Monitoring Centre, Cambridge, UK.

White, A.T., Martosubroto, P., Sadorra, M.S.M., 1989. The coastal environment profile of Segara Anakan – Cilacap, south Java, Indonesia. In: ICLARM. Association of Southeast Asian Nations. United states Coastal Resources Management Project. Technical Publication Series 4.

Wiryawan, B., Mous, P.J., 2003. Report on a Rapid Ecological Assessment of the Derawan Islands, Berau District, East Kalimantan, Indonesia. The Nature Conservancy, Bali, Indonesia.

Wiryawan, B., Khazali, M., Knight, M. (Eds.), 2005. Menuju Kawasan Konservasi Laut Berau, Kalimantan Timur: Status Sumberdaya Pesisir Dan Proses Pengembangannya. Program Bersama Kelautan Berau TNC-WWF-Mitra Pesisir/CRMP II USAID, Jakarta.

Woodroffe, C.D., 1992. Mangrove sediments and geomorphology. In: Robertson, A.I., Alongi, D.M. (Eds.), Tropical Mangrove Ecosystems. American Geophysical Union, Washington, DC, pp. 7–41.

World Weather Online, 2019. s, 4 December 2019. https://www.worldweatheronline.com/kalimantan-weather-averages/indonesia-general/id.aspx.

Yang, J., Gao, J., Liu, B., Zhang, W., 2014. Sediment deposits and organic carbon sequestration along mangrove coasts of the Leizhou Peninsula, southern China. Estuarine, Coastal and Shelf Science 136, 3–10.

Yuwono, E., Jennerjahn, T.C., Nordhaus, I., Ardli, E.R., Sastranegara, M.H., Pribadi, R., 2007. Ecological status of Segara Anakan, Java, Indonesia, a mangrove-fringed lagoon affected by human activities. Asian Journal of Water, Environment and Pollution 4, 61–70.

# 8

# Impact of megacities on the pollution of coastal areas—the case example Jakarta Bay

Andreas Kunzmann[1], Jan Schwarzbauer[2], Harry W. Palm[3], Made Damriyasa[4], Irfan Yulianto[5], Sonja Kleinertz[3], Vincensius S.P. Oetam[6,8,10], Muslihudeen A. Abdul-Aziz[6,7], Grit Mrotzek[6], Haryanti Haryanti[9], Hans Peter Saluz[6,10], Zainal Arifin[11], Gunilla Baum[2], Larissa Dsikowitzky[1], Dwiyitno[12], Hari Eko Irianto[12], Simon van der Wulp[13], Karl J. Hesse[13], Norbert Ladwig[13], Ario Damar[14]

*[1]LEIBNIZ CENTRE FOR TROPICAL MARINE RESEARCH (ZMT), BREMEN, GERMANY; [2]INSTITUTE OF GEOLOGY AND GEOCHEMISTRY OF PETROLEUM AND COAL, RWTH AACHEN UNIVERSITY, GERMANY; [3]AQUACULTURE AND SEA-RANCHING, FACULTY OF AGRICULTURAL AND ENVIRONMENTAL SCIENCES, UNIVERSITY OF ROSTOCK, GERMANY; [4]UDAYANA UNIVERSITY, FACULTY OF VETERINARY SCIENCES, BADUNG, BALI, INDONESIA; [5]FACULTY OF FISHERIES AND MARINE SCIENCES, BOGOR AGRICULTURAL UNIVERSITY, BOGOR, INDONESIA; [6]LEIBNIZ INSTITUTE FOR NATURAL PRODUCT RESEARCH AND INFECTION BIOLOGY, JENA, GERMANY; [7]AUSTRALIAN CENTRE FOR ANCIENT DNA, UNIVERSITY OF ADELAIDE, ADELAIDE, SA, AUSTRALIA; [8]MAX PLANCK INSTITUTE FOR CHEMICAL ECOLOGY, JENA, GERMANY; [9]GONDOL RESEARCH INSTITUTE FOR MARICULTURE GRIM, SINGARAJA, BALI, INDONESIA; [10]FRIEDRICH SCHILLER UNIVERSITY OF JENA, GERMANY; [11]RESEARCH CENTRE FOR OCEANOGRAPHY-LIPI, ANCOL TIMUR, JAKARTA, INDONESIA; [12]RESEARCH AND DEVELOPMENT CENTER FOR MARINE AND FISHERIES PRODUCT PROCESSING AND BIOTECHNOLOGY (BALAI BESAR RISET PENGOLAHAN PRODUK DAN BIOTEKNOLOGI KELAUTAN DAN PERIKANAN), JAKARTA, INDONESIA; [13]ECOLAB - GROUP COASTAL ECOSYSTEMS, RESEARCH AND TECHNOLOGY CENTRE WESTCOAST (FTZ), UNIVERSITY OF KIEL, BÜSUM, GERMANY; [14]CENTER FOR COASTAL AND MARINE RESOURCES STUDIES, BOGOR AGRICULTURAL UNIVERSITY (IPB UNIVERSITY), INDONESIA*

## Abstract

*The rapid development of Indonesia over the past 20 years and also its increasing negative impact on the environment are by far best to be seen at the metropolitan area and the corresponding coastal ecosystems of Jakarta. All the information and facts reported in this chapter regarding the environmental state of Jakarta Bay demonstrate impressively the huge anthropogenically induced pressure and stress this ecosystem is exposed to. Additionally, an intact ecosystem is a basic*

Science for the Protection of Indonesian Coastal Ecosystems (SPICE). https://doi.org/10.1016/B978-0-12-815050-4.00001-8

*precondition for many aspects directly impacting human beings around the Bay covering, e.g., seafood health, drinking water accessibility, or recreation needs. Noteworthy, the investigations reported here do not remain on the description of the environmental status but explored also the relevant processes behind inducing the pollution problems and, consequently, point to potential solutions.*

## Abstrak

*Pesatnya pembangunan di Indonesia selama 20 tahun terakhir, termasuk dampak negatifnya pada lingkungan dapat dilihat pada kawasan metropolitan Jakarta maupun ekosistem pesisir di sekitarnya. Data dan informasi yang ditampilkan pada chapter/bab ini menunjukkan adanya ancaman dan cekaman antropogenik yang serius pada lingkungan Teluk Jakarta. Sementara itu, sebuah ekosistem yang terpelihara dengan baik merupakan faktor penting bagi berbagai aspek yang berpengaruh langsung pada kehidupan masyarakat di sekitar pesisir, seperti keamanan produk perikanan, akses air bersih, maupun kegiatan rekreasi. Oleh karena itu, hasil penelitian ini tidak hanya menjelaskan kondisi lingkungan Teluk Jakarta, tetapi juga melihat lebih dalam faktor-faktor yang menyebabkan terjadinya pencemaran serta memberikan alternatif solusinya.*

## Chapter outline

**8.1 Introduction** ........ **287**
**8.2 Hydrological system and nutrient dispersion** ........ **289**
**8.3 Organic and inorganic pollution in Jakarta Bay** ........ **296**
8.3.1 Types, quantity, and distribution of pollutants ........ 297
*8.3.1.1 Trace hazardous elements* ........ 297
*8.3.1.2 Organic pollutants* ........ 297
8.3.2 Characterizing emission sources ........ 298
*8.3.2.1 Source apportionment of trace elements* ........ 298
*8.3.2.2 The insect repellent N,N-diethyl-m-toluamide as tracer for municipal sewage and the implications for coastal management* ........ 299
8.3.3 Industrial emissions in the Greater Jakarta area and their role for the contamination of the Jakarta Bay ecosystem ........ 300
8.3.4 The flushing-out phenomenon ........ 301
8.3.5 Accumulation in biota ........ 302
**8.4 Water quality and biological responses** ........ **303**
8.4.1 Water pollution in Jakarta Bay and the Thousand Islands ........ 303
8.4.2 Biological responses to anthropogenic stressors ........ 305
8.4.3 Impacts on the physiology of key coral reef organisms ........ 305
8.4.4 Impacts on reef composition ........ 308
8.4.5 Local versus regional stressors in Jakarta Bay and the Thousand Islands ........ 308
**8.5 Microbial diversity in Indonesian fish and shrimp: a comparative study on different ecological conditions** ........ **310**
8.5.1 Microbial diversity in *Epinephelus fuscoguttatus* ........ 312
8.5.2 Microbial diversity in feces of *Penaeus monodon* ........ 314
8.5.3 Comparison of microbial diversity between fish and shrimp from Jakarta and Bali ........ 316

**8.6 Fish parasites in Indonesian waters: new species findings, biodiversity patterns, and modern applications** ........ **317**
8.6.1 Species descriptions and taxonomic treatments ........ 318
8.6.2 Zoonotic *Anisakis* spp. in Indonesian waters ........ 319
8.6.3 Parasite biodiversity in wildlife and maricultured fish ........ 319
8.6.4 Fish parasites as biological indicators and new applications ........ 320
**8.7 Seafood consumption and potential risk** ........ **321**
**8.8 Implications** ........ **330**
**Acknowledgments** ........ **333**
**References** ........ **333**

# 8.1 Introduction

The rapid development of Indonesia over the past 20 years and also its increasing negative impact on the environment are by far best to be seen at the metropolitan area and the corresponding coastal ecosystems of Jakarta. The Indonesian capital is a center of ongoing industrialization, economic growth, and rapidly increasing population (Fig. 8.1). More than 10 million inhabitants are living in the central part of Jakarta City,

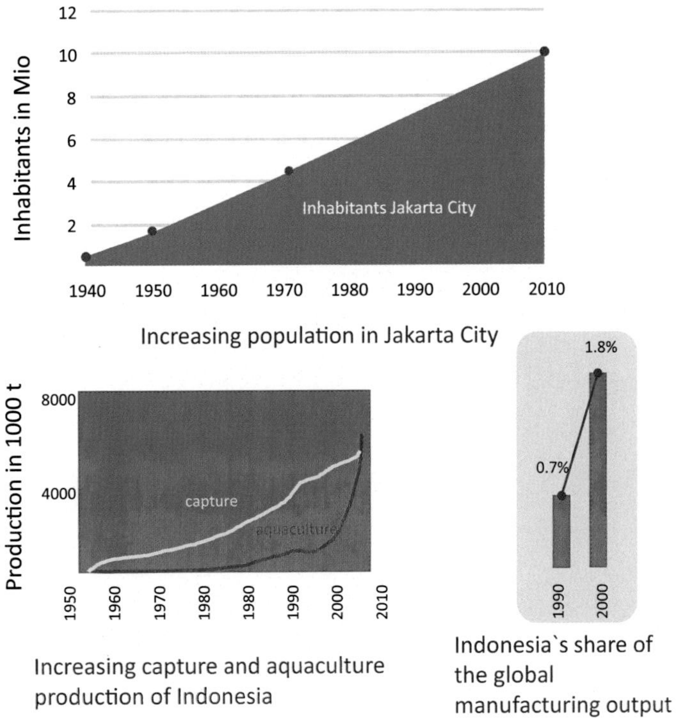


**FIGURE 8.1** Indications for the economic growth of Indonesia and its capital Jakarta. *Sources: FAO.*

and the currently much more expansive wider metropolitan area (Jabotabek region, conurbation representing the city area of Jakarta, Bogor, Depok, Tangerang, and Bekasi) covers an area of about 6700 $km^2$ with a population of some 28 million inhabitants according to the census in 2010 (BPS, 2010). As a result, Jakarta has become one of the 10 biggest megacities in Asia (Blackburn et al., 2014). In parallel, Jakarta faced in the past decades a tremendous increase of pollution by industrial and municipal emissions. Rough calculations point to a discharge of approximately 8.000 t of solid waste per day (BPLHD, 2012), from which a significant fraction is dumped into the rivers and canal systems due to the lack of organized rubbish collection systems. With respect to liquid waste discharged by sewage systems, it has to be noted that only a very low proportion of all households (ca. 15%–20%) are connected to a central or local sewage treatment installation. In addition, wastewater from more than 20,000 industrial facilities located in Jabotabek region is continuously discharged. All these sources contribute to the enormous waste emergence in Jakarta.

Notably, Jakarta Bay (JB) represents the receiving ecosystem of by far the largest proportion of sewage and waste via the urban riverine discharge. JB is bordered by the Thousand Islands in the North, some of which are part of the Kepulauan Seribu Marine National Park and the metropolitan area of Jakarta in the South (Fig. 8.2). The

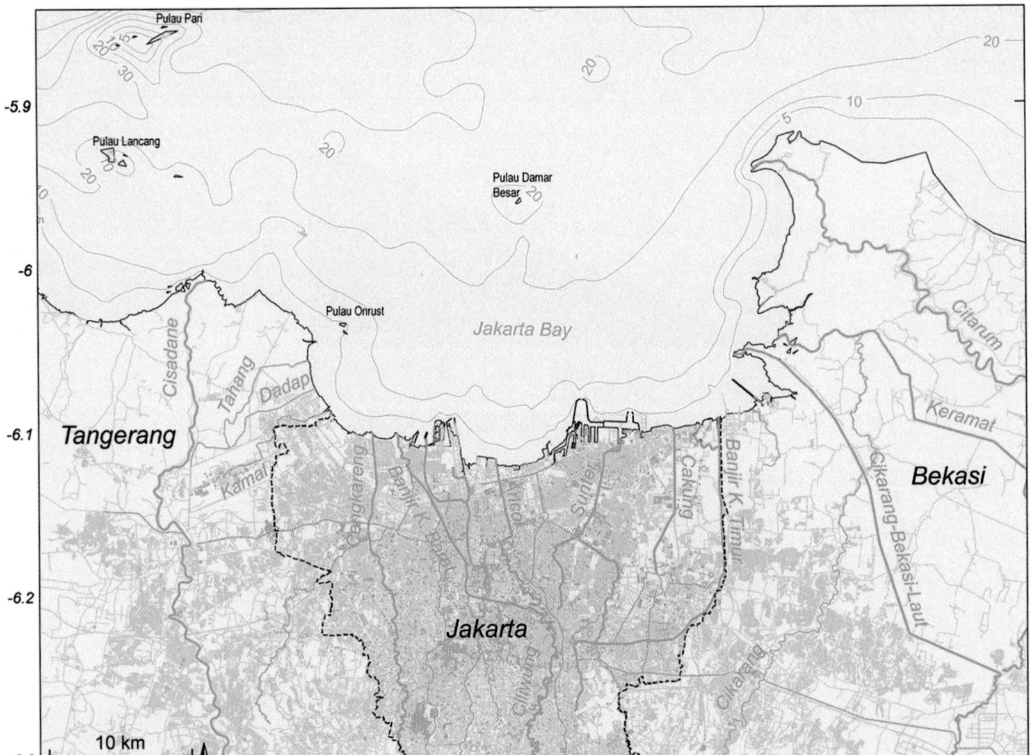


**FIGURE 8.2** Jakarta Bay and the Jakarta Metropolitan Area. The administrative districts of Tangerang, Jakarta, Bekasi, but also Bogor and Depok, situated more south, are drained by a ramified system of rivers and channels which ends up in Jakarta Bay.

corresponding shoreline faced a huge transition from natural conditions to an intensively used coastline. This is effective for industrial facilities, traffic infrastructure, urban settlements, aquafarming, agriculture, and many other developments.

The resulting pollution of industrial and municipal origin has an important impact on the environmental quality of these coastal ecosystems and affects in particular the aquatic biocoenosis. Since JB is also an important site for aquafarming as well as fishery and is used as recreation area for the metropolitan region, the pollution also impairs these relevant functions of the marine systems. Traditional fishery by more than 2500 fishermen is accompanied by modern aquafarming, in particular producing the green mussel *Perna viridis*. Finally, the aquatic pollution of riverine and coastal regions at Jakarta has also negative effects on drinking water resources, since riverine surface water and groundwater, highly affected by riverine infiltration, are the dominant sources for drinking water production, in particular for Jakarta city.

In summary, the aquatic ecosystems of JB represent an excellent example of a fundamental conflict between economic and ecological demands. As a consequence, this sensitive coastal region attracts not only political but also scientific attention. The described anthropogenic impacts and the complementary resource demands affect a complex system influenced by hydrological, geochemical, microbiological, hygienic, and many more parameters. Hence, understanding the principal mechanisms of pollution transport, dissemination and resulting effects on the ecosystems represents a huge scientific challenge.

It becomes dramatically obvious that there is a need for improvement of the environmental quality at JB. The enormous effects of anthropogenic pollution on the ecosystem JB have been continuously reported. As an example, in December 2015, a massive fish mortality along Ancol Beach has been chronicled in the *Jakarta Post*.

Many scientific studies also pointed to the disruption of the ecosystem functioning as the result of anthropogenic impact. Negative effects are visible at mangroves and coral reefs or by repetitive harmful algae blooms. Further observed ecological impairments cover enhanced fish mortality, harm to marine species by pollutants and reduced biodiversity (summarized and referenced in Dsikowitzky et al., 2016a,b).

For substantial changes of the ecological quality of JB, the extent and quality of pollution need to be determined as a basic precondition. However, in the past, studies focusing on contamination of JB and its linkage to long-term implications have been performed only more or less sporadically. As a consequence, Science for the Protection of Indonesian Coastal Ecosystems (SPICE) research activities have been focused inter alia on JB, its pollution, and the ecological implications. This chapter highlights a more comprehensive and complementary approach, considering the different geological, biological, microbial, and chemical aspects of anthropogenic impacts on the JB coastal ecosystems.

## 8.2 Hydrological system and nutrient dispersion

The susceptibility of a specific area to develop eutrophication symptoms such as algal blooms and hypoxia strongly depends on the scale of terrestrial nutrient loading relative to the local dilution capacity (Vollenweider, 1976; von Glasow et al., 2013). This, in turn, is a function of water volume and the rate of water exchange. Hence, it is important to understand the hydrodynamic processes of advective flux and turbulent diffusion, which control the transport of matter and the exposure time available for biological processes such as phytoplankton growth or microbial decomposition.

There are numerous studies predominantly from temperate regions focusing on the evaluation of nutrient discharge from point and nonpoint sources as well as on its impacts (e.g., De Jonge et al., 1994; EEA, 2001; Howarth et al., 2000; Schaub and Gieskes, 1991). Over the past few decades, a number of models have been developed to simulate nutrient fluxes in the catchment basins (e.g., Alexander et al., 2002; Behrendt et al., 2000; Marcé et al., 2004; Van Drecht et al., 2003) and the dispersion of nutrients and their effects in the coastal ecosystem (e.g., Moll, 1998; Lenhart, 2001; Skogen et al., 2004; Lancelot et al., 2005; Pätsch and Kühn, 2008). The quantification of nutrient transport to the coastal bays and estuaries is essential for understanding the consequences of nutrient pollution and for developing adequate nutrient management strategies.

This is not much different for Jakarta, as it is for many other coastal megacities. A ramified system of rivers and canals drains the Jakarta Metropolitan Area (Fig. 8.2). The most important rivers are the Citarum river discharging at the eastern flank of the bay with a catchment area of about 6600 $km^2$ and the Cisadane river at the periphery of the western cape with a catchment basin of 1400 $km^2$. Several smaller rivers, notably Angke, Ciliwung, Sunter, Bekasi, and Cikarang, with a combined total catchment area of ca. 2000 $km^2$, discharge into the central sector of the bay. These rivers drain Jakarta's urban center through mostly canalized downstream sections with several rivers sometimes merged into one canal. They form an integral part of the stormwater and sewage transport network within the city. The most important canals are the Cengkareng drain (dominated by Angke river), the West Banjir Canal (BKB, dominated by Ciliwung), the Sunter, the Cakung drain, the East Banjir Canal (BKT, diverting water from the Cipanang, Sunter, Cakung and Buaran river), and the Cikarang-Bekasi-Laut Canal (CBL, dominated by the Bekasi and Cikarang). In total, the coastline of JB comprises 16 river and canal mouths, which are the main gateways of pollutants from the JMA to the coastal ecosystem of the shallow JB.

Discharge and retention of water from the Jakarta area is regulated by a comprehensive network of barrages, weirs, retention basins, pumping stations, and artificial canals with their tributaries, mainly concentrated in the low lying plain of the densely populated Jakarta Province. The river water is deviated on demand for a variety of reasons, such as for flood prevention during times of high run-off in the rainy season as well as for irrigation (mostly paddy fields), domestic, municipal, and industrial water supply. As an example, the Citarum river provides up to 80% of Jakarta's raw surface water demand, which is channeled through the West Tarum canal to the tap water producing plants near the city (Fulazzaky, 2010).

Jakarta regularly experiences severe flooding events during the rainy season in January and February. To reduce flooding during times of heavy rainfall, river water from

**FIGURE 8.3** The rivers and canals of Jakarta receive a considerable amount of domestic and industrial wastewater and garbage due to the lack of proper solid and liquid waste management systems, leading to appearances such as the photo above, taken along the banks of the Cakung River.

the Ciliwung and Cengkareng drainage system is diverted to the West Banjir Canal, whereas water from the Cipinang, Sunter, Buaran, and Cakung rivers is bypassed to the recently finalized East Banjir canal. By contrast, during drought periods, water from the Citarum and the Ciliwung is drained to provide enough flushing to smaller canals and rivers, such as the Kali Baru river flowing through central Jakarta. Flow in the downstream area of the waterways is thus heavily influenced by interbasin transfer.

The rivers and canals receive a considerable amount of domestic and industrial wastewater and garbage due to the lack of proper solid and liquid waste management systems. This leads to a strong deterioration of these surface waters causing critical conditions for the natural ecosystem, public hygiene, and also esthetic values and blocking of the water drainage (see Fig. 8.3). The pollution received by the rivers and streams gradually finds its way to the river mouth and ends up in JB.

To date, less than 3% of Jakarta's population is connected to a centralized sewage treatment plant (the Setiabudi WWTP), which discharges the treated wastewater into the Ciliwung river (Apip Sagala and Luo, 2015). Instead, most households (>70%) rely on septic tanks, which are often poorly maintained because they are not emptied on a regular basis but overflow and leach into the soil. In addition, considering the high density of septic tanks in Jakarta, the draining fields are too small, resulting in the pollution of drinking water, groundwater, canals, and rivers (Vollaard et al., 2005). Besides, it is estimated that about 11% of Jakarta's population directly drain their wastes to neighboring watercourses (Apip Sagala and Luo, 2015). The lack of sufficient infrastructure to transport and treat domestic wastewater results in a flow of sewage into the public channels and rivers that cross the metropolitan area of Jakarta making them a major source for nutrients to the bay.

The impact of elevated nutrient levels becomes apparent as increasing frequencies of high biomass algae blooms (HBBs). Excessive blooms lead to undesired eutrophication effects where increased concentrations of biomass and subsequent microbial breakdown cause recurrent oxygen deficiencies in the water column and underlying sediments (Ladwig et al., 2016). Oxygen deficiencies on their turn are suspected to have triggered reoccurring fish kills (Wouthuizen et al., 2007). These high biomass blooms are reported to occur predominantly along the city shoreline of Jakarta (Damar, 2003; Mulyani Widiarti and Wardhana, 2012; Thoha et al., 2007; Yuliana, 2012).

Adverse eutrophication effects are driven not only by the level of nutrient concentrations but also by the physical characteristics of JB and its tributaries. A numerical model study by Koropitan et al. (2009) concluded that the influence of river discharge is limited to the coastal area with conformity to observed HBB occurrences. Quantification of nutrient loads to JB indicated a somewhat contradicting view where the largest rivers, such as the Cisadane and Citarum rivers, are the highest contributor to JB in terms of nutrient load. These rivers are situated along the northwest and northeast edges of the bay. By contrast, urban rivers and channels along the city shoreline are characterized by considerate, but relatively small loads of dissolved nutrients (Van der et al., 2016a). Also, the large rivers can be characterized by high river discharges with relatively low nutrient concentrations, whereas city bound rivers and channels have relatively low discharges, but very high concentrations of anthropogenic nutrient. More information on how river loads can be quantified can be found in "Determination of river loads for Jakarta Bay".

A numerical model for flow and dispersion of these land-based nutrients was set up within the framework of the SPICE project on impacts of marine pollution on biodiversity and coastal livelihoods. Simulations showed that nutrient loads from the Citarum and Cisadane resulted in a smaller elevation of nutrient concentrations, with respect to their urban counterparts. Favorable dispersion due to a larger interaction with offshore currents allowed these loads to be more rapidly assimilated than the rivers situated at the inner bay. Along the city shoreline, horizontal circulation was found to be limited, and pollutants are dispersed at a slower rate than along the outer ridge of JB, resulting in hot spots with elevated nutrient levels. In addition, dense water masses from the Java Sea in combination with less dense river discharges result in a profound horizontal and vertical density gradient. The vertical stratification leads to a stronger decoupling between surface and bottom flows allowing stronger turbulent mixing of the surface layers due to tides and wind-induced currents (Van der Wulp et al., 2016a). More information on the use of numerical models to simulate the dispersion of dissolved substances can be found in "Hydrodynamic and dispersion models to study pollution".

As discussed so far, quantified nutrient flux from rivers and channels do not reveal the actual source of pollution. No discrepancy can be made, whether nutrients originate from municipal wastewater, industrial wastewater, or other sources such as agriculture.

Elevated nutrient levels and localized eutrophication effects, however, can be attributed to those river inputs, which originate from catchment areas and which lie in predominantly urban areas and receive considerable amounts of wastewater. In addition, Dsikowitzky and Schwarzbauer (2014) found high levels of the insect repellent DEET (*N,N*-diethyl-*m*-toluamide) in urban rivers and nearshore coastal waters and proposed this substance as a molecular marker for municipal wastewater. The measurements of this study in combination with a simulated dispersion of molecular traces of DEET by Van der Wulp et al. (2016b) indicate that the distribution of municipal waste is limited to the city shoreline, making it more than likely that untreated municipal wastewaters are a key source for eutrophication effects.

The governmental authorities in Jakarta are well aware of this situation and set up a master plan to reinforce the city's infrastructure by expanding the sewerage system and constructing new treatment plants. It is planned to establish in a step-by-step procedure in total 15 sewerage zones with off-site treatment facilities expected to be ready to serve 80% of the population by 2050 (PD Pal Jaya, 2012). By 2020, off-site treatment capacity should cover 10% of the city's liquid waste mainly from the central area. In addition, on-site sanitation will be improved. New houses have to be equipped with modern septic tanks provided with filters, which promote bacterial degradation of organic matter in a closed system. Considering that currently up to 26% of human excreta are disposed of untreated into surface waters or gutters (Vollaard et al., 2005), it is further intended to strengthen enforcement of regulations on adequate desludging and proper sludge disposal. The target of a 100% coverage of regular desludging of on-site facilities and a complete conversion of conventional septic tanks into modern septic tanks shall be achieved by 2050 (PD Pal Jaya, 2012).

## Determination of river loads for Jakarta Bay

The river load of any given substance can be defined as the product of river discharge and concentration at a given time. This information is optimally provided by measurements and, where possible, at a frequent time interval to learn more about the temporal variation. Having all required data is more exception than the rule. For a complex hydrological system such as Jakarta, coinciding measurements of both river discharges and water quality measurements proved to be unavailable. Quantification of nutrient flux per individual tributary could only be done in combination with a numerical modeling approach. A hydrological model was used to approximate individual river discharges to quantify the nutrient flux, per tributary, into JB. The hydrological model was set up for the Jakarta Metropolitan Area providing river discharges entering JB (see the following figure).

Figure: Hydrodynamic model domain with drainage basins (*red lines*). Accumulation of runoff results in river discharges at the river mouths, indicated by *red dots*.

*Continued*

Determination of river loads for Jakarta Bay—cont'd

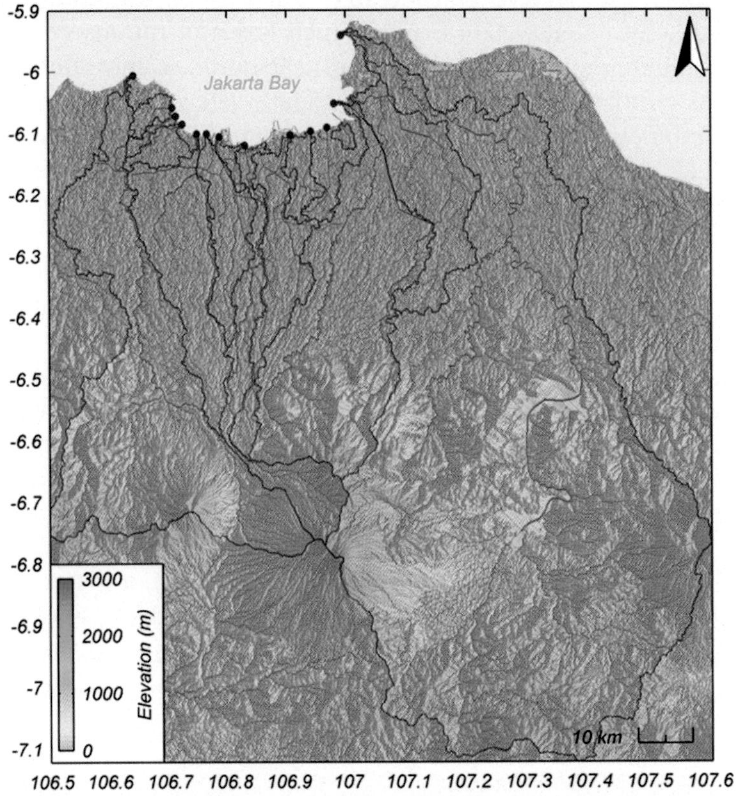


This grid-based model uses spatiotemporal precipitation and climate data to compute water balances considering various compartments including soil storage, groundwater storage, evapotranspiration, and potential runoff. The accumulation of runoff can be calculated spatially based on the topography, yielding an increased discharge with increasing downstream distance as illustrated in the illustrated figure. In addition, water samples were collected at the downstream end of selected rivers and channels and analyzed for nutrients. Among others, total nitrogen loads could be specified per tributary (see the following figure).

Figure: Based on modeled river discharges and field measurements, an approximation of river nutrient loads could be made as shown here for total nitrogen (TN).

*From Van der Wulp, S. A., Damar, A., Ladwig, N., & Hesse, K. J. (2016). Numerical simulations of river discharges, nutrient flux and nutrient dispersal in Jakarta Bay, Indonesia.* Marine Pollution Bulletin, 110*(2), 675–685.*

Determination of river loads for Jakarta Bay—cont'd

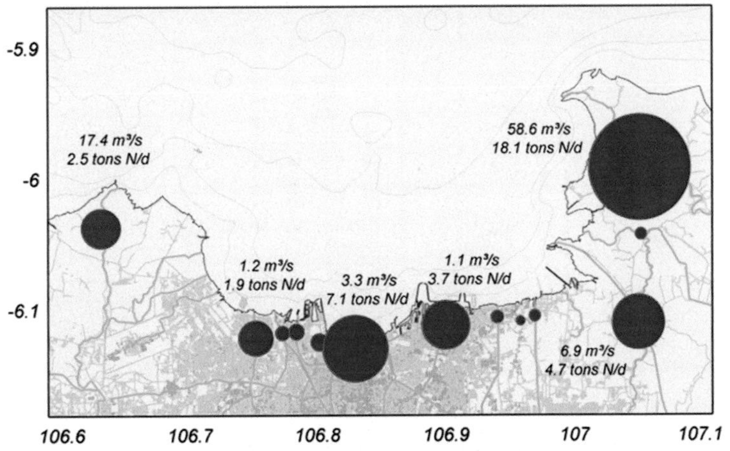


Hydrodynamic and dispersion models to study pollution

The study of pollution, anthropogenic sources, and fate of relevant substances requires a lot of field measurements to obtain an insight into the underlying mechanisms, which drive the transport cycle. Numerical models can complement field observations by its capability to reveal the underlying processes, which are difficult or not possible to observe through measurements. There are a variety of modeling systems (a.o. ROMS, POM, MIKE, and Delft3D) available to simulate flows, transport, and processes of decay of selected substances. To set up a model of a given region and with a given research objective, all relevant parameters and processes should be considered. For instance, the flow and dispersion of nutrients toward JB need a specification of river discharges, nutrient loads, bathymetry and forcing of tides + sea surface height, wind, sea temperature, and salinity enacting on the defined region. The following diagram shows the approach to simulate the flow and dispersion of total nitrogen (TN) and total phosphorus (TP) as conservative substances.

Figure: Diagram of the modeling approach for Jakarta Bay.

*Continued*

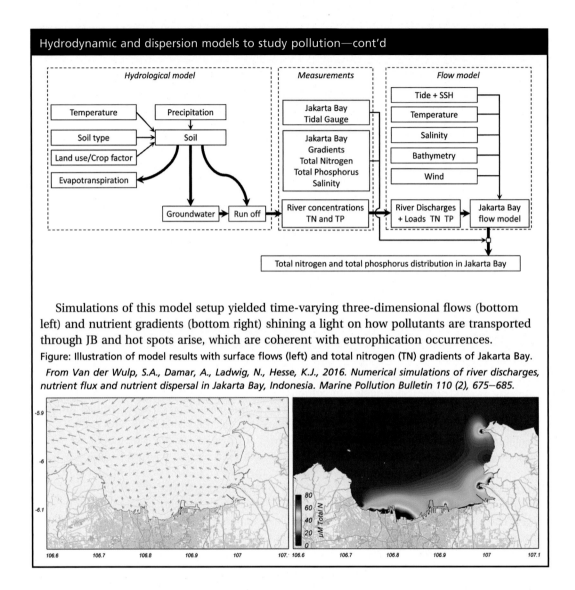

### Hydrodynamic and dispersion models to study pollution—cont'd

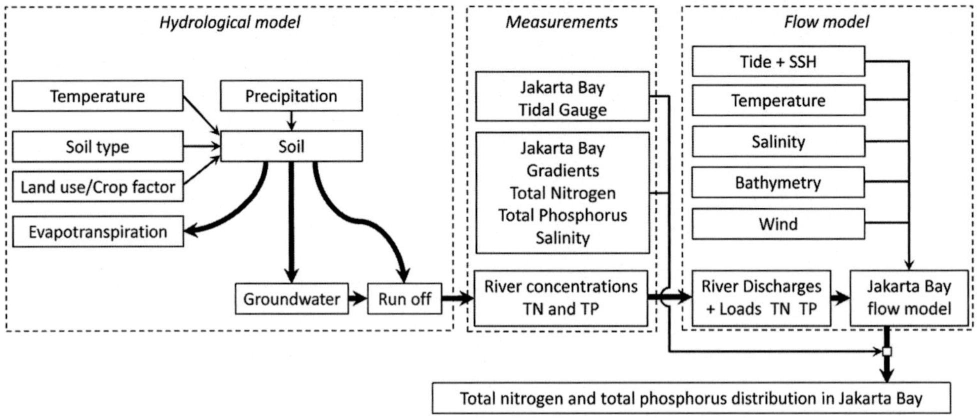


Simulations of this model setup yielded time-varying three-dimensional flows (bottom left) and nutrient gradients (bottom right) shining a light on how pollutants are transported through JB and hot spots arise, which are coherent with eutrophication occurrences.

Figure: Illustration of model results with surface flows (left) and total nitrogen (TN) gradients of Jakarta Bay.

*From Van der Wulp, S.A., Damar, A., Ladwig, N., Hesse, K.J., 2016. Numerical simulations of river discharges, nutrient flux and nutrient dispersal in Jakarta Bay, Indonesia. Marine Pollution Bulletin 110 (2), 675–685.*

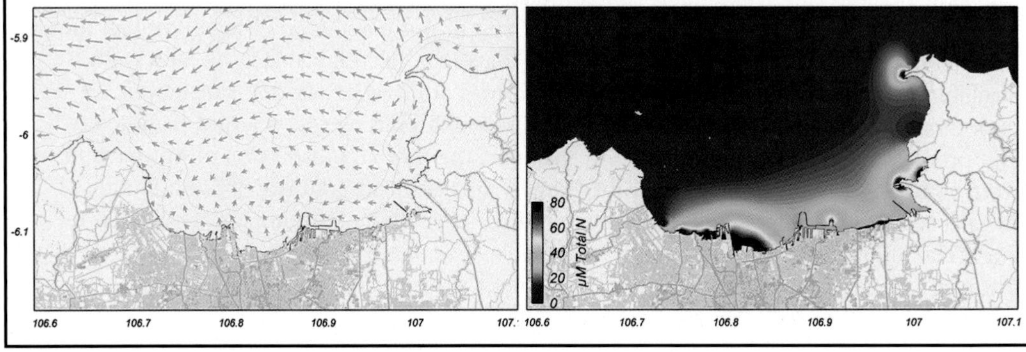


# 8.3 Organic and inorganic pollution in Jakarta Bay

The huge anthropogenic impact on the ecosystem is related to the emission of pollutants from the urban and riverine systems toward the marine environment at JB. The Jakarta river systems therefore receive enormous amounts of untreated or partially treated municipal wastewaters and transport these contaminant loads toward JB.

## 8.3.1 Types, quantity, and distribution of pollutants

### *8.3.1.1 Trace hazardous elements*

Trace hazardous elements build up a highly relevant pool of pollutants at JB. Therefore, their spatial distribution and seasonal variation is an important aspect for characterizing the state of pollution of JB (see also Arifin et al., 2012). Siregar et al. (2016) studied the dynamical distribution of the hazardous metals Hg, Pb, Cd, Cu, Cr, Co, and As. The levels of these selected trace hazardous elements in water, surface sediments, and animal tissues were determined in samples collected during two different seasons. A detailed interpretation of the data revealed two important aspects: trace hazardous element contamination in JB differed (1) between the considered metals and (2) for most of the metals also between premonsoon and postmonsoon time. Here, concentrations of most elements were lower after the wet season and higher at the end of the dry season. Further on, a quality assessment of the sediments showed that the concentrations of Hg, Cu, and Cr at some stations exceeded previously reported toxicity thresholds for benthic species. Noteworthy, not only river and canal sediments within Jakarta City but also sediments in canals of the industrial center Bekasi City are characterized by metal concentrations (mainly Zn) in excess of sediment quality guidelines. Consequently, adverse effects on benthic communities can be expected at all of these stations.

Contaminants in sediments and water are also the source for pollutants in biotic species living in the corresponding ecosystems. Hence, the level of pollution with hazardous elements in organisms is an essential criterion for environmental assessment of aquatic ecosystems. In JB, the order of element concentrations in tissue samples of economic important bivalve and fish species reflected very well the element concentrations found in water and sediment samples from the bay.

Generally, a thorough picture of the spatial distribution considering also the origin of the contamination, seasonal variations, and possible accumulation of the selected elements in economic important bivalve and fish species in JB was depicted for trace hazardous elements in JB. However, beside hazardous elements, organic contaminants play a major role for aquatic pollution. Organic pollutants cover a wide range of molecular structures and related physicochemical properties determining their environmental fate. Due to their partly unique structures, a clear linkage of occurrence to emission source is often possible remarking so-called marker or indicator substances.

### *8.3.1.2 Organic pollutants*

Applying a nontarget screening approach, numerous site-specific or indicative compounds have been identified in Jakarta river water samples (Dsikowitzky et al., 2016a,b). Noteworthy, pollutants detected in river waters from Jakarta area are also suspected to be main contaminants in JB as main receiving system for the urban discharge. Most of the identified organic contaminants can be linked with specific applications, e.g., usage in households or industry. As examples, the detected plasticizers, flame retardants, antioxidants, ingredients of personal care products, disinfectants, surfactant residues, pharmaceutical drugs, and stimulants were previously reported as constituents of municipal wastewaters (Dsikowitzky et al., 2014; Loraine and Pettigrove, 2006; Bueno et al., 2012; Rodil et al., 2010). In terms of concentrations and detection frequency, the flame retardant TCEP (tris(1-chloroethyl)phosphate), the disinfectant chloroxylenol, the personal care product ingredients oxybenzone, DEET (*N,N*-diethyl-*m*-toluamide), HHCB (1,3,4,6,7,8-hexahydro-4,6,6,7,8,8-hexamethylcyclopenta[g]-2-benzopyrane) and AHTN (7-acetyl-1,1,3,4,4,6-hexamethyl-1,2,3,4-tetrahydronaphthalene), the stimulants caffeine and nicotine, and the pain reliever ibuprofen and mefenamic acid were the most important source-specific compounds from municipal sources. These compounds occurred in exceptionally high concentrations as compared with other river systems across the globe. It is likely that these compounds are also relevant water contaminants in other Indonesian urban areas. This spectrum of pollutants reflects clearly the impact of partly untreated municipal discharge and its high implication for the water quality of Jakarta rivers and the adjunctive bay.

However, for some substances, an unambiguous discrimination between municipal and industrial sources is not possible, as these compounds may stem from both sources. This accounts, e.g., for selected plasticizers, flame retardants, and technical antioxidants that are used for paper and polymer manufacturing. The can derive not only from paper manufacturing but also from the leaching of solid waste (leaching of dumped waste papers and plastic materials).

Finally, only a few compounds could unequivocally be attributed to an application in agriculture or to the usage for industrial manufacturing. These include, e.g., the pesticides chlorpyrifos and carbofuran as well as the industrial derived pollutant triphenylphosphine oxide.

## 8.3.2 Characterizing emission sources

A main base for reducing pollution in ecosystems is the identification of the emission sources and their impact on the distribution and level of contamination. This knowledge is a key information for possible technical or political measures for reducing or mitigating the pollution. For inorganic and organic pollutants, two different approaches are followed: discrimination of influences from geogenic and anthropogenic sources as well as the usage of organic indicators.

### *8.3.2.1 Source apportionment of trace elements*

In more detail, to characterize the main sources of hazardous elements, two different sources have to be differentiated: the geogenic from the anthropogenic one. Analyses of major and trace elements in river and canal sediments of the Greater Jakarta area (in combination with published data on sediments of the JB as well as on volcanic rocks of the river catchment areas) showed that chemical characteristics of the river sediments are significantly controlled by the precursor volcanic rocks and the weathering in the catchment area (Sindern et al., 2016). The major element composition of river and canal sediments reflects the dominance of quartz, clay minerals, and Fe oxides/hydroxides. This marks a difference to the composition of the volcanic rocks that show a higher abundance of mafic minerals, which are most affected during weathering. Also trace metals and semimetals in the bay area are inherited from the volcanic rocks. The abundance of Cu and Cr is to various degrees, and the abundance of As is totally controlled by geogenic factors.

Beside these elements and metals of more geogenic origin, some elements such as Zn, Ni, Pb, and to lower degrees Cu are clearly emitted by anthropogenic sources, most of all in central Jakarta City. The marked contrast in enrichment of these elements points to the high variability of local sources, among which metal processing industries may be important, as well as fertilizers or untreated animal waste. In particular, the role of street dusts, which are transported to the rivers with rain water and which are characterized by extremely high Zn concentrations, has to be emphasized.

#### *8.3.2.2 The insect repellent N,N-diethyl-m-toluamide as tracer for municipal sewage and the implications for coastal management*

One of the most prominent organic compounds in terms of concentrations and detection frequency found in water samples from Jakarta rivers and from JB was DEET. DEET is the active component of most commercial insect repellents worldwide. Because there is concern about adverse effects on human health, it was replaced in most formulations sold in the European Union. This contaminant was frequently detected in surface waters in all areas of the world, indicating its mobility and persistence (Merel and Snyder, 2016). However, data from coastal areas are sparse, and data from tropical megacities have not been reported as yet.

Exceptionally high concentrations were analyzed in river water and seawater from Jakarta that exceeded by far all published concentrations in surface waters worldwide (Dsikowitzky and Schwarzbauer, 2014). This can be explained by its massive usage, lack of adequate wastewater treatment, and low average river flow. Due to its high source specificity, elevated concentrations, and its persistence, DEET is an ideal marker to trace the spatial distribution of municipal wastewater inputs into surface water systems.

Consequently, the distribution of DEET in JB mirrored the pattern in the rivers with highest DEET concentrations in the southern and western part of the bay—receiving the river discharges from the central part of Jakarta City—and lower concentrations in the eastern part. In the central part of the bay, which is $\sim 10$ km away from the coastline, still relatively high concentrations were found. These results show that the water quality of the whole inner JB is influenced by the municipal wastewater inputs from the metropolitan area.

As a continuing example, DEET was picked up as water-related indicator substance to estimate the impact of the phased construction of a giant seawall and large storage basins as idea to protect Jakarta City against floods from sea and rivers. A flow and mass tracer model was adapted to simulate scenarios similar to three phases of the construction of the Great Sea Wall to illustrate the fate of river-bound nutrients and municipal wastewater. DEET flux was introduced as an additional suitable tracer substance for municipal wastewaters (Van der Wulp et al., 2016c). The findings stressed that a phased construction should prioritize a parallel development of structural treatment of municipal wastewater to control the illustrated water quality deterioration.

### 8.3.3 Industrial emissions in the Greater Jakarta area and their role for the contamination of the Jakarta Bay ecosystem

The Greater Jakarta metropolitan area hosts the biggest Indonesian industrial manufacturing center. This center is located in Bekasi Regency and Bekasi City, a district with a population of ~5.7 Mio inhabitants (BPS, 2014). The industrial branches include in detail steel manufacturing, glass manufacturing, automotive industry, electrical industry, computer manufacturing, chemical industry (polymer synthesis), personal care product manufacturing, toy industry, and paper industry. The industrial wastewaters of the facilities are discharged into a system of small streams/canals that flow into the JB. These streams/canals are crossed by the West Tarum Canal, an artificial channel of 70 km length designed for irrigation, and serve the largest part of the public water supply of Jakarta City (Fares and Ikhwan, 2001).

As an example for the industrial impact on the aquatic system of Jakarta, Fig. 8.4 shows the concentrations of compounds used for paper manufacturing in the river system receiving discharges from the industrial area. The concentrations of these contaminants along the Western Tarum Canal were significantly lower than in the industrial area. The extremely high concentrations in the industrial area are striking. They were as high as in raw process waters from the paper industry (Dsikowitzky et al., 2015). An explanation for this extreme river pollution is the discharge of immense amounts of untreated or only partly treated wastewaters into the small river/channel system in the industrial area. Some of the contaminants from the industrial point sources are subject to transport into the coastal waters. DMPA (2,2-dimethoxy-2-phenylacetophenone), TMDD (2,4,7,9-tetramethyl-5-decyne-4,7-diol), DIPN (di-*iso*-propylnaphthalenes), and phenylmethoxynaphthalene were detectable in the water located ~30 km downstream the industrial discharges. These contaminants are obviously persistent enough to be transported in the aqueous phase over a distance of several kilometers and are therefore relevant for the water quality of the whole river section downstream the industrial area. All of them were also detectable in the seawater samples from JB, in the area of the river that transports the pollutant loads from the industry discharges into the bay. Hence, the paper industry wastewaters contribute to the contamination of the coastal ecosystem.

DIPN and phenylmethoxynaphthalene were present in the aqueous phase as well as in the sediments downstream the industrial area and in JB. The accumulation of these particle-associated contaminants in economic important mussels and fish species from JB was reported in Dwiyitno et al. (2016). From this it follows, that the paper industry

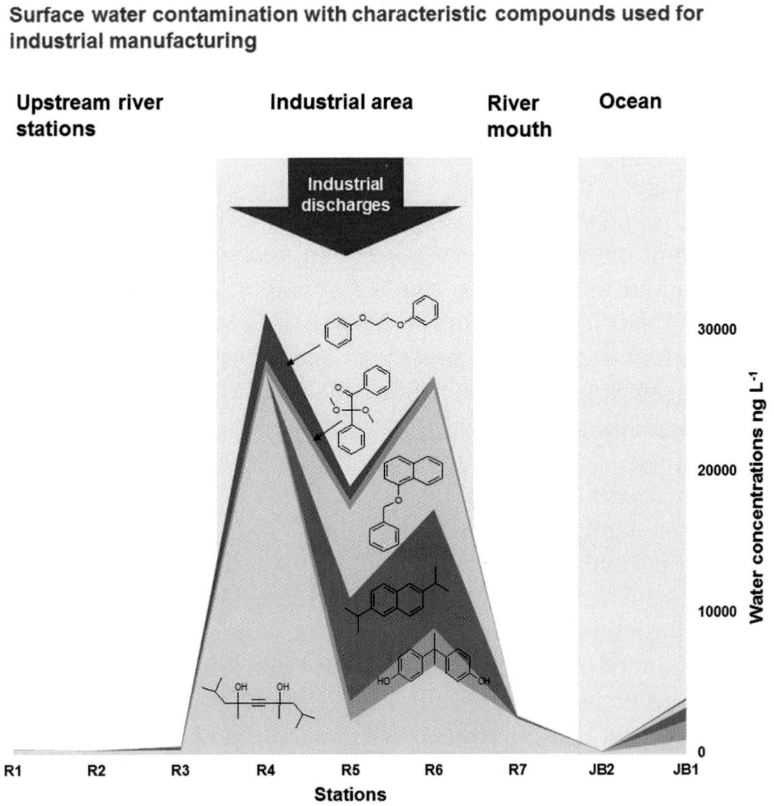


**FIGURE 8.4** Concentrations of chemicals that are used for paper production in water samples taken upstream the industrial area (R1–R3) and in the area receiving the industrial wastewaters (R4–R6). One station downstream was also sampled (R7) as well as two stations in JB, where the river receiving the industrial wastewaters discharges into the bay (JB1 and JB 2).

contributes to the contamination of fishery resources in the coastal waters. A comparison with toxicity thresholds shows that the bisphenol A concentrations in river water from the industrial area pose a threat to macrobenthic invertebrates. Exposure experiments with the freshwater snail *Marisa cornuarietis* revealed adverse effects of bisphenol A on reproduction and survival at $EC_{10}$ 13.9 ng $L^{-1}$ (concentration with response of 10% of the members of the tested population). At stations R5 and R6, the recorded maximum bisphenol A concentrations of 7500 and 8000 ng $L^{-1}$, respectively, were higher than the determined effect value of 998 ng $L^{-1}$.

## 8.3.4 The flushing-out phenomenon

Sewage contamination is a major cause for a deteriorated quality of surface waters, in particular in the rapidly growing coastal megacities of the developing and emerging economies (e.g., Peng et al., 2005; Phanuwan et al., 2006). Chemical marker compounds such as fecal steroids are useful to trace the water contamination by untreated municipal sewage (e.g., Furtula et al., 2012; Grimalt et al., 1990). Thereby, the concentrations of the chemical marker coprostanol show a good correlation with the number of fecal bacteria (e.g., Nichols et al., 1993).

This fecal marker approach was applied to water and sediments from the rivers and canals flowing through Jakarta as well as to figure out the spatial distribution of fecal pollution in JB as the coastal ecosystem that receives all urban river discharges (see Fig. 8.2). The concentrations of the fecal steroid coprostanol in river water ranged from 0.45 to 24.2 μg $L^{-1}$, and in sediments from 0.3 to 650 μg $g^{-1}$, reflecting the problem of inadequate sewage treatment capacities in Jakarta (Dsikowitzky et al., 2016a,b).

The spatial distribution of coprostanol in surface sediments from JB at different time periods is summarized in Fig. 8.5. In October 2012 and 2013, coprostanol was detected dominantly at nearshore samples, with a maximum concentration of 56 μg $g^{-1}$ (dry sediment). Interestingly, significant higher coprostanol concentrations up to 600 μg $g^{-1}$ (dry sediment) were found in May 2013. Here, coprostanol was detected even at stations in central JB, approximately 10 km offshore.

The steroid distribution in JB in May 2013 as compared with dry season data (October 2012 and 2013) indicates a flushing out of particle-associated pollutants from the urban rivers far offshore during the preceding rainy season, where the city experienced a severe flood. This flushing out of particle-associated pollutants during times of heavy rainfall as observed in this study is a discontinuous pollutant transport mechanism that is important for all tropical coastal systems. Overall, the pulsed pollutant transport into coastal areas that are normally not prone to urban pollution during flood events can strongly affect sensitive coastal habitats such as coral reefs.

## 8.3.5 Accumulation in biota

Uptake of particle-associated contaminants is one important exposure route for some aquatic organisms. Hence, an accumulation of organic contaminants by economic important fish and macrobenthic invertebrate species from JB might be the result of

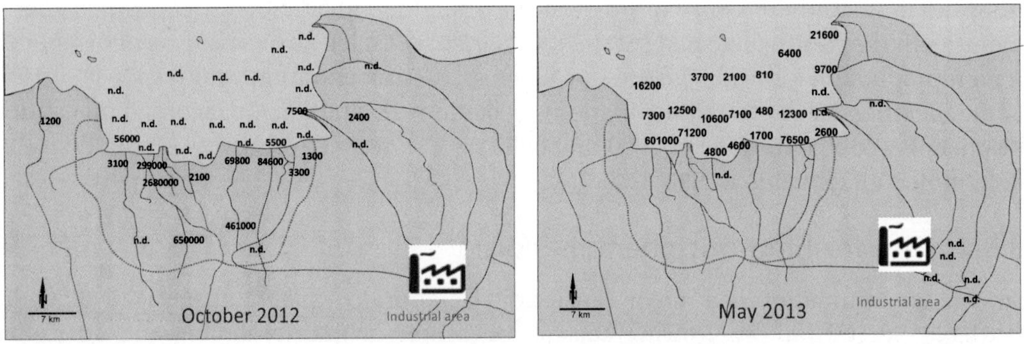


**FIGURE 8.5** Coprostanol distribution at two different times representing wet and dry season conditions. The concentrations of the fecal steroid in the sediments is given as ng g−$^1$ dw.

sediment contamination. Corresponding analyses of biota and sediments revealed those organic contaminants, which are highly relevant in terms of concentrations and detection frequency for the contamination of fisheries resources from JB (Dwiyitno et al., 2016).

High concentrations of DIPNs, linear alkylbenzenes (LABs) and polycyclic aromatic hydrocarbons (PAHs) were detected in all samples, whereas phenylmethoxynaphthalene (PMN), DDT, and DDT metabolites (DDX) were detected at lower concentrations. A comparison of the concentrations of DIPN, LABs, and PAHs in green mussels (*Perna viridis*) and selected fish species sampled in the bay revealed that the concentrations of all considered contaminant groups were significantly higher in the investigated mussel samples than in the fish samples. It was assumed that the higher concentration levels in mussels as compared with fish species can be attributed to higher exposure of the mussels to the contamination of the JB and the analyses of different tissue types. In addition, mussels might have a higher uptake rate of contaminants due to a different feeding mode. Mussels might also have a lower capacity to metabolize nonchlorinated aromatic hydrocarbons than fishes, as previously demonstrated for PAHs.

DIPNs, LABs, and PAHs are not single contaminants, but contaminant groups consisting in the case of DIPNs of eight different isomers and in the case of LABs of homologs with different chain lengths. PAHs are polycyclic aromatic compounds consisting of two or more condensed benzene rings. The different compounds within these contaminant groups exhibit different physicochemical properties. The processes organic contaminants undergo after their release into the environment such as distribution, uptake by organisms, accumulation, degradation, and transformation are strongly influenced by the physicochemical properties of compounds. Therefore, not only the concentration levels but also the patterns of the three contaminant groups DIPNs, LABs, and PAHs in sediments and animal tissue samples were considered.

DIPNs showed a low degree of degradation in the sediments, whereas an isomer-specific uptake or metabolization by the investigated species was evident. LABs in the sediments were more degraded than in animal tissue samples, suggesting the microbial degradation of LABs in the coastal sediments as predominant process. A preferential bioaccumulation of low-molecular weight PAHs as compared with high-molecular weight PAHs was observed. In addition, during the accumulation process, a shift in the proportion of parent PAHs to their methylated derivatives occurs, so that some common source-indicative PAH ratios cannot be applied to animal tissue samples.

In summary, different and compound-discriminating environmental processes are most relevant for organic contaminants on their way to bioaccumulation. Besides PAHs and LABs, also DIPNs are relevant for the contamination of marine fishery resources at JB.

## 8.4 Water quality and biological responses

### 8.4.1 Water pollution in Jakarta Bay and the Thousand Islands

JB has become one of the most polluted marine water bodies in Asia (Bengen et al., 2006). Various marine and coastal environmental impacts including decreased water

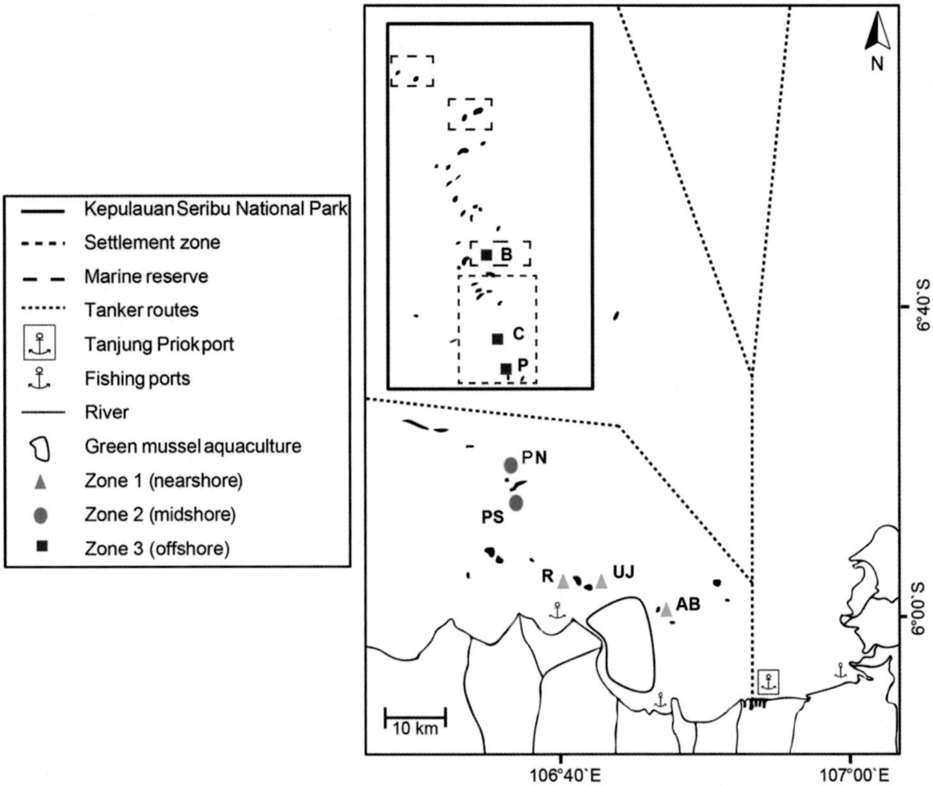


**FIGURE 8.6** Jakarta Bay and the Thousand Islands (Indonesian: Kepulauan Seribu). Map includes study sites (Baum et al., 2015) from nearshore reefs (within Jakarta Bay), as well as from the outer Thousand Islands (mid- and offshore): *AB*, Ayer Besar; *B*, Bira; *C*, Congkak; *P*, Panggang; *PN*, Pari North; *PS*, Pari South; *R*, Rambut; *UJ*, Untung Jawa.

quality, seafood contamination, depletion of fishery resources, land reclamation, coastal littering, land subsidence, loss of habitat as well as eutrophication and increased sedimentation rates are currently affecting the mega city of Jakarta. Directly to the north of JB is the island chain Kepulauan Seribu ("Thousand Islands") (Fig. 8.6). This island chain extends up to 80 km off the coast and is situated within the main impact area of anthropogenic stressors originating from Jakarta. Different ecosystems including coral reefs and mangroves that are crucial for the survival of marine organisms and that form the basis for the livelihoods of local communities can be found along the islands (Arifin, 2004). However, large amounts of untreated sewage and industrial effluents, with high pollutant levels, are transported by several rivers directly into JB (Rees et al., 1999). Many studies within the SPICE program have observed elevated concentrations of pollutants, especially within the bay (Dsikowitzky et al., 2016a,b).

Thousands of people such as fishermen in North Jakarta and along the Thousand Islands depend on the ecosystem goods and services provided by local coral reefs (Baum et al., 2016c). Coastal livelihoods, especially those that rely mainly on marine resources like in the JB/Thousand Islands complex, are vulnerable to long-term changes such as increasing pollution with toxic chemicals (Ferrol-Schulte et al., 2015). Here, we summarize findings on the impacts of declining water quality on reef organisms and communities, focusing on physiological impacts.

### 8.4.2 Biological responses to anthropogenic stressors

Local anthropogenic stressors such as pollution with toxic chemicals, eutrophication, and increased sedimentation are some of the most pressing stressors on coral reefs (Burke et al., 2012; Fabricius, 2005; Van Dam et al., 2011), affecting key functions such as community calcification (Silbiger et al., 2018). Chemicals enter the marine environment most commonly via terrestrial runoff from rivers or through urban runoff carrying large amounts of domestic wastes and industrial effluents. These pollutants can then accumulate in marine organisms such as fish or invertebrates like corals (bioaccumulation), which can lead to various physiological impairments varying from subcellular changes such as direct effects on DNA to metabolic stress (see reviews of Logan, 2007; Van Dam et al., 2011). Within the field of chemical stressors, the intensity and diversity of anthropogenic stressors has increased rapidly over the past decades. Organic contaminants such as hydrocarbons, surfactants, pesticides, and herbicides as well as inorganic pollutants such as sodium cyanide mixtures used in cyanide fishing (Arifin and Hindarti, 2006), metals, and organometallic compounds from industrial waste products are the most common groups (see review of Van Dam et al., 2011).

Scientists use a wide array of different response indicators, from subcellular to metabolic indicators, to determine stress responses in marine organisms (Logan, 2007). Stressed animals need additional energy to recover and maintain homeostasis (Calow and Forbes, 1998). By estimating the metabolic condition or fitness, i.e., the physiological status during changing environmental conditions, the stress level of an organism can be revealed (Fanslow et al., 2001; Lesser, 2013) by, e.g., respirometry (Fig. 8.7).

In general, organisms are able to tolerate stress to a certain extent; however, exposure to multiple stressors can pose additional threats to them and their ecosystems such as reefs. This in turn could lead to a higher sensitivity to other additional stressors (Beyer et al., 2014). So far, effects of multiple stressors have mainly been assumed to be additive (Halpern et al., 2007). However, recent literature indicates that multiple stressors tend to interact with each other (synergism, antagonism; Ban et al., 2014). Such combined effects can happen at the species level as well as on community or population levels (Fig. 8.7).

### 8.4.3 Impacts on the physiology of key coral reef organisms

Organic toxic pollutants are of growing concern to marine ecosystems (Arifin and Falahudin, 2017; Logan, 2007; Van Dam et al., 2011). PAHs are the most widespread class of organic pollutants, and some PAHs are considered to have mutagenic, carcinogenic, and endocrine-disrupting characteristics (Logan, 2007). Common sources for PAHs are

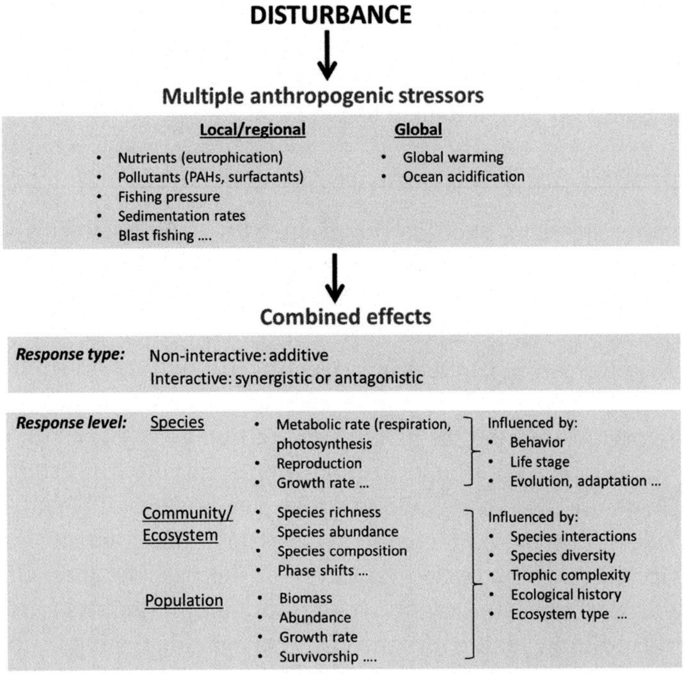


**FIGURE 8.7** General overview of multiple anthropogenic stressors (local and global) and combined effects of these stressors including biological response types and levels.

biomass burning, forest fires, internal combustion engines, and garbage incineration, as well as crude oil and petroleum products (Rinawati et al., 2012). Through the release of bilge and ballast water from boats, from both large tankers and small fishing boats alike, organic contaminants such as PAHs from diesel can enter marine waters as part of the water-accumulated fraction (WAF). This is of concern since many coral reefs are in close proximity to shipping lanes, where contaminated bilge water is disposed off (Halpern et al., 2007). Besides PAHs, another ubiquitous pollutant class is surfactants (LAS), which are contained in detergents and soaps and applied by households and industries in large amounts. Especially in untreated effluents, some surfactants can be present at concentrations that may be toxic to aquatic organisms (Ankley and Burkhard, 1992). The amount of LAS can be used as indicator for environments affected by sewage (Rinawati et al., 2012).

Other research groups within SPICE III found that higher concentrations of PAHs, as well as LABs, can be found within JB (Dsikowitzky et al., 2016a,b; Dwiyitno et al., 2016), and in green mussel and fish samples from JB (Dwiyitno et al., 2016).

Here, results from two different physiological studies (Baum et al., 2016a; Kegler et al., 2015) are presented in which the effects of PAHs and the surfactant LAS (linear alkyl benzene sulfonate) were analyzed on key coral reef organisms. Results from Baum et al. (2016a) show that both the surfactant LAS from sewage runoff and diesel-borne compounds such as PAHs from bilge water discharges are two very common local pollutants, not only within JB, but also along the outer Thousand Islands due to lack of sewage treatment and high boat traffic. Short-term exposure of sublethal concentrations of WAF-D (water-accumulated fraction from diesel) and LAS caused metabolic stress to various degrees in the commercially important herbivore coral reef fish *Siganus guttatus* (Baum et al., 2016a) and the hard coral *Pocillopora verrucosa* (Kegler et al., 2015). Exposure to WAF-D led to lower metabolic rates in *S. guttatus,* while no visible effect on hard coral metabolism of *P. verrucosa* could be found. In contrast, LAS exposure led to a significant increase in standard metabolic rates in *S. guttatus,* indicating an increased energy demand as a result of the higher stress (Sloman et al., 2000). The coral *P. verrucosa* reacted to LAS with a severe tissue loss and a decreased photosynthetic efficiency. Furthermore, the experiments could show that both pollutants interacted with each other. This highlights the need to account for stressor interactions in future management and conservation plans. Under combined exposure to both WAF-D and LAS, metabolic depression was observed in *S. guttatus.* LAS led to a significantly higher PAH concentration in the water, therefore suggesting that the effect of WAF-D (decrease in respiration) may have counteracted and neutralized the effect of LAS (increase in respiration).

Results also showed that a 3–4°C increase in temperature, reflecting predicted global warming effects for the end of this century (IPCC, 2013), caused more severe metabolic stress with regard to LAS and WAF-D toxicity for *S. guttatus* and *P. verrucosa.* However, in *S. guttatus,* a synergistic, i.e., amplified reduced (WAF-D) or increased (LAS) change in metabolic rates was not observed. Nonetheless effects were additive during combined exposure to high temperature and LAS, which further decreased the metabolic condition of *S. guttatus.* In the coral *P. verrucosa,* the combination of WAF-D and high temperature led to an increase in dark respiration, and the combination of LAS and high temperature to severe tissue loss and subsequent high mortality.

So far, effects of diesel or other oil products and of surfactants on both corals and fish seem to be ambiguous (DeLeo et al., 2015; Maki, 1979; Zaccone et al., 1985). The underlying physiochemical processes causing the toxicity of these pollutants are still barely understood. Even though the exact physiological mechanisms could not be discovered using only metabolic rates as indicators, the two studies presented here still indicate the severity of the toxicity.

While studies in the past focused primarily on pollutants such as heavy metals (e.g., Guzmán and Jiménez, 1992), single PAHs (e.g., Oliviera et al., 2008), and pesticides (e.g., Richmond, 1993), lesser studied pollutants such as surfactants and WAF-D should also be in the focus of future studies. Especially considering the frequency and amount of bilge water discharge and untreated sewage runoff in the JB/Thousand Islands complex and all over Indonesia, these two pollutants may be a regional rather than a local problem for marine organisms. Moreover, the two studies also highlight that to avoid long-term effects on fish and coral health, the import of these pollutants into coastal areas has to be reduced.

### 8.4.4 Impacts on reef composition

As a result of mounting anthropogenic stressors, the reefs in the JB/Thousand Islands reef complex have been significantly degraded, particularly within the bay. Here, historically rich coral communities declined to around 10% coral cover in the 1980s and to less than 5% by 2011 (Cleary et al., 2014). By differentially affecting the physiology of key benthic organisms, stressors can result in shifts in the overall composition of benthic communities. In JB and the Thousand Islands, the composition of benthic communities is strongly structured by environmental factors, resulting in a marked inshore/offshore gradient (Cleary et al., 2016). By assessing the physiological response of soft corals in the JB/Thousand Islands reef complex to levels of key pollutants and assessing benthic community composition and levels of water pollution, Baum et al. (2016b) concluded that water quality may control abundance and physiology of dominant soft corals (*Sarcophyton* sp. and *Nephthea* sp.) in JB. Water quality, mostly inorganic nutrient concentrations and sedimentation rates, affected photosynthetic yield and electron transport system activity of the two soft coral species, indicating that metabolic condition in both species is affected by reduced water quality. The abundances of both species were moreover directly linked to declining water quality. This in turn was hypothesized to have facilitated phase shifts from hard to soft coral dominance. The findings highlight the need to improve management of water quality to prevent or reverse phase shifts.

### 8.4.5 Local versus regional stressors in Jakarta Bay and the Thousand Islands

Both local and regional factors interact and have caused severe reef degradation in the JB/Thousand Islands reef complex, including shifts to soft coral dominance in the bay. A little over a decade ago, Cleary et al. (2016) concluded that large-scale environmental gradients played the strongest role in structuring benthic communities, noting no effect of local factors related to land use on the Thousand Islands on coral cover or diversity. In November 2012, during the transition time between northwest and southeast monsoon, a large coral reef survey was performed as part of the SPICE III project, and eight sites across the Thousand Islands chain were visited (see Fig. 8.6). Results from that study (Baum et al., 2015) confirm that the bay is facing extreme eutrophication coupled with increased primary production and turbidity. $PO_4$ levels in the upper layer of JB reached 4 $\mu M\ L^{-1}$ and DIN (dissolved inorganic nutrient) levels up to 13 $\mu M\ L^{-1}$. Similarly, Ladwig et al. (2016) showed that inorganic and organic nutrient concentrations in the nearshore area of JB are highly increased. The river discharges from the urban area of Jakarta were identified as the largest contributors to nutrient concentrations within the nearshore area of JB (Van der Wulp et al., 2016a). During the survey in 2012, at all sites in JB, Chl a levels were between 5 and 15 $\mu g\ L^{-1}$, far above the Eutrophication Threshold Concentration for Chl a of 0.2–0.3 $\mu g\ L^{-1}$ (Bell et al., 2007), indicating high primary productivity. Phytoplankton bloom formations are fostered within JB, and an oxygen

**FIGURE 8.8** A highly degraded reef in Jakarta Bay at the island Rambut (A) and a still relatively intact reef further offshore at Pari Island (B).

deficiency area of 20 $km^2$ was found in the eastern part of JB (Ladwig et al., 2016). In addition, sites within JB had significantly higher sedimentation rates compared with offshore sites in the Thousand Islands, with up to 30 g $m^{-2}$ $d^{-1}$ (Baum et al., 2015).

This decline in water quality, especially in JB, went hand in hand with severe changes in reef communities (Fig. 8.8). The reef condition along the Thousand Islands has dramatically declined since the first scientists conducted investigations in the area in the beginning of the 20th century and described reef systems with high species diversity (Umgrove, 1939). In 2012, hard coral cover was around 2% at sites within JB. But also along the outer Thousand Islands, the overall reef condition is poor, since total coral cover at most sites was <25%. Furthermore, shifts to soft coral dominance were found in the bay. Even though shifts to soft coral dominance are far less common than those to macroalgae dominance, such shifts have been reported for other degraded reefs in the Indo Pacific (Chou and Yamazato, 1990; Fox et al., 2003). Severe changes were also found for fish communities in the area, with currently 80% lower fish abundance in the bay compared with sites from the outer Thousand Islands (Baum et al., 2015). A subsequent study assessing traits of fish species along the JB/Thousand Islands reef complex concluded that eutrophication and pollution have resulted in depauperated fish communities on inshore reefs, selecting for fast-growing and short-lived species (Cleary, 2017).

The results from Baum et al. (2015) furthermore showed a clear difference in benthic and fish communities between sites in JB and the outer Thousand Islands (Fig. 8.9). A direct impact on shallow coral reefs may be restricted to within the bay itself. In contrast to findings from earlier studies, localized effects of anthropogenic stressors rather than regional gradients appear to have gained in importance and now shape the spatial structure of reefs in the outer Thousand Islands. Furthermore, results showed that over 80% of variation in benthic community composition was linked to factors related to terrestrial runoff and eutrophication, especially $NO_3$, sedimentation, turbidity, $PO_4$, and Chl a. Local anthropogenic stressors can become the dominant factors shaping benthic

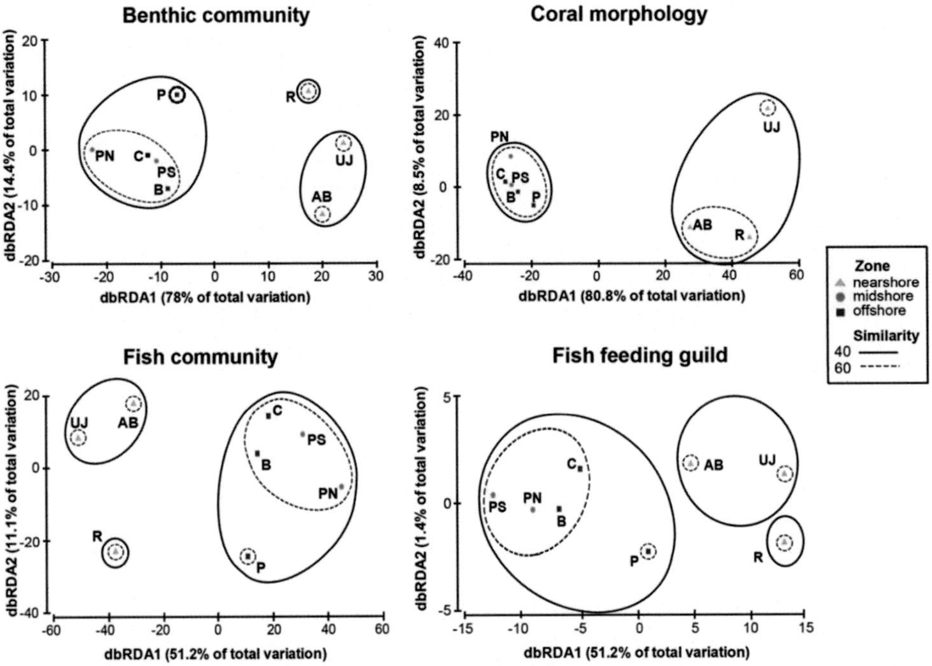


FIGURE 8.9 Visualization of fish and benthic community composition based on distance-based redundancy analysis (dbRDA). Benthic community composition (A), coral morphology composition (B), fish community taxonomic composition (C), and fish feeding guild composition (D) are shown. Study sites: *AB*, Ayer Besar; *B*, Bira; *C*, Congkak; *P*, Panggang; *PN*, Pari North; *PS*, Pari South; *R*, Rambut; *UJ*, Untung Jawa (Baum et al., 2015).

reef communities (Williams et al., 2015), and this trend appears to be reflected in the Thousand Islands. Overall, the spatial structure of reefs in the JB/Thousand Islands reef complex is directly related to both local and regional anthropogenic sources.

In summary, the degradation of coral reefs in JB and the Thousand Islands is caused by a combination of multiple anthropogenic stressors acting in concert. Especially within JB, pollution and sedimentation have become disastrous, leading to extremely degraded coral reefs. Further offshore along the Thousand Islands, reefs are in a slightly improved condition; however, the increasing threat due to overfishing, global warming, eutrophication due to highly densely populated islands and a lack of sewage treatment, and chemicals released from increasing shipping traffic and urban runoff from islands will pose a severe challenge for those reefs in the future.

## 8.5 Microbial diversity in Indonesian fish and shrimp: a comparative study on different ecological conditions

Indonesia, as an archipelago country has numerous and different coastal ecosystems, which are of great importance concerning natural resources for people's livelihoods. Therefore, sea products play a crucial role as a source of income and foreign exchange.

This is a major reason why the Indonesian government strives to sustain marine and coastal affairs. Consequently, this chapter focuses on metagenomes and microbiomes of two food industry-relevant marine organisms, i.e., *Epinephelus fuscoguttatus*, a marine grouper and *Penaeus monodon*, well known as giant black tiger shrimp, respectively. There were two broad objectives using functional metagenomics: (1) elucidating the genes of different community members that affect host–microbe and microbe–microbe interactions and (2) identifying relevant functions regarding pathogenicity in different microbiome communities (Mandal et al., 2015). Here, detailed information on cultivatable and noncultivatable prokaryotic and eukaryotic organisms, derived from food and environment, i.e., pathogenic, nonpathogenic microbes, parasites, and others will be discussed. Furthermore, this chapter provides insights into the microbiome, the function of complex microbial communities, and their role in host health (Srivasta, 2007). The elucidation of the bacterial community composition further leads to the recognition of bacterial pathogens that might be harmful for hosts and consumers. Moreover, knowing the composition of the gut microbiome, the presence of parasites and feeding habits under different environmental conditions lays the groundwork for future application of, for example, probiotics to improve fish and shrimp immunity to recurrent bacterial infections that are critical for aquaculture in Indonesia.

As reported in various studies, there are many different factors that affect the gut microbiome (Asplund, 2013; Chaiyapechara et al., 2012; Sullam et al., 2012; Xia et al., 2014; Zhang et al., 2014). Apart from the factors that can alter the composition of bacterial communities, the core microbiome remains stable. A core is typically defined as the suite of members shared among microbial consortia from similar habitats. Discovering a core microbiome is important for understanding the stability, i.e., consistent components across complex microbial assemblages (Shade and Handelsman, 2012). The host phylogeny shapes core microbiome predominantly (Hennersdorf et al., 2016a,b). *Proteobacteria* are the dominant phylum, whereas *Vibrionales* are the dominant order (Chaiyapechara et al., 2012; Liu et al., 2011; Rungrassamee et al., 2014). Diet also plays a significant role in shaping the gut microbiome under starvation or under rapid variations in food sources and have been shown to have altered gut bacterial communities (Xia et al., 2014). Ecological and environmental conditions have also been reported to influence the stability of gut microbiome composition (Ronnback, 2002; Sullam et al., 2012). In fish and shrimp aquaculture, the gut microbiome plays an important role in host immunity and as a defense mechanism against pathogenic infections (Balcázar et al., 2006; Balcázar et al., 2007; Guarner and Malagelada, 2003). Invertebrates, such as shrimp, and fish have only innate immunity and, therefore, require the support of commensal bacteria that are part of the gut microbiome to defend against pathogenic infection. In developing countries such as Indonesia, where a large number of traditional aquaculture facilities do not have proper aeration, food control, and effective disease management, improving host metabolism and immunity through understanding and controlling the gut microbiome is necessary to ensure success in the aquacultural industry.

### 8.5.1 Microbial diversity in *Epinephelus fuscoguttatus*

*E. fuscoguttatus* was chosen as a model fish, because it is widespread over the Indo-Pacific Ocean. It is of great economic value and a common protein resource in Indonesia. However, this species has been added to the IUCN Red List of Threatened Species due to overfishing and increasing pollution leading to destruction of seagrass and coral reefs crucial for juvenile growth. Free-living and mariculture samples were collected in two different regions, respectively. Free-living samples originated from the Thousand Islands (Pulau Seribu) Marine National Park, whereas the mariculture samples were collected from the open water mariculture facility Nusa Karamba Aquaculture. Using both 16S amplicon sequencing and whole metagenome analysis, it was observed that free-living and mariculture showed similar overall gut microbiomes. They consisted predominantly of *Proteobacteria, Firmicutes, Spirochaetes,* and *Actinobacteria* at phyla level. At order level (Fig. 8.10A), free-living samples were dominated by *Vibrionales* and *Bacillales,* whereas the mariculture samples were dominated by *Pseudomonadales* and *Enterobacteriales.* The mariculture samples were found to have a low intragroup relative abundance variability compared with free-living samples, which strongly varied among samples.

The microbial diversity has been explored using three statistical methods, i.e., observed taxonomic richness (Gotelli and Colwell, 2011), Chao1 method (Chao, 1984), and Shannon–Wiener diversity index (Shannon, 1948) (Fig. 8.10B–D). Overall, the free-living samples revealed higher observed taxonomic richness in comparison with the aquaculture samples. Similarly, the unobserved taxonomic richness using the Chao1 method (nonparametric richness estimator, providing a statistical estimation of true species richness, including unobserved species within a community) showed similar results. The Chao1 method showed a greater taxonomic richness in free-living samples when compared with mariculture samples. Shannon–Wiener diversity index demonstrated that free-living samples have a greater index, i.e., a greater microbial biodiversity than mariculture samples. Finally, the Bray–Curtis distances method, a statistical analysis, which is used to quantify the compositional dissimilarity between two different sites, based on counts at each site was used to measure the differences between bacterial compositions under the two environmental conditions (Beals, 1984). As shown in Fig. 8.10E, the mariculture samples constructed one cluster revealing their low intragroup relative abundance, whereas the free-living samples spread around this cluster pointing toward the large variability between samples.

Metagenomic analysis allowed to assess the eukaryotic composition (Hennersdorf et al., 2016a,b). The eukaryotic composition showed most reads belonging to the host species. Reads corresponding to nine fish taxa endemic to the Java island region were found. In addition, the reads corresponding to parasitic phyla such as *Platyhelminthes, Arthropoda,* and *Acanthocephala* were found. The results showed that the eukaryotic content of both environments was not significantly different. Moreover, metagenome analyses revealed functional components of the observed microbial composition. This was done using the gene ontology database (Gene and Consortium, 2000), which designates a series of biological events performed by a number of organized assemblies of

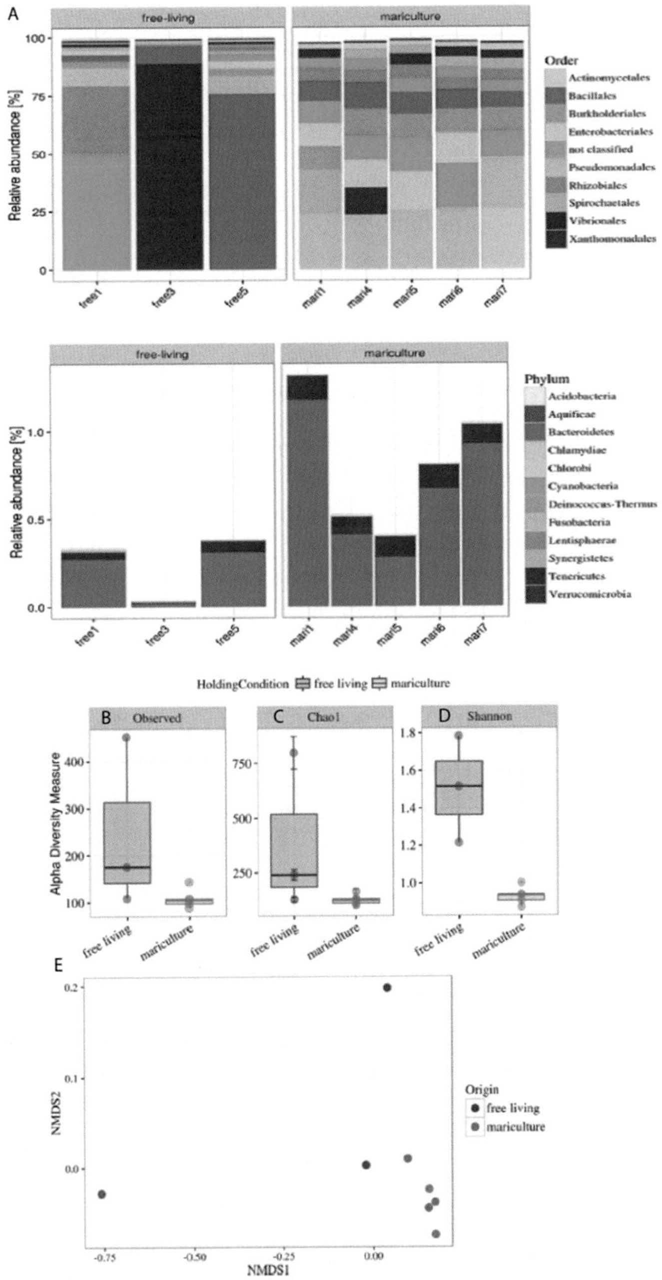


**FIGURE 8.10** (A): Relative abundance of bacterial communities in order level from free-living and mariculture *Epinephelus fuscoguttatus* samples (Hennersdorf et al., 2016a,b). Alpha diversity observations from different conditions (red: free-living samples, blue: mariculture samples; (B): observed OTUs richness; (C): Chao1 richness estimator; (D): Shannon indices; boxes represent the quartile, bars represent the interquartiles, dots represent single sample, and solid lines represent median from all samples; E: NMDS (nonmetric multidimensional scaling) measurement aims to ordinate the between-sample dissimilarities. *OTU*, operational taxonomical unit.

molecular functions. This was evident as most of the functional reads were linked to the phylum *Proteobacteria* (Hennersdorf et al., 2016a,b). The results revealed an increased enrichment of genes corresponding with DNA damage and other DNA repair functions specifically in the free-living samples. These functions were obtained from genes assigned to *Vibronales, Pseudomondales,* and *Enterobacteriales,* while samples from mariculture were enriched for metabolic processing mostly assigned to *Proteobacteria.* The link observed between potentially pathogenic bacteria and the enrichment of functions relating to DNA repair point toward the role of the environment on microbial composition as opposed to that of the controlled mariculture facility where it was absent (Hennersdorf et al., 2016a,b).

### 8.5.2 Microbial diversity in feces of *Penaeus monodon*

The microbial diversity in feces of *P. monodon* was investigated in Oetama et al. (2016). *P. monodon* is a high-value shrimp species that is widespread in the Indo-Pacific. It is a worldwide consumed seafood and is exported mainly to Japan, Europe, and the United States (FAO, 2012). Free-living shrimp samples were collected from two different locations: JB and Bali. Concurrently, the aquaculture shrimp samples were collected from traditional aquaculture in Pejarakan, Bali. Abiotic factors that could influence microbial diversity and shrimp health such as salinity, temperature, pH, nitrate, and phosphate concentration were measured (Sullam et al., 2012). The Illumina Miseq sequencing platform was occupied. On average, there were 304,800 obtained mapped reads and assigned to 935 operational taxonomical units (OTUs), which is an operational definition used to classify groups of closely related individuals (Oetama et al., 2016). The most abundant phyla are *Proteobacteria* (96.08%), *Bacteriodetes* (2.32%), *Fusobacteria* (0.96%), and *Firmicutes* (0.53%). On the order level of taxonomy, *Vibrionales* (66.20%) was the most detected order among the samples and was followed by the order of *Alteromonadales* (24.81%) (Fig. 8.11A). The complete sequence information can be found in NCBI's Short Read Archive under accession number SRP059721.

Observed OTUs in taxonomical richness showed a high median number of taxa in both Bali free-living samples (295 taxa) and aquaculture samples (269 taxa), whereas Jakarta free-living samples included only 122 taxa as a median (Fig. 8.11B). The Chao1 method showed similar trends under three environmental and regional conditions: Bali free-living samples included 329.29 taxa, aquaculture samples 303.23, and Jakarta free-living samples 140.47 samples (Fig. 8.11C). The Shannon–Wiener diversity index revealed similar results as obtained by the other two methods. Bali free-living (2.57) and aquaculture samples (2.58) showed a higher index in comparison with Jakarta free-living samples (0.93) (Fig. 8.11D). However, the individual samples of Jakarta free-living shrimp resulted in a higher variation of diversity. Conclusively, Bray–Curtis distance was applied to access the cluster based on dissimilarities. Three out of five samples from Jakarta free-living shrimp are in one cluster, independently from another cluster of Bali free-living and aquaculture samples (Fig. 8.11E). Finally, 4 out of 17 samples from Bali free-living and aquacultured shrimps were observed as outliers, i.e., not belonging to any cluster (Oetama et al., 2016).

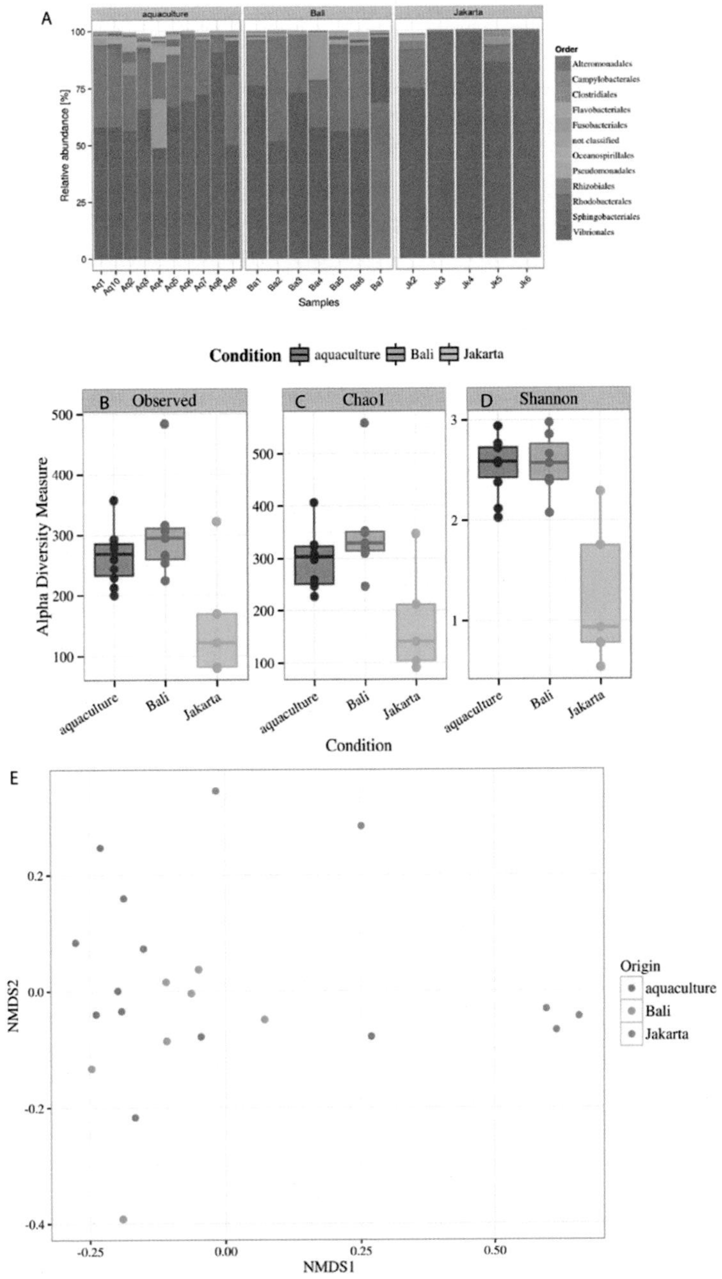


**FIGURE 8.11** (A): Relative abundances of bacterial communities in order level of free-living and aquacultured *Penaeus monodon* samples (Oetama et al., 2016). (Aq: shrimp samples collected from aquaculture; Ba: free-living shrimp samples derived from Bali; Jk: free-living shrimp samples derived from Jakarta bay). Alpha diversity observations under different conditions (red: samples derived from aquaculture, blue: free-living samples from Bali bay, green: free-living samples from Jakarta Bay). (B): observed OTUs richness; (C): Chao1 richness estimator; (D): Shannon indices; E: NMDS (nonmetric multidimensional scaling) measurement. *OTU*, operational taxonomical unit.

On the species level, nonpathogenic *Vibrio alginolyticus* bacteria and the pathogens *Vibrio vulnificus* and *Photobacterium damselae* were detected. *V. alginolyticus* is commonly used in probiotics for aquaculture (Gomez-Gil et al., 2002; Krupesha Sharma et al., 2010). The coculture of *V. alginolyticus* C7b and phytoplankton (microalgae) produces polyunsaturated fatty acids and vitamins, which are crucial for shrimp health (Krupesha Sharma et al., 2010). It has been described that healthy and robust gut microbiomes avail the host immunity against bacterial and viral pathogens (Gomez-Gil et al., 2002). Thus, nowadays, farmers apply probiotics to improve production rates. Regarding a healthy gut microbiome, the infections by opportunist pathogens could be inhibited.

The attempt to determine the presence of common pathogenic *Vibrio* species, such as *Vibrio cholerae* and *Vibrio parahaemolyticus*, was made with quantitative PCR, although they were absent in all samples. *V. cholerae* secrete cholera toxin that causes diarrhea, which can lead to death due to dehydration.

As mentioned earlier, *V. vulnificus* and *P. damselae* were detected almost in all samples. The reason might be due to frequent alterations in environmental parameters such as pH conditions, temperature, salinity, and so on resulting in stress and thus increasing probability of infection (Jones and Oliver, 2009; Lee and Rangdale, 2008; Venkateswara-Rao, 1998). Therefore, the findings regarding *V. vulnificus* and *P. damselae* are essential, not only for aquaculture farmers, but also for seafood consumers (Oetama et al., 2016). *V. vulnificus* causes systemic infection starting with fever, chills, and nausea, especially patients suffering from chronic liver disease or diabetes, and is dangerous for immuno-compromised patients. Infection occurs not only via consumption of corresponding marine organisms, but also via direct contact with open wounds. For example, there have been many known cases in the United States, especially during periods of strongly increased water temperature (Jones and Oliver, 2009).

*P. damselae* is found in a broad variety of marine organisms. Its virulence factors have not yet been elucidated. It has been hypothesized that phospholipase-D Dly (damselysin) and the pore-forming toxins HlyApl and HylAch are relevant virulence factors in *P. damselae* (Hundenborn et al., 2013; Vaseeharan et al., 2007). In our studies, more than 50% of the aquaculture and Jakarta free-living samples were infected by *P. damselae*, and all free-living samples from Bali were infected as well. However, larger sample sizes are needed to verify these findings.

### 8.5.3 Comparison of microbial diversity between fish and shrimp from Jakarta and Bali

Most predominant phyla observed in the gut microbiome of both *E. fuscoguttatus* and *P. monodon* from all sampling sites were *Proteobacteria*. In addition, three other phyla with more than 1% abundance were determined. *Firmicutes, Spirochaetes,* and *Actinobacteria* were found in *E. fuscoguttatus*, whereas *Bacteriodetes, Fusobacteria,* and *Firmicutes* were present in *P. monodon*. As reported in other studies, *Proteobacteria* and *Firmicutes* are the two phyla, which are found in all marine organisms (Liu et al., 2011; Rungrassamee et al., 2014; Sanchez et al., 2012). Therefore, these two phyla belong to the core microbiome. Surprisingly, we could not detect any common pathogens in marine organisms, such as *V. cholerae* and *V. parahaemolyticus*. However, we detected *V. vulnificus* and *P. damselae* in all samples of cultured shrimp. These pathogens have virulence factors that could initiate severe diseases in host and consumers. The presence of well-known probiotics, e.g., *V. alginolyticus* was also determined. Moreover, the composition of the gut microbiome in free-living and aquaculture samples is very similar, but with a marked decrease in diversity in the aquaculture samples. With respect to the environmental factors that influence the gut microbiome, traditional aquacultures share similar conditions with natural habitats. Conjointly, traditional aquaculture does not involve food additives, and therefore, diet of aquacultured shrimp and fish is composed of microalgae and phytoplankton present in the ponds. Thus, diet and ecological conditions do not play a big role in shaping the bacterial populations in all samples of our study.

In summary, the diversity in free-living fish and shrimp is greater than in the corresponding aquacultures. Location is a crucial factor, which leads to differences in the origin of evolutionary ancestors and genetic diversity among *P. monodon*, although the impact of environmental conditions on alterations of the microbiome cannot be dismissed. The substantial distinction of microbial diversity under different conditions could be explained, because of less environmental changes in aquaculture than in natural habitats; hence, it shapes more stable host microbiomes (Ronnback, 2001, 2002; Sullam et al., 2012).

## 8.6 Fish parasites in Indonesian waters: new species findings, biodiversity patterns, and modern applications

As already pointed out in Chapter 6.5, the fisheries industry with its valuable food products is a driver for the future economic development of the maritime nation Indonesia. Consequently, food safety and security are consumer's main interest. Marine fishes can be a source for foodborne, parasitic human diseases (zoonoses), primarily when larval helminths are ingested through the consumption of semicooked or uncooked fisheries products (e.g., Petersen et al., 1993). Anisakid nematodes of the genus Anisakis have been reported to cause the Anisakiasis, an inflammation of the human gastrointestinal tract (Ishikura and Kikuchi, 1990; Klimpel and Palm, 2011). Other parasite taxa such as the trypanorhynch cestodes are commonly found inside the fish musculature (Palm, 2004, p. 724) and can offend consumers (Palm and Overstreet, 2000) or cause allergic reactions (Ivanovic et al., 2015).

The Indonesian marine habitats host one of the highest biodiversity's on earth (Palm, 2011; Roberts et al., 2002; Yuniar et al., 2007). This includes the fish parasite fauna, though only few local studies have been recognized outside the country (e.g., Yamaguti, 1952, 1953, 1954). After the first record of the potential zoonotic Anisakis in 1954, several

Indonesian researchers studied the Anisakidae based on morphology (e.g., Asmanelli and Muchari, 1993; Burhanuddin and Djamali, 1978; Humoto and Burhanuddin, 1978; Ilahude et al., 1978; Koesharyani et al., 2001; Martosewojo, 1980). Comprehensive taxonomic treatments (e.g., Palm and Bray, 2014) as well as fish ecological studies on fish parasites from Indonesia occurred only recently mainly as a result of SPICE I–III (Kleinertz and Palm, 2015).

The biodiversity of fish parasites and their complex life cycles allow them to be used as indicators for a wide range of biological and environmental applications (Palm, 2011; Palm and Bray, 2014). Because of the direct linkage and dependence of parasites with multiple-host life cycles to the surrounding animal communities (Hechinger et al., 2007), these organisms have been considered as sensitive bioindicators for aquatic ecosystem health (Dzikowski et al., 2003; Overstreet, 1997). They are useful to indicate food web relationships especially in unaffected marine habitats (e.g., Klimpel et al., 2006; Lafferty et al., 2008a; Palm, 1999), where the full range of their required hosts is present (Lafferty et al., 2008b). Consequently, they can serve as indicators for environmental change and pollution (Diamant et al., 1999; Dzikowski et al., 2003; Kleinertz and Palm, 2015; Palm, 2011) or environmental stress (Landsberg et al., 1998). While the occurrence of endoparasites often decreases in polluted waters (Nematoda, Kiceniuk and Khan, 1983), ectoparasites can increase (Monogenea, Khan and Kiceniuk, 1988; Trichodina, Khan, 1990; Ogut and Palm, 2005). Some ectocommensals with direct life cycles such as trichodinid ciliates favor polluted waters and can additionally indicate high bacterial load (Palm and Dobberstein, 1999).

Many methods have been used to assess real and potential hazards of contaminants by using target organisms (Kang et al., 2014). They involve evaluation of ecological health by using key bioindicators at the population and community levels (de la Torre et al., 2007). Other approaches aim at analyzing biomarkers on their molecular, physiological, and cellular levels, which are sensitive to ecosystem degradation (Chèvre et al., 2003) and contamination. As an example, according to Sures (2008) and Sures and Reimann (2003), acanthocephalans can be used as accumulation indicators for heavy metals, because they accumulate 1000 times higher amounts of heavy metals compared with their host tissues.

This chapter summarizes the German–Indonesian joint research effort on fish parasites from Indonesian coastal waters during the third phase of SPICE, also summarizing the earlier phases (e.g., Kleinertz, 2010; Rückert, 2006; Theisen, 2009; Yuniar, 2005).

## 8.6.1 Species descriptions and taxonomic treatments

Comprehensive collections of fish parasites during the 10 years of SPICE revealed the discovery of new species and increased our taxonomic knowledge. The most comprehensive treatment of the tropical fish parasitic Trypanorhyncha Diesing, 1863 by Palm (2004), already utilized material sampled under this program and added several new species from Indonesian waters, totaling about 20% of the then known biodiversity in that group. Kuchta et al. (2009) studied the four then known bothriocephalidean cestodes from Indonesia and provided redescriptions of species that were earlier reported from the Indo-Pacific region.

Bucephalids are the most common fish trematode family in commercially important groupers. Bray and Palm (2009) provided a taxonomic treatment including two new species, *Rhipidocotyle danai* and *Rhipidocotyle jayai.* Dewi and Palm (2013) described two new species of philometrid nematodes, *Spirophilometra endangae,* and *Philometra epinepheli,* from the orange spotted grouper *Epinephelus coioides,* and Dewi and Palm (2017) added *Philometra damriyasai* from the tetraodontiform *Tylerius spinosissimus.* The papers described the first opercula-infecting species of *P. damriyasai* from the Epinephelinae, with a total of 14 philometrids that so far have been identified from marine fishes in Indonesia. Theisen et al. (2017) for the first time identified a new endoparasitic monogenean from Indonesian marine fishes (*Nibea soldado, Otolithes ruber*; both Sciaenidae). Monogeneans are mainly ectoparasitic, and the infection of inner organs is scarce. In all these studies, identification keys and a summary of the current state of knowledge on these parasite taxa from Indonesia were provided.

### 8.6.2 Zoonotic *Anisakis* spp. in Indonesian waters

A high prevalence of infestation (97%–100%) of the economically important *Trichiurus lepturus* as well as other oceanic and pelagic fish with the potentially zoonotic *Anisakis* spp. at the southern coast of Java demonstrated a high risk of Indonesian predatory fish getting infested (Jakob and Palm, 2006). Because that study did not use molecular identification, the real species identity and potential risk for the human consumer remained unclear. Palm et al. (2008) genetically identified *Anisakis* spp. from Bali and recorded the distribution of Anisakis larvae in Indonesia, based on the available literature and a sample from five fish species from Kedonganan, Bali, and Pelabuhan Ratu, South Java. The larvae mainly belonged to *Anisakis typica.* Because the musculature infection in *Auxis rochei rochei* was low (2.5%), no major risk for the fish consumers was concluded.

Palm et al. (2017) genetically identified 118 *Anisakis* spp. and established 16 new host records. To date, 53 Indonesian teleosts harbor *Anisakis* spp., 32 of them with known sequence data. The analyses identified three specimens of *Anisakis* sp. HC-2005 and 39 (16%) *A.* typica (s.s.). *Anisakis berlandi* and *Anisakis pegreffii* were reported for the first time from teleosts in the equatorial region and *Anisakis physeteris* from the Pacific Ocean. 193 worms (~79%) belonged to the already-detected genotype by Palm et al. (2008) and were nominated as a new *Anisakis aff. typica* var. *indonesiensis* until the description of the adults. The musculature infection was very low, resulting in minor risk of Anisakiasis in Indonesia.

### 8.6.3 Parasite biodiversity in wildlife and maricultured fish

These studies during SPICE focused on sampling of commercially important fish from Segara Anakan lagoon, the Java Sea, and Balinese waters. Jakob and Palm (2006) examined five oceanic fish species from the southern Java coast. An overlapping infestation pattern in fish from entirely different families underlined a low specificity of many helminths in their second intermediate hosts and their ability to infest fishes without respect to their host phylogeny.

Yuniar et al. (2007) carried out a first thorough investigation on ectoparasites of commercial important fish from Segara Anakan. Eight economically important marine fish species were examined for crustaceans. A diverse copepod fauna consisting of 23 different species and two isopods was found. Rückert et al. (2009a) reported the metazoan fish parasites also from Segara Anakan lagoon. Again, a highly diverse parasite fauna was found, consisting of 43 species/taxa. Kleinertz et al. (2014) examined *Epinephelus areolatus* off the anthropogenic influenced Segara Anakan lagoon and a relatively undisturbed reference site in Balinese waters. Kleinertz and Palm (2015) studied *E. coioides* from Segara Anakan and Bali. Regional differences for *E. coioides* were found in terms of different used parameters. Neubert et al. (2016a) provided the first comprehensive information on the parasites of the white-streaked grouper *Epinephelus ongus* from Karimunjawa, Java Sea. For comparison, the parasite community of *E. areolatus, E.coioides,* and *E. fuscoguttatus* from previous studies was analyzed. The ectoparasite fauna was predominated by the monogenean *Pseudorhabdosynochus quadratus*. The endoparasite fauna was predominated by generalists, which were already known from Indonesia, demonstrating the potential risk of parasite transmission through *E. ongus* into grouper mariculture and *vice versa*. Rückert et al. (2008) stated that fish parasites have been repeatedly reported to be a major threat to the developing industry of finfish mariculture. They sampled the metazoan parasite fauna and trichodinid ciliates from *Lates calcarifer* in a representative mariculture farm in Lampung Bay, South Sumatra.

Rückert et al. (2009b) studied differently fed groupers *E. coioides* from an Indonesian finfish mariculture farm. Pellet-fed *E. coioides* were infested with 13 parasite species/taxa. The use of pellet food significantly reduced the transfer of heteroxenous endohelminths. Trash fish was held responsible for the transmission of these parasites, though the endohelminth infestation of pellet fed fish demonstrated that parasite transfer also occurred via organisms that naturally live in, on, and in the surroundings of the net cages. Rückert et al. (2010) examined *E. fuscoguttatus* during three consecutive seasons from floating net cages of the National Sea Farming Development Centre (Balai Budidaya Laut) and from wild catches in Lampung Bay, South Sumatra. The parasite findings contrasted wild grouper, where heteroxenous parasites occurred at a similar prevalence compared with the fairly abundant *Pseudorhabdosynochus* spp. No seasonality of infestation was observed for both cultured and wild fish. Palm et al. (2015) summarized the results from *E. fuscoguttatus* from four mariculture facilities in Lampung Bay (South Sumatra) and one in Pulau Seribu (North of Jakarta). Their results demonstrated that one of the major future tasks in Indonesian mariculture is the search for alternative feed sources and feeding strategies to prevent parasite spread and pathogenic outbreaks.

### 8.6.4 Fish parasites as biological indicators and new applications

The use of fish parasites as biological indicators required adaptation of a stargraph to visualize different parasite metrics on a single figure and to provide a "holistic view in sustainable development" (Bell and Morse, 2003). Palm and Rückert (2009) used three different indicators to visualize ecosystem health by using marine fish parasites. Palm et al. (2011) added further characters and applied this system to indicate long-term changes in a grouper mariculture facility. Kleinertz et al. (2014) used reef-associated grouper *E. areolatus* to demonstrate regional differences. The authors included further parameters to the system such as the hepatosomatic index to indicate high enzymatic activity of stressed fish (Munkittrick et al., 1994). Finally, Kleinertz and Palm (2015) adjusted this system to *E. coioides*. For the first time, it was possible to visualize regional differences between Indonesian coastal waters (Kleinertz et al., 2014; Palm and Rückert, 2009) and long-term annual change (Palm et al., 2011) by using fish parasites.

During the final SPICE Phase, Neubert et al. (2016b) were able to assess the environmental conditions of a heavily polluted Indonesian marine habitat (JB as well as off JB, star graph, Fig. 8.12), and compared it with already existing data. The data were normalized and transferred into a colored traffic light, assessing the environmental conditions in the range from poor (= red), medium (= yellow), and good (= green) (pollution light, Fig. 8.13).

Kleinertz et al. (2016) for the first time applied nontargeted Py-FIMS analyses on fish parasites, using the acanthocephalan *Rhadinorhynchus zhukovi* from *A. rochei* and *Auxis thazard* as a potential accumulation indicator for organic pollutants, such as fuel residues like diesel as indicated by the Py-FI mass spectra within this study.

In summary, from an estimate of about 10.500–14.000 marine fish parasite species occurring in Indonesia, so far, not more than about 500 species have been scientifically reported and/or correctly identified, demonstrating the little knowledge in this field. This is astonishing considering the increasing importance of aqua- and mariculture in Indonesia, necessitating a thorough knowledge on potential harmful organisms that threaten the cultivated fish. At our main sampling sites in Indonesia, the parasite biodiversity was expectedly high. One surprising result was the different species richness according to the fish ecology. The comparison of wild and maricultured fish demonstrated a high risk to gain and accumulate fish parasites also from the wild into the cultivated fish.

## 8.7 Seafood consumption and potential risk

Nowadays, aquaculture contributes predominantly to seafood production, compared with capture/wild fishery. In JB, approximately 35,000 tons year$^{-1}$ are produced from capture fishery (mollusk species, pelagic fish, demersal species, and crustaceans) and 2500 tons from marine/coastal culture (green mussel, milkfish, grouper and shrimp).

Seafood is the main source of protein (54%) for the majority of Indonesian people. Currently, national fish consumption is between 20 and 40 kg person$^{-1}$ year$^{-1}$, with an average 36 kg person$^{-1}$ year$^{-1}$ (MMAF, 2015). The consumption level in megapolitan Jabodetabek (Jakarta, Bogor, Depok, Tangerang and Bekasi) is around 25–30 kg person$^{-1}$ year$^{-1}$. Governmental campaigns such as the "*Gemar Makan Ikan*" ("Eating Fish")

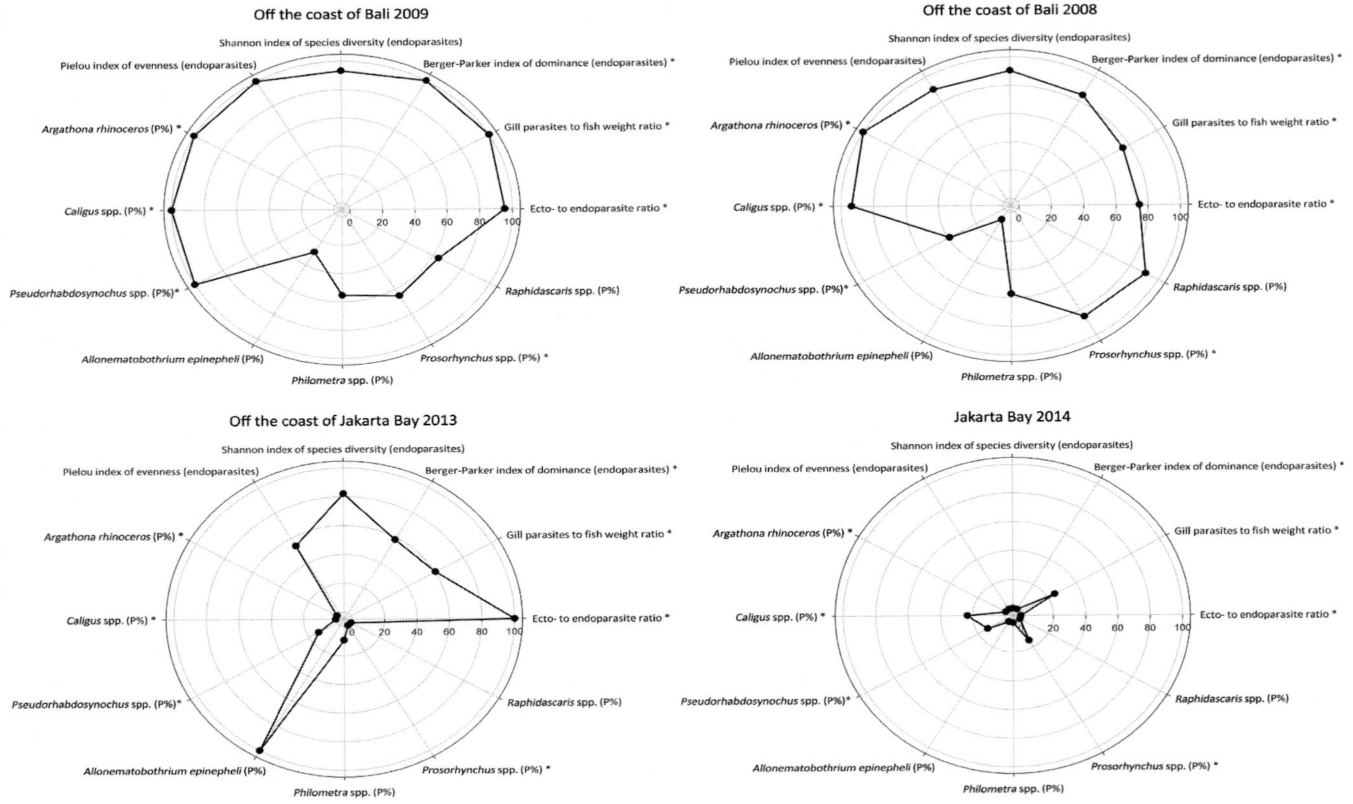


**FIGURE 8.12** Visual integration of normalized parasitological parameters from *Epinephelus coioides* for Indonesian coastal waters. Large areas reflect near-natural conditions, small areas unnatural conditions. Comparison is made between the coast off Bali, Jakarta Bay, and off Jakarta Bay. * Inverse parameter (Neubert et al., 2016b). *Modified after Kleinertz (2010) and Kleinertz and Palm (2015).*

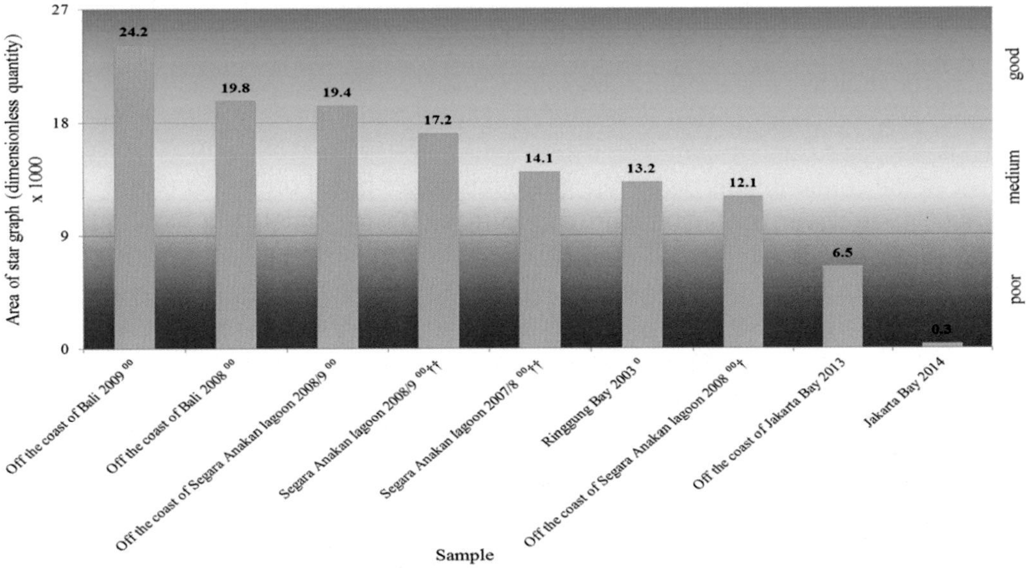


**FIGURE 8.13** Pollution light: Areas of star graphs calculated from normalized parasitological parameters of *Epinephelus coioides*. ° data from Rückert (2006, p. 240), °° data from Kleinertz (2010, p. 263) and Kleinertz and Palm (2015), † identified as sample from inside Segara Anakan lagoon, †† identified as sample from off the coast of Segara Anakan lagoon (Neubert et al., 2016b).

campaign have contributed in elevating seafood consumption. Furthermore, fishery and aquaculture are important livelihoods for coastal inhabitants of JB and the Thousand Islands, with few available livelihood alternatives to local fishers (Fauzi and Buchary, 2001).

Contrarily to the health benefit of seafood consumption (e.g., Al et al., 2000; Clandinin, 1999; Kawarazuka and Béné, 2011; Otto et al., 2001), there are possible opposite effects related to harmful contaminants. For example, there is evidence of risk of coronary heart disease associated with methyl-mercury contamination (FAO/WHO, 2007, 2010) or neurodevelopment disorder (Lynch et al., 2011). Potential cancer risks associated with persistent organic pollutants (POPs) such as dioxins (PCDD/F) and dioxin-like PCBs (polychorinated biphenyls) as well as carcinogenic PAHs may negate the coronary heart disease benefits from fish consumption (EC, 2011).

Due to their possible adverse effects, harmful contaminants in seafood have received increasing awareness. This includes organic contaminants (mainly residue of pesticides and PAHs), inorganic pollutants (heavy metals), and organometallic pollutants such as butyl tin derivatives (e.g., Monirith et al., 2003; Rumengan et al., 2008; Sudaryanto et al., 2007; Williams et al., 2000). Additionally, biological contaminants such as pathogenic microbes, biotoxins, and biogenic amines also potentially contaminate seafood from JB (Andayani and Sumartono, 2012; Makmur et al., 2014). Finally, illegal additives and preservatives frequently misused during handling and processing of seafood need to be considered (Dwiyitno et al., 2009).

Concern of adverse effects due to chemical residues in seafood is reflected in several exposure assessments conducted recently for the JB region. Agusa et al. (2007) estimated exposure of 15 heavy metals (V, Cr, Mn, Co, Cu, Zn, Se, Sr, Mo, Ag, Cd, Sn, Pb, Hg, and Ba) from seafood consumption based on the concentration in 12 seafood species from JB, Lada Bay, and Cirebon Bay. The assessment revealed an average of estimated daily intakes of the contaminants below the guideline values based on JECFA (2003) and US-EPA (2005). However, the maximum exposure of Hg (detected in Talang queenfish, *Scomberoides commersonnianus*, from JB) was found to be 15 μg $day^{-1}$, which is 136% and 118% over these limits, respectively. Noteworthy, dietary intake of heavy metals from seafood in Indonesia is lower than that, e.g., of Malaysia and Cambodia, but higher than in Thailand (Agusa et al., 2007).

Sudaryanto et al. (2007) studied the potential exposure of organochlorine residues in the JB region. The results showed that the residues were dominated by PCBs and total DDT in all fish samples, followed by hexachlorocyclohexanes (HCHs), chlordane compounds (CHLs), and hexachlorobenzene (HCB). Based on the concentration in mussels, DDT and HCHs were particularly higher in samples from suburban area, but PCBs and CHLs were more abundant in samples from JB (Monirith et al., 2003). The calculated daily intake of these organochlorines did not reach the TDIs, suggesting a low risk. The mean daily intake of PCBs was 0.81 μg $person^{-1}$ $day^{-1}$ via fish consumption, which was less than 2% of the FAO/WHO (2007) values. For total DDT, the dietary intake by Indonesians was 1.1 μg $person^{-1}$ $day^{-1}$, less than 1% of the FAO/WHO guideline.

Potential exposure of similar organochlorines from seafood, collected in traditional markets in Jakarta, Bogor, and Yogyakarta, has been also studied by Shoiful and Colleagues (2013). They estimated the daily intake of HCB and DDT contaminants through milkfish (*Chanos chanos*) to be 0.35 and 0.5 ng kg $BW^{-1}$ $day^{-1}$, respectively. However, these exposures also were far below the guideline of acceptable daily intake (ADI) (according to FAO-WHO, 2010).

Recently, Dwiyitno et al. (2015) have identified more than 40 organic contaminants in selected seafood species in JB, including persistent pollutants and emerging contaminants. A survey on the corresponding exposure in four districts around JB (Cilincing, Penjaringan, Untung Jawa, and Tanjung Pasir) showed that the potential risk of organic contaminants from seafood was below the thresholds of moderate and serious risk limits suggested by RIVM (2001) and ATSDR (2002), indicating that they may not cause any serious health risk (Dwiyitno et al., 2017).

Dwiyitno et al. (2017) estimated a daily intake of total DDT and dichlorobenzenes by residents of around 2 ng $day^{-1}$ (maximum limit is 10 μg kg $BW^{-1}$ according to FAO/WHO, 2000), much higher than $PAH_4$ (0.8 ng $day^{-1}$). DDT-related contamination of green mussels and the majority of fish samples was related mainly to *p,p'*-DDE. This result is in line with an earlier study reporting *p,p'*-DDE in fish samples from several locations in Indonesia (Shoiful et al., 2013; Sudaryanto et al., 2007).

Furthermore, the exposure of $PAH_4$ (sum of the four most carcinogenic, mutagenic, and estrogenic isomers, see Fig. 8.14) was dominated by BaA and BbF in green mussels

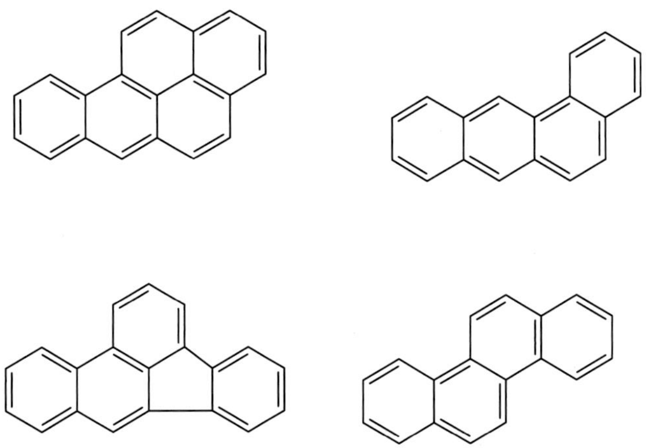

**FIGURE 8.14** $PAH_4$—(benzo[a]pyrene; benzo[a]anthracene; benzo[f]fluoranthene; chrysene).

as well as certain pelagic and benthic fish such as milk fish, Spanish mackerel, and mullet (Dwiyitno et al., 2017). Maximum concentrations of $PAH_4$ in green mussel and fish species were 7 and 3 μg $kg^{-1}$, which were below the threshold of 30 and 12 μg $kg^{-1}$, respectively (EC, 2011). The calculated dietary intake of PAHs was comparable to that of various fish species in Mumbai, India (1.8–10.7 ng kg $BW^{-1}$ $day^{-1}$), Korea (13.8–16.7 ng kg $BW^{-1}$ $day^{-1}$), Kuwait (231 ng $day^{-1}$), and Spain (627–712 ng $day^{-1}$) (Dhananjaya and Muralidharan, 2012; Falcó et al., 2003; Saeed et al., 1995).

A tolerable daily intake (TDI) of 1,4-dichlorobenzene (DCB), used as deodorizer and disinfectant, is suggested to be 107 μg kg $BW^{-1}$ $day^{-1}$ (WHO, 2010). RIVM (2001) adopted this level to define TDI level of *1,4-* and *1,2-*DCB as 100 and 430 μg kg $BW^{-1}$ $day^{-1}$, respectively. Accordingly, Dwiyitno et al. (2017) estimated that DCB intake via seafood consumption from JB is below the TDI. Important species captured in JB and the risk of potential contaminants are presented in Table 8.1.

According to a survey of 84 households on the Thousand Islands and 140 in coastal communities of JB, the main target species differ between JB and Thousand Island households (Table 8.2). The species ranked highest in importance on average on the islands, and caught by the majority (around 90%) of households there, is the fusilier, *Caesio cuning*. For coastal households in JB, however, this species ranks last in mean importance. The most important species in JB, caught in about 85% of households, is *Rastrelliger kanagurta*. This species ranks third in importance on the islands and is caught by 60% of households there, indicating that it is the most important species overall for coastal households in the JB/Thousand Islands area (Baum et al., 2016c). The important role of green mussels, *Perna viridis*, for coastal households in JB underlines their particular exposure to pollution risks (Table 8.1), as well as their economic exposure in case culture and sale of these mussels becomes difficult or illegal due to high contaminant levels. This economic exposure of marine resource-dependent households to pollution risks is further underlined by a significantly lower level of livelihood diversity

**Table 8.1** Important seafood species in Jakarta Bay and the risk of potential contaminants.

| Seafood group | Common name/*Scientific name* | Potential[a] contaminants | Risk Level[b] | References |
|---|---|---|---|---|
| Mollusks | Green mussel | OCPs | L | Monirith et al. (2003) |
| | *Perna viridis* | PCBs | L | Sudaryanto et al. (2009) |
| | | PAHs | M | Dwiyitno et al. (2015) |
| | | BFRs | L | |
| | | HMs | M | |
| | | OTs | L | |
| | | STX | M | |
| | Blood cockle | HMs | M | Andayani and Sumartono |
| | *Anadara granosa* | STX | L | (2012) |
| | Feathers cockle | STX | L | Andayani and Sumartono |
| | *Anadara antiquata* | | | (2012) |
| Pelagic fish | Indian mackerel | OCPs | L | Agusa et al. (2007) |
| | *Rastrelliger kanagurta* | PAHs | L | Dwiyitno et al. (2015) |
| | | HMs | L | |
| | Spanish mackerel | OCPs | L | Sudaryanto et al. (2005) |
| | *Scomberomorus commerson* | PCBs | L | Sudaryanto et al. (2007) |
| | | PAHs | L | Dwiyitno et al. (2015) |
| | | OTs | L | |
| | Slender shad | OCPs | L | Dwiyitno et al. (2015) |
| | *Ilisha elongate* | PAHs | L | |
| | Milkfish | OCPs | L | Shoiful et al. (2013) |
| | *Chanos chanos* | PCBs | L | Dwiyitno et al. (2015) |
| | | PAHs | L | Sudaryanto et al. (2007) |
| | Talang queenfish | HMs | M | Agusa et al. (2007) |
| | *Scomberoides* | OTs | L | Sudaryanto et al. (2005) |
| | *commersonnianus* | | | |
| Demersal fish | Mahogany snapper | OCPs | L | Dwiyitno et al. (2015) |
| | *Lutjanus johnii* | PAHs | L | Sudaryanto et al. (2009) |
| | | BFRs | L | |
| | Croaker | OCPs | L | Dwiyitno et al. (2015) |
| | *Argyrosomus amoyensis* | PAHs | L | |
| | White emperor | OCPs | L | Dwiyitno et al. (2015) |
| | *Lethrinus lentjan* | PAHs | L | |
| | Blue-tail mullet | OCPs | L | Dwiyitno et al. (2015) |
| | *Valamugil buchanani* | PCBs | L | Sudaryanto et al. (2007) |
| | | PAHs | L | |
| | Rabbitfish | OCPs | L | Dwiyitno et al. (2015) |
| | *Siganus javus* | PAHs | L | Sudaryanto et al. (2009) |
| | | BFRs | L | |
| | Sea catfish | OCPs | L | Dwiyitno et al. (2015) |
| | *Netuma thalassina* | PAHs | L | Sudaryanto et al. (2009) |
| | | BFRs | L | |
| | White-spotted spinefoot | OCPs | L | Sudaryanto et al. (2005) |
| | *Siganus canaliculatus* | PCBs | L | Sudaryanto et al. (2007) |
| | | PAHs | L | Dwiyitno et al. (2015) |
| | | OTs | L | |

**Table 8.1** Important seafood species in Jakarta Bay and the risk of potential contaminants.—cont'd

| Seafood group | Common name/*Scientific name* | Potential[a] contaminants | Risk Level[b] | References |
|---|---|---|---|---|
| | Telkara perchlet | OCPs | L | Sudaryanto et al. (2005) |
| | *Ambassis vachelli* | OTs | L | Sudaryanto et al. (2007) |
| | Slender ponyfish | OTs | L | Sudaryanto et al. (2005) |
| | *Leiognathus elongatus* | | | |
| | Jarbua terapon | OCPs | L | Sudaryanto et al. (2007) |
| | *Terapon jarbua* | PCBs | L | |
| | Pugnose ponyfish | OCPs | L | Sudaryanto et al. (2007) |
| | *Secutor ruconius* | PCBs | L | |
| | Little jaw fish | OCPs | L | Sudaryanto e al. (2007) |
| | *Johnius vogleri* | PCBs | L | |
| Crustaceans | Banana shrimp | OCPs | L | Dwiyitno et al. (2015) |
| | *Penaeus marguiensis* | PAHs | L | |
| | Blue crab | OCPs | L | Dwiyitno et al. (2015) |
| | *Callinectes sapidus* | PAHs | L | |

[a]*BFRs*, brominated flame retardants; *HMs*, heavy metals; *OCPs*, organochlorine pesticides; *OTs*, organotins; *PAHs*, polyaromatic hydrocarbons; *PCBs*, polychlorinated biphenyl; *STX*, saxitoxin.
[b]L (low) < 50% of maximum residue limit (MRL); M (moderate): 50%–100% of MRL; H (high) >MRL.

and lack of alternative income: While 80% of surveyed island households relied on fishing as primary source of income, about half of these households had secondary sources of income, notably in tourism. In contrast, although fishing was the primary source of income for only 60% of surveyed coastal households in JB (with another 20% depending on fish processing), only about one-seventh of JB coastal households had secondary sources of income (Baum et al., 2016c). These observations underline that pollution of marine resources poses not only an immediate risk to the health of coastal households and consumers, but also furthermore that threats to the safety of marine food resources pose significant risks to marine resource-dependent households.

There are several other contaminants that can harm the consumer through seafood consumption. They include organotins (OTs), brominated flame retardants (BFRs), and marine biotoxins. Among OTs, tributyltin (TBT) is essential due to its former use in various industrial purposes such as slime control in paper mills, as biocide in antifouling paints for ships and boats, for aquaculture nets and in wood protection. Total BT residues have been observed in six fish species collected from JB at concentrations of 21–84 ng $g^{-1}$ wet weight (Sudaryanto et al., 2005). Based on recent estimates of average daily seafood consumption of 77.71 g $person^{-1}$ $day^{-1}$ in Jakarta (Dwiyitno, 2017), the estimated maximum daily intake of BTs is about 6.5 μg $person^{-1}$ $day^{-1}$, which is lower than the TDI of 15 μg 60 kg $person^{-1}$ $day^{-1}$ (0.25 μg TBT kg body $weight^{-1}$ $day^{-1}$). Based on the study of Sudaryanto et al. (2005), BT concentrations in seafood from JB were higher compared with those collected from Lada Bay (4.2–18 ng g wet $wt^{-1}$) and Cirebon

**Table 8.2** List of target species of island (Tab. 2A, $n = 84$) and coastal households in Jakarta Bay (Tab. 2B, $n = 140$) with local and latin names, mean rank of importance to the surveyed households, as well as number of households, in which the respective species is important. Table 8.2A.

**(A)**

| Local name | Latin name | Rank | Number of households | % of all households |
|---|---|---|---|---|
| Ekor kuning | *Caesio cuning* | 1.29 | 75 | 89.3 |
| Tongkol | *Euthynnus affinis* | 1.91 | 66 | 78.6 |
| Banyar | *Rastrelliger kanagurta* | 1.95 | 58 | 69.0 |
| Kerapu | *Epinephelus* sp./*Plectropomus* sp. | 2.06 | 32 | 38.1 |
| Selar kuning | *Selaroides leptolepis* | 2.33 | 51 | 60.7 |
| Baronang tompel | *Siganus guttatus* | 2.69 | 48 | 57.1 |
| Tembang | Clupeidae: *Sardinella albella, Sardinella brachysoma, Sardinella fimbriata* | 2.74 | 38 | 45.2 |
| Lemuru | *Sardinella lemuru (Amblygaster leiogaster?)* | 2.80 | 49 | 58.3 |
| Tenggiri | *Acanthocybium solandri* | 2.95 | 38 | 45.2 |
| Kerang hijau | *Perna viridis* | 3.00 | 7 | 8.3 |

**(B)**

| Local name | Latin name | Rank | Number of households | % of all households |
|---|---|---|---|---|
| Banyar | *Rastrelliger kanagurta* | 1.41 | 118 | 84.3 |
| Lemuru | *Sardinella lemuru (Amblygaster leiogaster?)* | 1.68 | 102 | 72.9 |
| Kerang hijau | *Perna viridis* | 1.75 | 85 | 60.7 |
| Tembang | Clupeidae: *Sardinella albella, Sardinella brachysoma, Sardinella fimbriata* | 1.87 | 94 | 67.1 |
| Selar kuning | *Selaroides leptolepis* | 1.97 | 94 | 67.1 |
| Tenggiri | *Acanthocybium solandri* | 2.60 | 91 | 65.0 |
| Tongkol | *Euthynnus affinis* | 2.82 | 88 | 62.9 |
| Kerapu | *Epinephelus* sp./*Plectropomus* sp. | 2.85 | 13 | 9.3 |
| Baronang tompel | *Siganus guttatus* | 2.94 | 36 | 25.7 |
| Ekor kuning | *Caesio cuning* | 2.98 | 53 | 37.9 |

Bay (3.3–25 ng g wet wt$^{-1}$), indicating a heavier boat traffic and ship building activities in JB than in other areas in Indonesia. Concentrations of BTs in fish from Indonesia were generally similar, or slightly higher than those from Vietnam, Thailand, Taiwan, and Oceanian countries, but lower than those from the Malacca Strait, Malaysia (Kannan et al., 1995; Sudaryanto et al., 2005).

BFRs have been detected in seafood samples collected from JB at concentrations of 0.42–42 ng g$^{-1}$ lipid (Sudaryanto, 2009). BFR residues were dominated by polybrominated diphenyl ethers (PBDEs) and hexabromocyclododecanes (HBCDs). Compared with fish species, mussels tends to accumulate more congeners that may be due to the different feeding behavior. PBDEs and HBCDs have received global concern due to their persistency, bioaccumulative nature, and possible adverse effects on wildlife and humans.

Saxitoxin (STX) is a common biotoxin contaminant present in shellfish species due to accumulation from saxitoxin-producing dinoflagellates (Dam et al., 2009). Consumption of STX contaminated shellfish could promote paralytic shellfish poisoning (PSP) with various symptoms from mild tingling, numbness, headache, dizziness, nausea, vomiting, and diarrhea to respiratory paralysis and possible death (EFSA, 2009; FAO, 2004). Determination of STX of green mussels collected from JB has been conducted by Kusnoputranto and Colleagues (2013). The results showed that STX concentrations of shellfish samples ranged from 6.92 to 17.34 μg STXeq 100 $g^{-1}$, which is below the maximum residue limit of STX in most countries, i.e., 80 μg STXeq 100 $g^{-1}$ (BSN, 2009; FAO, 2004). The concentration was higher than that reported earlier by Andayani and Sumartono (2012), which was 0.87–5.39 μg STXeq 100 $g^{-1}$. With reference to the general mussel consumption in Jakarta of approximately 185 g $portion^{-1}$ (Makmur et al., 2014), the maximum exposure of STX is approximately 31.45 μg $person^{-1}$, which is above the acute Rf dose of EFSA (2009) 30 μg $person^{-1}$ and 75% of FAO/WHO (2007) guideline of 42 μg $person^{-1}$. However, the exposure is lower than the guideline of the Australia New Zealand Food Authority, which suggests 120–180 μg STX can produce moderate symptoms and 400–1060 μg can cause death (ANZFA, 2001).

A number of regulations have been established by the Indonesian government to support the quality and safety assurance of fish products distributed either inside or outside the country. Based on law No.18/2012, Chapter 4, the central and local governments are obliged to assure the safety of food items along all supply chains. Additionally, Chapter 7 of the law No.45/2009 asks the Ministry of Marine Affairs and Fisheries (MMAF) to prevent the contamination and destruction of marine and fishery resources, including the environment. Furthermore, decree No.52A-/KEPMEN-KP/2013 deals with the requirements of quality assurance and safety of fishery products in production, processing, and distribution. The decree elaborates the general structure and hygiene requirements of the whole chain including during fishing, landing, storage, fish markets, as well as food security and health standards. Furthermore, monitoring of chemical residues, biological material in aquaculture, including marine aquaculture, has been mandated by the ministerial decree No. PER.02/MEN/2007. With reference to persistent organic pollutants, Indonesia has adopted the Stockholm Convention on POPs in September 2009 as law No.19/2009. The law is followed by related regulations, such as the management of toxic and hazardous waste in law No.101/2014. This law regulates many aspects including transporting, dumping, mitigating, and monitoring of the waste. Generally, the levels of maximum residue limit (MRL) of certain contaminants have been established by the National Standardization Board (BSN) and implemented as national standard.

To monitor the quality and safety of Indonesia's aquaculture products, the MMAF has implemented a management system to control residues, namely the National Residue

Monitoring Plan (NRMP). In 2013, for example, Indonesia had cooperated with the European Commission on a mitigation program through the Commission Decision 2011/163/EU, resulting in aquaculture products free of residue and the inclusion of Indonesia in the list of countries to export aquaculture products to the EU. Quality inspections of fishery products are conducted by official laboratories both of the end products and during the production process.

With regard to the potential contamination of seafood from JB, the general level is not yet seen to generate public health problems. However, the increasing level of certain contaminants in recent years, such as heavy metals, hexachlorobenzene, and biotoxin, should alarm the authorities to mitigate the anthropogenic pollution load. Notably, the potency of multicontaminants in certain seafood could indicate that consumption of those products might be hazardous to residents around JB due to either additive or synergistic effects among the toxicants.

## 8.8 Implications

All the information and facts reported in this chapter regarding the environmental state of JB demonstrate impressively the huge anthropogenically induced pressure and stress this ecosystem is exposed to. Additionally, an intact ecosystem is a basic precondition for many aspects directly impacting human beings around the Bay covering, e.g., seafood health, drinking water accessibility, or recreation needs. Noteworthy, the investigations reported here do not remain on the description of the environmental status, but explored also the relevant processes behind inducing the pollution problems and, consequently, point to potential solutions.

These scientific efforts are not only relevant for JB or, more general, for Indonesia. Most big cities around the globe evolved from settlements, located close to rivers, which in past and present times function as water suppliers, transport routes, and dumping sites (such as it is the case for Bangkok, Shanghai, Manila, Hanoi or even Jakarta). And especially in rapidly developing countries, urbanization and industrial development is progressing faster than the development of environmental services, infrastructure, and emission controls. As a result, the quality of surface water systems significantly declines along with population growth and economic development in these regions in general (e.g., He et al., 2014; Hosono et al., 2009).

Consequently, the gained knowledge about the anthropogenic impact on JB as derived from the SPICE project is not solely of regional interest, but has relevance for many vulnerable ecosystems near coastal megacities worldwide.

The degradation of coral reefs in JB and the Thousand Islands is caused by a combination of multiple anthropogenic stressors acting in concert. Especially within JB, pollution and sedimentation have become disastrous, leading to extremely degraded coral reefs. Further offshore along the Thousand Islands, reefs are in a slightly improved condition; however, the increasing threat due to overfishing, global warming, eutrophication due to highly densely populated islands and a lack of sewage treatment, and chemicals released from increasing shipping traffic and urban runoff from islands will pose a severe challenge for those reefs in the future.

Increased stress, as seen in the experiments by Baum et al. (2016a) and Kegler et al. (2015), may culminate in responses at population and ecosystem levels and reduce reef resilience and the potential for recovery (Hoegh-Guldberg et al., 2007). Therefore, there is a need for more controlled factorial experiments in the future, investigating physiological and ecological stress responses of marine organisms to different pollutants simultaneously and together with other environmental stressors to detect possible interactions.

Combined effects of multiple stressors are still barely understood (Ban et al., 2014), and studies as in Section 8.4 are needed to determine responses of organisms and ecosystems to these stressors and understand potential interactions against a background of ongoing environmental change. Similarly, ocean management can no longer focus on individual stressors (Halpern et al., 2007), but must account for combined stressor effects, as shown in Section 8.4.

Section 8.5 provided detailed information about the microbiome and its functional and metabolic diversity in Indonesian fish and shrimps. In addition, metagenomics and metatranscriptomics combined with corresponding functional analyses should be also performed to address differential expression under different ecological and environmental conditions in future. A metaproteome-based analysis is a further approach to elucidate functionality of the proteins within the environments under different conditions (Mandal et al., 2015; Prakash and Taylor, 2012; Srivasta, 2007).

Section 8.6 summarized the current state of knowledge on Indonesian mariculture fish. It is evident that all thoroughly studied fish species have a diverse parasite fauna and many of these records still require further taxonomic work. Detailed overviews on the potential parasites infecting these fish are given, guiding the farmers to safeguard their aquaculture activities also with new coming parasite species and threats in future. Rückert et al. (2010) compared free-living and maricultured *E. fuscoguttatus* and demonstrated the transmission pathways into the net cages (e.g., fouling organisms, crustaceans, trash fish feed). The fish pathogens *Vibrio* sp., *Flavobacterium* sp., and *Photobacteria* sp. were found, especially in those specimens that were parasite free and kept in mariculture. This opens new possibilities for a disease prevention and treatment management of maricultured finfish in future. Palm et al. (2015) recommended that one of the major future tasks is the search for alternative feed sources and feeding strategies to prevent parasite spread and pathogenic outbreaks. Finally, a new method to use grouper fish parasites as biological indicators was demonstrated to be a useful tool to compare different aquaculture sites as well as ecosystems (Neubert et al., 2016a). It is evident that this methodology will be useful not only in Indonesia, but also in other southeast Asian regions, allowing recommendations on the carrying capacity at different locations.

With regard to the potential contamination of seafood from JB (Section 8.7), the general level is not yet seen to generate public health problems. However, the increasing level of certain contaminants in recent years, such as heavy metals, hexachlorobenzene, and biotoxin, should alarm the authorities to mitigate the anthropogenic pollution load. Notably, the potency of multicontaminants in certain seafood could indicate that consumption of those products might be hazardous to residents around JB due to either additive or synergistic effects among the toxicants.

In conclusion, marine spatial planning adjusted to local conditions is an alternative to current management strategies (MPAs and NTAs, Wilson et al., 2010). Monitoring key biological and environmental parameters continuously over several years and across seasons is crucial for the establishment of successful management and conservation plans. The involvement of local communities in reef management is needed (Ferse et al., 2010), and marine awareness and local education campaigns could aid the enforcement of protection areas (Breckwoldt et al., 2016). Any conservation and management plan, however, will only be successful if pollution in Jakarta is reduced, e.g., by implementing sewage treatment and waste disposal plans (Clara et al., 2007).

#### Knowledge gaps and directions of future research

Jakarta faced in the past decades a tremendous increase of pollution by industrial and municipal emissions, and consequently, the JB as the receiving ecosystem is affected by a large proportion of sewage and waste via the urban riverine discharge. Thereby, the bay ecosystem faced a huge transition from natural conditions to an intensively used coastline as effective for industrial facilities, traffic infrastructure, urban settlements, aquafarming, agriculture, and many other developments.

Since the aquatic ecosystems of JB represent an excellent example of a fundamental conflict between economic and ecological demands, this chapter deals with the following key research questions related to the overall consequences:

- How do the anthropogenic modifications of the hydrological system at JB impact the local nutrient dispersion?
- What are the most important organic and inorganic pollutants? What are the dominant emission sources?
- How is water pollution affecting sensitive ecosystems at JB? What are the biological answers to anthropogenic stressors?
- Can microbial diversity in aquatic organisms reflect ecological conditions on different quality levels?
- Does the pollution at JB impact the quality of products of the fishery industry—as exemplified for fish parasites?
- What is the overall impact of the JB pollution on the quality of seafood quality and the corresponding risk of consumption?

Implications/recommendations for policy and society

Based on the discussed relation between pollution and ecological impact, an overall aim of management at JB should be a significant decrease of anthropogenic stress. Main stressors are hereby complex mixtures of pollutants derived from sewage effluents and industrial emissions. Effective management can imply the following:

- Implementation of sewage treatment and waste disposal plans
- Local marine spatial planning
- Continuous monitoring of key biological and environmental parameters
- Local education campaigns

# Acknowledgments

The financial and administrative support of the reported research by the German Federal Ministry of Education and Research (Grant No. 03F0641), the Indonesian Ministry for Research and Technology (RISTEK), the Indonesian Ministry for Maritime Affairs and Fisheries (KKP), and the German Academic Exchange Service (DAAD) is highly appreciated. Further on, the highly motivated practical support of many students from both Germany and Indonesia is gratefully acknowledged.

# References

Agency for Toxic Substances and Disease Registry [ATSDR], 2002. Toxicological Profile for Polycyclic Aromatic Hydrocarbons. U.S. Department of Health and Human Services, Public Health Services, Atlanta, GA.

Agusa, T., Kunito, T., Sudaryanto, A., Monirith, I., Kan-Atireklap, S., Iwata, H., et al., 2007. Exposure assessment for trace elements from consumption of marine fish in South East Asia. Environ Pollut 145 (3), 766–777. https://doi.org/10.1016/j.envpol.2006.04.034.

Al, M.D.M., van Houwelingnen, A.C., Hornstra, G., 2000. Long chain polyunsaturated fatty acids, pregnancy and pregnancy outcome. American Journal of Clinical Nutrition 71, 285S–291S.

Alexander, R.B., Johnes, P.J., Boyer, E.W., Smith, R.A., 2002. A comparison of models for estimating the riverine export of nitrogen from large watersheds. Biogeochemistry 57/58, 295–339.

Andayani, W., Sumartono, A., 2012. Saxitoxin in green mussels (Perna viridis, Mytiliae), blood cockle (Anadara granosa) and feathers cockle (Anadara antiquata, Arcidae) using high pressure liquid chromatography. Journal of Coastal Development 15 (3), 252–259.

Ankley, G.T., Burkhard, L.P., 1992. Identification of surfactants as toxicants in a primary effluent. Environmental Toxicology and Chemistry 11, 1235–1248.

Apip Sagala, S., Luo, P., 2015. Overview of Jakarta water-related environmental challenges. In: Water and Urban Initiative Working Paper Series, vol. 4. United Nations University, p. 5.

Arifin, Z., 2004. Local Millennium Ecosystem Assessment: Condition and Trend of the Greater Jakarta Bay Ecosystem. The Ministry of Environment, Jakarta, Republic of Indonesia.

Arifin, Z., Falahudin, D., 2017. Contribution of fish consumption to cadmium and lead intakes in coastal communities of west Kalimantan, Indonesia. Marine Research in Indonesia 42 (1), 1–10.

Arifin, Z., Hindarti, D., 2006. Effects of cyanide on ornamental coral fish (*Chromis viridis*). Marine Research in Indonesia 30, 15–20.

Arifin, Z., Puspitsari, R., Miyazaki, N., 2012. Heavy metal contamination in Indonesian coastal marine ecosystems: a historical perspective. Coastal Marine Science 35, 227–233.

Asmanelli, Y., Muchari, H., 1993. Penyakit ikan laut di lokasi Keramba Jaring Apung di Kepulauan Riau. [Marine fish diseases in floating net cages in Riau Archipelago.]. In: Prosiding Seminar Hasil Penelitian Perikanan Budidaya Pantai, 16-19 July 1993, vol. 11. Maros, Indonesia, pp. 13–24.

Asplund, M., 2013. Ecological Aspects of marine Vibrio Bacteria-Exploring Relationships to Other Organisms and a Changing Environment. Retrieved from https://130.241.16.4/handle/2077/31813.

Australia New Zealand Food Authority [ANZFA], 2001. Shellfish Toxins in Food. A Toxicological Review and Risk Assessment. Technical Report Series No. 14. November 2001. Australia New Zealand Food Authority (ANZFA). www.nicnas.gov.au/australia/fsanz.htm.

Balcázar, J.L., De Blas, I., Ruiz-Zarzuela, I., Cunningham, D., Vendrell, D., Múzquiz, J.L., 2006. The role of probiotics in aquaculture. Veterinary Microbiology 114 (3–4), 173–186. https://doi.org/10.1016/j.vetmic.2006.01.009.

Balcázar, J.L., Rojas-Luna, T., Cunningham, D.P., 2007. Effect of the addition of four potential probiotic strains on the survival of pacific white shrimp (*Litopenaeus vannamei*) following immersion challenge with Vibrio parahaemolyticus. Journal of Invertebrate Pathology 96 (2), 147–150.

Ban, S.S., Graham, N.A.J., Connolly, S.R., 2014. Evidence for multiple stressor interactions and effects on coral reefs. Global Change Biology 20, 681–697.

Baum, G., Januar, H.I., Ferse, S.C., Kunzmann, A., 2015. Local and regional impacts of pollution on coral reefs along the Thousand Islands North of the Megacity Jakarta, Indonesia. PloS One 10 (9), e0138271.

Baum, G., Januar, I., Ferse, S.C., Wild, C., Kunzmann, A., 2016a. Abundance and physiology of dominant soft corals linked to water quality in Jakarta Bay, Indonesia. PeerJ 4, e2625.

Baum, G., Kegler, P., Scholz-Böttcher, B.M., Alfiansah, Y.R., Abrar, M., Kunzmann, A., 2016b. Metabolic performance of the coral reef fish *Siganus guttatus* exposed to combinations of water borne diesel, an anionic surfactant and elevated temperature in Indonesia. Marine Pollution Bulletin 110 (2), 735–746.

Baum, G., Kusumanti, I., Breckwoldt, A., Ferse, S.C., Glaser, M., Dwiyitno Adrianto, L., et al., 2016c. Under pressure: investigating marine resource-based livelihoods in Jakarta Bay and the Thousand islands. Marine Pollution Bulletin 110 (2), 778–789.

Beals, E.W., 1984. Bray-curtis ordination: an effective strategy for analysis of multivariate ecological data. Advances in Ecological Research 14 (C), 1–55. https://doi.org/10.1016/S0065-2504(08)60168-3.

Behrendt, H., Huber, P., Kornmilch, M., Opitz, D., Schmoll, O., Scholz, G., et al., 2000. Nutrient Emissions into River Basins of Germany. UBA-Texte 23/00, Umweltbundesamt Berlin.

Bell, S., Morse, S., 2003. Measuring Sustainability: Learning by Doing. Earthscan, London, Sterling, VA.

Bell, P.R., Lapointe, B.E., Elmetri, I., 2007. Reevaluation of ENCORE: support for the eutrophication threshold model for coral reefs. AMBIO: A Journal of the Human Environment 36 (5), 416–424.

Bengen, D.G., Knight, M., Dutton, I., 2006. Managing the port of Jakarta bay: overcoming the legacy of 400 years of adhoc development. In: The Environment in Asia Pacific Harbours. Springer, Dordrecht, pp. 413–431.

Beyer, J., Petersen, K., Song, Y., Ruus, A., Grung, M., Bakke, T., et al., 2014. Environmental risk assessment of combined effects in aquatic ecotoxicology: a discussion paper. Marine Environmental Research 96, 81–91.

Blackburn, S., Marques, C., de Sherbinin, A., Modesto, F., Ojima, R., Oliveau, S., et al., 2014. Mega-urbanisation on the coast. In: Pelling, M., Blackburn, S. (Eds.), Megacities and the Coast: Risk, Resilience and Transformation. Routledge, Oxon, US, pp. 1–21.

BPLHD (Badan Pengelolaan Lingkungan Hidup), 2012. Solid waste handling management: a case of Jakarta. In: Presented at Review and Planning Workshop on Eco-Town 11–13 December 2012. Environment Management Board — Jakarta Provincial Government, Penang — Malaysia.

BPS (Badan Pusat Statistik), 2010. Statistical Yearbook of Indonesia 2010. Badan Pusat Statistik (BPS) – Statistics, Indonesia, Jakarta.

BPS (Badan Pusat Statistik), 2014. Jumlah Penduduk Menurut Jenis Kelamin Kabupaten Bekasi Tahun 2014 (Population of Bekasi Regency Based on Gender 2014). Statistic of Bekasi Regency 2014. http://bekasikab.bps.go.id.

Bray, R.A., Palm, H.W., 2009. Bucephalids (Digenea: Bucephalidae) from marine fishes off the south-western coast of Java, Indonesia, including the description of two new species of Rhipidocotyle and comments on the marine fish digenean fauna of Indonesia. Zootaxa 2223 (1), 1–24.

Breckwoldt, A., Dsikowitzky, L., Baum, G., Ferse, S.C., van der Wulp, S., Kusumanti, I., et al., 2016. A review of stressors, uses and management perspectives for the larger Jakarta Bay Area, Indonesia. Marine Pollution Bulletin 110, 790–794.

Bueno, M.M., Gomez, M.J., Herrera, S., Hernando, M.D., Agüera, A., Fernandez-Alba, A.R., 2012. Occurrence and persistence of organic emerging contaminants and priority pollutants in five sewage treatment plants of Spain: two years pilot survey monitoring. Environ Pollut 164, 267–273.

Burhanuddin, D.A., 1978. Parasit Anisakis sebagai petunjuk perbedaan populasi ikan laying, Decapterus russelli Ruppell, di laut Jawa. Osean. Indonesia 9, 1–11.

Burke, L., Reytar, K., Spalding, M.D., Perry, A., 2012. Reefs at Risk Revisited in the Coral triangle. World Resources Institute, Washington DC.

Calow, P., Forbes, V.E., 1998. How do physiological responses to stress translate into ecological and evolutionary processes? Comparative Biochemistry and Physiology A 120, 11–16.

Chaiyapechara, S., Rungrassamee, W., Suriyachay, I., Kuncharin, Y., Klanchui, A., Karoonuthaisiri, N., et al., 2012. Bacterial community associated with the intestinal tract of *P. monodon* in commercial farms. Microbial Ecology 63 (4), 938–953. https://doi.org/10.1007/s00248-011-9936-2.

Chao, A., 1984. Nonparametric estimation of the number of classes in a population. Scandinavian Journal of Statistics 11, 265–270.

Chèvre, N., Gagné, F., Blaise, C., 2003. Development of a biomarker-based index for assessing the ecotoxic potential of aquatic sites. Biomarkers 8 (3–4), 287–298.

Chou, L.M., Yamazato, K., 1990. Community structure of coral reefs within the vicinity of Motobu and Sesoko (Okinawa) and the effects of human and natural influences. Galaxea 9, 9–75.

Clandinin, M.T., 1999. Brain development and assessing the supply of polyunsaturated fatty acid. Lipids 34, 131–137.

Clara, M., Scharf, S., Scheffknecht, C., Gans, O., 2007. Occurrence of selected surfactants in untreated and treated sewage. Water Research 41, 4339–4348.

Cleary, D.F.R., 2017. Linking fish species traits to environmental conditions in the Jakarta Bay-Pulau Seribu coral reef system. Marine Pollution Bulletin 122, 259–262.

Cleary, D.F.R., Polónia, A.R.M., Renema, W., Hoeksema, B.W., Wolstenholme, J., Tuti, Y., et al., 2014. Coral reefs next to a major conurbation: a study of temporal change (1985–2011) in coral cover and composition in the reefs of Jakarta, Indonesia. Marine Ecology: Progress Series 501, 89–98.

Cleary, D.F.R., Polónia, A.R.M., Renema, W., Hoeksema, B.W., Rachello-Dolmen, P.G., Moolenbeek, R.G., et al., 2016. Variation in the composition of corals, fishes, sponges, echinoderms, ascidians, molluscs, foraminifera and macroalgae across a pronounced in-to-offshore environmental gradient in the Jakarta Bay–Thousand Islands coral reef complex. Marine Pollution Bulletin 110, 701–717.

Dam, H., Colin, S., Haley, S., Avery, D., Chen, L., Zhang, H., 2009. Copepod Resistance to Toxic Pyhtoplankton. http://www.pices.int/publications/presentations/Zoopl%202007/Zoop%202007%20S3/S3_Dam.pdf.

Damar, A., 2003. Effects of enrichment on nutrient dynamics, phytoplankton dynamics and productivity in Indonesian tropical waters: a comparison between Jakarta Bay, Lampung Bay and Semangka Bay. In: Berichte aus dem Forschungs- und Technologiezentrum Westküste der Universität Kiel, Nr. 29. Büsum.

De Jonge, V.N., Boynton, W., D'Elia, C.F., Elmgren, R., Welsh, B.L., 1994. Responses to developments in eutrophication in four different North Atlantic estuarine systems. In: Dyer, K.R., Orth, R.J. (Eds.), Changes in Fluxes in Estuaries. ECSA/ERF Symposium plymouth, September 1992. Olsen & Olsen Fredensborg, pp. 179–196.

de la Torre, F.R., Salibián, A., Ferrari, L., 2007. Assessment of the pollution impact on biomarkers of effect of a freshwater fish. Chemosphere 68, 1582–1590.

DeLeo, D.M., Ruiz-Ramos, D.V., Baums, I.B., Cordes, E.E., 2015. Response of Deep-Water Corals to Oil and Chemical Dispersant Exposure. . Deep-Sea Research Part II. https://doi.org/10.1016/j.dsr2.2015.02.028.

Dewi, K., Palm, H.W., 2013. Two new species of philometrid nematodes (Nematoda: Philometridae) in *Epinephelus coioides* (Hamilton, 1822) from the south Bali Sea, Indonesia. Zootaxa 3609 (1), 49–59.

Dewi, K., Palm, H.W., 2017. Philometrid nematodes (Philometridae) of marine teleosts from Balinese waters, Indonesia, including the description of Philometra damriyasai sp. nov. Zootaxa 4341 (4), 577–584.

Dhananjayan, V., Muraldidharan, S., 2012. Polycyclic aromatic hydrocarbons in various species of fishes from Mumbai Harbour, India and their dietary intake concentration to human. The International Journal of Ocean 9, 1–7. https://doi.org/10.1155/2012/645178.

Diamant, A., Banet, A., Paperna, I., Westernhagen, H.V., Broeg, K., Krüner, G., et al., 1999. The use of fish metabolic, pathological and parasitological indices in pollution monitoring. Helgoland Marine Research 53 (3), 195–208.

Dsikowitzky, L., Schwarzbauer, J., 2014. Industrial organic contaminants: identification, toxicity and fate in the environment. Environmental Chemistry Letters 12, 371–386.

Dsikowitzky, L., Dwiyitni, D., Heruwati, E., Ariyani, F., Irianto, H.E., Schwarzbauer, J., 2014. Exceptionally high concentrations of the insect repellant N,N-diethyl-m-toluamide (DEET) in surface waters from Jakarta, Indonesia. Environmental Chemistry Letters 12, 407–411.

Dsikowitzky, L., Botalova, O., Illgut, S., Bosowski, S., Schwarzbauer, J., 2015. Identification of characteristic organic contaminants in wastewaters from modern paper production sites and subsequent tracing in a river. Journal of Hazardous Materials 300, 254–262.

Dsikowitzky, L., Ferse, S., Schwarzbauer, J., Vogt, T.S., Irianto, H.E., 2016a. Impacts of megacities on tropical coastal ecosystems-the case of Jakarta, Indonesia. Marine Pollution Bulletin 110 (2), 621.

Dsikowitzky, L., Sträter, M., Ariyani, F., Irianto, H.E., Schwarzbauer, J., 2016b. First comprehensive screening of lipophilic organic contaminants in surface waters of the megacity Jakarta, Indonesia. Marine Pollution Bulletin 110 (2), 654–664.

Dwiyitno, Priyanto, N., Wulanjari, W.A., Atmawidjaya, 2009. Sub-chronic toxicity of green mussel (Perna viridis) treated with synthetic colorant on liver of mouse. JPBKP 4 (2), 113–120.

Dwiyitno, Dsikowitzky, L., Andarwulan, N., Irianto, H.E., Lioe, H.N., Ariyani, F., et al., 2015. Non-target screening method for the identification of persistent and emerging organic contaminants related to seafood and sediment samples from Jakarta Bay. Squalen Bulletin 10 (3), 141–157.

Dwiyitno, Dsikowitzky, L., Nordhaus, I., Andarwulan, N., Irianto, H.E., Lioe, H.N., et al., 2016. Accumulation patterns of lipophilic organic contaminants in surface sediments and in economic important mussel and fish species from Jakarta Bay, Indonesia. Marine Pollution Bulletin 101, 767–777.

Dwiyitno, A.N., Irianto, H.E., Lioe, H.N., Ariyani, F., Dsikowitzky, L., et al., 2017. Potential risk of organic contaminants to the coastal population through seafood consumption from Jakarta Bay. Squalen Bull 12 (3).

Dzikowski, R., Paperna, I., Diamant, A., 2003. Use of fish parasite species richness indices in analyzing anthropogenically impacted coastal marine ecosystems. Helgoland Marine Research 57 (3–4), 220–227.

EEA, European Environment Agency, 2001. Eutrophication in Europe's Coastal Waters. Topic Report No. 7, Copenhagen.

Environmental Protection Agency [US-EPA], 2005. Risk-based Concentration Table, April, 2005. U.S. EPA, region 3, Philadelphia, PA. http://www.epa.gov/reg3hwmd/risk/human/index.htm.

European Commission [EC], 2011. Commission Regulation (EU) No 835/2011 of 19 August 2011 amending Regulation (EC) No 1881/2006 as regards maximum levels for polycyclic aromatic hydrocarbons in foodstuffs. The Official Journal of the European Union 215, 4–8.

European Food Safety Agency [EFSA], 2009. Scientific Opinion of the Panel on Contamination in the Food Chain. Marine biotoxin in shellfish-saxitoxin Group. http://www.efsa.europa.eu/EFSA/Scientific_Opinion/contam-op-ej1019-saxitoxinmarine-saxitoxin.pdf.

Fabricius, K.E., 2005. Effects of terrestrial runoff on the ecology of corals and coral reefs: review and synthesis. Marine Pollution Bulletin 50, 125–146.

Falcó, G., Domingo, J.L., Llobet, J.M., Teixidó, A., Casas, C., Müller, L., 2003. Polycyclic aromatic hydrocarbons in foods: human exposure through the diet in Catalonia. Spain J Food Prot 66, 2325–2331.

Fanslow, D.L., Nalepa, T.F., Johengen, T.H., 2001. Seasonal changes in the respiratory electron transport system (ETS) and respiration of the zebra mussel, *Dreissena polymorpha* in Saginaw Bay, Lake Huron. Hydrobiologia 448, 61–70.

FAO (Food & Agriculture Organisation), 2012. The State of World Fisheries and Aquaculture 2012. Sofia. https://doi.org/10.5860/CHOICE.50-5350.

FAO/WHO, 2000. Pesticide Residue in Food and Feed. Codex Alimentarius-International Food Standard.

FAO/WHO, 2007. Evaluation of Certain Food Additives and Contaminants: Sixty-Seventh Report of the Joint FAO/WHO Expert Committee on Food Additives. World Health Organization (WHO Technical Report Series, Geneva. No. 940.

FAO/WHO, 2010. Report of the Joint FAO/WHO Expert Consultation on the Risks and Benefits of Fish Consumption, 25–29 January. 2010. Italy, Rome.

Fares, Y.R., Ikhwan, M., 2001. Conceptual modeling for management of the Citarum/Ciliwung basins, Indonesia. Journal of Environmental Hydrology 9.

Fauzi, A., Buchary, E.A., 2001. An overview of socioeconomic aspects of an Indonesian marine protected area: a perspective from Kepulauan Seribu Marine Park. Fisheries Centre Research Reports 9 (8), 62–69.

Ferrol-Schulte, D., Gorris, P., Baitoningsih, W., Adhuri, D.S., Ferse, S.C., 2015. Coastal livelihood vulnerability to marine resource degradation: a review of the Indonesian national coastal and marine policy framework. Marine Policy 52, 163–171.

Ferse, S.C., Costa, M.M., Manez, K.S., Adhuri, D.S., Glaser, M., 2010. Allies, not aliens: increasing the role of local communities in marine protected area implementation. Environmental Conservation 23–34.

Food and Agriculture Organization [FAO], 2004. Marine Toxin Food and Agriculture Organization of the United Nations. http://www.fao.org/docrep/.

Fox, H.E., Pet, J.S., Dahuri, R., Caldwell, R.L., 2003. Recovery in rubble fields: long-term impacts of blast fishing. Marine Pollution Bulletin 46, 1024–1031.

Fulazzaky, M.A., 2010. Water quality evaluation system to assess the status and the suitability of the Citarum river water to different uses. Environmental Monitoring and Assessment 168 (1), 669–684.

Furtula, V., Liu, J., Chambers, P., Osachoff, H., 2012. Sewage treatment plants efficiencies in removal of sterols and sterol ratios as indicators of faecal contamination sources. Water, Air, and Soil Pollution 223, 1017–1031.

Gene & Consortium, 2000. Gene ontology: tool for the. Nature Genetics 25 (5), 25–29. https://doi.org/10.1038/75556.

Gomez-Gil, B., Roque, A., Velasco-Blanco, G., 2002. Culture of Vibrio alginolyticus C7b, a potential probiotic bacterium, with the microalga *Chaetoceros muelleri*. Aquaculture 211 (1–4), 43–48.

Gotelli, N., Colwell, R., 2011. Chapter 4: estimating species richness. Biological diversity. Frontiers in Measurement and Assessment (2), 39–54. https://doi.org/10.2307/3547060.

Grimalt, J.O., Fernandez, P., Bayona, J.M., Albaiges, J., 1990. Assessment of faecal sterols and ketones as indicators of urban sewage inputs to coastal waters. Environmental Science and Technology 24, 357–363.

Guarner, F., Malagelada, J.-R., 2003. Gut flora in health and disease. The Lancet 360 (9356), 512–519. https://doi.org/10.1016/S0140-6736(03)12489-0.

Guzmán, H.M., Jiménez, C.E., 1992. Contamination of coral reefs by heavy metals along the Caribbean coast of Central America (Costa Rica and Panama). Marine Pollution Bulletin 24, 554–561.

Halpern, B., Selkoe, K., Micheli, F., Kappel, C., 2007. Evaluating and ranking the vulnerability of global marine ecosystems to anthropogenic threats. Conservation Biology 21, 1301–1315.

He, Q., Bertness, M.D., Bruno, J., Li, B., Chen, G., Coverdale, T.C., et al., 2014. Economic development and coastal ecosystem change in China. Scientific Reports 4. Article number: 5995.

Hechinger, R.F., Lafferty, K.D., Huspeni, T.C., Brooks, A.J., Kuris, A.M., 2007. Can parasites be indicators of free-living diversity? Relationships between species richness and the abundance of larval trematodes and of local benthos and fishes. Oecologia 151 (1), 82–92.

Hennersdorf, P., Kleinertz, S., Theisen, S., Abdul-Aziz, M.A., Mrotzek, G., Palm, H.W., et al., 2016. Microbial diversity and parasitic load in tropical fish of different environmental conditions. PLoS One 11 (3), e0151594.

Hennersdorf, P., Mrotzek, G., Abdul-Aziz, M.A., Saluz, H.P., 2016. Metagenomic analysis between free-living and cultured Epinephelus fuscoguttatus under different environmental conditions in Indonesian waters. Marine Pollution Bulletin 110 (2), 726–734. https://doi.org/10.1016/j.marpolbul.2016.05.009.

Hoegh-Guldberg, O., Mumby, P.J., Hooten, A.J., Steneck, R.S., Greenfield, P., Gomez, E., et al., 2007. Coral reefs under rapid climate change and ocean acidification. Science 318, 1737–1742.

Hosono, T., Umezawa, Y., Onodera, S., Wang, C.H., Siringan, F., Buapeng, S., et al., 2009. Comparative study on water quality among Asian megacities based on major ion concentrations. In: Taniguchi, M., Burnett, W.C., Fukushima, Y., Haigh, M., Umezawa, Y. (Eds.), From Headwaters to the Ocean: Hydrological Changes and Watershed Management. CRC Press/Balkema, Leiden, The Netherlands, pp. 295–300.

Howarth, R.W., Anderson, D., Cloern, J., Elfring, C., Hopkinson, C., Lapointe, B., et al., 2000. Nutrient pollution of coastal rivers, bays, and seas. Issues Ecol 7, 1–15.

Hundenborn, J., Thurig, S., Kommerell, M., Haag, H., Nolte, O., 2013. Case Report Severe Wound Infection with Photobacterium Damselae Ssp. Damselae and Vibrio Harveyi, Following a Laceration Injury in marine Environment: A Case Report and Review of the Literature.

Hutomo, M., Burhanuddin, H.P., 1978. Observations on the incidence and intensity of infection of nematode larvae (Fam. Anisakidae) in certain marine fishes of waters around Panggang Island, Seribu Islands. Marine Research in Indonesia 21, 49–60.

Ilahude, H.D., Hadidjaja, P., Mahfudin, H., 1978. Survey on Anisakid larvae in marine fish from fish markets in Jakarta. The Southeast Asian Journal of Tropical Medicine and Public Health 9 (1), 48–50.

IPCC, 2013. Summary for policymakers. In: Stocker, T.F., Qin, D., Plattner, G.-K., Tignor, M., Allen, S.K., Boschung, J. (Eds.), Climate Change 2013: The Physical Science Basis. Contribution of Working Group I to the Fifth Assessment Report of the Intergovernmental Panel on Climate Change. Cambridge University Press, New York, pp. 3–29.

Ishikura, H., Kikuchi, K., 1990. Intestinal Anisakiasis in Japan. Springer, Tokyo.

Ivanovic, J., Baltic, M.Z., Boskovic, M., Kilibarda, N., Dokmanovic, M., Markovic, R., et al., 2015. Anisakis infection and Allergy in humans. Procedia Food Science 5, 101–104.

Jakob, E., Palm, H.W., 2006. Parasites of commercially important fish species from the southern Java coast, Indonesia, including the distribution pattern of trypanorhynch cestodes. Verhandlungen der Gesellschaft für Ichthyologie 5, 165–191.

Joint FAO/WHO Expert Committee on Food Additives [JECFA], 2003. Summary and Conclusions of the 61st Meeting of the Joint FAO/WHO Expert Committee on Food Additives (JECFA). JECFA/61/SC, Rome, Italy.

Jones, M.K., Oliver, J.D., 2009. Vibrio vulnificus: disease and pathogenesis. Infection and Immunity 77 (5), 1723–1733. https://doi.org/10.1128/IAI.01046-08.

Kang, N., Kang, H.I., An, K.G., 2014. Analysis of fish DNA biomarkers as a molecular-level approach for ecological health assessments in an urban stream. Bulletin of Environmental Contamination and Toxicology 93 (5), 555–560.

Kannan, K., Tanabe, S., Iwata, H., Tatsukawa, R., 1995. Butyltins in muscle and liver of fish collected from certain Asian and Oceanian countries. Environmental Pollution 90, 279–290.

Kawarazuka, N., Béné, C., 2011. The potential role of small fish species in improving micronutrient deficiencies in developing countries: building evidence. Public Health Nutrition 14, 1927–1938.

Kegler, P., Baum, G., Indriana, L.F., Wild, C., Kunzmann, A., 2015. Physiological response of the hard coral Pocillopora verrucosa from Lombok, Indonesia, to two common pollutants in combination with high temperature. PloS One 10 (11), e0142744.

Khan, R.A., 1990. Parasitism in marine fish after chronic exposure to petroleum hydrocarbons in the laboratory and to the Exxon Valdez oil spill. Bulletin of Environmental Contamination and Toxicology 44 (5), 759–763.

Khan, R.A., Kiceniuk, J.W., 1988. Effect of petroleum aromatic hydrocarbons on monogeneids parasitizing Atlantic cod, *Gadus morhua* L. Bulletin of Environmental Contamination and Toxicology 41 (1), 94–100.

Kiceniuk, J.W., Khan, R.A., 1983. Toxicology of chronic crude oil exposure: sublethal effects on aquatic organisms. Advances in Environmental Science and Technology 13, 425–436.

Kleinertz, S., 2010. Fischparasiten als Bioindikatoren: Zum Umweltstatus von Küstenökosystemen und einer Zackenbarschmarikultur in Indonesien Dissertation der Naturwissenschaften. Fachbereich 2 (Biologie/Chemie) der Universität Bremen.

Kleinertz, S., Palm, H.W., 2015. Fish parasites of *Epinephelus coioides* (Serranidae) as potential environmental indicators in Indonesian coastal ecosystems. Journal of Helminthology. https://doi.org/10.1017/S0022149X1300062X.

Kleinertz, S., Damriyasa, I.M., Hagen, W., Theisen, S., Palm, H.W., 2014. An environmental assessment of the parasite fauna of the reef-associated grouper Epinephelus areolatus from Indonesian waters. Journal of Helminthology 88, 50–63. https://doi.org/10.1017/S0022149X12000715.

Kleinertz, S., Eckhardt, K.U., Theisen, S., Palm, H.W., Leinweber, P., 2016. Acanthocephalan fish parasites (Rhadinorhynchidae Lühe, 1912) as potential biomarkers: molecular–chemical screening by pyrolysis-field ionization mass spectrometry. Journal of Sea Research 113, 51–57.

Klimpel, S., Palm, H.W., 2011. Anisakid nematode (Ascaridoidea) life cycles and distribution: increasing zoonotic potential in the time of climate change? In: Mehlhorn, H. (Ed.), Progress in Parasitology, Parasitology Research Monographs 2. Springer-Verlag, Berlin Heidelberg, pp. 201–222. https://doi.org/10.1007/978-3-642-21396-0_11.

Klimpel, S., Rückert, S., Piatkowski, U., Palm, H.W., Hanel, R., 2006. Diet and metazoan parasites of silver scabbard fish *Lepidopus caudatus* from the Great Meteor Seamount (North Atlantic). Marine Ecology Progress Series 315, 249–257.

Koesharyani, I., Roza, D., Mahardika, K., Johnny, F., Zafran, Y.K., 2001. Manual for Fish Disease Diagnosis. II. Marine Fish and Crustacean Diseases in Indonesia. Gondol Research Institute for Mariculture. Central Research Institute for Sea Exploration and Fisheries, Departement of Marine Affairs and Fisheries and Japan International Cooperation Agency.

Koropitan, A.F., Ikeda, M., Damar, A., Yamanaka, Y., 2009. Influences of physical processes on the ecosystem of Jakarta bay: a coupled physical-ecosystem model experiment. ICES Journal of Marine Science 66, 336–348. https://doi.org/10.1093/icesjms/fsp011.

Krupesha Sharma, S.R., Shankar, K.M., Sathyanarayana, M.L., Sahoo, A.K., Patil, R., Narayanaswamy, H. D., et al., 2010. Evaluation of immune response and resistance to diseases in tiger shrimp, *Penaeus monodon* fed with biofilm of Vibrio alginolyticus. Fish and Shellfish Immunology 29 (5), 724–732.

Kuchta, R., Scholz, T., Vlčková, R., Řhía, M., Walter, T., Yuniar, A.T., et al., 2009. Revision of tapeworms (Cestoda: Bothriocephalidae) from lizardfish (Saurida: Synodontidae) from the Indo-Pacific region. Zootaxa 1977, 55–67.

Kusnoputranto, H., Moersidik, S.S., Makmur, M., Dwiyitno, 2013. Determination of saxitoxin concentration of green mussel collected from markets around Jakarta, and from original sources at Lampung Bay and Panimbang Bays. JPBKP 8 (2), 115–123.

Ladwig, N., Hesse, K.J., van der Wulp, S.A., Damar, A., Koch, D., 2016. Pressure on oxygen levels of Jakarta bay. Marine Pollution Bulletin 110 (2), 665–674.

Lafferty, K.D., Allesina, S., Arim, M., Briggs, C.J., De Leo, G., Dobson, A.P., et al., 2008a. Parasites in food webs: the ultimate missing links. Ecology Letters 11, 533–546.

Lafferty, K.D., Shaw, J.C., Kuris, A.M., 2008b. Reef fishes have higher parasite richness at unfished Palmyra Atoll compared to fished Kiritimati Island. Eco Health 5, 338–345.

Lancelot, C., Spitz, Y., Gypens, N., Ruddick, K., Becquevort, S., Rousseau, V., et al., 2005. Modelling diatom-Phaeocystis blooms and nutrient cycles in the Southern Bight of the North Sea: the MIRO model. Marine Ecology Progress Series 289, 63–78.

Landsberg, J.H., Blakesley, B.A., Reese, R.O., McRae, G., Forstchen, P.R., 1998. Parasites of fish as indicators of environmental stress. Environmental Monitoring and Assessment 51, 211–232.

Lee, R.J., Rangdale, R.E., 2008. In Improving Seafood Products for the Consumer 11. Bacterial Pathogens in Seafood.

Lenhart, H.-J., 2001. Effects of river nutrient load reductions on the eutrophication of the North Sea, simulated with the ecosystem model ERSEM. Senckenbergiana Maritima 31 (2), 299–312.

Lesser, M.P., 2013. Using energetic budgets to assess the effects of environmental stress on corals: are we measuring the right things? Coral Reefs 32, 25–33.

Liu, H., Wang, L., Liu, M., Wang, B., Jiang, K., Ma, S., et al., 2011. The intestinal microbial diversity in Chinese shrimp (Fenneropenaeus chinensis) as determined by PCR-DGGE and clone library analyses. Aquaculture 317 (1–4), 32–36. https://doi.org/10.1016/j.aquaculture.2011.04.008.

Logan, D.T., 2007. Perspective on ecotoxicology of PAHs to fish. Human and Ecological Risk Assessment: An International Journal 13, 302–316.

Loraine, G.A., Pettigrove, M.E., 2006. Seasonal variations in concentrations of pharmaceuticals and personal care products in drinking water and reclaimed wastewater in southern California. Environmental Science and Technology 40, 687–695.

Lynch, M.L., Huang, L.S., Cox, C., Strain, J.J., Myers, G.J., Bonham, M.P., et al., 2011. Varying coefficient function models to explore interactions between maternal nutritional status and prenatal methylmercury toxicity in the Seychelles Child Development Nutrition Study. Environmental Research 111 (1), 75–80.

Maki, A.W., 1979. Respiratory Activity of Fish as a Predictor of Chronic Fish Toxicity Values for Surfactants, vol. 667. Special Technical Publ, Philadelphia, pp. 77–95.

Makmur, M., Moersidik, S.S., Wisnubroto, D.S., Kusnoputranto, H., 2014. Consumer health risk assessment of green mussell containing saxytoxyn in Cilincing, North Jakarta. Journals of Health Ecology 13 (2), 165–178.

Mandal, R.S., Saha, S., Das, S., 2015. Metagenomic surveys of gut microbiota. Genomics, Proteomics and Bioinformatics 13 (3), 148–158. https://doi.org/10.1016/j.gpb.2015.02.005.

Marcé, R., Comerma, M., García, J.C., Armengol, J., 2004. A neuro-fuzzy modeling tool to estimate fluvial nutrient loads in watersheds under time-varying human impact. Limnology and Oceanography: Methods 2, 342–355.

Martosewojo, S., 1980. Cacing Anisakis yang hidup sebagai parasit pada ikan laut. Pewarta Oseana 6 (2), 1–5.

Merel, S., Snyder, S.A., 2016. Critical assessment of the ubiquitous occurrence and fate of the insect repellent N,N-diethyl-m-toluamide in water. Environment International 96, 98–117.

Ministry of Marine Affairs and Fisheries (MMAF), 2015. Marine and Fisheries in Figure 2011-2015. http://statistik.kkp.go.id/sidatik-dev/Publikasi/src/kpda2015.pdf.

Moll, A., 1998. Regional distribution of primary production in the North Sea simulated by a three-dimensional model. Journal of Marine Systems 16 (1–2), 151–170.

Monirith, I., Ueno, D., Takahashi, S., Nakata, H., Sudaryanto, A., Subramanian, A.N., et al., 2003. Asia-Pacific mussel watch: monitoring contamination of persistent organochlorine compounds in coastal waters of Asian countries. Marine Pollution Bulletin 46, 281–300.

Mulyani Widiarti, R., Wardhana, W., 2012. Sebaran spasial spesies penyebab harmful algal bloom (HAB) di Lokasi Budidaya Kerang Hijau (Perna viridis) Kamal Muara, Jakarta Utara Pada Bulan Mei 2011. Jurnal Akuatika III (1), 28–39.

Munkittrik, K.R., van der Kraak, G.J., McMaster, M.E., Portt, D.C.B., van den Heuval, M.R., Servos, M.R., 1994. Survey of receiving-water environmental impacts associated with discharges from pulp mills. II. Gonad size, liver size, hepatic EROD activity and plasma sex steroid levels in white sucker. Environmental Toxicology and Chemistry 13, 1089–1101.

Neubert, K., Yulianto, I., Kleinertz, S., Theisen, S., Wiryawan, B., Palm, H.W., 2016a. Parasite fauna of white-streaked grouper, *Epinephelus ongus* (Bloch, 1790) (Epinephelidae) from Karimunjawa, Indonesia. Parasitology Open 2 (e12), 1–11. https://doi.org/10.1017/pao.2016.6.

Neubert, K., Yulianto, I., Theisen, S., Kleinertz, S., Palm, H.W., 2016b. Parasites of *Epinephelus coioides* (Epinephelidae) from Jakarta Bay: missing link in the development of an environmental indicator system for Indonesian coastal waters. Marine Pollution Bulletin 110, 747–756.

Nichols, P.D., Leeming, R., Rayner, M.S., 1993. Comparison of the abundance of the faecal sterol coprostanol and faecal bacterial groups in inner-shelf waters and sediments near Sydney, Australia. Journal of Chromatography A 643, 189–195.

Oetama, V.S.P., Hennersdorf, P., Abdul-Aziz, M.A., Mrotzek, G., Haryanti, H., Peter, H., 2016. Microbiome analysis and detection of pathogenic bacteria of *Penaeus monodon* from Jakarta Bay and Bali. MPB. https://doi.org/10.1016/j.marpolbul.2016.03.043.

Ogut, H., Palm, H.W., 2005. Seasonal dynamics of Trichodina spp. on whiting (*Merlangius merlangus*) in relation to organic pollution on the Eastern Black Sea coast of Turkey. Parasitology Research 96, 149–153.

Oliveira, M., Pacheco, M., Santos, M.A., 2008. Organ specific antioxidant responses in golden grey mullet (*Liza aurata*) following a short-term exposure to phenanthrene. The Science of the Total Environment 396, 70–78.

Otto, S.J., van Houwelingen, A.C., Badart-Smook, A., Hornstra, G., 2001. Comparison of the peripartum and postpartum phospholipids polyunsaturated fatty acid profiles of lactating and nonlactating women. American Journal of Clinical Nutrition 73, 1074–1079.

Overstreet, R.M., 1997. Parasitological data as monitors of environmental health. Parassitologia 39, 169–175.

Pal Jaya, P.D., 2012. Concept and Strategy for Wastewater Management of Jakarta. Jakarta.

Palm, H.W., 1999. Ecology of *Pseudoterranova decipiens* (Krabbe, 1878) (Nematoda: Anisakidae) from Antarctic waters. Parasitology Research 85, 638–646.

Palm, H.W., 2004. The Trypanorhyncha Diesing, 1863. IPB-PKSPL Press.

Palm, H.W., 2011. Fish parasites as biological indicators in a changing world: can we monitor environmental impact and climate change?. In: Mehlhorn, H. (Ed.), Progress in Parasitology, vol. 2. Parasitology Research Monographs, pp. 223–250. https://doi.org/10.1007/978-3-642-21396-0_12.

Palm, H.W., Bray, R.A., 2014. Marine Fish Parasitology in Hawaii. Hohenwarsleben. Westarp & Partner Digitaldruck, p. XII.

Palm, H.W., Dobberstein, R.C., 1999. Occurrence of trichodinid ciliates (Peritricha: Urceolariidae) in the Kiel Fjord, Baltic sea, and its possible use as a biological indicator. Parasitology Research 85, 726–732.

Palm, H.W., Overstreet, R., 2000. New records of trypanorhynch cestodes from the Gulf of Mexico, including Kotorella pronosoma (Stossich, 1901) and Heteronybelinia palliata (Linton, 1924) comb. n. Folia Parasitologica 47, 293–302.

Palm, H.W., Rückert, S., 2009. A new method to visualize fish parasites as biological indicators for ecosystem health. Parasitology Research 105, 539–553.

Palm, H.W., Damriyasa, I.M., Linda, Oka, I.B.M., 2008. Molecular genotyping of Anisakis Dujardin, 1845 (Nematoda: Ascaridoidea: Anisakidae) larvae from marine fish of Balinese and Javanese waters, Indonesia. Helminthologia 45, 3–12.

Palm, H.W., Kleinertz, S., Rückert, S., 2011. Parasite diversity as an indicator of environmental change? – an example from tropical grouper (*Epinephelus fuscoguttatus*) mariculture in Indonesia. Parasitology 138, 1–11. https://doi.org/10.1017/S0031182011000011.

Palm, H.W., Yulianto, I., Theisen, S., Rückert, S., Kleinertz, S., 2015. Epinephelus fuscoguttatus mariculture in Indonesia: implications from fish parasite infections. Regional Studies of Marine Science. https://doi.org/10.1016/j.rsma.2015.07.003.

Palm, H.W., Theisen, S., Damriyasa, I.M., Kusmintarsih, E.S., Oka, I.B.M., Setyowati, E.A., et al., 2017. Anisakis (Nematoda: Ascaridoidea) from Indonesia. Diseases of Aquatic Organisms 123, 141–157.

Pätsch, J., Kühn, W., 2008. Nitrogen and carbon cycling in the North Sea and exchange with the North Atlantic – a model study, Part I. Nitrogen budget and fluxes. Continental Shelf Research 28 (6), 767–787.

Peng, X., Zhang, G., Mai, B., 2005. Tracing anthropogenic contamination in the Pearl River estuarine and marine environment of South China Sea using sterols and other organic molecular markers. Marine Pollution Bulletin 50, 856–865.

Petersen, F., Palm, H.W., Möller, H., Cuzi, M.A., 1993. Flesh parasites of fish from central Philippine waters. Diseases of Aquatic Organisms 15, 81–86.

Phanuwan, C., Takizawa, S., Oguma, K., Katayama, H., Yunika, A., Ohgaki, S., 2006. Monitoring of human enteric viruses and coliform bacteria in waters after urban flood in Jakarta, Indonesia. Water Science and Technology 54, 203–210.

Prakash, T., Taylor, T.D., 2012. Functional assignment of metagenomic data: challenges and applications. Briefings in Bioinformatics 13 (6), 711–727. https://doi.org/10.1093/bib/bbs033.

Rees, J.G., Setiapermana, D., Sharp, V.A., Weeks, J.M., Williams, T.M., 1999. Evaluation of the impacts of land-based contaminants on the benthic faunas of Jakarta Bay, Indonesia. Oceanologica Acta 22, 627–640.

Richmond, R.H., 1993. Coral reefs: present problems and future concerns resulting from anthropogenic disturbance. American Zoologist 33, 524–536.

Rinawati, Koike, T., Koike, H., Kurumisawa, R., Ito, M., Sakurai, S., et al., 2012. Distribution, source identification, and historical trends of organic micropollutants in coastal sediment in Jakarta Bay, Indonesia. Journal of Hazardous Materials 217, 208–216.

Roberts, C.M., McClean, C.J., JEN, V., Hawkins, J.P., Allen, G.R., McAllister, D.E., et al., 2002. Marine biodiversity hotspots and conservation priorities for tropical reefs. Science 295, 1280–1284.

Rodil, R., Quintana, J.B., Basaglia, G., Pietrogrande, M.C., Cela, R., 2010. Determination of synthetic phenolic antioxidants and their metabolites in water samples by downscaled solid-phase extraction, silylation and gas chromatography–mass spectrometry. Journal of Chromatography A 1217, 6428–6435.

Rönnbäck, P., 2001. Shrimp Aquaculture: State of the Art. Swedish Environmental Impact Assessment Centre for the Swedish International Development Cooperation Agency (Sida).

Ronnback, P., 2002. Environmentally Sustainable Shrimp Aquaculture.

Rückert, S., 2006. Marine Fischparasiten in Indonesien: Befallssituation und Bedeutung für die Marikultur von Zackenbarschen. Dissertation, Mathematisch-Naturwissenschaftliche Fakultät. Heinrich-Heine-Universität Düsseldorf.

Rückert, S., Palm, H.W., Klimpel, S., 2008. Parasite fauna of seabass (*Lates calcarifer*) under mariculture conditions in Lampung Bay, Indonesia. Journal of Applied Ichthyology 24, 321–327.

Rückert, S., Hagen, W., Yuniar, A.T., Palm, H.W., 2009a. Metazoan parasites of Segara Anakan Lagoon, Indonesia, and their potential use as biological indicators. Regional Environmental Change 9, 315–328.

Rückert, S., Klimpel, S., Al-Quraishy, Mehlhorn, H., Palm, H.W., 2009b. Transmission of fish parasites into grouper mariculture (Serranidae: *Epinephelus coioides* (Hamilton, 1822)) in Lampung Bay, Indonesia. Parasitology Research 104, 523–532.

Rückert, S., Klimpel, S., Palm, H.W., 2010. Parasites of cultured and wild grouper (*Epinephelus fuscoguttatus*) in Lampung Bay, Indonesia. Aquaculture Research 41, 1158–1162. https://doi.org/10.1111/j.1365-2109.2009.02403.x.

Rumengan, I., Ohiji, M., Arai, T., Harino, H., Arifin, Z., Miyazaki, N., 2008. Contamination status of butyltin compounds in Indonesian coastal waters. Coastal Marine Science 32, 116–126.

Rungrassamee, W., Klanchui, A., Maibunkaew, S., Chaiyapechara, S., Jiravanichpaisal, P., Karoonuthaisiri, N., 2014. Characterization of intestinal bacteria in wild and domesticated adult black tiger shrimp (*Penaeus monodon*). PLoS One 9 (3). https://doi.org/10.1371/journal.pone.0091853.

Saeed, T., Al-Yakoob, S., Al-Hashash, H., Al-Bahloul, M., 1995. Preliminary exposure assessment for Kuwaiti consumers to polycyclic aromatic hydrocarbons in seafood. Environment International 21, 255–263.

Sanchez, L.M., Wong, W.R., Riener, R.M., Schulze, C.J., Linington, R.G., 2012. Examining the fish microbiome: vertebrate-derived bacteria as an environmental niche for the discovery of unique marine natural products. PLoS One 7 (5). https://doi.org/10.1371/journal.pone.0035398.

Schaub, B.E.M., Gieskes, W.W.C., 1991. Eutrophication of the North Sea: the relation between Rhine river discharge and chlorophyll-a concentration in Dutch coastal waters. In: Elliott, M., Ducrotoy, J.-P. (Eds.), Estuaries and Coasts: Spatial and Temporal Intercomparisons. Olsen & Olsen, Fredensborg, pp. 85–90.

Shade, A., Handelsman, J., 2012. Beyond the Venn diagram: the hunt for a core microbiome. Environmental Microbiology 14 (1), 4–12. https://doi.org/10.1111/j.1462-2920.2011.02585.x.

Shannon, C.E., 1948. A mathematical theory of communication. The Bell System Technical Journal 27 (July 1928), 379–423. https://doi.org/10.1145/584091.584093.

Shoiful, A., Fujita, H., Watanabe, I., Honda, K., 2013. Concentrations of Organochlorine Pesticides (OCPs) residues in foodstuffs collected from traditional markets in Indonesia. Chemosphere 90 (5), 1742–1750. https://doi.org/10.1016/j.chemosphere.2012.10.022.

Silbiger, N.J., Nelson, C.E., Remple, K., Sevilla, J.K., Quinlan, Z.A., Putnam, H.M., et al., 2018. Nutrient pollution disrupts key ecosystem functions on coral reefs. Proceedings of the Royal Society 285.

Sindern, S., Tremöhlen, M., Dsikowitzky, L., Schwarzbauer, J., Siregar, T.H., Ariyani, F., 2016. Heavy metals in river and coast sediments of the Jakarta Bay region (Indonesia) – geogenic versus anthropogenic sources. Marine Pollution Bulletin 110, 624–633.

Siregar, T.H., Priyanto, N., Putri, A.K., Rachmawati, N., Triwibowo, R., Dsikowitzky, L., et al., 2016. Spatial distribution and seasonal variation of the trace hazardous element contamination in Jakarta Bay, Indonesia. Marine Pollution Bulletin 110 (2), 634–646.

Skogen, M.D., Søiland, H., Svendsen, E., 2004. Effects of changing nutrient loads to the North Sea. Journal of Marine Systems 46 (1), 23–38.

Sloman, K.A., Motherwell, G., O'Conner, K.I., Taylor, A.C., 2000. The effect of social stress on the standard metabolic rate (SMR) of brown trout, *Salmo trutta*. Fish Physiology and Biochemistry 23, 49–53.

Srivasta, K., 2007. Comparative metagenomics revealed commonly enriched gene sets in human gut microbiomes. DNA Research: An International Journal for Rapid Publication of Reports on Genes and Genomes 2007 14 (4), 169–181.

Sudaryanto, A., Takahashi, S., Iwata, H., Tanabe, S., Muchtar, M., Razak, H., 2005. Organotin residues and the role of anthropogenic tin sources in the coastal marine environment of Indonesia. Marine Pollution Bulletin 50 (2), 227–236. https://doi.org/10.1016/j.marpolbul.2004.11.041.

Sudaryanto, A., Monirith, I., Kajiwara, N., Takahashi, S., Hartono, P., Muawanah Omori, K., et al., 2007. Levels and distribution of organochlorines in fish from Indonesia. Environment International 33, 750–758.

Sudaryanto, A., Setiawan, I.E., Ilyas, M., Soeyanto, E., Riadi, A.S., Isobe, T., 2009. Levels of brominated flame retardants in sediments and their bioaccumulation potential in biota from Jakarta Bay and its surroundings, Indonesia. In: Obayashi, Y., Isobe, T., Subramanian, A., Suzuki, S., Tanabe, S. (Eds.), Interdisciplinary Studies on Environmental Chemistry — Environmental Research in Asia. TERRAPUB, 126 A, pp. 125–131.

Sullam, K.E., Essinger, S.D., Lozupone, C., O'Connor, M.P., Rosen, G.L., Knight, R., et al., 2012. Environmental and ecological factors that shape the gut bacterial communities of fish: a meta-analysis. Molecular Ecology 21 (13), 3363–3378. https://doi.org/10.1111/j.1365-294X.2012.05552.x.

Sures, B., 2008. Host-parasite interactions in polluted environments. Journal of Fish Biology 73, 2133–2142.

Sures, B., Reimann, N., 2003. Analysis of trace metals in the Antarctic host-parasite system Notothenia coriiceps and Aspersentis megarhynchus (Acanthocephala) caught at King George Island, South Shetland Islands. Polar Biology 26, 680–686.

The National Institute of Public Health and the Environment [RIVM], 2001. Re-evaluation of Human-Toxicological Maximum Permissible Risk Level. RIVM Report 711701025. Netherland.

Theisen, S., 2009. Fischparasiten von der Südküste Javas, Indonesien. Diplomarbeit, Mathematisch-Naturwissenschaftliche Fakultät. Heinrich-Heine-Universität Düsseldorf.

Theisen, S., Palm, H.W., Al-Jufaili, S.H., Kleinertz, S., 2017. Oesophagus infecting Pseudempleurosoma indonesiensis sp. nov. (Monogenoidea: Ancyrocephalidae) from croakers (Teleostei: Sciaenidae) in Indonesia. PLoS One 12 (9), e0184376.

Thoha, H., Adnan, Q., Sidabutar, T., Sugestiningsih, 2007. Note on the occurrence of phytoplankton and its relation with mass morality in the Jakarta Bay, May and November 2004. MAKARA of Science Series 11, 63–67.

Umbgrove, J.H.F., 1939. Madreporaria from the Bay of Batavia. Zoologische Mededelingen 22, 1–64.

Van Dam, J.W., Negri, A.P., Uthicke, S., Mueller, J.F., 2011. Chemical pollution on coral reefs: exposure and ecological effects. In: Sánchez-Bayo, F., van den Brink, P.J., Mann, R.M. (Eds.), Ecological Impacts of Toxic Chemicals. Bentham Science Publishers Ltd, pp. 187–211.

Van der Wulp, S.A., Damar, A., Ladwig, N., Hesse, K.J., 2016a. Numerical simulations of river discharges, nutrient flux and nutrient dispersal in Jakarta Bay, Indonesia. Marine Pollution Bulletin 110 (2), 675–685.

Van der Wulp, S.A., Dsikowitzky, L., Hesse, K.J., Schwarzbauer, J., 2016b. Master Plan Jakarta, Indonesia: the Giant Seawall and the need for structural treatment of municipal waste water. Marine Pollution Bulletin 110 (2), 686–693.

Van der Wulp, S.A., Hesse, K.-J., Ladwig, N., Damar, A., 2016c. Numerical simulations of river discharges, nutrient flux and nutrient dispersal in Jakarta Bay, Indonesia. Marine Pollution Bulletin Special Issue Jakarta Bay Ecosystem. https://doi.org/10.1016/j.marpolbul.2016.05.015.

Van Drecht, G., Bouwman, A.F., Knoop, J.M., Beusen, A.H.W., Meinardi, C.R., 2003. Global modeling of the fate of nitrogen from point and nonpoint sources in soils, groundwater, and surface water. Global Biogeochemical Cycles 17, 1115. https://doi.org/10.1029/2003GB002060.

Vaseeharan, B., Sundararaj, S., Murugan, T., Chen, J.C., 2007. Photobacterium damselae ssp. damselae associated with diseased black tiger shrimp *Penaeus monodon* Fabricius in India. Letters in Applied Microbiology 45 (1), 82–86. https://doi.org/10.1111/j.1472-765X.2007.02139.x.

Venkateswara-Rao, a, 1998. Vibriosis in Shrimp Aquaculture, (Over 280).

Vollaard, A.M., Ali, S., Smet, J., Van Asten, H., Widjaja, S., Visser, L.G., et al., 2005. A survey of the supply and bacteriologic quality of drinking water and sanitation in Jakarta, Indonesia. Southeast Asian Journal of Tropical Medicine and Public Health 36 (6), 1552–1561.

Vollenweider, R.A., 1976. Advances in defining critical loading levels for phosphorus in lake eutrophication. Memories Istituto Italiano Idrobiologie 33, 53–83.

Von Glasow, R., Jickells, T.D., Baklanov, A., Carmichael, G.R., Church, T.M., Gallardo, L., et al., 2013. Megacities and Large Urban Agglomerations in the coastal zone: interactions between atmosphere, land, and marine ecosystems. AMBIO 42 (1), 13–28. https://doi.org/10.1007/s13280-012-0343-9.

Williams, T.M., Rees, J.G., Setiapermana, D., 2000. Metals and trace organic compounds in sediments and waters of Jakarta Bay and the Pulau Seribu complex, Indonesia. Marine Pollution Bulletin 40, 277–285.

Williams, G.J., Gove, J.M., Eynaud, Y., Zgliczynski, B.J., Sandin, S.A., 2015. Local human impacts decouple natural biophysical relationships on Pacific coral reefs. Ecography 38, 001–011.

Wilson, S.K., Adjeroud, M., Bellwood, D.R., Berumen, M.L., Booth, D., Bozec, et al., 2010. Crucial knowledge gaps in current understanding of climate change impacts on coral reef fishes. Journal of Experimental Biology 213, 894–900.

World Health Organization] [WHO], 2010. Inventory of IPCS and Other WHO Pesticide Evaluations and Summary of Toxicological Evaluations Performed by the Joint Meeting on Pesticide Residues (JMPRs) through 2010. http://www.who.int/foodsafety/chem/jmpr/publications/jmpr_pesticide/en/index.html.

Wouthuizen, S., Tan, C.K., Ishizaka, J., Son, T.P.H., Ransi, V., Tarigan, S., et al., 2007. Monitoring of algal blooms and massive fish kill in the Jakarta Bay, Indonesia using satellite imageries. In: Proceedings of the First Joint PI Symposium of ALOS Data Nodes for ALOS Science Program, vol. 19. Japan Aerospace Exploration Agency (JAXA), Kyoto, p. 4.

Xia, J.H., Lin, G., Fu, G.H., Wan, Z.Y., Lee, M., Wang, L., et al., 2014. The intestinal microbiome of fish under starvation. BMC Genomics 15 (1), 266. https://doi.org/10.1186/1471-2164-15-266.

Yamaguti, S., 1952. Parasitic worms mainly from Celebes. Part 1. New digenetic trematodes of fishes. Acta Medica Okayama 8, 146–198.

Yamaguti, S., 1953. Parasitic worms mainly from Celebes. Part 3. Digenetic Trematoda of fishes II. Acta Medica Okayama 8, 257–295.

Yamaguti, S., 1954. Parasitic worms mainly from Celebes. Part 9. Nematodes of fishes. Acta Medica Okayama 9, 122–133.

Yuliana, Y., 2012. Implikasi Perubahan Ketersediaan Nutrien Terhadap Perkembangan Pesan (Blooming) Fitoplankton di Perairan Teluk Jakarta Ph.D. Dissertation. IPB Graduate School.

Yuniar, A., 2005. Parasites of marine fish from Segara Anakan, Java, Indonesia and their potential use as biological indicators. In: Master of Science Thesis in International Studies in Aquatic Tropical Ecology. ISATEC) Universität Bremen, p. 118.

Yuniar, A., Palm, H.W., Walter, T., 2007. Crustacean fish parasites from Segara Anakan Lagoon, Java Indonesia. Parasitology Research 100, 1193–1204.

Zaccone, G., Cascio, P.L., Fasulo, S., Licata, A., 1985. The effect of an anionic detergent on complex carbohydrates and enzyme activities in the epidermis of the catfish Heteropneustes fossilis (Bloch). The Histochemical Journal 17, 453–466.

Zhang, M., Sun, Y., Chen, K., Yu, N., Zhou, Z., Chen, L., et al., 2014. Characterization of the intestinal microbiota in Pacific white shrimp, Litopenaeus vannamei, fed diets with different lipid sources. Aquaculture 434, 449–455. https://doi.org/10.1016/j.aquaculture.2014.09.008.

9

# Late quaternary environmental history of Indonesia

Mahyar Mohtadi[1], Andreas Lückge[2], Stephan Steinke[3], Haryadi Permana[4], Susilohadi Susilohadi[5], Rina Zuraida[6], Tim C. Jennerjahn[7,8]

[1]*MARUM-CENTER FOR MARINE ENVIRONMENTAL SCIENCES, UNIVERSITY OF BREMEN, GERMANY;* [2]*FEDERAL INSTITUTE FOR GEOSCIENCES AND NATURAL RESOURCES, HANNOVER, GERMANY;* [3]*DEPARTMENT OF GEOLOGICAL OCEANOGRAPHY AND STATE KEY LABORATORY OF MARINE ENVIRONMENTAL SCIENCE (MEL), COLLEGE OF OCEAN AND EARTH SCIENCES, XIAMEN UNIVERSITY, CHINA;* [4]*RESEARCH CENTER FOR GEOTECHNOLOGY, INDONESIA INSTITUTE OF SCIENCES (LIPI), BANDUNG, INDONESIA;* [5]*MARINE GEOLOGICAL INSTITUTE (MGI), BANDUNG, INDONESIA;* [6]*CENTER FOR GEOLOGICAL SURVEY, BANDUNG, INDONESIA;* [7]*LEIBNIZ CENTRE FOR TROPICAL MARINE RESEARCH (ZMT), BREMEN, GERMANY;* [8]*FACULTY OF GEOSCIENCES, UNIVERSITY OF BREMEN, BREMEN, GERMANY*

## Abstract

*Paleoclimate research in Indonesia has become increasingly important with growing awareness of the critical role of this region for global climate. By reconstructing the past climate trend and variability, paleoclimate research contributes to a more reliable simulation of future Indonesian climate, which has proven difficult owing to the lack of temporally long instrumental records and model deficiencies in reproducing the complex regional setting. After two decades of paleoclimate research, there is strong evidence that melting of the ice sheets and slowdown of the ocean circulation in the North Atlantic, e.g., during Heinrich Events, result in a cooler Northern Hemisphere, and warmer and drier conditions over Sumatra, Java, and the Lesser Sunda Islands. On the other hand, this region experiences a cooler windy season and a wetter rainy season during periods of a generally warmer Northern Hemisphere summer, e.g., during precession minima, with environmental changes being more severe in monsoonal south Indonesia compared with the ever-wet equatorial regions.*

## Abstrak

*Penelitian iklim purba di Indonesia telah menjadi semakin penting dengan meningkatnya kesadaran akan peran kritis wilayah ini dalam mempengaruhi iklim global. Dengan merekonstruksi kecenderungan iklim masa lalu dan variasinya, penelitian iklim purba memberikan kontribusi dalam simulasi iklim Indonesia masa depan yang lebih dapat diandalkan, hal tersebut sebelumnya telah terbukti sulit karena kurangnya rekaman pengukuran parameter iklim secara teratur berjangka panjang dan kekurang-mampuan model dalam menggambarkan wilayah regional yang kompleks. Setelah dua dekade penelitian iklim purba, terdapat bukti kuat bahwa pencairan tudung es dan pelambatan sirkulasi lautan di Atlantik Utara, seperti selama Peristiwa Heinrich, menghasilkan iklim yang lebih sejuk di belahan bumi utara dan kondisi lebih hangat dan kering di*

**Science for the Protection of Indonesian Coastal Ecosystems (SPICE). https://doi.org/10.1016/B978-0-12-815050-4.00012-2** 

*Sumatra, Jawa dan Kepulauan Sunda Kecil. Sebaliknya daerah-daerah ini mengalami musim angin sejuk dan musim hujan basah selama periode yang umumnya hangat di belahan bumi utara pada musim panas, misalnya selama presesi minimal, yang disertai dengan perubahan lingkungan yang drastis selama musim hujan di selatan Indonesia dibandingkan dengan daerah khatulistiwa yang selalu basah.*

**Chapter outline**

**9.1 Introduction** 348
**9.2 Tools for reconstructing the environmental history of Indonesia** 349
**9.3 Environmental history of western and southern Indonesia** 352
9.3.1 Glacial–interglacial and orbital timescales 352
9.3.2 Millennial-scale climate variability 359
9.3.3 Holocene and high-frequency climate variations 360
**Acknowledgments** 364
**References** 365

# 9.1 Introduction

The Indonesian Archipelago lies in the heart of the Indo-Pacific Warm Pool (IPWP), a key region for deep atmospheric convection characterized by strong precipitation and low-level wind convergence. Deep atmospheric convection in this region mainly arises from the high sea surface temperature (SST) in the Indonesian seas, which provides a significant source of latent heat release to the atmosphere and strongly influences the Walker cell and the Australasian monsoon systems (e.g., Barsugli and Sardeshmukh, 2002; Webster et al., 1998). Regional SST variations cause changes in the surface winds and the atmospheric deep convection and alter precipitation and ocean circulation within the entire Indo-Pacific region (Sprintall et al., 2019). Similarly, model simulations suggest that a more realistic simulation of El Niño–Southern Oscillation (ENSO) requires a proper representation of this dynamical ocean–atmosphere link within the Indonesian Archipelago (e.g., Koch-Larrouy et al., 2010; Neale et al., 2008). In addition, the Indonesian Archipelago is home to the Indonesian Throughflow (ITF), the only low-latitude connection between the oceans that plays a critical role for the global ocean heat and freshwater balance as well as the (sub)surface circulation (e.g., Gordon, 1986; Sprintall et al., 2019).

Anomalous coastal upwelling offshore Java and Sumatra is capable of triggering the Indian Ocean Dipole Mode (IOD), an ocean–atmosphere coupled climate phenomenon that causes temperature and rainfall anomalies across the Indian Ocean realm and worldwide (Saji et al., 1999; Webster et al., 1999). Thus, assessing the thresholds and tipping points of such a system will help to test the skills and sensitivity of climate models in projecting tropical climate in the coming centuries. However, our understanding of climate dynamics in this region is limited by the temporal coverage of instrumental data of only a few decades, which are too short to detect forced changes against the background of natural variability. Moreover, the effect of anthropogenic forcing on regional climate and environmental variability in Indonesia is largely unknown, to a large extent owing to the scarcity of data indicating the natural variability on the relevant historical timescales. Such knowledge is essential to develop appropriate strategies in managing marine and terrestrial ecosystem changes in Indonesia in times of accelerating global climate change. To this end, temperature and rainfall reconstructions from geological past beyond the historical records are required to understand the climate response to various forcing mechanisms (Mohtadi et al., 2016). The Indonesian–German research and education program "Science for the Protection of Indonesian Coastal Marine Ecosystems" (SPICE) provided the frame for the collection and evaluation of this type of information in its cluster "Marine Geology and Biogeochemistry." Building on data evaluations derived from sediment archives in Indonesian repositories and collected during SPICE research cruises, this chapter synthesizes the available information to increase awareness of the relevance of climate change for Indonesia among stakeholders.

## 9.2 Tools for reconstructing the environmental history of Indonesia

Sample material that served in the SPICE project for reconstructing the environmental history of Indonesia were collected mainly during the seagoing expeditions SO-184 and SO-189 with the German research vessel R/V SONNE in 2005 and 2006, respectively (Fig. 9.1). During these expeditions, the Indian Ocean side of the Indonesian Archipelago was sampled systematically from the Simeulue Basin, Nias Basin, and Mentawai Basin offshore west Sumatra to Java Basin, Lombok Basin, and Savu Basin south of the Lesser Sunda Islands. Before these expeditions, this part of the Indonesian Archipelago was mostly uncharted territory with only a handful of studies on sediment cores with a low temporal resolution from Dutch, French, and Australian expeditions in the 1980s and 1990s. Most notably, collecting water column and surface sediment samples using a CTD-rosette water sampler, a sediment trap, and a multicorer during the expeditions SO-184 and SO-189 resulted in an unprecedented set of data that allowed for relating measured parameters in marine archives to modern environmental conditions and calibrating the different proxies used for paleoenvironmental reconstructions (Mohtadi et al., 2007, 2009, 2011a,b; Chen et al., 2014, 2016; Gibbons et al., 2014).

To avoid a tenuous and diluted discussion in this chapter, studies on reconstructing past variation of the ITF by utilizing marine sedimentary archives mainly from the Timor Sea and Makassar Strait and without a connection to the SPICE program are excluded

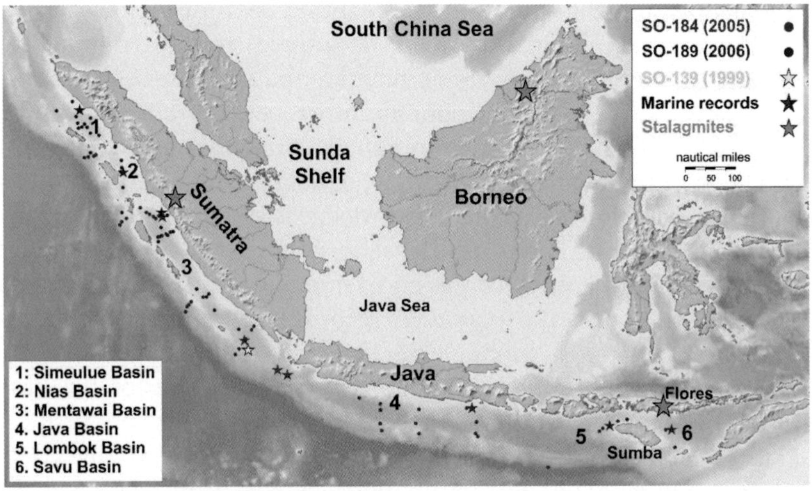


**FIGURE 9.1** Map of the sites discussed in this chapter. Numbers indicate the different basins along the Indian Ocean coast of Indonesia. Dots represent water and surface sediment sampling stations during the expeditions SO-184 (red, Mohtadi et al., 2007, 2009, 2011a,b) and SO-189 (blue, Mohtadi et al., 2011a,b). Orange stars indicate the stalagmite sites in Borneo (Carolin et al., 2013; Partin et al., 2007), Sumatra (Wurtzel et al., 2018), and Flores (Ayliffe et al., 2013; Griffiths et al., 2009). Blue/red stars show the position of the published gravity corer and piston corer data in the Simeulue Basin (Mohtadi et al., 2014, 2017), Nias Basin (Niedermeyer et al., 2014), Mentawai Basin (Mohtadi et al., 2010, 2014, 2017; Kwiatkowski et al., 2015), offshore the Sunda Strait (Mohtadi et al., 2010a,b; Setiawan et al., 2015), Java Basin (Mohtadi et al., 2011; Mohtadi et al., 2017; Setiawan et al., 2015; Ruan et al., 2019), Lombok Basin (Steinke et al., 2014a,b), and Savu Basin (Dubois et al., 2014; Gibbons et al., 2014). Yellow star marks the position of site SO139-74KL offshore the Sunda Strait (Lückge et al., 2009; Wang et al., 2018).

(Xu et al., 2006, 2008, 2010; Dürkop et al., 2008; Hendrizan et al., 2017; Holbourn et al., 2005, 2012; Kawamura et al., 2006; Kuhnt et al., 2015; Schröder et al., 2016; Zhang et al., 2018; Zuraida et al., 2009). The reader is referred to the original publications for more information.

For reconstructing past ocean temperature, one of the well-established methods is the magnesium (Mg) to calcium (Ca) ratio in fossils of surface dwelling planktic foraminifera. Numerous culture and field studies have shown an exponential relationship of this ratio to temperature, with Mg being more incorporated in the calcite shell of planktic foraminifera with increasing temperatures (see, e.g., Mohtadi et al., 2011a,b). Generally, for SST reconstructions, the mixed layer dwelling species such as *Globigerinoides ruber* and *Trilobatus sacculifer* are used, while *Neogloboquadrina dutertrei* and *Pulleniatina obliquiloculata* are commonly used for thermocline temperature reconstructions. However, since the living- and calcification depth and season of planktic foraminifera might vary regionally, depending on upper ocean conditions such as nutrient content, turbidity, and stratification, the Mg/Ca-based thermometry requires a regional calibration based on field and instrumental data (e.g., Groeneveld et al., 2019). Studies on sediment trap and surface sediment samples from SO-184 and SO-189 expeditions have confirmed the aforementioned concept and suggest that the shell geochemistry of *G. ruber* and *T. sacculifer* (*Globigerina bulloides*) in each sample can be used to reconstruct mean annual (boreal summer) conditions in the upper 50 m of the mixed-layer, and of *P. obliquiloculata* and *N. dutertrei* for the upper thermocline conditions around 70 and 100 m, respectively (Mohtadi et al., 2007, 2009, 2011; Gibbons et al., 2014).

The other established method for SST reconstruction is the alkenone paleotemperature thermometry, which is based on the unsaturation index of long-chain ketones ($U^{K'}{}_{37}$) produced by coccolithophores of the family Gephyrocapsaceae in lipid extracts of marine sediments (e.g., Prahl and Wakeham, 1987). Similar to the Mg/Ca thermometry, there are a number of calibration equations available to convert measured $U^{K'}{}_{37}$ in the sample to SST. Surface sediment samples from SO-184 and SO-189 expeditions suggest that alkenones reflect SST during the upwelling season south of the Lesser Sunda Islands and in the Southern Mentawai Basin (Chen et al., 2014; Mohtadi et al., 2011a,b). In the Northern Mentawai, Nias and Simeulue Basins that are characterized by a deep mixed layer throughout the year and no seasonal upwelling, the alkenones possibly underestimate the mean annual SST being at the upper limit of this method (Chen et al., 2014; Mohtadi et al., 2011a,b).

Past rainfall changes over Indonesia have been reconstructed by using the compound-specific isotopic signature of hydrogen and carbon ($\delta D$ and $\delta^{13}C$) in leaf waxes of higher plants from marine archives offshore Sumatra and Java (Mohtadi et al., 2017; Niedermeyer et al., 2014; Ruan et al., 2019) and Sumba (Dubois et al., 2014). The first presumption here is that riverine transport of leaf waxes is the major transport process from the continent to the ocean in wet areas, where they are constantly eroded by rain either directly from plants or from the top soils and transported as riverine suspended materials before being deposited in marine sediments (Bird et al., 1995). Hence, the isotopic composition of plant waxes in marine sediments is thought to reflect climate conditions on the adjacent land mass. Accordingly, the $\delta^{13}C$ in leaf waxes from marine sediments reflects the relative distribution of plants over land, mainly the proportion of the $C_3$ and $C_4$ plants (Chikaraishi et al., 2004). In tropical regions, the $C_4$ photosynthetic pathway typically occurs in warm-season grasses and sedges, while the $C_3$ pathway is utilized by trees and shrubs (e.g., Eglinton and Eglinton, 2008). Therefore, higher $\delta^{13}C$ values can be used simplistically to detect periods with a higher contribution of $C_4$ plants reflecting open canopy vegetation, while periods of lower $\delta^{13}C$ values indicate more $C_3$ plants and a close canopy vegetation.

For hydrogen isotope values (deuterium, $\delta D$), an additional presumption is that in convectively active regions like the IPWP, where air mass trajectories have a large vertical component and temperature variations are limited, the hydrogen isotope composition is mainly influenced by the amount of rainfall (Lee et al., 2012; Risi et al., 2008). Terrestrial plants subsequently incorporate this signal in their leaf waxes without a significant fractionation (Sachse et al., 2012). In fact, studies of both living plants and lake-surface sediments show a strong linear correlation between leaf wax $\delta D$ values and source water (precipitation) δD values along climatic gradients, implying that precipitation δD is the primary control of leaf wax δD (Sachse et al., 2012 and references therein). This relationship indicates that a lower (higher) δD value in fossil remains of plant waxes reflects an increase (a decrease) in precipitation during the period covered by the measured sample. It is noteworthy that no regional calibration or ground-truthing for the leaf wax data based on field studies or modern samples exists, owing to the relatively high amount of sediment material required for the analysis. Therefore, one of the major shortcomings of this method is the qualitative nature of the results, contrary to the quantitative results when reconstructing ocean temperatures in the past.

Other studies have used the element composition in marine sediments as an indirect measure to reconstruct past rainfall changes over Sumatra (Kwiatkowski et al., 2015; Mohtadi et al., 2010), Java (Mohtadi et al., 2011a,b; Setiawan et al., 2015), and Sumba (Steinke et al., 2014a). The presumption is that higher rainfall over land results in a higher riverine runoff, as shown for modern runoff variations of the Brantas River in east Java (Jennerjahn et al., 2004). An increased runoff, in turn, increases the relative contribution of terrestrially derived elements such as iron (Fe) or titanium (Ti) in marine sediments compared with those of marine origin, mainly calcium (Ca).

Finally, the $\delta^{18}O$ of seawater has been used to infer past rainfall changes over western and southern Indonesia (Gibbons et al., 2014; Mohtadi et al., 2014), calculated by subtracting the temperature- (estimated with Mg/Ca) and ice volume-related changes from the measured $\delta^{18}O$ in calcite shells of planktic foraminifera (see method in Mohtadi et al., 2014). The presumptions for this approach are (1) the linear relationship between salinity and $\delta^{18}O$ of seawater in the modern IPWP (Hollstein et al., 2017) existed also in the past, and (2) surface ocean salinity is primarily controlled by rainfall and freshwater input from the continent, rather than by upwelling or advection by surface ocean currents. The latter presumption limits the use of this proxy to regions where salinity changes are primarily controlled by varying amounts of rainfall in hinterland, such as the semienclosed, nonupwelling basins off west Sumatra (Sprintall and Tomczak, 1992; Du et al., 2005; Qu and Meyers, 2005). Notwithstanding, the great similarity among records across Indonesia (see next subchapter and Fig. 9.2) implies that this proxy rather reflects large-scale changes in the hydrological cycle (Gibbons et al., 2014).

# 9.3 Environmental history of western and southern Indonesia

## 9.3.1 Glacial–interglacial and orbital timescales

Rainfall reconstructions based on element composition and $\delta^{18}O$ of seawater show a different pattern across timescales and regions than the isotope-based terrestrial rainfall reconstructions (Figs. 9.2 and 9.3). Generally, $\delta^{18}O$ of seawater and element ratios in bulk sediments offshore Sumatra, Java, and Sumba do not display any glacial–interglacial variability and imply no changes in the rainfall-related runoff on this timescale (Figs. 9.2 and 9.3B). Only offshore the Sunda Strait, there is a notable change in the Ti/Ca at the

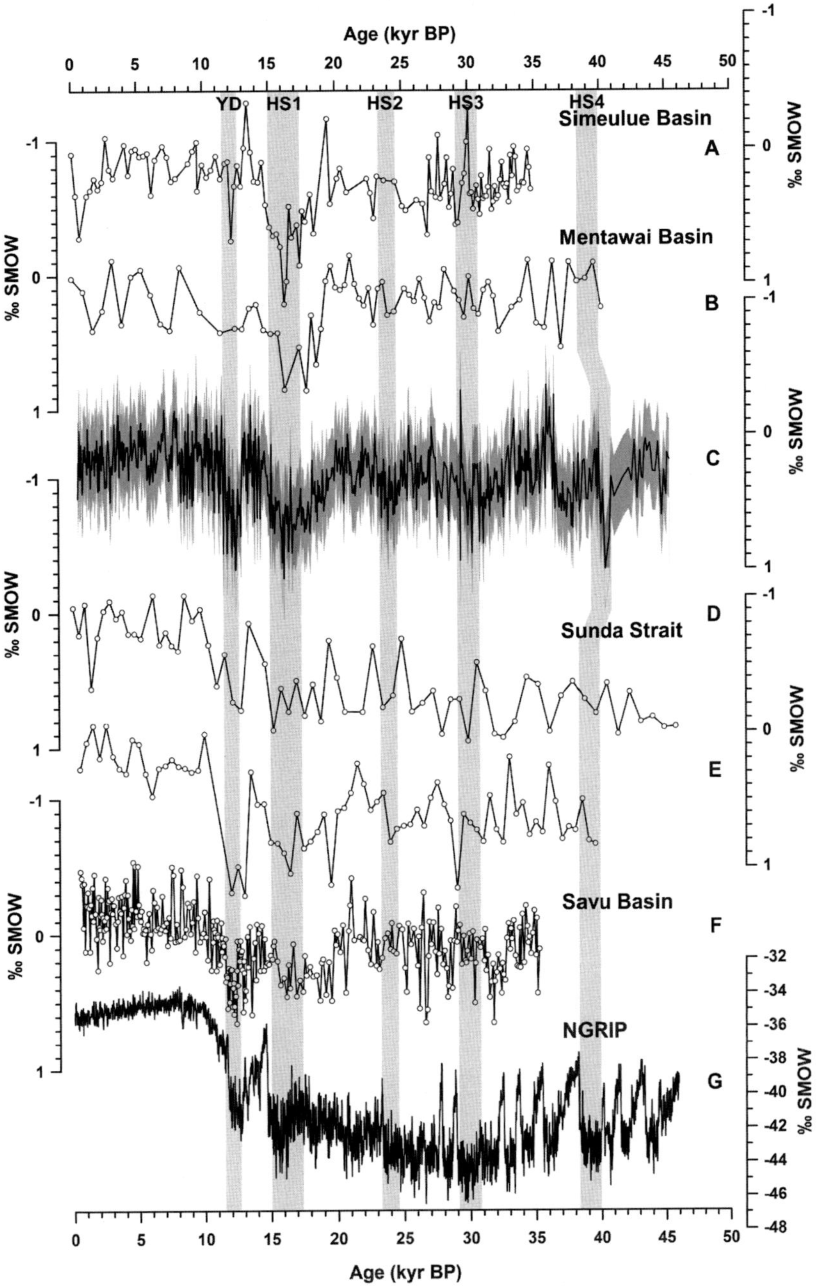


**FIGURE 9.2** Seawater $\delta^{18}O$ records from the Indian Ocean side of Indonesia. For comparability, all the records are corrected for changes in global sea level and plotted on the same scale, $\pm$ 1‰ of the Vienna Standard Mean Ocean Water (VSMOW). The 1$\sigma$ error of the method is exemplarily given as gray shading in (C). Vertical gray bars indicate the North Atlantic cold climate anomalies, the Younger Dryas (YD), and the Heinrich Stadials (HS). All ages are in 1000 years before present (kyr BP, present: CE1950). From top to bottom: (A) Site SO189-119KL from the Simeulue Basin (Mohtadi et al., 2014, 2017). (B) Site GeoB10029-4 (Mohtadi et al., 2010) and (C) site SO189-39KL (Mohtadi et al., 2014, 2017) from the northern Mentawai Basin. (D) Site GeoB10038-4 (Mohtadi et al., 2010; Mohtadi et al., 2010; Mohtadi et al., 2017) and (E) spliced record of sites GeoB10042-1 and GeoB10043-3 (Setiawan et al., 2015) offshore the Sunda Strait. (F) Site GeoB10069-3 (Gibbons et al., 2014) from the Savu Basin. (G) $\delta^{18}O$ of the Greenland ice core NGRIP (Svensson et al., 2008).

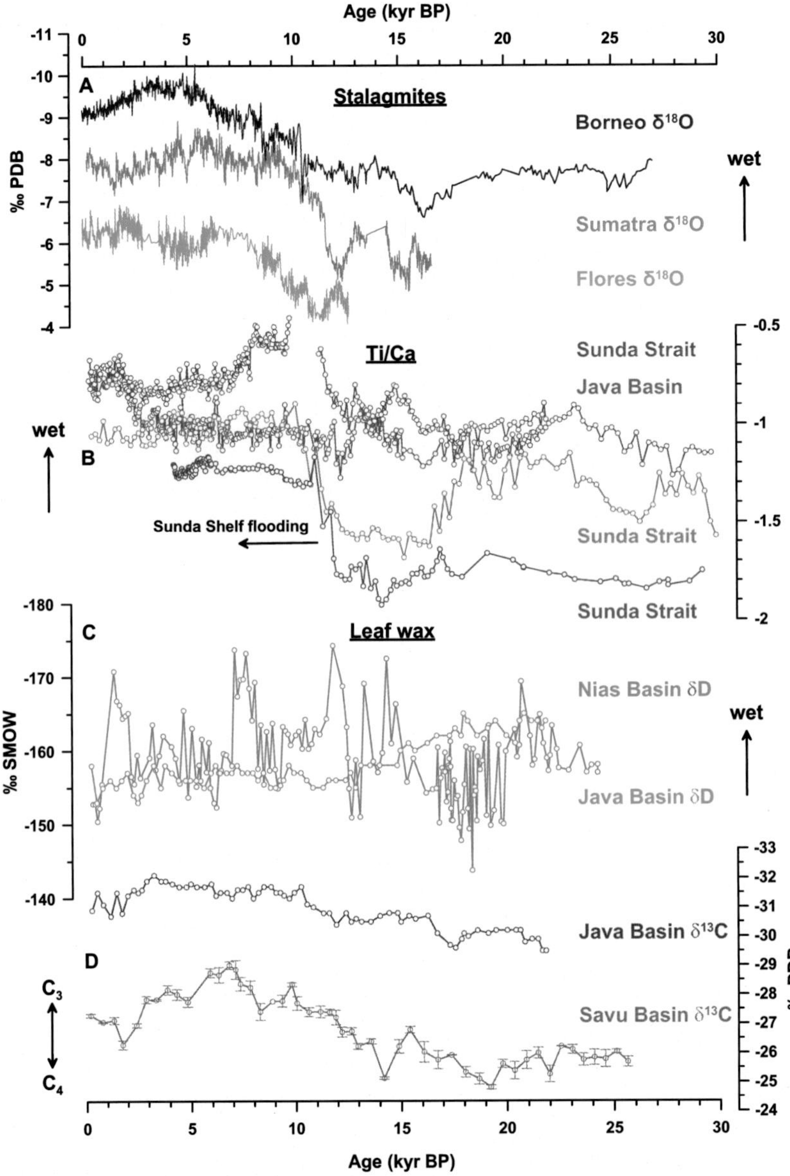


**FIGURE 9.3** Rainfall proxy records from cave stalagmites and the Indian Ocean side of Indonesia. Vertical gray bars indicate the North Atlantic cold climate anomalies, the Younger Dryas (YD), and the Heinrich Stadials (HS). From top to bottom: (A) Stalagmite $\delta^{18}O$ records from Borneo in dark blue (Carolin et al., 2013; Partin et al., 2007), Sumatra in blue (Wurtzel et al., 2018), and Flores in cyan (Ayliffe et al., 2013; Griffiths et al., 2009). Values are in ‰ versus Vienna PeeDee Belemnite (VPDB) with more negative values (upward) indicating wetter conditions. (B) Logarithmic ratios (dimensionless) between titanium (Ti) as a terrestrial endmember and calcium (Ca) as a marine endmember in bulk sedimentary deposits indicating qualitative changes in the rainfall-related terrestrial runoff. Records offshore the Sunda Strait are at sites GeoB10042-1/GeoB10043-3 in magenta (Setiawan et al., 2015), site GeoB10038-4 in orange (Mohtadi et al., 2010), and site SO139-74KL in red (Lückge et al., 2009). The record from Java Basin at site GeoB10053-7 is shown in gray (Mohtadi et al., 2011a,b). The horizontal arrow indicates the start of the Sunda Shelf flooding around 11,000 years ago, when records offshore the Sunda Strait show an abrupt increase in terrigenous supply. (C) Compound-specific hydrogen isotopes (δD) of higher terrestrial plants at site SO189-144KL from Nias Basin in deep green (Niedermeyer et al., 2014) and at site GeoB10053-7 from Java Basin in pale green (Ruan et al., 2019), corrected for global sea level changes and given in ‰ VSMOW. More negative values (upward) indicate wetter conditions. (D) Compound-specific carbon isotopes ($\delta^{13}C$, in ‰ VPDB) of higher terrestrial plants at site GeoB10053-7 from Java Basin in deep brown (Ruan et al., 2019) and at site GeoB10069-3 from Savu Basin in brown (Dubois et al., 2014). More negative values (upward) indicate a more C3-dominated vegetation. The 1σ error of the method is exemplarily given as vertical bars.

end of the last deglaciation (Fig. 9.3B) that has been related to the opening of the Sunda Strait around 11,000 years ago (Setiawan et al., 2015). The gradual flooding of the previously exposed Sunda Shelf during the deglacial sea level rise culminated in the opening of the Sunda Strait and flushing of the Java Sea sediments into the Indian Ocean (Setiawan et al., 2015). This finding is supported by the $\delta^{18}O$ of seawater records offshore the Sunda Strait (Fig. 9.2 D–E) suggesting that fresher conditions coincided with the opening of the Sunda Strait and the onset of the outflow of fresher Java Sea waters into the Indian Ocean (Mohtadi et al., 2010; Setiawan et al., 2015).

Leaf wax isotope records show a more heterogeneous pattern across Indonesia than the sediment element composition and the $\delta^{18}O$ of seawater. No glacial–interglacial shifts in the $\delta D$ values were observed in a sediment core from the Nias Basin (Fig. 9.3C, Niedermeyer et al., 2014), which conform with the $\delta^{18}O$ of seawater records from the Simeulue Basin and the Northern Mentawai Basin (Fig. 9.2A–C, Mohtadi et al., 2014). Together, these data suggest that the glacial climate of west Sumatra at and north of the equator was as wet as, if not wetter than, today. Similarly, a stalagmite $\delta^{18}O$ record from northern Borneo suggests that the hydroclimate was only weakly affected by glacial–interglacial changes in global climate boundary conditions (Carolin et al., 2013). The lack of glacial–interglacial variability in rainfall and vegetation cover in the ever-wet regions of Sumatra and Borneo is probably due to the minimum Coriolis force at the equatorial band that causes a sluggish atmospheric and oceanic circulation on the one hand, and the absence of a distinct dry or wet season on the other hand, making these regions less sensitive to changing seasonality, windiness, or hemispheric temperature gradient on glacial–interglacial scale.

Offshore south Java, the $\delta D$ values are lower and imply wetter conditions in the montane region during the last glacial period compared with the present interglacial (Fig. 9.3C, Ruan et al., 2019). However, the leaf wax $\delta^{13}C$ values of the same archive (Fig. 9.3D) suggest that glacial lowland vegetation included more $C_4$ grasses and sedges, implying that during the last glacial, lowland and montane vegetations grew under different climate conditions (Ruan et al., 2019). A longer and drier dry season during the last glacial was also inferred from a leaf wax $\delta^{13}C$ record offshore east Sumba/south Flores (Fig. 9.3D, Dubois et al., 2014). Together, leaf wax isotope data from monsoonal Indonesia suggest that the glacial climate was characterized by a stronger seasonality with a shorter but rainier wet season and a longer and drier dry season (Ruan et al., 2019). Alternatively, $\delta D$ records from interior Indonesia suggest that glacial $\delta D$ values are biased by changes in moisture convection and source (Konecky et al., 2016; Wicaksono et al., 2017, see also next paragraph).

A comparative time-slice analysis of $\delta D$ records from offshore northwest Sumatra to southeast Java suggests relatively wetter conditions during the Last Glacial Maximum (LGM), around 21,000 years ago, compared with the past 2000 years (Mohtadi et al., 2017). This is in stark contrast to drier glacial conditions inferred from cave stalagmites in Flores (Griffiths et al., 2009; Ayliffe et al., 2013) and Borneo (Partin et al., 2007; Carolin et al., 2013), as shown in Fig. 9.3A. Climate simulations indicate that during the LGM, when the Sunda Shelf was exposed due to sea level lowstand, the atmospheric convection center was shifted toward the margins of the Maritime Continent over the ocean, resulting in higher rainfall over the Indian Ocean and Pacific coasts of Indonesia on the one hand (Hollstein et al., 2018; Mohtadi et al., 2017), and relatively drier conditions over central Indonesia on the other hand, where the stalagmite sites reside.

As for the different rainfall reconstruction methods, one of the consistent features in SST reconstructions is the mismatch between Mg/Ca-based and alkenone-based paleothermometry, both in their absolute values and evolution (Figs. 9.4 and 9.5). This feature has been observed globally and attributed to differences in the seasonality and ecology of their producing organisms, i.e., coccolithophorids and foraminifera (e.g., Leduc et al., 2010). As mentioned in the previous subchapter, regional studies on the modern SST proxy carriers also support this view and indicate that alkenone-based SST represents the upwelling SST from June to October, while Mg/Ca-based SST reflects mean annual SST of the Indonesian seas (Mohtadi et al., 2011a,b; Chen et al., 2014; Gibbons et al., 2014). Not surprisingly, alkenone-based SST reconstructions from offshore southwest Sumatra (Lückge et al., 2009; Mohtadi et al., 2010) vary at the precessional band (19,000–23,000 years) that also controls local insolation and monsoon-related seasonal upwelling intensity on orbital timescales (Fig. 9.4A–B). Axial and apsidal precession are the wobble of Earth's axis and rotation of Earth's elliptical orbit over time, with periods of about 19 and 23 kyr, respectively (see, e.g., Mohtadi et al., 2016 and references therein). Precession is considered as a major control on changes in monsoon intensity since it modulates the temporal and spatial distribution of insolation and affects the seasonal cycle of incoming solar radiation and its hemispheric distribution (Kutzbach, 1981). During low precession, the summer hemisphere receives more energy (Fig. 9.4A) that is compensated by the atmosphere through stronger cross-equatorial energy transport by winds, which in turn results in a stronger Ekman transport and upwelling in monsoonal Indonesia (see Mohtadi et al., 2016 and references therein).

Contrary to the alkenone-based SST records, the Mg/Ca-based SST reconstructions from the eastern tropical Indian Ocean do not show a precession control but vary on glacial–interglacial timescale akin to the Antarctic temperatures (Figs. 9.4 and 9.5, Mohtadi et al., 2010; Setiawan et al., 2015; Wang et al., 2018). This pattern is consistent from the nonupwelling, northern Mentawai and Simeulue Basins (Mohtadi et al., 2014) to as far east as in the Savu Basin (Gibbons et al., 2014), implying a consistent response of the mean annual SST in the eastern tropical Indian Ocean to changing global ice volume and atmospheric carbon dioxide on glacial–interglacial timescales.

Thermocline temperatures offshore the Sunda Strait appear to vary on precessional band (Fig. 9.4E), since thermocline depth and temperature in this region are mainly controlled by monsoon-related seasonal upwelling (Wang et al., 2018). Similar to the mechanisms responsible for variations in the alkenone-based SST records, periods of low precession are accompanied by a higher interhemispheric energy transfer that is

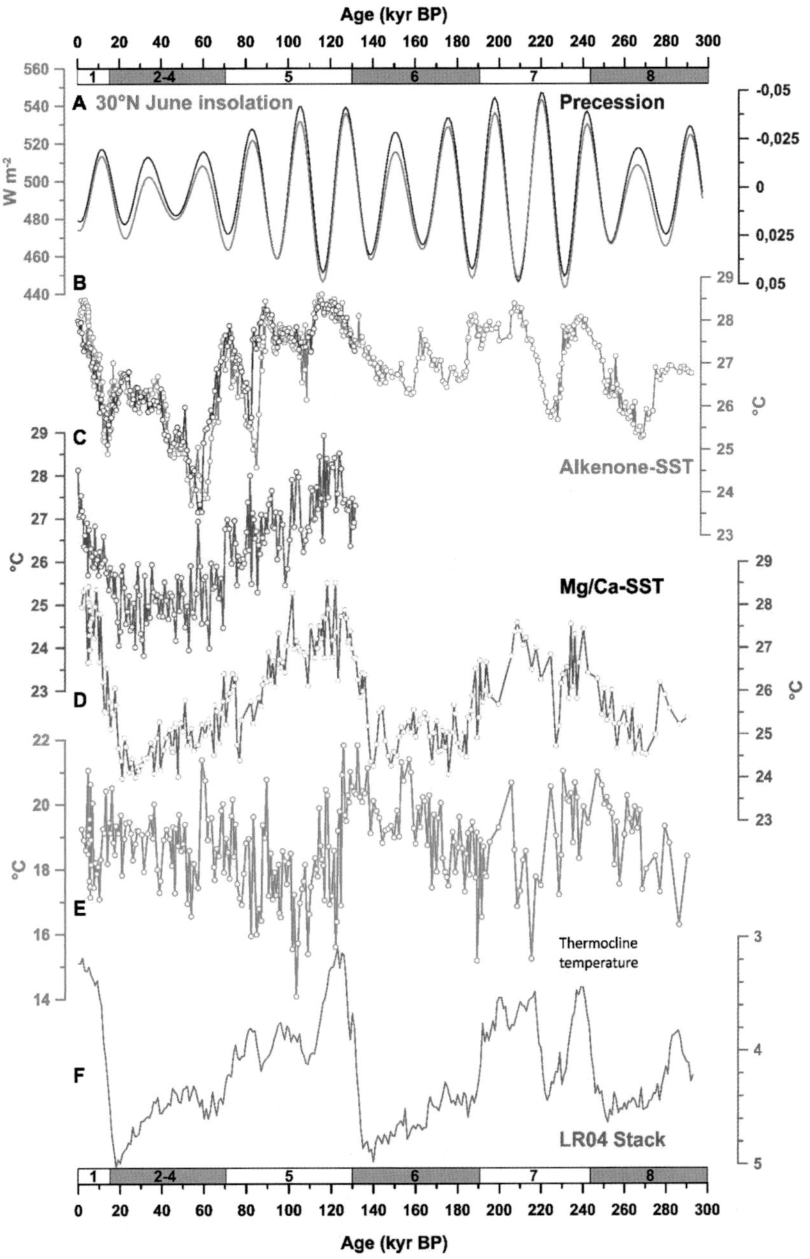


**FIGURE 9.4** Glacial–interglacial and orbital-scale ocean temperature variability offshore the Sunda Strait. Glacial and interglacial periods are indicated by gray and white bars, respectively, with numbers indicating marine isotopic stages. From top to bottom: (A) Precession index in blue (reverse scale, dimensionless) and Northern Hemisphere June insolation at 30°N in orange (Watt per square meter, W m−2). (B) Alkenone-based SST reconstructions at site SO139-74KL in gray (Lückge et al., 2009) and at site GeoB10038-4 in purple (Mohtadi et al., 2010). (C) and (D) Mg/Ca-based SST reconstructions at site GeoB10038-4 in magenta (Mohtadi et al., 2010) and at site SO139-74KL in red (Wang et al., 2018). For comparability, the SST records are plotted on the same temperature range between 23 and 29°C. (E) Mg/Ca-based thermocline temperature reconstruction at site SO139-74KL in green (Wang et al., 2018). Note that the temperature range of the y-axis is larger (8°C) compared with the SST records (6°C). (F) Relative changes in global sea level (LR04-stack, dimensionless) during the past glacial–interglacial cycles (Lisiecki and Raymo, 2005). *SST*, sea surface temperature.

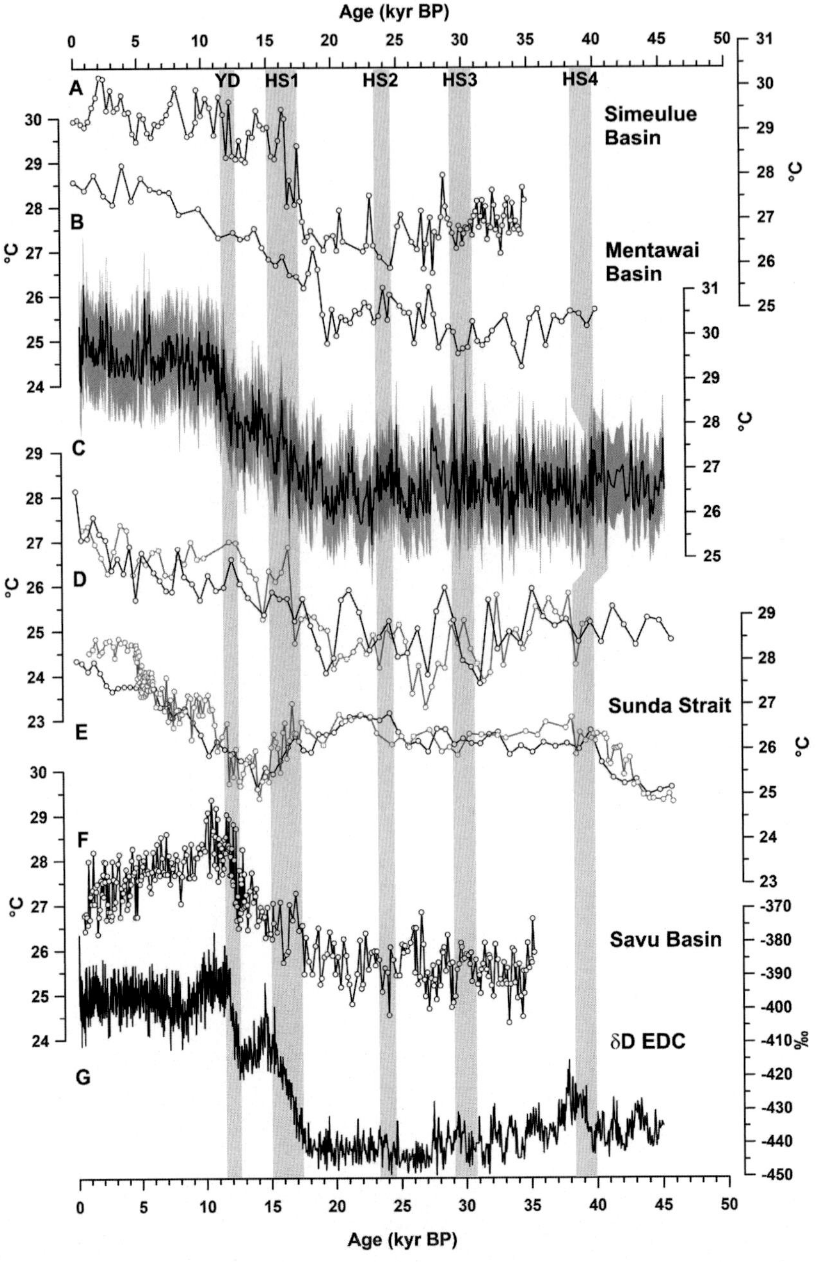


**FIGURE 9.5** SST records from the Indian Ocean side of Indonesia. For comparability, all the records are plotted at the same temperature range of 6°C. The 1σ error of the Mg/Ca-based method is exemplarily given as gray shading in (C). Vertical gray bars and abbreviations as in Fig. 9.2. From top to bottom: (A) Site SO189-119KL (Mohtadi et al., 2014, 2017) from the Simeulue Basin. (B) Site GeoB10029-4 (Mohtadi et al., 2010) and (C) site SO189-39KL (Mohtadi et al., 2014, 2017) from the northern Mentawai Basin. (D) Site GeoB10038-4 in black (Mohtadi et al., 2010; Mohtadi et al., 2017) and the spliced record of sites GeoB10042-1 and GeoB10043-3 in gray (Setiawan et al., 2015) offshore the Sunda Strait. (E) Alkenone-based SST reconstruction at sites GeoB10038-4 in black (Mohtadi et al., 2010) and site SO139-74KL in gray (Lückge et al., 2009) offshore the Sunda Strait. (F) Site GeoB10069-3 (Gibbons et al., 2014) from the Savu Basin. (G) δD record of ice core EPICA Dome C (EDC) from Antarctica (Parrenin et al., 2013). SST, seas surface temperature.

accomplished by cross-equatorial winds, which in turn facilitate coastal upwelling and shoaling/cooling of the thermocline in monsoonal Indonesia.

Taken together, evidence from orbital-scale climate variability suggests that a stronger interhemispheric temperature gradient, e.g., during precession minima, results in stronger cross-equatorial winds that facilitate upwelling and marine productivity in the eastern tropical Indian Ocean and a stronger seasonality of rainfall over monsoonal Indonesia. These findings have important implications for the future considering that the Northern Hemisphere is projected to warm at a faster and higher rate than the Southern Hemisphere in the 21st century, resulting in a higher interhemispheric temperature gradient (e.g., Wang et al., 2013).

## 9.3.2 Millennial-scale climate variability

On multicentennial to millennial timescales, the $\delta^{18}O$ of seawater offshore Sumatra covaries with abrupt climate events in the North Atlantic over the past 40,000 years (Fig. 9.2A–C, Mohtadi et al., 2014). Similar millennial-scale variability with the same sign of change was observed in the element composition of sediments from offshore south Java (Fig. 9.3B, Mohtadi et al., 2011a,b) and in $\delta^{18}O$ of seawater offshore east Sumba (Fig. 9.2F) that resembles the first principal component of all $\delta^{18}O$ records of the IPWP (Gibbons et al., 2014). The consistency among different locations and methods suggests a North Atlantic control on the IPWP climate on millennial timescales, which is mainly related to changes in the strength of the global thermohaline circulation forced by variations in the North Atlantic deep water formation (Mohtadi et al., 2014): Periods of a less vigor thermohaline circulation, e.g., during Heinrich Stadials and the Younger Dryas, are associated with cooler conditions in the North Atlantic and less rainfall over the entire monsoonal Asia including western and southern Indonesia, while periods of stronger deep water formation in a relatively warmer North Atlantic result in generally wetter conditions in Asia (Mohtadi et al., 2014). This scenario is supported by the sole stalagmite $\delta^{18}O$ record from Sumatra (Wurtzel et al., 2018) that shows the same timing and sign of rainfall change (Fig. 9.3A) as inferred from marine records from the Mentawai and Simeulue Basins (Mohtadi et al., 2014). Mechanistically, a weaker or a disrupted thermohaline circulation results in a reduced northward heat transport by the surface ocean in the Northern Hemisphere. Consequently, the atmosphere compensates for the reduced oceanic heat transport by transferring warm air from the warmer Southern Hemisphere to the cooler Northern Hemisphere, which is accomplished by a southward displacement of the Intertropical Convergence Zone (ITCZ, Donohoe et al., 2013; Marshall et al., 2014). A more southerly position of the ITCZ results in a general drying of the Northern Hemisphere and the equatorial ocean including the Maritime Continent, and wetter conditions in the Southern Hemisphere, e.g., over northern Australia (Gibbons et al., 2014; Mohtadi et al., 2011, 2014).

On millennial timescales, there is only a subtle response of surface and thermocline temperature offshore Indonesia ($\sim 1°C$ increase, Fig. 9.5) to abrupt climate changes in the North Atlantic during Heinrich Stadials and the Younger Dryas (Gibbons et al., 2014; Mohtadi et al., 2014). In the upwelling regions offshore southwest Sumatra and south Java, these cold climate events are associated with an increased austral winter upwelling due to stronger cross-equatorial winds (Mohtadi et al., 2011a,b). Akin to orbital-paced variations, cross-equatorial winds compensate for the interhemispheric energy imbalance during Heinrich Stadials and the Younger Dryas by transferring energy from the anomalously warm Southern Hemisphere to the anomalously cool Northern Hemisphere, hence increasing the austral winter upwelling of cooler subsurface waters that should cool the austral winter SST. This scenario is supported by alkenone-based SST reconstructions showing a slight decrease in the upwelling SST offshore the Sunda Strait (Fig. 9.5E). However, both data and model results suggest a net mean annual SST warming offshore Sumatra and the Lesser Sunda Islands during Heinrich Stadials and the Younger Dryas (Gibbons et al., 2014; Mohtadi et al., 2014; Setiawan et al., 2015). It appears that the subtle increase in mean annual (Mg/Ca-based) SST is mainly a result of the large and abrupt drying over the IPWP during these periods caused by a southward shift in the position of the ITCZ (see Section 9.4), anomalous atmospheric descent, and high pressure over the entire northern Indian Ocean and a higher sensible heat gain of the ocean (Mohtadi et al., 2014). In addition, the global rise in the atmospheric $CO_2$ during these periods should have contributed to the net SST warming of the entire IPWP (Moffa-Sanchez et al., 2019).

Taken together, millennial-scale variability in rainfall over Sumatra, Java, and the Lesser Sunda Islands, and in the ocean temperature and productivity of the eastern tropical Indian Ocean, appears to be controlled by changes in the strength of the oceanic global thermohaline circulation. Freshwater perturbations in the North Atlantic are capable of greatly reducing this circulation (Alley and Clark, 1999), as evidenced by iceberg release and freshwater discharge during Heinrich Stadials and the Younger Dryas. The subsequent cooling of the North Atlantic provokes rapid changes in the atmospheric circulation by displacing the ITCZ to the warmer Southern Hemisphere, a weaker Southern Hemisphere Hadley and Northern Hemisphere monsoon circulation, and a drier and warmer western and southern Indonesia (Mohtadi et al., 2014). Thus, evidence from the past provides clues for the Indonesian climate in face of the projected rapid reduction in the northern high-latitude ice sheets and slowdown of the thermohaline circulation in the 21st century (Christensen et al., 2013; Flato et al., 2013).

### 9.3.3 Holocene and high-frequency climate variations

Leaf wax $\delta^{13}C$ records from offshore east and west Sumba show an increasing trend during the early and mid-Holocene (Fig. 9.3D), which has been interpreted as an overall drying of the monsoonal Indonesia during the Holocene (Dubois et al., 2014). However, the interpretation of leaf wax $\delta^{13}C$ in terms of rainfall changes is not straightforward, and the $\delta^{13}C$ record of *n*-alkanes offshore south Java lacks any trend during the Holocene (Fig. 9.3D), casting doubt on a continuous drying of the monsoonal Indonesia during the

Holocene (Ruan et al., 2019). The latter study argues that different homologs are not equally sensitive to $C_3/C_4$ vegetation-type shifts, with $\delta^{13}C$ of $n$-$C_{29}$ being the most sensitive homolog in a tropical ecosystem and lacking any notable change during the Holocene (Ruan et al., 2019). This view is supported by the δD record at the same site offshore south Java (Fig. 9.3C) deprived of a trend during the Holocene (Ruan et al., 2019). Holocene δD values from the Nias Basin also lack a trend during the Holocene and are punctuated by millennial-scale changes indicating wetter conditions between 8000–6500 years ago, 4500–2500 years ago, and 2000–1000 years ago (Fig. 9.3C, Niedermeyer et al., 2014). These intervals coincide roughly with changes in the stalagmite $\delta^{18}O$ record from Flores (Fig. 9.3A, Griffiths et al., 2009) and have been interpreted as reflecting changes in the mean state of the IOD (Niedermeyer et al., 2014).

Holocene records of rainfall-related runoff paint a different picture across Indonesia. Offshore south Java, the Ti/Ca values and the terrigenous component of the sediment suggest higher rainfall during the past 3000 years compared with the rest of the Holocene (Fig. 9.3B, Mohtadi et al., 2011a,b). Likewise, a high-resolution, 6000-year runoff record from the Lombok Basin offshore west Sumba (Fig. 9.6A) indicates lower riverine detrital supply and hence weaker summer monsoon rainfall before 3000 years ago followed by wetter conditions during the Late Holocene (Steinke et al., 2014a). The last century is characterized by highest riverine runoff at this site, which is probably related to increasing human (agriculture) impact on vegetation and land surface. Using a climate model simulation, this study suggests that the abrupt shift to wetter conditions about 2800 years ago coincided with a grand solar minimum, and short-term variability in runoff over the past 6000 years was modulated by solar activity, whereas ENSO did not exert a significant control on rainfall variability at multidecadal to multimillennial timescales (Steinke et al., 2014a).

There is no ubiquitous trend in the Holocene evolution of SST in the eastern tropical Indian Ocean (Fig. 9.5). Records based on Mg/Ca in *G. ruber* offshore Sumatra show little variability mostly within the error of the method (±1°C, Mohtadi et al., 2014) and partly with opposing long-term trends, in both the nonupwelling and upwelling regions (Fig. 9.5), and remain inconclusive throughout the Holocene (Mohtadi et al., 2010; Mohtadi et al., 2010; Mohtadi et al., 2014; Setiawan et al., 2015). However, offshore east Sumba in the Savu Basin (Fig. 9.5F, Gibbons et al., 2014) and along the ITF path in the Indonesian seas, there is a significant overall decrease during the course of the Holocene that follows the decline in the Northern Hemisphere insolation (see compilations in Gibbons et al., 2014; Linsley et al., 2010; Mohtadi et al., 2010a,b; Stott et al., 2004). A sole control of precession-paced insolation changes on the Holocene SST in this region is rather unlikely since the decreasing trend is evident in records from 6°N to 11°S, i.e., in both hemispheres (Stott et al., 2004). It has been speculated that the decreasing trend in these SST records stems from an eastward shift of the IPWP during the mid- and late Holocene, though this view is not supported by climate models (Linsley et al., 2010).

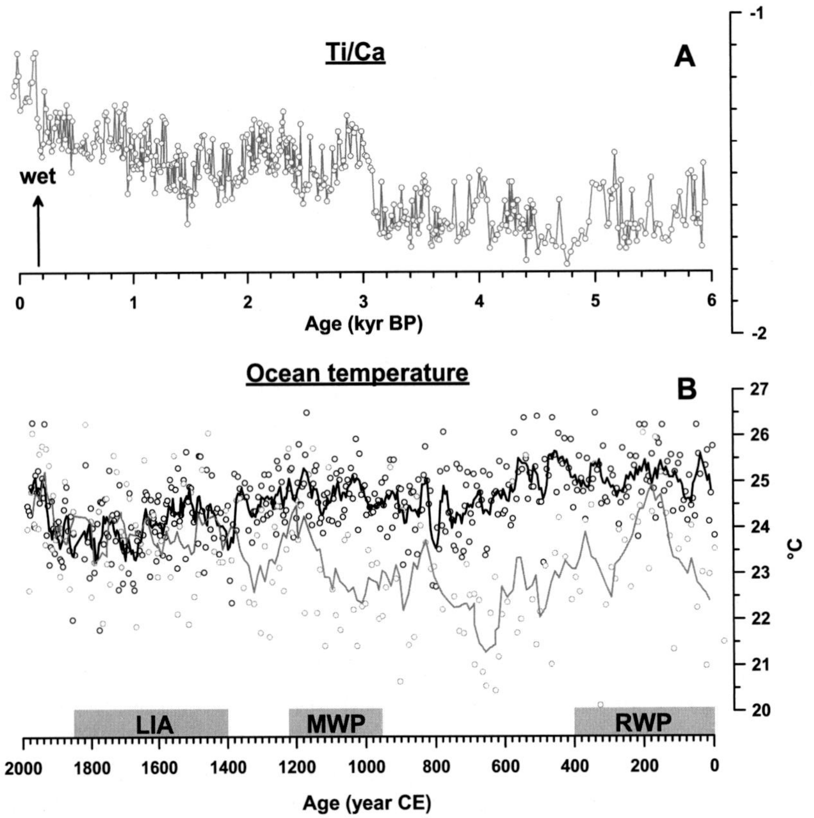


**FIGURE 9.6** Holocene records of rainfall and ocean temperature from the Lombok Basin. All the data are generated at site GeoB10065-7 in the Lombok Basin offshore western Sumba Island (see Fig. 9.1). From top to bottom: (A) Natural logarithmic ratio between Ti and Ca (Ti/Ca) indicating changes in the rainfall-related runoff over the past 6000 years (Steinke et al., 2014a). More positive values (upward) indicate wetter conditions. (B) Reconstructed ocean temperature during the upwelling season at surface (black circles) and thermocline (gray circles) and their 5-point running averages (lines) for the past 2000 years (Steinke et al., 2014b). Note that ages are in year CE. Gray boxes frame the Roman Warm Period (RWP), Medieval Warm Period (MWP), and the Little Ice Age (LIA).

Tidal mixing within the Indonesian seas is capable of changing the ITF properties and SST (Sprintall et al., 2014), but the cooling trend is also evident in further upstream records from offshore Mindanao (Stott et al., 2004), indicating that the Holocene cooling along the ITF path resides in the North Pacific Subtropical Waters, the main source water of the ITF (Gordon, 2005).

A study from the northern Mentawai Basin suggests a gradual cooling and shoaling of the thermocline caused by more frequent and severe upwelling events toward the late Holocene (Kwiatkowski et al., 2015). An orbital-forced climate model indicates a shift from a westerly dominated wind regime to an easterly dominated wind regime during the upwelling season in the mid-Holocene, resulting in a gradual shift from a more negative IOD-like mean state to a more positive IOD-like mean state (Kwiatkowski et al., 2015). However, the two alkenone-based SST records offshore the Sunda Strait show a continuous increase during the course of the Holocene (Fig. 9.5E, Lückge et al., 2009; Mohtadi et al., 2010a,b). Assuming that alkenones record the upwelling season SST (Mohtadi et al., 2010, 2011; Chen et al., 2014, 2016), their increasing trend can be only explained by a continuous decline in the upwelling of cooler subsurface waters during the Holocene. These sites lie within the area of highest upwelling and temperature anomalies during the IOD events and thus indicate a more negative IOD-like mean state toward the late Holocene. It is unclear to what extent proxy records used for paleoclimate reconstructions are affected by meridional changes in the atmospheric circulation, i.e., the position and strength of the Hadley cell and the ITCZ with a seasonal cycle, and zonal changes in the atmospheric circulation, i.e., the strength of the Walker circulation and the state of ENSO and IOD that operate on interannual scale. Therefore, the interpretation of proxy records remains limited and speculative as long as long-term sampling and monitoring efforts across the Maritime Continent are missing.

Ocean temperature variability on shorter timescales is reported from one site in the Lombok Basin offshore west Sumba covering the past 2000 years (Steinke et al., 2014a,b). Based on Mg/Ca-based surface and thermocline temperature reconstructions during the upwelling season, this study shows an overall SST cooling during the upwelling season toward the end of the Little Ice Age (LIA, 1400–1850 CE, Fig. 9.6B). Thermocline temperatures and SST converge after the Medieval Warm Period (MWP, 950–1250 CE, Fig. 9.6B) and suggest a stronger upwelling during the LIA and a warmer thermocline and weaker upwelling during the MWP and Roman Warm Period (250 BCE–400 CE). While ocean temperature reconstructions from this site indicate a strong control of ENSO on upwelling intensity during the late Holocene (Steinke et al., 2014a,b), rainfall-related runoff reconstruction at the same site (Fig. 9.6A) shows no connection to ENSO but varies with solar activity (Steinke et al., 2014a,b). These findings suggest that different proxies in the same archive might be affected by several climate forcings that operate at different timescales and underscore the great potential of paleoclimate studies in reconstructing the frequency, amplitude, and trend of various climate phenomena.

Taken together, there are only a handful of Holocene rainfall and ocean temperature records with a sufficient temporal resolution from the Indian Ocean side of Indonesia. The observed changes, if any, have been interpreted to reflect variability in the ENSO, IOD, ITCZ, or solar output yet remain inconsistent among proxies. This might stem from different sensitivity of different proxies to environmental conditions and their forcing mechanisms on various timescales and requires a deeper understanding of these proxies and how they preserve different climate signals.

**Knowledge gaps and directions of future research**

Despite the efforts during the past two decades, one of the main issues in understanding the climate evolution of Indonesia is the lack of continuous and reliable sampling on land and in the ocean. Satellites, meteorological stations, and buoys have delivered invaluable data during the past decades, but to extend the instrumental data back in time and improve the "climate memory" for future predictions of such a complex and diverse setting, it is pivotal to understand how the environment is being "recorded" by the biosphere and geosphere. There are presently no efforts in sampling proxy carriers for temperature and productivity, such as marine plankton (foraminifera, algae) and benthos (corals, bivalves), and measuring their chemical composition on weekly to annual basis. Likewise, no systematic effort exists in monitoring water isotopes of rainfall in terrestrial plants, stalagmites, and riverine discharge to understand the hydrological cycle and its fingerprint on proxy carriers. This modern ground-truthing of the proxies is critical to decipher changes related to seasonal (monsoon, ITCZ) or interannual variability (ENSO, IOD) in the past.

Another major gap in paleoclimate research in Indonesia is the lack of long records that cover several glacial–interglacial cycles. Only one stalagmite record from Borneo and two marine records from offshore the Sunda Strait do not suffice to understand the spatiotemporal response and sensitivity of the Indonesian climate to sea level and temperature changes in the past. Records that ideally extend back to periods of higher-than-today SST and sea level, e.g., the marine isotope stage 11 or Pliocene, are required to compare paleodata and future projections.

**Implications/recommendations for policy and society**

Climate change greatly affects the life and livelihood of the Indonesian people, but understanding and predicting natural climate variability in this climatically complex region have been an everlasting challenge for decision-makers and the scientific community. Monitoring and sampling endeavors in Indonesia have been historically difficult mainly due to political and logistical impediments for sampling campaigns. However, such endeavors are key to understanding the trend and variability of the Indonesian climate. There is an increasing awareness for the consequences of climate change in Indonesia and worldwide, yet climate sciences are not a focal point of the Indonesian educational system. Setting up scholarship programs for Indonesian undergraduate and postgraduate students to attend training courses and workshops at foreign universities and research institutions might help to improve the domestic expertise. However, for a deep understanding of the complex dynamics of the Indonesian climate, a clear emphasis on climate sciences should be included in studies at both bachelor and master levels.

# Acknowledgments

We are grateful to the Indonesian Institute of Sciences (LIPI) and the Marine Geological Institute (MGI) in Bandung, the Agency for the Assessment and Application of Technology (BPPT), and the Indonesian

Ministry for Research and Technology (RISTEK) in Jakarta for great support in preparing and conducting the research expeditions. The studies presented here were funded by the Federal Ministry for Education and Research (BMBF) through grants 03G0139A (GINCO3), 03G0184A (PABESIA), 03G0189A (SUMATRA), 03F0645A, B (SPICE), and the German Science Foundation (DFG) grants HE3412/15-1, STE1044/4-1, and JE281/4-1.

# References

Alley, R.B., Clark, P.U., 1999. The deglaciation of the northern hemisphere: a global perspective. Annual Review of Earth and Planetary Sciences 27, 149–182.

Ayliffe, L.K., Gagan, M.K., Zhao, J.-X., Drysdale, R.N., Hellstrom, J.C., Hantoro, W.S., et al., 2013. Rapid interhemispheric climate links via the Australasian monsoon during the last deglaciation. Nature Communications 4. https://doi.org/10.1038/ncomms3908.

Barsugli, J.J., Sardeshmukh, P.D., 2002. Global atmospheric sensitivity to tropical SST anomalies throughout the Indo-Pacific basin. Journal of Climate 15 (23), 3427–3442. https://doi.org/10.1175/1520-0442(2002)015<3427:gastts>2.0.co;2.

Bird, M.I., Chivas, A.R., Brunskill, G.J., 1995. Carbon-isotope composition of sediments from the Gulf of Papua. Geo-Marine Letters 15 (3–4), 153–159. https://doi.org/10.1007/bf01204457.

Carolin, S.A., Cobb, K.M., Adkins, J.F., Clark, B., Conroy, J.L., Lejau, S., et al., 2013. Varied response of western Pacific hydrology to climate forcings over the last glacial period. Science 340 (6140), 1564–1566. https://doi.org/10.1126/science.1233797.

Chen, W., Mohtadi, M., Schefuß, E., Mollenhauer, G., 2014. Organic-geochemical proxies of sea surface temperature in surface sediments of the tropical eastern Indian Ocean. Deep Sea Research Part I: Oceanographic Research Papers 88 (0), 17–29. https://doi.org/10.1016/j.dsr.2014.03.005.

Chen, W., Mohtadi, M., Schefuß, E., Mollenhauer, G., 2016. Concentrations and abundance ratios of long-chain alkenones and glycerol dialkyl glycerol tetraethers in sinking particles south of Java. Deep Sea Research Part I: Oceanographic Research Papers 112, 14–24. https://doi.org/10.1016/j.dsr.2016.02.010.

Chikaraishi, Y., Naraoka, H., Poulson, S.R., 2004. Hydrogen and carbon isotopic fractionations of lipid biosynthesis among terrestrial (C3, C4 and CAM) and aquatic plants. Phytochemistry 65 (10), 1369–1381. https://doi.org/10.1016/j.phytochem.2004.03.036.

Christensen, J.H., Kumar, K.K., Aldrian, E., An, S.-I., Cavalcanti, I.F.A., Castro, M., et al., 2013. Climate phenomena and their relevance for future regional climate change. In: Stocker, T.F., Qin, D., Plattner, G.-K., Tignor, M., Allen, S.K., Boschung, J., et al. (Eds.), Climate Change 2013: The Physical Science Basis. Contribution of Working Group I to the Fifth Assessment Report of the Intergovernmental Panel on Climate Change. Cambridge University Press, Cambridge, UK, New York, NY, USA.

Donohoe, A., Marshall, J., Ferreira, D., McGee, D., 2013. The relationship between ITCZ location and cross-equatorial atmospheric heat transport: from the seasonal cycle to the last glacial maximum. Journal of Climate 26 (11), 3597–3618. https://doi.org/10.1175/jcli-d-12-00467.1.

Dubois, N., Oppo, D.W., Galy, V.V., Mohtadi, M., van der Kaars, S., Tierney, J.E., et al., 2014. Indonesian vegetation response to changes in rainfall seasonality over the past 25,000 years. Nature Geoscience 7 (7), 513–517. https://doi.org/10.1038/ngeo2182.

Du, Y., Qu, T., Meyers, G., Masumoto, Y., Sasaki, H., 2005. Seasonal heat budget in the mixed layer of the southeastern tropical Indian Ocean in a high-resolution ocean general circulation model. Journal of Geophysical Research 110, C04012. https://doi.org/04010.01029/02004JC002845.

Dürkop, A., Holbourn, A., Kuhnt, W., Zuraida, R., Andersen, N., Grootes, P.M., 2008. Centennial-scale climate variability in the Timor Sea during marine isotope stage 3. Marine Micropaleontology 66 (3–4), 208–221.

Eglinton, T.I., Eglinton, G., 2008. Molecular proxies for paleoclimatology. Earth and Planetary Science Letters 275 (1), 1–16. https://doi.org/10.1016/j.epsl.2008.07.012.

Flato, G., Marotzke, J., Abiodun, B., Braconnot, P., Chou, S.C., Collins, W., et al., 2013. Evaluation of climate models. In: Stocker, T.F., Qin, D., Plattner, G.-K., Tignor, M., Allen, S.K., Boschung, J., et al. (Eds.), Climate Change 2013: The Physical Science Basis. Contribution of Working Group I to the Fifth Assessment Report of the Intergovernmental Panel on Climate Change. Cambridge University Press, Cambridge, United Kingdom, New York, NY, USA.

Gibbons, F.T., Oppo, D.W., Mohtadi, M., Rosenthal, Y., Cheng, J., Liu, Z., et al., 2014. Deglacial $\delta^{18}O$ and hydrologic variability in the tropical Pacific and Indian Oceans. Earth and Planetary Science Letters 387 (0), 240–251. https://doi.org/10.1016/j.epsl.2013.11.032.

Gordon, A.L., 1986. Interocean exchange of thermocline water. Journal of Geophysical Research 91, 5037–5046.

Gordon, A.L., 2005. Oceanography of the Indonesian seas and their throughflow. Oceanography 18 (4), 14–28.

Griffiths, M.L., Drysdale, R.N., Gagan, M.K., Zhao, J., Ayliffe, L.K., Hellstrom, J.C., et al., 2009. Increasing Australian-Indonesian monsoon rainfall linked to early Holocene sea-level rise. Nature Geoscience 2 (9), 636–639.

Groeneveld, J., Ho, S.L., Mackensen, A., Mohtadi, M., Laepple, T., 2019. Deciphering the variability in Mg/Ca and stable oxygen isotopes of individual foraminifera. Paleoceanography and Paleoclimatology 34 (5), 755–773. https://doi.org/10.1029/2018pa003533.

Hendrizan, M., Kuhnt, W., Holbourn, A., 2017. Variability of Indonesian throughflow and Borneo runoff during the last 14 kyr. Paleoceanography 32 (10), 1054–1069. https://doi.org/10.1002/2016PA003030.

Holbourn, A., Kuhnt, W., Kawamura, H., Jian, Z., Grootes, P., Erlenkeuser, H., et al., 2005. Orbitally paced paleoproductivity variations in the Timor Sea and Indonesian Throughflow variability during the last 460 kyr. Paleoceanography 20, PA3002. https://doi.org/3010.1029/2004PA001094.

Holbourn, A., Kuhnt, W., Xu, J., 2012. Indonesian throughflow variability during the last 140 ka: the Timor Sea outflow. In: Hall, R., Cottam, M.A., Wilson, M.E.J. (Eds.), The SE Asian Gateway: History and Tectonics of the Australia–Asia Collision, vol. 355. Geological Society, London, pp. 283–303.

Hollstein, M., Mohtadi, M., Rosenthal, Y., Moffa Sanchez, P., Oppo, D., Martínez Méndez, G., et al., 2017. Stable oxygen isotopes and Mg/Ca in planktic foraminifera from modern surface sediments of the western Pacific warm pool: implications for thermocline reconstructions. Paleoceanography 32 (11), 1174–1194. https://doi.org/10.1002/2017PA003122.

Hollstein, M., Mohtadi, M., Rosenthal, Y., Prange, M., Oppo, D.W., Martínez Méndez, G., et al., 2018. Variations in Western Pacific Warm Pool surface and thermocline conditions over the past 110,000 years: forcing mechanisms and implications for the glacial Walker circulation. Quaternary Science Reviews 201, 429–445. https://doi.org/10.1016/j.quascirev.2018.10.030.

Jennerjahn, T.C., Ittekkot, V., Klöpper, S., Adi, S., Purwo Nugroho, S., Sudiana, N., et al., 2004. Biogeochemistry of a tropical river affected by human activities in its catchment: Brantas River estuary and coastal waters of Madura Strait, Java, Indonesia. Estuarine, Coastal and Shelf Science 60 (3), 503–514.

Kawamura, H., Holbourn, A., Kuhnt, W., 2006. Climate variability and land-ocean interactions in the Indo Pacific warm pool: a 460-ka palynological and organic geochemical record from the Timor Sea. Marine Micropaleontology 59 (1), 1–14.

Koch-Larrouy, A., Lengaigne, M., Terray, P., Madec, G., Masson, S., 2010. Tidal mixing in the Indonesian Seas and its effect on the tropical climate system. Climate Dynamics 34 (6), 891–904.

Konecky, B., Russell, J., Bijaksana, S., 2016. Glacial aridity in central Indonesia coeval with intensified monsoon circulation. Earth and Planetary Science Letters 437, 15–24.

Kuhnt, W., Holbourn, A., Xu, J., Opdyke, B., De Deckker, P., Röhl, U., et al., 2015. Southern Hemisphere control on Australian monsoon variability during the late deglaciation and Holocene. Nature Communications 6. https://doi.org/10.1038/ncomms6916.

Kutzbach, J.E., 1981. Monsoon climate of the early Holocene: climate experiment with the Earth's orbital parameters for 9000 years ago. Science 214, 59–61.

Kwiatkowski, C., Prange, M., Varma, V., Steinke, S., Hebbeln, D., Mohtadi, M., 2015. Holocene variations of thermocline conditions in the eastern tropical Indian Ocean. Quaternary Science Reviews 114, 33–42. https://doi.org/10.1016/j.quascirev.2015.01.028.

Leduc, G., Schneider, R., Kim, J.H., Lohmann, G., 2010. Holocene and Eemian sea surface temperature trends as revealed by alkenone and Mg/Ca paleothermometry. Quaternary Science Reviews 29 (7–8), 989–1004. https://doi.org/10.1016/j.quascirev.2010.01.004.

Lee, J.-E., Risi, C., Fung, I., Worden, J., Scheepmaker, R.A., Lintner, B., et al., 2012. Asian monsoon hydrometeorology from TES and SCIAMACHY water vapor isotope measurements and LMDZ simulations: implications for speleothem climate record interpretation. Journal of Geophysical Research: Atmospheres 117 (D15). https://doi.org/10.1029/2011jd017133.

Linsley, B.K., Rosenthal, Y., Oppo, D.W., 2010. Holocene evolution of the Indonesian throughflow and the western Pacific warm pool. Nature Geoscience 3 (8), 578–583. https://doi.org/10.1038/ngeo920.

Lisiecki, L.E., Raymo, M.E., 2005. A Pliocene-Pleistocene stack of 57 globally distributed benthic $\delta^{18}O$ records. Paleoceanography 20, PA1003. https://doi.org/1010.1029/2004PA001071.

Lückge, A., Mohtadi, M., Rühlemann, C., Scheeder, G., Vink, A., Reinhardt, L., et al., 2009. Monsoon versus ocean circulation controls on paleoenvironmental conditions off southern Sumatra during the past 300,000 years. Paleoceanography 24, PA1208. https://doi.org/10.1029/2008PA001627.

Marshall, J., Donohoe, A., Ferreira, D., McGee, D., 2014. The ocean's role in setting the mean position of the Inter-Tropical Convergence Zone. Climate Dynamics 42 (7), 1967–1979. https://doi.org/10.1007/s00382-013-1767-z.

Moffa-Sanchez, P., Rosenthal, Y., Babila, T.L., Mohtadi, M., Zhang, X., 2019. Temperature evolution of the Indo-Pacific warm pool over the Holocene and the last deglaciation. Paleoceanography and Paleoclimatology 34 (7), 1107–1123. https://doi.org/10.1029/2018pa003455.

Mohtadi, M., Lückge, A., Steinke, S., Groeneveld, J., Hebbeln, D., Westphal, N., 2010a. Late Pleistocene surface and thermocline conditions of the eastern tropical Indian Ocean. Quaternary Science Reviews 29 (7–8), 887–896. https://doi.org/10.1016/j.quascirev.2009.12.006.

Mohtadi, M., Max, L., Hebbeln, D., Baumgart, A., Krück, N., Jennerjahn, T., 2007. Modern environmental conditions recorded in surface sediment samples off W and SW Indonesia: planktonic foraminifera and biogenic compounds analyses. Marine Micropaleontology 65, 96–112.

Mohtadi, M., Oppo, D.W., Lückge, A., DePol-Holz, R., Steinke, S., Groeneveld, J., et al., 2011a. Reconstructing the thermal structure of the upper ocean: insights from planktic foraminifera shell chemistry and alkenones in modern sediments of the tropical eastern Indian Ocean. Paleoceanography 26 (3), PA3219. https://doi.org/10.1029/2011pa002132.

Mohtadi, M., Oppo, D.W., Steinke, S., Stuut, J.-B.W., De Pol-Holz, R., Hebbeln, D., et al., 2011b. Glacial to Holocene swings of the Australian-Indonesian monsoon. Nature Geoscience 4 (8), 540–544. https://doi.org/10.1038/ngeo1209.

Mohtadi, M., Prange, M., Oppo, D.W., De Pol-Holz, R., Merkel, U., Zhang, X., et al., 2014. North Atlantic forcing of tropical Indian Ocean climate. Nature 509 (7498), 76–80. https://doi.org/10.1038/nature13196.

Mohtadi, M., Prange, M., Schefuß, E., Jennerjahn, T.C., 2017. Late Holocene slowdown of the Indian ocean Walker circulation. Nature Communications 8 (1), 1015. https://doi.org/10.1038/s41467-017-00855-3.

Mohtadi, M., Prange, M., Steinke, S., 2016. Palaeoclimatic insights into forcing and response of monsoon rainfall. Nature 533 (7602), 191–199. https://doi.org/10.1038/nature17450.

Mohtadi, M., Steinke, S., Groeneveld, J., Fink, H.G., Rixen, T., Hebbeln, D., et al., 2009. Low-latitude control on seasonal and interannual changes in planktonic foraminiferal flux and shell geochemistry off south Java: a sediment trap study. Paleoceanography 24, PA1201. https://doi.org/1210.1029/2008PA001636.

Mohtadi, M., Steinke, S., Lückge, A., Groeneveld, J., Hathorne, E.C., 2010b. Glacial to Holocene surface hydrography of the tropical eastern Indian Ocean. Earth and Planetary Science Letters 292, 89–97.

Neale, R.B., Richter, J.H., Jochum, M., 2008. The impact of convection on ENSO: from a delayed oscillator to a series of events. Journal of Climate 21 (22), 5904–5924. https://doi.org/10.1175/2008jcli2244.1.

Niedermeyer, E.M., Sessions, A.L., Feakins, S.J., Mohtadi, M., 2014. Hydroclimate of the western Indo-Pacific warm pool during the past 24,000 years. Proceedings of the National Academy of Sciences of the United States of America 111 (26), 9402–9406. https://doi.org/10.1073/pnas.1323585111.

Parrenin, F., Masson-Delmotte, V., Köhler, P., Raynaud, D., Paillard, D., Schwander, J., et al., 2013. Synchronous change of atmospheric $CO_2$ and Antarctic temperature during the last deglacial warming. Science 339 (6123), 1060–1063. https://doi.org/10.1126/science.1226368.

Partin, J.W., Cobb, K.M., Adkins, J.F., Clark, B., Fernandez, D.P., 2007. Millennial-scale trends in west Pacific warm pool hydrology since the last glacial maximum. Nature 449 (7161), 452–455.

Prahl, F.G., Wakeham, S.G., 1987. Calibration of unsaturation patterns in long-chain ketone compositions for palaeotemperature assessment. Nature 330 (26), 367–370.

Qu, T., Meyers, G., 2005. Seasonal variation of barrier layer in the southeastern tropical Indian Ocean. Journal of Geophysical Research 110, C11003. https://doi.org/10.11029/12004JC002816.

Risi, C., Bony, S., Vimeux, F., 2008. Influence of convective processes on the isotopic composition ($\delta^{18}O$ and $\delta D$) of precipitation and water vapor in the tropics: 2. Physical interpretation of the amount effect. Journal of Geophysical Research: Atmospheres 113 (D19). https://doi.org/10.1029/2008jd009943.

Ruan, Y., Mohtadi, M., van der Kaars, S., Dupont, L.M., Hebbeln, D., Schefuß, E., 2019. Differential hydroclimatic evolution of East Javanese ecosystems over the past 22,000 years. Quaternary Science Reviews 218, 49–60. https://doi.org/10.1016/j.quascirev.2019.06.015.

Sachse, D., Billault, I., Bowen, G.J., Chikaraishi, Y., Dawson, T.E., Feakins, S.J., et al., 2012. Molecular Paleohydrology: Interpreting the hydrogen-isotopic composition of lipid biomarkers from Photosynthesizing organisms. Annual Review of Earth and Planetary Sciences 40, 221–249. https://doi.org/10.1146/annurev-earth-042711-105535.

Saji, N.H., Goswami, B.N., Vinayachandran, P.N., Yamagata, T., 1999. A dipole mode in the tropical Indian Ocean. Nature 401, 360–363.

Schröder, J.F., Holbourn, A., Kuhnt, W., Küssner, K., 2016. Variations in sea surface hydrology in the southern Makassar Strait over the past 26 kyr. Quaternary Science Reviews 154, 143–156. https://doi.org/10.1016/j.quascirev.2016.10.018.

Setiawan, R.Y., Mohtadi, M., Southon, J., Groeneveld, J., Steinke, S., Hebbeln, D., 2015. The consequences of opening the Sunda Strait on the hydrography of the eastern tropical Indian Ocean. Paleoceanography 30 (10), 1358–1372. https://doi.org/10.1002/2015pa002802.

Sprintall, J., Gordon, A.L., Koch-Larrouy, A., Lee, T., Potemra, J.T., Pujiana, K., et al., 2014. The Indonesian seas and their role in the coupled ocean-climate system. Nature Geoscience 7 (7), 487–492. https://doi.org/10.1038/ngeo2188.

Sprintall, J., Gordon, A.L., Wijffels, S.E., Feng, M., Hu, S., Koch-Larrouy, A., et al., 2019. Detecting change in the Indonesian seas. Frontiers in Marine Science 6 (257). https://doi.org/10.3389/fmars.2019.00257.

Sprintall, J., Tomczak, M., 1992. Evidence of the barrier layer in the surface layer of the tropics. Journal of Geophysical Research 97, 7305–7316. https://doi.org/10.1029/92jc00407.

Steinke, S., Mohtadi, M., Prange, M., Varma, V., Pittauerova, D., Fischer, H.W., 2014a. Mid- to Late-Holocene Australian–Indonesian summer monsoon variability. Quaternary Science Reviews 93 (0), 142–154. https://doi.org/10.1016/j.quascirev.2014.04.006.

Steinke, S., Prange, M., Feist, C., Groeneveld, J., Mohtadi, M., 2014b. Upwelling variability off southern Indonesia over the past two millennia. Geophysical Research Letters 41 (21). https://doi.org/10.1002/2014gl061450. 2014GL061450.

Stott, L., Cannariato, K., Thunell, R., Haug, G.H., Koutavas, A., Lund, S., 2004. Decline of surface temperature and salinity in the western tropical Pacific Ocean in the Holocene epoch. Nature 431 (7004), 56–59.

Svensson, A., Andersen, K.K., Bigler, M., Clausen, H.B., Dahl-Jensen, D., Davies, S.M., et al., 2008. A 60 000 year Greenland stratigraphic ice core chronology. Climate of the Past 4, 47–57. https://doi.org/10.5194/cp-4-47-2008.

Wang, X., Jian, Z., Lückge, A., Wang, Y., Dang, H., Mohtadi, M., 2018. Precession-paced thermocline water temperature changes in response to upwelling conditions off southern Sumatra over the past 300,000 years. Quaternary Science Reviews 192, 123–134. https://doi.org/10.1016/j.quascirev.2018.05.035.

Wang, B., Liu, J., Kim, H.-J., Webster, P.J., Yim, S.-Y., Xiang, B., 2013. Northern Hemisphere summer monsoon intensified by mega-El Niño/southern oscillation and Atlantic multidecadal oscillation. Proceedings of the National Academy of Sciences of the United States of America 110 (14), 5347–5352. https://doi.org/10.1073/pnas.1219405110.

Webster, P.J., Magaña, V.O., Palmer, T.N., Shukla, J., Tomas, R.A., Yanai, M., et al., 1998. Monsoons: processes, predictability, and the prospects for prediction. Journal of Geophysical Research 103 (C7), 14451–14510.

Webster, P.J., More, A.M., Loschnigg, J.P., Leban, R.R., 1999. Coupled ocean-atmosphere dynamics in the Indian Ocean during 1997–1998. Nature 401, 356–360.

Wicaksono, S.A., Russell, J.M., Holbourn, A., Kuhnt, W., 2017. Hydrological and vegetation shifts in the Wallacean region of central Indonesia since the last glacial Maximum. Quaternary Science Reviews 157, 152–163. https://doi.org/10.1016/j.quascirev.2016.12.006.

Wurtzel, J.B., Abram, N.J., Lewis, S.C., Bajo, P., Hellstrom, J.C., Troitzsch, U., et al., 2018. Tropical Indo-Pacific hydroclimate response to North Atlantic forcing during the last deglaciation as recorded by a speleothem from Sumatra, Indonesia. Earth and Planetary Science Letters 492, 264–278. https://doi.org/10.1016/j.epsl.2018.04.001.

Xu, J., Holbourn, A., Kuhnt, W., Jian, Z., Kawamura, H., 2008. Changes in the thermocline structure of the Indonesian outflow during Terminations I and II. Earth and Planetary Science Letters 273 (1–2), 152–162.

Xu, J., Kuhnt, W., Holbourn, A., Andersen, N., Bartoli, G., 2006. Changes in the vertical profile of the Indonesian throughflow during termination II: evidence from the Timor Sea. Paleoceanography 21, PA4202. https://doi.org/4210.1029/2006PA001278.

Xu, J., Kuhnt, W., Holbourn, A., Regenberg, M., Andersen, N., 2010. Indo-pacific warm pool variability during the Holocene and last glacial maximum. Paleoceanography 25 (4), PA4230. https://doi.org/10.1029/2010pa001934.

Zhang, P., Xu, J., Schröder, J.F., Holbourn, A., Kuhnt, W., Kochhann, K.G.D., et al., 2018. Variability of the Indonesian throughflow thermal profile over the last 25-kyr: a perspective from the southern Makassar Strait. Global and Planetary Change 169, 214–223. https://doi.org/10.1016/j.gloplacha.2018.08.003.

Zuraida, R., Holbourn, A., Nürnberg, D., Kuhnt, W., Dürkop, A., Erichsen, A., 2009. Evidence for Indonesian throughflow slowdown during Heinrich events 3-5. Paleoceanography 24, PA2205. https://doi.org/10.1029/2008PA001653.

10

# Decision tool for assessing marine finfish aquaculture sites in Southeast Asia

Roberto Mayerle[1], Ketut Sugama[2], Simon van der Wulp[4], Poerbandono[3], Karl-Heinz Runte[1]

[1]RESEARCH AND TECHNOLOGY CENTRE WESTCOAST (FTZ), UNIVERSITY OF KIEL, BÜSUM, GERMANY; [2]CENTRE FOR AQUACULTURE RESEARCH AND DEVELOPMENT (CARD), MINISTRY OF MARINE AFFAIRS AND FISHERIES, JAKARTA, INDONESIA; [3]INSTITUTE OF TECHNOLOGY BANDUNG (ITB), BANDUNG, JAWA BARAT, INDONESIA; [4]ECOLAB - GROUP COASTAL ECOSYSTEMS, RESEARCH AND TECHNOLOGY CENTRE WESTCOAST (FTZ), UNIVERSITY OF KIEL, BÜSUM, GERMANY

## Abstract

*This chapter summarizes the development of a decision support system (DSS) for spatial planning of marine finfish aquaculture sites in Southeast Asia. The DSS implements methods for site selection and estimation of carrying capacities. These methods are based primarily on results of dynamic models supplemented with in situ measurements and remote sensing information. They follow the FAO's Ecosystem Approach to Aquaculture for ensuring the proper planning and operation of aquaculture sites. In this study, the effectiveness of the DSS is demonstrated for a target aquaculture site in the northwest of Bali in Indonesia. Results of the study led to recommendations for relocation of farms to suitable areas and for reduction of fish production in farms operating above carrying capacity. Recommendations to increase fish production while ensuring sustainable ecological operation in the site were made. The DSS has broad applicability and will help to make decisions for aquaculture sites around the world.*

## Abstrak

*Makalah ini menjelaskan tentang pengembangan sistem pendukung keputusan (decision support system/DSS) tentang penentuan tata ruang dalam budidaya laut di Asia Tenggara. Penerapan metoda DSS untuk pemilihan lokasi dan penentuan daya dukung perairan utamanya berdasarkan model dinamis yang dilengkapi dengan pengukuran mutu perairan in-situ dan informasi penginderaan jauh. Metode ini termasuk prosedur umum dalam kontek pendekatan ekosystem akuacultur oleh FAO untuk memastikan perencanaan yang tepat dan operasional yang benar di lokasi budidaya laut. Dalam studi ini efektivitas DSS didemontrasikan di lokasi budidaya laut yang terletak di perairan Utara-Barat Bali-Indonesia. Hasil dari penilaian lokasi ternyata keramba jaring apung (KJA) pembudidaya diletakkan diluar area perairan yang sesuai untuk budidaya laut dan produksi ikannya berlebih. Rekomendasi untuk penempatan KJA yang sesuai untuk meningkatkan produksi berkelanjutan telah dilakukan. Dengan demikian dapat disimpulkan bahwa DSS mempunyai penerapan yang sangat luas dan dapat membantu dalam penentuan lokasi budidaya di seluruh dunia.*

Science for the Protection of Indonesian Coastal Ecosystems (SPICE). https://doi.org/10.1016/B978-0-12-815050-4.00003-1 

**Chapter outline**

**10.1 Introduction** .......... **372**
**10.2 Methods for sustainable management of marine finfish aquaculture** .......... **374**
10.2.1 Site selection .......... 374
10.2.2 Production carrying capacity .......... 374
10.2.3 Ecological carrying capacity (ECC) .......... 375
**10.3 Decision support system for sustainable management of marine finfish aquaculture** ... **375**
**10.4 Marine finfish aquaculture site in Bali, Indonesia** .......... **376**
**10.5 Surveys and monitoring at the aquaculture site in Bali** .......... **378**
**10.6 Dynamic model of the aquaculture site in Bali** .......... **379**
**10.7 Assessment of operation of the marine finfish aquaculture site in Bali** .......... **380**
10.7.1 Site selection .......... 380
10.7.2 Production carrying capacity .......... 381
10.7.3 Ecological carrying capacity .......... 382
**10.8 Conclusion** .......... **384**
**Acknowledgments** .......... **385**
**References** .......... **385**

## 10.1 Introduction

In the past decades, average fish consumption globally has increased significantly. As a result, aquaculture nowadays surpasses global capture fisheries in seafood production (FAO, 2016). Asia accounts currently for nearly 90% of global aquaculture production (FAO, 2018). China will remain the world's leading producer for the next decades, while several Southeast Asia countries and India are expected to intensify their production (FAO, 2018; World Bank, 2013, p. 80). Expansion of mariculture worldwide is rapid and will require significant increase of the cultivated area both onshore and offshore and exert considerable strain on aquatic and terrestrial resources, as well as on the environment. Several countries in Southeast Asia have been increasing fish production to meet food security and job creation targets. The Government of Indonesia, e.g., is currently drafting an ambitious new-term development plan to meet production target of 31.4 million tons by 2027 and 37.6 million tons by 2030 (Bone et al., 2018). Implementation of this plan will substantially expand the activity in the country and will require spatial planning approaches to account for constraints and interactions. As the current adopted spatial planning of aquaculture facilities in place is insufficient or nonexistent, and data for decision-making are scarce, there is a need for systematic procedures that allow for the growth and development of the aquaculture industry, within working toward minimal impacts to the marine ecosystems.

To this end, the Food and Agricultural Organization of the United Nations (FAO) has proposed the Ecosystem Approach to Aquaculture (EAA) as a strategy for sustainable development of aquaculture sites (FAO, 2010; Soto et al., 2008). The adoption of environmental sustainable practices and managerial schemes following the lines of the proposed EAA is essential for enhancing sustainable development and expansion of the aquaculture industry; moreover, it is a prerequisite to assure compliance with the existing regulatory framework. Carrying capacity is the primary concept of EAA (Inglis et al., 2000; McKindsey et al., 2006). Although production plays a significant role in the procedure, a more comprehensive approach that also considers physical, ecological, and social carrying capacity was proposed by Ross et al. (2013). Physical carrying capacity or site selection (SS) comprises the identification of sites or potential aquaculture zones from which specific SS can be made for actual farm development. Production carrying capacity (PCC) corresponds to the maximum aquaculture production of individual fish farms and is typically considered at the farm level. Ecological carrying capacity (ECC) is defined as the magnitude of aquaculture production that can be supported by the entire site without leading to significant changes to ecological processes, services, species, populations, or communities in the environment. In this study, emphasis is given to the development of methods for SS, PCC, and ECC of marine finfish aquaculture sites following FAO's EAA.

Since the implementation of the EAA about 10 years ago, there has been increased awareness of the need for comprehensive and participatory approaches. Despite the positive developments and increasing consciousness of farm developers and policy-makers, the practical implementation of the EAA has been slow (Brugère et al., 2018). The lack of specific guidelines for the adoption of tools and models and complexity of the assessments remain the main challenge toward the fulfillment of the EAA in emerging countries. In view of the limitations and absence of guidelines, a general procedure for assessing coastal finfish aquaculture in sites with limited data has been developed in this study. The procedure implements cost-effective methods for the estimation of SS, PCC, and ECC based essentially on results of dynamic models. As the cost involved in collecting field data and developing dynamic models is high and requires skillful personnel, this study emphasizes model development using data from available databases, global oceanic models, and remote sensing information. Bearing in mind the complexity in developing dynamic models particularly to data-poor sites, model development for specific sites is done by experts and the results of applications imported to a decision support system (DSS) to facilitate decision-making.

This chapter summarizes the development of a computerized DSS to facilitate spatial planning of marine finfish aquaculture sites in Southeast Asia. The DSS has been developed primarily for assessing operating aquaculture sites and potential new sites selected for expansion of the activity. A description of the methods for SS, PCC, and ECC embedded within the DSS is provided. The investigations have been carried out in a target aquaculture site in the northwest of Bali in Indonesia. Information on the in situ measurements and details of the dynamic models for the site under investigation are provided in the paper. Results of the application of the DSS for selecting suitable locations and estimating carrying capacities at farm level and site scale are presented for the site in Bali. Recommendations to promote sustainable ecological operation of the aquaculture site are made.

## 10.2 Methods for sustainable management of marine finfish aquaculture

A description of the methods developed in this study for selecting suitable locations and estimating carrying capacities of marine finfish aquaculture facilities is summarized in the following.

### 10.2.1 Site selection

SS comprises the identification of sites or potential aquaculture zones from which a subsequent more selection of fish farming locations can be made for actual farm development (Kutty, 1987). It is probably the most relevant step for the sustainable operation of coastal finfish aquaculture sites. SS is relatively straightforward to determine but requires substantial amount of data with good spatial coverage. Many potential environmental impacts and risks can be avoided with prudent farm siting. The method proposed for defining suitable locations is based on criteria and threshold values usually adopted for target finfish species listed in Mayerle et al. (2017). In the selection of suitable locations, thematic maps of the most relevant criteria for marine finfish aquaculture facilities are prepared. For each thematic map, templates are built according to the threshold values. Templates are then overlaid for generating suitability maps for marine finfish facilities (Windupranata, 2007; Windupranata and Mayerle, 2009). As data in remote sites in Asia are limited, dynamic models for circulation and waves are developed and applied to deliver relevant spatial information for SS particularly in the early stages. This information is supplemented with on-site measurements and satellite imagery information. An overview of satellites providing remote sensing data for SS in Indonesia is listed in Mayerle et al. (2018).

### 10.2.2 Production carrying capacity

PCC is defined as the environmentally sustainable stocking densities of individual fish farms (Ross et al., 2013). PCC limits are estimated using primarily current velocity values at fish farming locations. A dimensionless relationship was derived to estimate PCC at fish farming locations on the basis of the flow Reynolds number (Re), the settling velocity of particulate waste (ws), and a specified limit for benthic deterioration underneath fish farms (Mayerle et al., 2020). By combining results of dynamic models for circulation with the derived empirical equation, maximum stocking densities and/or cage depths of individual fish farms can be determined at any location. Predictions take note of farming conditions into consideration, such as daily feeding rate, proportion of wasted feed, and excrete feces, as well as the percentage of carbon in feed and feces. A correction coefficient is introduced to account for the simplifications and assumptions made in the derivation of the empirical relationship and the estimation of the waste emissions from the fish farms. In situ assessments of the benthic conditions underneath farms, in conjunction with specific farming conditions, are used to determine this coefficient. For further details of the method, the reader is referred to Mayerle et al. (2020).

### 10.2.3 Ecological carrying capacity (ECC)

ECC is defined as the magnitude of aquaculture production that can be supported at site scale without leading to significant changes to ecological processes, species, populations, or communities in the environment (Gibbs, 2007; Byron and Costa-Pierce, 2013). The method proposed for estimating ECC of marine finfish aquaculture sites relies primarily on results of high-resolution water quality models accounting for emissions of multiple farms and their interactions. Simulations are done for scenarios taken into consideration background nutrient concentrations as recommended by Ferreira et al. (2013) in conjunction with patterns of circulation within the aquaculture site. To deliver conservative estimates of ECC, scenarios considering farm emissions at the end of the grow-out period are considered. Modeled concentrations of nutrients in places of highest cumulative concentrations are used in conjunction with water quality standards to determine ECC. Dose–response curves following the Driver–Pressure–State–Impact–Response (DPSIR) environmental assessment framework are adopted for this purpose (Tett et al., 2011). Several production-level scenarios are simulated to obtain continuous variation of nutrient concentrations for the entire range of farm biomass. Diagrams displaying nutrient concentrations against production-level scenarios are constructed for locations within the aquaculture site, most impacted by the cumulative effects due to fish farming. ECC is defined by the magnitude of aquaculture production, by which there is a balance between the maximum nutrient concentration and the applicable environmental quality standard. The latter should not be exceeded when aquaculture is included into the system. In this study, the parameters adopted for assessing the impacts of dissolved waste from fish farms on the water column are nitrate and ammonia. The standards defined by ASEAN marine water quality criteria for aquatic life protection are considered in the assessment of impacts due to marine finfish aquaculture operations (Secretariat ASEAN, 2008). On the basis of the results, regions within the site with the highest cumulative nutrient concentrations are identified and locations for monitoring defined.

## 10.3 Decision support system for sustainable management of marine finfish aquaculture

A DSS has been developed to facilitate applications and support planners in the assessment of marine finfish cage clusters. The interactive computer-based DSS was designed to control a sequence of interactions and the operations performed in the selection of suitable sites and estimations of carrying capacities. The system provides guidance on planning and supports the identification of potential areas for expansion of the aquaculture activity. The DSS is modular and implements the methods developed in this study for SS, PCC, and ECC. Thematic maps based on in situ measurements, simulation models, and remote sensing information are imported to the DSS. The DSS is also equipped with facilities for importing and visualizing operational data from global oceanic models and monitoring systems (Mayerle et al., 2018). Data from the various sources are embedded with a graphical user interface. The DSS has been successfully adopted to several sites selected by the Government of Indonesia for expansion of the activity. The system is currently being used by several institutions and research centers under the Indonesian Ministry of Marine Affairs and Fisheries (MMAF).

## 10.4 Marine finfish aquaculture site in Bali, Indonesia

The investigations were done in Pegametan Bay (8.13°S, 114.6°E) in the northwest of Bali The area of interest (model area) covers about 35 $km^2$ along a coastal stretch of around 10 km. Fig. 10.1 shows an overview of the study area with a satellite image from 2014.

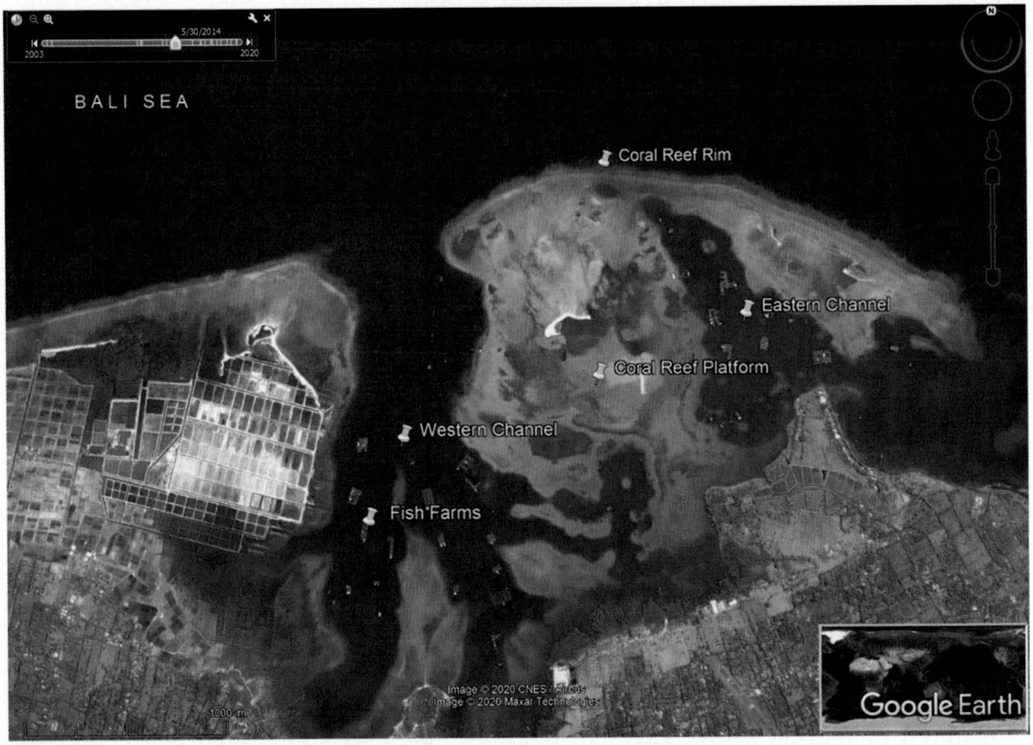


**FIGURE 10.1** Satellite image of Pegametan Bay, North Bali, Indonesia in 2014. *Modified from Mayerle et al. (2017). Copyright Google Earth.*

The prominent characteristics of the site are a coral reef system embedding two large tidal channels that embrace a central coral reef platform. Water depths are less than 1 m in the coral reef flats and greater than 50 m at the reef slope facing the Bali Sea. The reefs divide the inner bay into the two main channels, whose water depths reach about 20–25 m. Circulation patterns within the bay are strongly dependent on the tidal conditions. The tidal range varies between about 1 and 1.8 m throughout the year. Modeled velocities are generally smaller than about 0.02–0.03 m/s but can reach about 0.15–0.20 m/s during spring tides.

According to the intersectorial zoning scheme under the umbrella of the MMAF (see Fig. 10.3), Pegametan Bay has been dedicated exclusively to mariculture activity, whereas the adjacent coastal areas have been dedicated mainly for tourism and natural conservation (Mayerle et al., 2017). Finfish mariculture using floating net cages has been practiced in Pegametan Bay since 2001 and has grown to a total of 30 farms in 2015. The bulk of the standing stock consists of tiger grouper and Asian seabass. Operative fish farms in Pegametan Bay in 2015 have a wide range of sizes (Mayerle et al., 2017). The floating net cages of most farms are of the type most common in Indonesia: the fish farms consist of wooden rafts, kept afloat by plastic drums. Each cage typically measures 3 m × 3 m × 3 m. They are connected to form a floating raft to reduce the effect of waves and currents. The stocking density of these farms for cultivation of grouper is about 10–20 kg/m$^3$. Following the trends recently observed elsewhere in Asia, there are two bigger fish farms in the eastern channel with seven to eight circular floating units of high-density polyethylene cages for nursery and on-growing. The cages are 20 m in diameter, have a depth of about 6–7 m, and are located relatively close to each other. The stocking densities of these larger fish farms for the cultivation of Asian seabass reach values of about 25–30 kg m$^{-3}$. Total fish production in Pegametan Bay reached about 1200 tonnes in 2015 (Pusat Data Statistik dan).

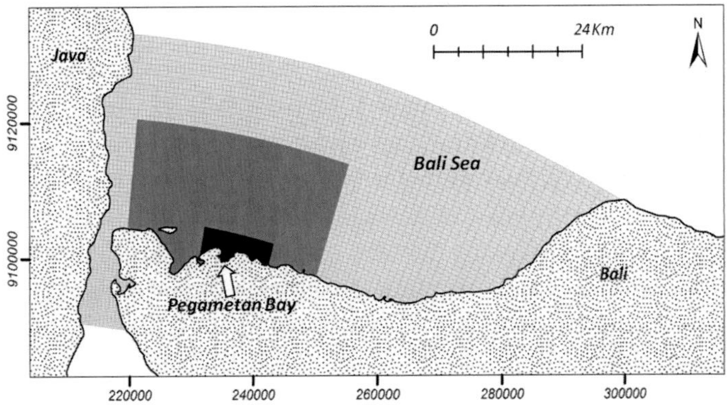


**FIGURE 10.2** Downscaling modeling sequence for the aquaculture site in Pegametan Bay.

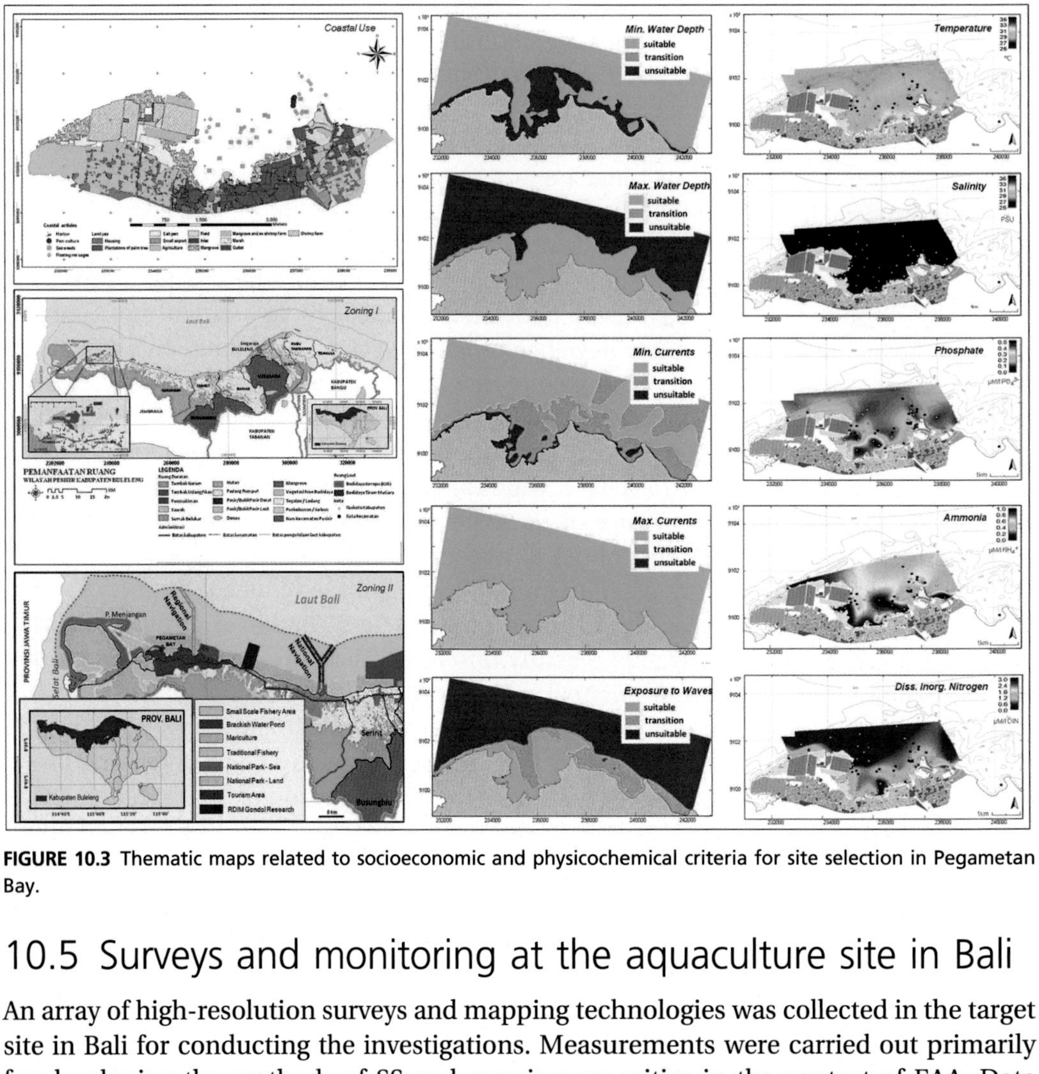


**FIGURE 10.3** Thematic maps related to socioeconomic and physicochemical criteria for site selection in Pegametan Bay.

## 10.5 Surveys and monitoring at the aquaculture site in Bali

An array of high-resolution surveys and mapping technologies was collected in the target site in Bali for conducting the investigations. Measurements were carried out primarily for developing the methods of SS and carrying capacities in the context of EAA. Data from existing databases and large-scale global models were supplemented with remote sensing information of farming locations and places of diffusive sources of turbidity along the coastline (Mayerle et al., 2018). Gaps of information were identified, and several measuring campaigns were carried out. Emphasis was given to the chemical and sediment data for defining suitable locations for installation of fish farms and for estimating carrying capacities at farm level and site scale. Bathymetry in the bay for the setup of the model grid within Pegametan Bay was measured with a vessel mounted echo sounder. A tidal gauge was installed within the bay for measurement of water levels for assessment of performance of the circulation model. Vessel mounted CTDs were deployed in conjunction with Niskin bottles throughout the entire bay for water quality.

Measurements were carried out for delivering concentrations of nutrients, salinity, and water temperature at the open sea boundaries of the model and within the bay, respectively, for forcing and assessment of performance of the water quality model. To obtain high-resolution information on the spatial variability of nutrients, a large number of surficial water samples were collected throughout the entire bay under different environmental conditions. Measurements covered periods with different tidal conditions during rainy and dry seasons. Appraisals of benthic impacts due to the operation of fish farms were carried out underneath the 12 largest farms that are responsible for about 80% of the production in Pegametan Bay (Mayerle et al., 2020). The aquaculture site has been equipped with operational monitoring systems. Multiparameter sensors for water temperature, turbidity, and water quality parameters are currently in operation at three fish farms in Pegametan Bay (Mayerle et al., 2018). Data are transferred in real time to a data center at the Gondol Research Institute of Mariculture (GRIM) of MMAF, and to the Research and Technology Centre Westcoast (FTZ) of the University of Kiel in Germany. The aim is to alert farmers of adverse conditions relevant to the operation of fish farms.

## 10.6 Dynamic model of the aquaculture site in Bali

A three-dimensional process–based model was developed for the aquaculture site in Bali. The model covering Pegametan Bay comprises of submodels for circulation, waves, and water quality. The modeling software used in this study is based on the freely available Delft3D modeling suite developed by Deltares in the Netherlands (https://www.deltares.nl/en/software/delft3d-4-suite, Deltares, 2014a,b). Computation was performed on three curvilinear grids with increasing grid resolution toward the coast. Fig. 10.2 shows the horizontal setup of the model grid. Subdomain decomposition was adopted to permit grid refinements from the coarse model covering parts of the Bali Sea and Bali Strait to the more refined model covering Pegametan Bay. Horizontal grid resolution ranges from 800 m in the larger-scale model to about 25 m near the coast. Bathymetry data stems from different sources and embeds information from several surveys and databases. Bathymetry offshore of the bay was compiled using data from GEBCO (IOC, IHO, BODC, 2003). Near-shore bathymetric data stem from own surveys done in 2008 complemented with information from nautical charts from Geospatial Information Agency, Indonesia (BIG). To resolve the flow and fate of dissolved and particulate waste in three dimensions, the grid of Pegametan Bay model was vertically divided into five sigma layers, each covering 20% of the water depth. For further details of the circulation model, the reader is referred to Niederndorfer (2017).

The performance of the circulation model was assessed by comparing computed with measured water level time series registered at the tidal gauge installed within the model area. A simulation period of 31 days from mid-March to mid-April 2008 was considered for validation. Results of assessment of performance of the circulation model are summarized in Niederndorfer (2017). The water levels at Pegametan Bay are predicted very well with the model. The mean and standard deviations of computed water levels from measured water levels are found to be 0.05 and 0.07 m, respectively. Waves are simulated using the fully spectral model SWAN developed by Booij et al. (1999) (see Mayerle et al., 2018). The circulation (tide-flow) model was coupled to the water quality model for simulation of the fate of dissolved nutrients and of transport and accumulation of solid particulate waste from fish farms. Details of the water quality model are provided in Van der Wulp (2015). The performance of the water quality of Pegametan Bay model was assessed by comparing predicted with measured nutrient concentrations at several locations of Pegametan Bay. The assessment was done for three periods, respectively, during the wet season on January 2008 and December 2008 and by the end of the dry season on September 2012.

## 10.7 Assessment of operation of the marine finfish aquaculture site in Bali

In this study, the marine finfish aquaculture site in Pegametan Bay was assessed regarding the suitability of fish farming locations, carrying capacity of individual fish farms and overall fish farm production cultivated in the site. The assessment was done for the fish farming conditions in 2015. Fish farming sizes and species cultivated are provided in Mayerle et al. (2017).

### 10.7.1 Site selection

The array of spatial data from different sources collected during the project was used to assess the site suitability of existing farms following the SS method developed in this study. Fig. 10.3 presents the set of thematic maps used in the assessment. In addition to the zoning scheme for the northwest of Bali and maps of coastal uses, results of on-site measurements and dynamic models are adopted for SS.

Fig. 10.4 shows the resulting suitability map of marine finfish farming in Pegametan Bay. Values in the suitability map are rated according to a suitability index (Windupranata, 2007; Windupranata and Mayerle, 2009). Indices below zero refer to areas in the bay unsuitable for the activity; values ranging between 0 and 50 indicate moderately suitable areas; and indices higher than 50 refer to areas well suited for marine finfish cage facilities. The location of the 30 fish farms in operation in 2015 is also indicated in Fig. 10.4. It can be seen that most farms are located within suitable areas, reflecting the experience of fish farmers and planners. Based on the results, there are six farms on the western channel located outside suitable areas (see farms numbered 5, 6, 7, 10, 11, and 12 in Fig. 10.4). Farms located outside suitable areas were spotted, and recommendations for their relocation were made to the responsible authorities (Mayerle et al., 2017; Bone et al., 2018). Areas within Pegametan Bay suitable for expansion of the activity, particularly to the east of the bay, were also identified (see Fig. 10.4). Validation of the SS method was done on the basis of on-site fish health assessments. It was found that farms located outside suitable areas are generally subject to higher risks of disease and that fish growth was generally lower as compared with those farms located within suitable areas.

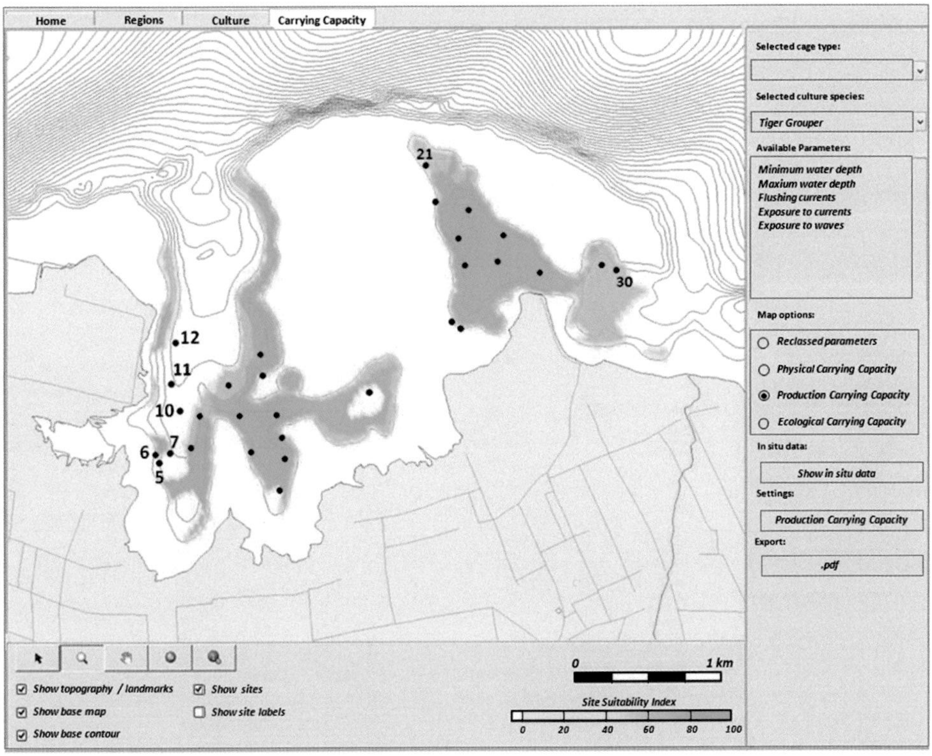


**FIGURE 10.4** Overall suitability map for the aquaculture site in Pegametan Bay. Dots indicate locations of existing fish farms.

## 10.7.2 Production carrying capacity

The proposed PCC method was applied to the aquaculture site in Bali. Fig. 10.5 shows predicted cage depths over the entire suitable area for stocking density values equal to 20 kg $m^{-3}$ typical to the cultivation of grouper. According to the results, the majority of traditional fish farms with cage depths of 3 m are operating within acceptable limits. On the other hand, farms numbered 21 and 30 with cage depths of 6–7 m are found to be unsuitable to operate in the area. In situ assessments of the benthic conditions underneath the 12 largest fish farms were used to validate the method (Mayerle et al., 2020). It was found that the sediment cores collected under the farms 21 and 30 consisted mainly of pure mud with significant enrichment of particulate organic matter and particulate organic nitrogen, confirming model predictions. Underneath the remaining farms, sediment samples resulted comparable with those samples collected at undisturbed locations away from fish farms. Key biogeochemical parameters for assessment and monitoring operations were identified (Mayerle et al., 2020). On the basis of the results, recommendations for reducing the size of fish farms 21 and 30 and for regular monitoring of sediment quality have been made.

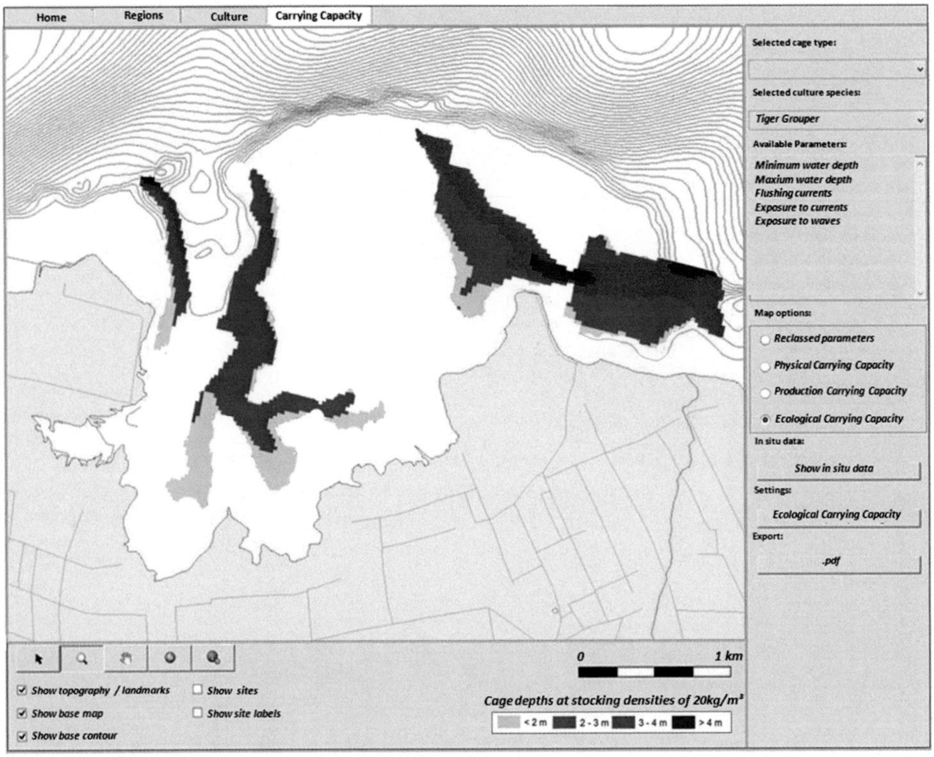


**FIGURE 10.5** Predicted cage depths for stocking densities of 20 kg/m³ in Pegametan Bay.

## 10.7.3 Ecological carrying capacity

Furthermore, the ECC of all the fish farms in the bay was estimated on the basis of the ECC method proposed in this study. The three-dimensional water quality model for simulation of organic loading due to fish farming covering Pegametan Bay is applied. Predictions considered the 30 fish farms in operation in 2015. The location of the fish farms is shown in Fig. 10.4. Stocking densities of traditional fish farms cultivating grouper were assumed equal to 20 kg $m^{-3}$. For the two larger farms (farms 21 and 30) deploying circular cages for cultivation of Asian seabass, stocking densities were considered equal to 30 kg $m^{-3}$. Farm emissions corresponding to the end of the grow-out period leading to maximum emissions are considered. The assessment was done for three scenarios in terms of tidal conditions and background nutrient concentrations for which measurements are available. Simulations covered conditions in January 2008, December 2008, and September 2012. ECC in Pegametan Bay is estimated with dose–response curves following the DPSIR environmental assessment framework under the consideration of environmental water quality standards defined by ASEAN. Limiting concentration values for ammonia and nitrate, respectively, equal to 70 and 60 μg/L, as defined by the ASEAN environmental water quality standards, apply.

To be on the safe side, the analysis focused on locations within Pegametan Bay with highest levels of nutrient concentrations as a result of fish farming. Based on model simulations, higher cumulative nutrient concentrations occur in parts of the western channel near to the coast, along the eastern channel and on the coral reefs between the two tidal channels. The data for construction of the diagrams stem from results of model simulations for production-level scenarios of 10%, 25%, 50%, 75%, and 100% of the capacity of each individual fish farm corresponding to total fish biomass values, respectively, of 340, 850, 1700, 2560, and 3400 tons.

Dose–response curves of the amount of biomass versus nutrient concentrations are constructed for the conditions at the monitoring points in the bay with the highest cumulative concentrations of ammonia and nitrate. According to the results, conditions in December 2008 led to the highest nutrient concentrations in Pegametan Bay. Concentrations of nitrate over the site for a production-level scenario of 100% at the time of peak values in December 2008 are shown in Fig. 10.6. The location of the selected monitoring point is also indicated in the figure. Predicted time average maximum and minimum nutrient concentrations values at the monitoring location are plotted in the diagrams for the results of the five production-level scenarios. Values are interpolated leading to continuous variation of maximum and minimum nutrient concentrations for the full range of biomass values. ECC of the aquaculture site in Pegametan Bay is defined by the cumulative farm loading, by which there is a balance between the maximum nutrient concentration and the applicable environmental standard.

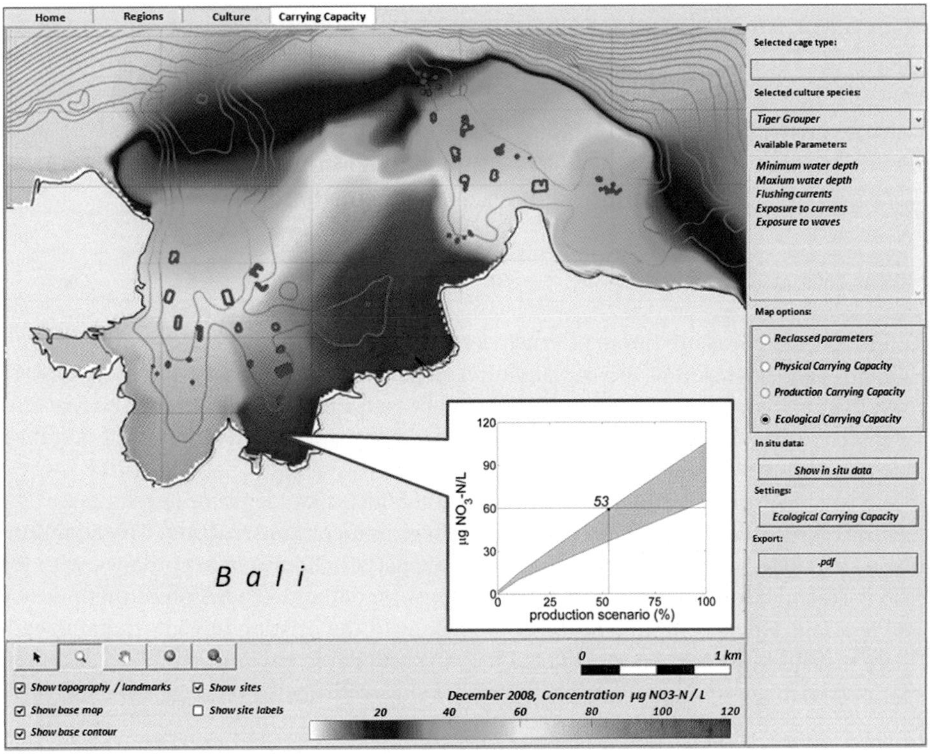


**FIGURE 10.6** Map of nitrate concentrations in Pegametan Bay at the time of peak values for a production-level scenario of 100% in December 2008.

The sketch in Fig. 10.6 shows the resulting DPSIR diagram at the monitoring point with the overall highest concentrations of nitrate. The limiting concentration value for nitrate equal to 60 μg/L is also plotted in the figure. Based on the conditions of December 2008, ECC values resulted in the order of 1800 tons corresponding to 53% of fish production in the site. As the total fish production in the site was about 1200 tons in 2015, and fish is harvest twice a year, Pegametan Bay is operating well within ECC. Hence, increase in fish production is feasible. Bearing in mind that currently most farms in Pegametan Bay are located in the western and eastern channels, it is recommended to expand the activity to the suitable areas identified to the east of the bay (see Fig. 10.4). Recommendations for regular monitoring of water quality in locations identified in the course of the investigations have been made.

## 10.8 Conclusion

In this project, a DSS for enhancing spatial planning of marine finfish aquaculture facilities was developed for data-poor sites in Southeast Asia. The decision tool implements methods for identifying suitable locations and estimating the limits of fish production at farm level and site scale. Results of the application of the DSS to an aquaculture site in Bali, Indonesia, show its effectiveness for assessing operating conditions. Recommendations for relocation of farms and reduction of fish production of certain farms while expanding overall fish production in the aquaculture site in Bali have been made. Regular monitoring of sediment and water quality should also be emphasized. The DSS has wider applicability and will be helpful in making decisions for mariculture development at the feasibility stage of projects. In association with our partners in Indonesia, the DSS has been applied to several aquaculture sites selected by the Government of Indonesia for the envisaged expansion of the activity.

### Knowledge gaps and directions of future research

- The Indonesian Seas are home of sensitive ecosystems; they may be jeopardized by the uncontrolled expansion of fish aquaculture. In particular, the excessive release of waste from the growing aquaculture production might cause eutrophication with severe effects on the benthic community, consequently impacting the fish cultivation itself and the entire marine environment.
- There is a need to foster environmental sustainability of aquaculture development ensuring both economic benefit and marine environment conservation. The adoption of environmentally sustainable practices and managerial schemes in accordance with the EAA is essential for enhancing sustainable development and expansion of the industry and is also a prerequisite to assure compliance with the existing regulatory framework.
- There is need for user-friendly systems for the sustainable environmental management of marine finfish aquaculture facilities in remote sites with scarce data.

Implications/recommendations for policy and society

The strength of aquaculture in Indonesia is available space, high biodiversity, favorable geography, and climate as well as human resources. Environmentally sustainable aquaculture reduces exploitation of marine natural resources and is regarded to contribute to the four pillars of Indonesian policy, viz., economic growth, creation of job opportunities, reduction of poverty by generating wealth for the people living along the coast, and environmental recovery and pollution mitigation. The Indonesian Directorate General for Fisheries and Aquaculture estimates that the potentially suitable area for aquaculture development is approximately 12 Mio. ha (DJB, 2017). However, only ca. 326,000 ha, corresponding to less than 3% of the available area, is in use so far; the potential for expansion is significant.

## Acknowledgments

We thank the Ministry for Education and Research (BMBF), the Project Management Jülich (PtJ), and the Leibniz Centre for Tropical Marine Research (ZMT) GmbH in Germany, as well as the State Ministry of Research and Technology (RISTEK) and the Ministry of Marine Affairs and Fisheries (KKP) in Indonesia for coordination, administration, and financial support of the projects SPICE I, Cluster 3.2, FK 03F0393A and SPICE II, Cluster 3.1, FK 03F0469A.

## References

Bone, J., Clavelle, T., Ferreira, J.G., Grant, J., Ladner, I., Immink, A., et al., 2018. Best practices for aquaculture management, guidance for implementing the ecosystem approach in Indonesia and beyond. In: Amy Sweeting (Sustainable Fisheries Partnership), Conservation International, Sustainable Fisheries Partnership. University of California Santa Barbara, p. 55.

Booij, N., Ris, R.C., Holthuijsen, L.H., 1999. A third-generation wave model for coastal regions 1 - model description and validation. Journal of Geophysical Research 104 (C4), 7649–7666.

Brugère, C., Aguilar-Manjarrez, J., Beveridge, M., Soto, D., March 1, 2018. The ecosystem approach to aquaculture 10 years on - a critical review and consideration of its future role in blue growth. Reviews in Aquaculture. https://doi.org/10.1111/raq.12242.

Byron, C.J., Costa-Pierce, B.A., 2013. Carrying capacity tools for use in the implementation of an ecosystems approach to aquaculture. In: Ross, L.G., Telfer, T.C., Falconer, L., Soto, D., Aguilar-Manjarrez, J. (Eds.), Site Selection and Carrying Capacities for Inland and Coastal Aquaculture. FAO, Rome, pp. 87–101.

Deltares, 2014a. Delft3D FLOW User Manual, Version: 3.15.34158. May 2014. Delft, Netherlands.

Deltares, 2014b. Delft3D-Water Quality User Manual. Version: 4.99.34158, 28 May 2014. Delft, Netherlands.

DJB (Directorat Jenderal Perikanan Budidaya), 2017. KKP Genjot Pemanfaatan Potensi Budiaya Laut (KKP - Potential for Sea Culture Utilization) [WWW Document]. URL. http://www.djpb.kkp.go.id/index.php/arsip/c/492/KKP-GENJOT-PEMANFAATAN-POTENSI-BUDIDAYA-LAUT/?category_id=9.

FAO, 2010. Aquaculture development. 4. Ecosystem approach to aquaculture. In: Technical Guidelines for Responsible Fisheries, vol. 5, 53. Suppl. 4. Rome.

FAO, 2016. The State of World Fisheries and Aquaculture 2016. Contributing to Food Security and Nutrition for All (Rome).

FAO, 2018. The State of World Fisheries and Aquaculture 2018 – Meeting the Sustainable Development Goals. Rome. Licence: CC BY-NC-SA 3.0 IGO. http://www.fao.org/fishery/affris/species-prAofiles/barramundi/feeding-formulation/en/.

Ferreira, J.G., Grant, J., Verner-Jeffreys, D.W., Taylor, N.G.H., 2013. Carrying capacity for aquaculture, modeling frameworks for determination of. In: Christou, P., Savin, R., Costa-Pierce, B.A., Misztal, I., Whitelaw, C.B.A. (Eds.), Sustainable Food Production. Springer, New York, NY.

Gibbs, M.T., 2007. Sustainability performance indicators for suspended bivalve aquaculture activities. Ecological Indicators 7 (1), 94–107.

Inglis, G.J., Hayden, B.J., Ross, A.H., 2000. An overview of factors affecting the carrying capacity of coastal embayments for mussel culture. In: NIWA Client Report; CHC00/69 Project No. MFE00505. National Institute of Water and Atmospheric Research, Ltd, Christchurch, New Zealand, p. 31.

IOC, IHO, BODC, 2003. Centenary Edition of the GEBCO Digital Atlas, Published on CD-ROM on Behalf of the Intergovernmental Oceanographic Commission and the International Hydrographic Organization as Part of the General Bathymetric Chart of the Oceans.

Kutty, M.N., 1987. Site Selection for Aquaculture: Physical Features of Water.

Mayerle, R., Niederndorfer, K.R., Fernandez Jaramillo, J.M., Runte, K.-H., 2020. Hydrodynamic method for estimating production carrying capacity of coastal finfish cage aquaculture in Southeast Asia. Aquacultural Engineering 88, 102038.

Mayerle, R., Sugama, K., Hesse, K., Lehner, S., Pramono, G., Niedendorfer, K., 2018. Spatial technologies for early warning and preparedness of marine fish cage culture in Indonesia. In: Aguilar-Manjarrez, J., Wickliffe, L.C., Dean, A. (Eds.), Guidance on Spatial Technologies for Disaster Risk Management in Aquaculture. Full Document. FAO, Rome, pp. 227–297, 312 pp.

Mayerle, R., Sugama, K., Runte, K.-H., Radiarta, N., Maris Vellejo, S., 2017. Spatial planning for marine finfish aquaculture facilities in Indonesia. In: Aguilar-Manjarrez, J., Soto, D., Brummett, R. (Eds.), Aquaculture Zoning, Site Selection and Area Management under the Ecosystem Approach to Aquaculture. Report ACS113536. FAO and World Bank Group, Rome, p. 395.

McKindsey, C.W., Thetmeyer, H., Landry, T., Silvert, W., 2006. Review of recent carrying capacity models for bivalve culture and recommendations for research and management. Aquaculture 261 (2), 451–462.

Niederndorfer, K.R., 2017. Proposal of a Practical Method to Estimate the Ecological Carrying Capacity for Finfish Mariculture with Respect to Particulate Organic Carbon Deposition to the Seafloor (Ph.D. Thesis). Christian-Albrechts University of Kiel.

Pusat Data Statistik dan Informasi (Data Centre of Statistics and Information), 2015. Analysis Data Pokok Kemeterian Kelautan Dan Perikanan 2015 (Analysis Data of the Ministry of Marine Affairs and Fisheries 2015 (In Indonesian).

Ross, L.G., Telfer, T.C., Falconer, L., Soto, D., Aguilar-Manjarrez, J., 2013. Site selection and carrying capacities for inland and coastal aquaculture. In: Ross, L.G., Telfer, T.C., Falconer, L., Soto, D., Aguilar-Manjarrez, J. (Eds.), FAO Fisheries and Aquaculture Proceedings No. 21. Rome, p. 282.

Secretariat, A.S.E.A.N., 2008. ASEAN Marine Water Quality Management Guidelines and Monitoring Manual. Australia Marine Science and Technology Ltd.(AMSAT), Australia.

Soto, D., Aguilar-Manjarrez, J., Hishamunda, N. (Eds.), 2008. Building an Ecosystem Approach to Aquaculture. FAO/Universitat de les Illes Balears Expert Workshop, 7–11 May 2007, Palma de Mallorca, Spain. FAO Fisheries and Aquaculture Proceedings. No. 14. Rome, FAO. 2008. 221 pp.

Tett, P., Portilla, E., Gillibrand, P.A., Inall, M., 2011. Carrying and assimilative capacities: the ACExRLESV model for sea-loch aquaculture. Aquaculture Research 42, 51–67.

Van der Wulp, S.A., 2015. A Strategy to Optimize the Arrangement of Multiple Floating Net Cage Farms to Efficiently Accommodate Dissolved Nitrogenous Wastes. Ph.D. thesis. Christian-Albrechts University Kiel.

Windupranata, W., 2007. Development of a Decision Support System for Suitability Assessment of Mariculture Site Selection. Ph.D. thesis. Research and Technology Centre of the University of Kiel, Kiel Germany, p. 125.

Windupranata, W., Mayerle, R., 2009. Decision support system for selection of suitable mariculture site in the western part of Java sea, Indonesia. ITB Journal of Engineering Science 41b (1), 77–96.

World Bank, 2013. Fish to 2030: Prospects for Fisheries and Aquaculture. World Bank Group, Washington, DC.

11

# Decision tool for estimating energy potential from tidal resources

Roberto Mayerle[1], Kadir Orhan[1], Wahyu W. Pandoe[2], Poerbandono[3], Peter Krost[4]

[1]RESEARCH AND TECHNOLOGY CENTRE WESTCOAST (FTZ), UNIVERSITY OF KIEL, BÜSUM, GERMANY; [2]AGENCY FOR THE ASSESSMENT AND APPLICATION OF TECHNOLOGY (BPPT), JAKARTA, INDONESIA; [3]INSTITUTE OF TECHNOLOGY BANDUNG (ITB), BANDUNG, JAWA BARAT, INDONESIA; [4]COASTAL RESEARCH & MANAGEMENT (CRM), KIEL, GERMANY

**Abstract**

*This chapter presents the results of the investigations leading to the development and application of a general procedure for the characterization of tidal current energy resources in data-poor sites. It implements methods to identify suitable sites for tidal power extraction, estimate effective power potentials, and assess impacts of tidal turbine arrays on the environment. The process relies primarily on results from high-resolution, three-dimensional circulation models. Guidelines for the development of circulation models in data-poor sites are provided. Furthermore, a decision support system embedding the entire resource characterization procedure that has been developed to facilitate analysis and support authorities in the assessment and spatial planning of tidal turbine arrays is presented. Results of the application of the proposed procedure to the target sites selected by the Indonesian government led to the improvement of information base regarding the potential of tidal energy in Indonesia.*

**Abstrak**

*Makalah ini menyajikan hasil investigasi untuk pengembangan dan penerapan prosedur umum dalam melakukan karakterisasi terhadap sumber daya energi arus pasut di lokasi-lokasi yang datanya sangat sedikit. Proses yang dilakukan utamanya didasarkan pada hasil dari model sirkulasi tiga dimensi beresolusi tinggi. Model yang digunakan menerapkan metode untuk mengidentifikasi lokasi yang sesuai untuk ekstraksi tenaga pasut, memperkirakan potensi daya efektifnya, dan menilai dampak dari susunan turbin pasutnya terhadap lingkungan. Selain itu, di makalah ini disajikan pula sistem pendukung keputusan yang melibatkan seluruh prosedur karakterisasi sumber daya yang telah dikembangkan untuk memfasilitasi analisis dan mendukung pemerintah dalam penilaian dan perencanaan tata ruang dari penempatan rangkaian turbin pasut. Hasil dari prosedur yang diterapkan di lokasi-lokasi yang telah dipilih oleh pemerintah Indonesia memberikan gambaran tentang potensi energi arus pasut secara nasional.*

Science for the Protection of Indonesian Coastal Ecosystems (SPICE). https://doi.org/10.1016/B978-0-12-815050-4.00006-7 

**Chapter outline**

**11.1 Introduction** .......... **390**
**11.2 Investigation sites in Indonesia** .......... **392**
**11.3 Guidelines for setting up circulation models in data-poor sites** .......... **393**
**11.4 Procedure for assessing the energy potential from tidal currents** .......... **397**
11.4.1 Site suitability .......... 397
11.4.2 Technically extractable power under natural conditions .......... 399
11.4.3 Technically extractable power accounting for the flow interactions with turbines .......... 399
11.4.4 Environmental impact assessment .......... 400
**11.5 Decision support system for assessing the energy potential from tidal currents** .......... **401**
**11.6 Assessing the energy potential from tidal currents within the target sites in Indonesia** .......... **401**
**11.7 Conclusion** .......... **402**
**Acknowledgments** .......... **404**
**References** .......... **404**

## 11.1 Introduction

The sun, wind, waves, and tides provide vast resources of renewable energy, which may be capable of satisfying the present and future global demands for electricity with ease and without inflicting considerable damage on the global ecosystem (Asif and Muneer, 2007). While the technology for the extraction of solar and wind power is well established, the development of ocean renewable energy (ORE) technologies to harness waves, tidal currents, ocean thermal resources, and salinity gradients is facing various challenges. Nevertheless, in the short to medium term (2025–30), it is expected that particularly wave and tidal current energy conversion systems will make significant contributions to global energy supply mainly due to their advanced technology readiness levels (Sannino and Cavicchioli, 2013; Magagna and Uihlein, 2015).

Indonesia has great potential for electricity production from renewable resources. The ocean covers as much as 70% of the archipelago and possesses a substantial amount of clean energy in the form of wind, wave, ocean current/thermal, and tidal power. However, so far, the portion of the renewable energy contribution to the country's primary energy mix is meager, and utilizing renewable resources might be of great help to meet the continually increasing energy demand of the growing Indonesian population. To this end, the Indonesian government aims to increase the share of renewable energy to at least 23% by 2025 and 31% by 2050 (Boer et al., 2018; UNEP, 2019). ORE can help to reach this target, but so far, there is only limited information on the country's resources.

This study emphasizes the development of a general procedure for assessing the energy potential of tidal currents in data-poor sites. Tidal currents are primarily driven by lunar and solar gravitational forces, and the astronomic nature of the underlying driving mechanism results in an inherently predictable source (Bryden et al., 2004). Due to its high predictability and quality, tidal current energy might prove highly valuable in the future high-penetration renewable energy electricity grids (Lewis et al., 2019). Currently, the global tidal resources are estimated at 3 TW (Kempener and Neumann, 2014), with the actual technically harvestable part in the vicinity of the coast being in the order of 1 TW (Carbon Trust, 2011; Lewis et al., 2011; Kempener and Neumann, 2014). There are over 200 companies operating in the tidal energy sector, most of them involved in the design of energy converters (Magagna and Uihlein, 2015). Tidal current turbines have reached the technology readiness level 8 (TRL 8) out of a total of nine levels and have made significant progress toward commercialization (Kempener and Neumann, 2014; Magagna and Uihlein, 2015; Magagna et al., 2016).

Despite the advanced TRL, there are still significant challenges hindering the development of tidal energy conversion systems. In addition to the lack of accurate information regarding suitable locations and effective power potentials, there is general agreement that pilot commercial-scale installations should be accompanied by research investigations evaluating performances and the associated impacts of the devices on the environment. To date, the majority of assessments are inaccurate, as they are based on models driven mainly by tides. Other relevant forcing phenomena such as density variations, meteorology, and ocean currents such as the Indonesian Throughflow are mostly disregarded. Moreover, most assessments are incomplete, as they are restricted to single turbines and limited to short periods. Little is known on the effects of tidal turbine arrays on the patterns of circulation in straits. Comprehensive life cycle assessments and cost–benefit analyses of converter arrays are scarce. Studies concerning the integration of the different ORE resources, accounting for the fluctuation of power output, storage, and grid integration, are limited. Furthermore, socioeconomic impacts of ORE installations on issues such as well-being, quality of life, job creation, household income, public budget, and gross domestic product (GDP) are not yet well understood (Uihlein and Magagna, 2016). Research and development efforts must be intensified to include the aspects mentioned earlier and develop a holistic approach for the assessment of ORE resources (Magagna and Uihlein, 2015; Magagna et al., 2016).

This study focuses on some of the fundamental research gaps regarding the characterization of tidal current energy resources and the estimation of effective power from tidal currents with a focus on Indonesia. The aim is to develop a reliable information base and tools for high-resolution analysis alongside advanced strategies/techniques for resource characterization. The main objective of the present study is the development and application of a general procedure for assessing the potential of tidal current resources in data-poor sites. As the assessment relies mainly on the results of model simulations, guidelines for the development of dynamic models are provided. The investigations are done in several straits connecting the inner Indonesian Seas with the Indian Ocean in Indonesia. Due to their advanced technological status, horizontal-axis tidal turbines were taken into account throughout the investigations (Magagna and Uihlein, 2015). The proposed procedure is embedded in a decision support system (DSS) to facilitate the analysis and decision-making process. Results of the application of the DSS to estimate the power potential of tidal currents in the most promising straits selected by the Government of Indonesia are presented.

## 11.2 Investigation sites in Indonesia

The investigations were conducted in several straits connecting the inner Indonesian Seas with the Indian Ocean. Tidal flows constrained by land, islands, and shallows cause high tidal current velocities, making the straits promising zones for renewable electricity production from tidal currents. Fig. 11.1 shows the location of potential sites identified by the Government of Indonesia to increase the energy mix in the country. Accurate estimates demonstrating the power potentials of these straits are of utmost importance for the decision-making process. There is a need for standardized assessments capable of delivering estimates in sites with limited data. To this end, preliminary investigations targeted the straits of Larantuka in the far east, Alas between Bali and Lombok islands, and Sunda between Java and Sumatra islands (see Fig. 11.1). These straits were selected to cover a wide range of conditions in terms of size, current speeds, and driving mechanisms.

The Strait of Larantuka connecting the Flores Sea with the Indian Ocean was previously studied by Erwandi et al. (2011). With current speeds well above 3 m $s^{-1}$, it is a promising candidate for the deployment of second-generation tidal stream technologies, which are assumed to be capable of exploiting currents with peak spring flows above 2 m $s^{-1}$ (Lewis et al., 2015). The strait, with a width of ca. 650 m and water depths up to about 35 m in the bottleneck, is relatively narrow. Hence, it is well suited for investigating the

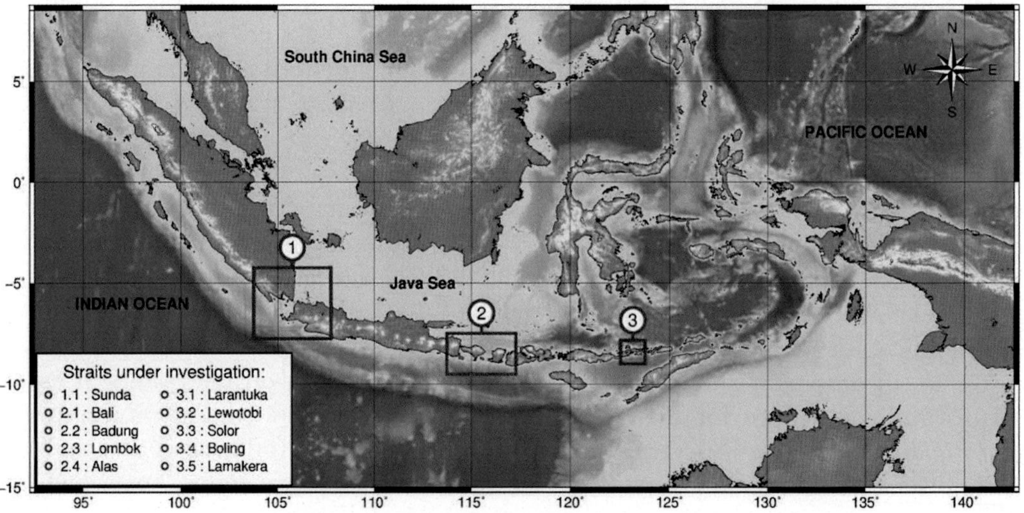

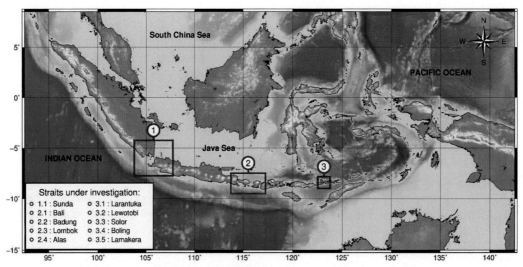


**FIGURE 11.1** Location of target sites selected by the Government of Indonesia for the extraction of tidal current energy. *Source for bathymetry/topography: SRTM15_PLUS.*

hydrodynamic impacts of tidal turbines in bounded channels. Additional investigations were done in the Alas Strait between the Bali Sea and the Indian Ocean. Although peak currents are not as high as in the Strait of Larantuka (Blunden et al., 2013), Alas is characterized by a 16 km width and water depths around 180 m, which provides a vast area for the deployment of turbines. The third site chosen is the Strait of Sunda, linking the Java Sea with the Indian Ocean. Flow conditions in this strait are strongly affected by the regional excess of freshwater entering into Pacific waters through the Sunda Shelf and the monsoon and other sea-air-land interactions such as El Niño and Dipole Mode Event (Gordon, 2005; Mayer et al., 2010; Putri, 2005). With a width of about 30 km and water depths up to 60 m in its narrowest section, it is well suited to investigate the relevance of regional atmospheric and oceanographic dynamics on tidal currents. This strait is also characterized by a steep bottom slope. Water depths increase gradually from about 35 m at the mouth to the Java Sea to ca. 500 m depths at the Indian Ocean.

## 11.3 Guidelines for setting up circulation models in data-poor sites

The development of circulation models requires essentially two components, namely the modeling software for the representation of physics and the data for setting up the model to a specific site. The modeling software adopted in this study is based on the open-source Delft3D modeling suite developed by Deltares in the Netherlands (Hydraulics, 2006; Lesser et al., 2004). As the cost of measurements for development and assessment of the performance of models is high and difficult to obtain, particularly in remote sites, dynamic models are developed using data primarily from global databases and global operational models established mainly by governmental agencies.

Bathymetric data present the main obstacle for the setup of circulation models in remote sites. Data for model development are usually obtained from a variety of sources, and a full model bathymetry typically stems from several databases and surveys. In Indonesia, model bathymetries have been constructed by integrating offshore bathymetric information from the SRTM15_PLUS land and sea topography global data set (Tozer et al., 2019), with a spatial resolution in the order of 0.5 km, with data from surveys and nautical charts provided by the Geospatial Information Agency (BIG).

Much attention has been given to the construction of the model grids. Experience from the development of models for characterization of tidal energy resources in Indonesia showed that while horizontal grid resolution can be in the order of a few hundred meters for the initial site assessments, it should typically be in the order of few tenths of meters within the identified suitable areas to simulate the effect of tidal turbine arrays on flow conditions. Horizontal grid resolutions are gradually increased from the open sea boundaries toward regions suitable for the installation of tidal turbines to reduce computing time. Grid development is done in several steps. Preliminary models are set up with uniform horizontal grid spacing. The initial grid resolution is determined based on the size of the model, the bathymetry available, and the computing time. The results are then used to gradually adjust horizontal grid spacing in locations of higher current velocities. Fig. 11.2 shows the model grid and bathymetry of the selected sites. Based on the size of computational domains, bathymetries, and results of the

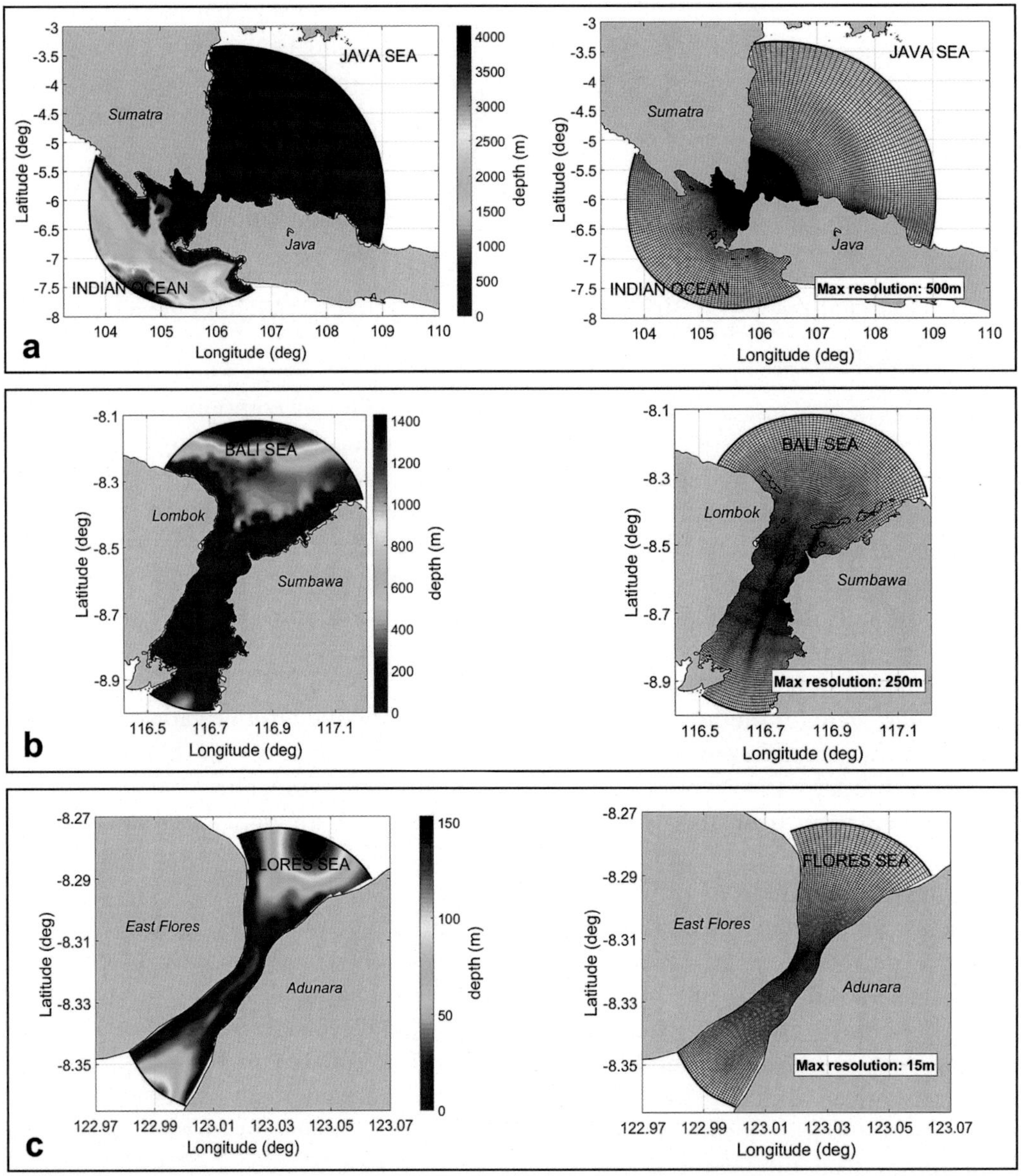


**FIGURE 11.2** Bathymetries and computational grids of the models of the straits of (A) Sunda, (B) Alas, and (C) Larantuka.

preliminary assessments, Sunda Strait computational grid was constructed with a maximum resolution of 500 m around the strait's bottleneck, whereas the Alas Strait grid had a maximum resolution of 250 m. In the case of the strait of Larantuka, after the initial site assessment, a detailed model was nested in the vicinity of the identified suitable area to match the rotor size of the individual turbines and simulate their impacts. The computational grid of the detailed model was constructed with a maximum grid resolution of 15 m.

Three-dimensional models in baroclinic mode have been developed for the straits in question. Over the vertical, several layers are implemented to take into consideration the stratifications and the three-dimensional effect of tidal turbines on flow. Z-layers fixed in the vertical direction are preferred for enhancing the accuracy of the vertical distribution of salinity and temperature (Cornelissen, 2004). In this study, the k-ε model is adopted to include the effects of the turbulent kinetic energy and turbulent kinetic energy dissipation in simulations (Burchard and Baumert, 1995; Postma et al., 1998).

Models of straits have at least two open sea boundaries to the adjacent coastal seas or oceans (see Fig. 11.2). Open sea boundaries are placed in deep water away from the mouths to capture the external forcing more adequately. Common forcing phenomena are tides, winds, offshore waves, river discharges, salinity and water temperature, and their combination. Specified at these boundaries are usually modeled values. TPXO is a global model for the estimation of ocean tides, providing tidal information drawn from tidal constituents (Egbert and Erofeeva, 2002). The latter is supplemented with sea surface heights, current velocities, salinities, and water temperature from global circulation models. Profiles of salinity and the temperature imposed at the open sea boundaries are supplied from the 3D free surface baroclinic hydrodynamic HAMSOM Model (HAMburg Shelf Ocean Model) developed at the Institute of Oceanography at the University of Hamburg (Backhaus, 1985; Pohlmann, 2006). Additionally, winds and atmospheric pressure are specified at the free surface. There are several weather-related systems developed by governmental agencies and institutes from all around the world that characterize meteorology, tides, circulation, and sea conditions. Data from the Global Forecast System (GFS) developed at the Environmental Modeling Center (2003) have been used for delivering forcing in terms of winds and atmospheric pressure.

In this study, the relevance of the main forcing processes on current velocities was assessed in the selected straits. The studies show that while phenomena such as wind-induced circulation and resulting density gradients can significantly amplify tidal currents in Indonesian straits (Orhan et al., 2019), the influence of the Indonesian Throughflow can increase tidal current speeds around 40% (Goward Brown et al., 2019). Hence, the tide-only approach was thought to be inadequate for the target domains. In addition to tidal forcing, profiles of salinity and temperature were taken into account while defining open sea boundary conditions. Effects of air cloudiness, air temperature, atmospheric pressure, relative air humidity, and wind have been considered as meteorological inputs in simulations. The ocean heat flux module of Delft3D, which typically applies to large water bodies, is activated to account for the exchange of heat through the

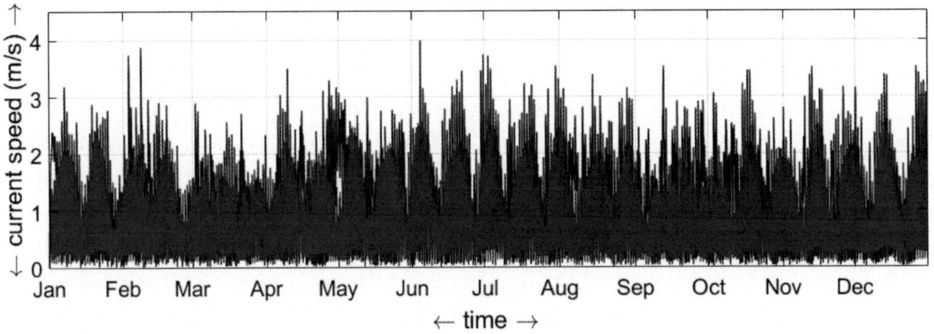


**FIGURE 11.3** Relevance of forcing mechanisms on velocities in the Sunda Strait. Modeled velocities for a year considering tide only (red) and the main forcing phenomena (blue).

free surface (Lane, 1989). Sensitivity analyses were conducted in the Sunda Strait to illustrate the relevance of driving forces on current velocities (Orhan et al., 2019). Fig. 11.3 shows the results of model simulations considering tide only and all the forcing mechanisms for a monitoring point located in the suitable areas of the Sunda Strait. It could be concluded that particularly in larger straits, atmospheric effects are of utmost importance for adequately capturing current velocities in straits. In the Sunda Strait, the discrepancies between current velocities resulted up to 100% during some periods (see Fig. 11.3).

Assessment of the accuracy of circulation models in predicting water levels and current velocities is essential for the development of reliable models. It is recommended to measure water levels at several locations along the straits. The roughness coefficient and wind drag coefficients are adjusted against water level measurements. This usually provides reasonable water level predictions in the early stages of development when a site is assessed regarding its potential for tidal current resources. Later on, once the site is selected for expansion of the activity, predictions based on current velocity measurements using acoustic doppler current profilers (ADCPs) are recommended to improve the accuracy of the circulation model. Due to the difficulty and costs involved in conducting on-site measurements in high-speed flows, so far, measurement data concerning energetic tidal currents are scarce and predominantly restricted to commercial tidal sites (Goward Brown et al., 2019). Especially the in situ measurements concerning the performance and hydrodynamic impacts of tidal turbines are limited with the turbine arrays located in the United Kingdom (Lewis et al., 2019).

The ability of the Larantuka Strait model to simulate water level elevations was verified through measurements recorded at the southern entrance of the strait via a stationary acoustic device manufactured by General Acoustics, Germany. Fig. 11.4 displays the comparison of simulated and measured tidal water levels for July 2014. The agreement between modeled and measured values was checked with Pearson's product–moment correlation coefficient r and root mean square deviation (RMSD).

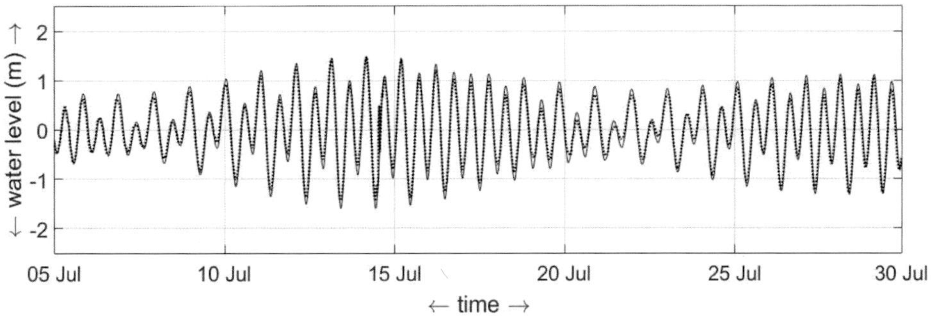


**FIGURE 11.4** Modeled (*continuous line*) versus measured (*dotted line*) tidal elevations during July 2014 at the southern entrance of the Strait of Larantuka.

RMSD and r values resulted, respectively, equal to 0.11 m and 0.99, indicating a strong, positive linear relationship (Weeser et al., 2019).

Porous plates are used to account for the interactions of turbines with the flow. Drag forces are introduced at turbine locations to represent the velocity deficit caused by turbines. Tidal energy extraction is parameterized in 3D as an enhanced momentum sink to accurately evaluate the impacts and performances of individual turbines (Brown et al., 2017; Piano et al., 2017). Momentum sinks in the form of porous plates have been located at the grid cell faces over several vertical layers to represent the tidal turbines (Defne et al., 2011; Yang and Wang, 2015; Chen et al., 2014). The effect of turbines on the flow of the Strait of Larantuka is illustrated in Fig. 11.5D.

## 11.4 Procedure for assessing the energy potential from tidal currents

A general procedure for resource characterization and estimation of energy potential from tidal currents in straits has been proposed. The standardized procedure comprises basically four methods, namely for site selection, estimation of tidal power considering natural conditions, estimation of tidal power accounting for the flow interaction with turbines, and environmental impact assessment (EIA). Horizontal axis turbines are adopted in the assessments. Fig. 11.5 illustrates the Strait of Larantuka, the main deliverables of the various methods. A detailed description follows.

### 11.4.1 Site suitability

Site suitability (SS) is relatively straightforward to determine but requires a substantial amount of data with good spatial coverage to assess site function and define suitable locations for tidal turbines. Several criteria and threshold values are considered. Defne et al. (2011) introduced a geographic information system (GIS)-based multicriteria assessment accounting for physical, environmental, and socioeconomic constraints. Because of the scarce data in remote sites, in this study, preliminary assessments are

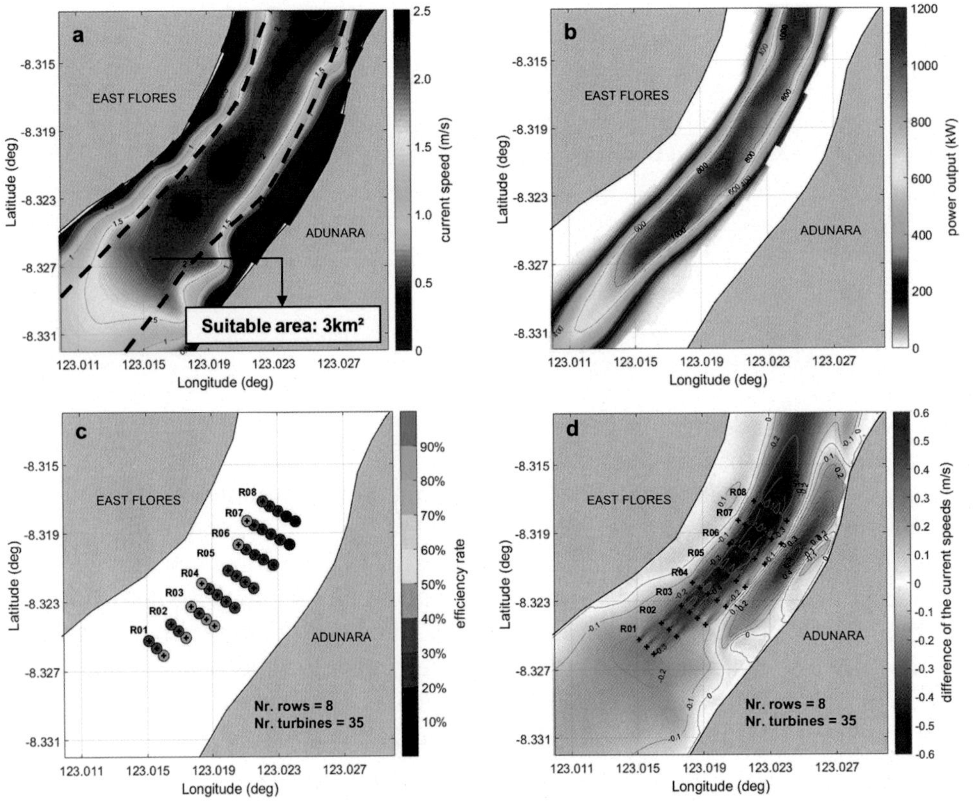


**FIGURE 11.5** Proposed stepwise procedure for assessing the tidal energy potential in the Strait of Larantuka: (A) Spatial variation of the average current velocities and suitable area (between *dashed lines*), (B) average power output under natural conditions, (C) turbine efficiencies for a tidal energy extraction scenario, (D) difference between depth-averaged current speeds of reference state and tidal energy extraction scenario.

based exclusively on physical criteria. Essentially three criteria are considered, namely current velocity, water depth, and space. Typically, commercial-grade horizontal axis tidal turbines require a minimum current velocity, i.e., cut-in velocity, to start operating. Depending on the design of the turbines, cut-in velocity values ranging between ca. 0.5 and 1 m $s^{-1}$, corresponding to hydrokinetic power densities, respectively, of 0.0625 and 0.5 kW $m^{-2}$, have been used (EPRI, 2006). Water depths must be sufficiently high to accommodate clearances above and below the turbines. Near the free surface, a 5 m clearance applies to avoid conflicts with recreational activities, minimize turbulence and wave loading effects on turbines, and reduce damage from floating materials. On the seabed, 10% of the mean lower low water depth is eliminated to escape from the low-speed benthic boundary layer (EPRI, 2006; EMEC, 2009). Additionally, a minimum area of about 0.1–0.5 $km^2$ is required for the development of tidal energy resources using arrays of horizontal axis turbines (Haas et al., 2011). For more details of the adopted criteria and threshold values, the reader is referred to Orhan et al. (2017) and Orhan and Mayerle (2017).

In the selection of suitable locations, thematic maps based primarily on results of simulation models and in situ measurements are prepared. For each thematic map, templates are built and overlaid for generating suitability maps. Based on the properties of the energy conversion device, marine areas with adequate hydrokinetic power densities to ensure efficient operation and adequate water depths are defined. Fig. 11.5A shows the spatial variation of the current velocity at the energy hot spot within the Strait of Larantuka. According to the results in the deeper and more energetic main channel, average current speeds are up to ca. 2.5 $m\ s^{-1}$, and average hydrokinetic power densities exceed about 9 $kW\ m^{-2}$. An area of ca. 3 $km^2$ in the center of the strait is found suitable for deployment of up to about 500 turbines with rotor diameters ranging from 1.5 to 20 m.

### 11.4.2 Technically extractable power under natural conditions

The method delivers approximate estimates of the technically extractable power generated by individual turbines for undisturbed flow conditions. Calculations taking into account swept areas, current speeds, turbine efficiencies, and a series of water to wire efficiencies are carried out. The method proposed by Blunden et al. (2013) for providing an approximate estimation of the total technically extractable electric power from the energy hot spots is adopted. The wake effects of turbines are avoided by setting the spacing among generators at sufficient distances to minimize the interferences (EMEC, 2009). The total extractable power is then calculated considering the flow fields in their natural (undisturbed) states alongside the applicable size, performance, and packing density of the devices within the energy hot spots. Fig. 11.5B illustrates the spatial variation of the estimated electric power that can be delivered to the local electrical grid by individual turbines from the identified energy hot spot. The figure shows, within the suitable area in the center of the strait, average extractable electric power by individual turbines of over 1.2 MW. This is compatible with the rated power production of commercial tidal turbines (Roberts et al., 2016).

### 11.4.3 Technically extractable power accounting for the flow interactions with turbines

In this step, a more realistic estimate of the extractable power accounting for arrays of turbines and their interactions with the flow is provided. For a thorough assessment, several array layouts with varying spacing and installed capacities are tested, and the corresponding changes in hydrodynamics and performances of individual devices and arrays are evaluated (Chen et al., 2014; Orhan and Mayerle, 2020). The interaction of turbines with the flow might cause changes in hydrodynamics and sedimentation patterns, which in turn might lead to erosion/degradation of the coastline and deterioration of water quality in adjacent marine areas (Ahmadian and Falconer, 2012; Neill et al., 2009). Decreasing current velocity magnitudes both upstream and downstream of the turbine arrays and increases in velocities above and below the turbine rotors are observed in conjunction with stronger velocities along the margins of the strait (see Fig. 11.5). The installation of additional rows of turbines causes clear decreases in the efficiency of turbines. Fig. 11.5C depicts the estimated efficiencies of the turbines for a given array layout with eight rows and a total of 35 turbines. Marker colors indicate the efficiency rate of each turbine. Notice that toward the behind of the array, turbine efficiencies decrease down to around 20%–30%. The maximum average operational efficiency rates of the devices resulted in the order of 50%–60% as compared with the undisturbed conditions (Fig. 11.5D). The method delivers realistic estimates of the effective power potential and on the effects of the turbines on flow and sediment transport. Issues such as the flow asymmetry and turbine misalignment (Piano et al., 2017), as well as power variability of the tidal current energy (Lewis et al., 2019), should also be addressed in future investigations. Emphasis should be given to the optimization of the turbine array layouts to maximize power output while minimizing environmental impacts in the strait.

### 11.4.4 Environmental impact assessment

A method based on the DPSIR (Driver, Pressure, State, Impact, and Response) for assessing environmental impacts was proposed. In this study, the method was adopted for the characterization, and evaluation of environmental pressures imposed by tidal power converters on sensitive tropical marine environments was done. Tidal power converters are rather large structures built within potentially sensitive and vulnerable environments. In addition to the interaction with the flow, the operation of tidal turbines is also associated with certain risks due to emissions in the form of exhaust gas, wastewater, light, and noise. Hence, turbine installations can put pressure on vulnerable habitats and species (Orhan et al., 2015).

Application to the Strait of Larantuka led to the identification of vulnerable marine ecosystems and species in the strait. The results indicated that the development of tidal turbine arrays in the Strait of Larantuka might affect adjacent beaches, coral reefs, mangroves, seagrass meadows, and coastal wetlands as well as species cetacea, dugong, whales, migratory fish species, and sea turtles. It was concluded that the method should be shifted to SS to identify relevant interactions at the feasibility stage of projects.

In this study, the emphasis was given to the far-field impacts of multiturbine arrays on current patterns in the Strait of Larantuka. Fig. 11.5.D shows differences in depth-averaged current speeds between the undisturbed conditions and the tidal energy extraction scenario with 8 rows and 35 turbines. An overall change in the flow patterns with respect to the reference state led to deficits in current speeds of about 0.5–0.6 m $s^{-1}$. Especially along the margins of Adunara Island, a drastic increase in currents speeds, which equals around 80%, resulted mainly because of the proximity of turbines to the nearshore shallow waters of the island. As a result, changes in the sediment patterns leading to erosion and sedimentation are expected to occur. Measures aiming at mitigating such impacts and seeking alternative sites/layouts for tidal arrays are required.

## 11.5 Decision support system for assessing the energy potential from tidal currents

The proposed procedure has been embedded into a DSS to facilitate the analysis and support planners in the assessment, spatial planning, and estimation of energy potential from tidal resources. The DSS has been designed to process, analyze, and evaluate the results from circulation models as well as to estimate the tidal stream power potentials to identify the best sites for the installation of ocean renewable energy. The DSS is equipped with an integrated GIS database for storage, processing, and visualization of measurement data, results of dynamic models, and calculations concerning the electric power yield in the specific sites. Data from various sources and results from dynamic models are embedded with a graphical user interface. The system provides guidance on planning and identification of potential areas for the installation of tidal converters. The DSS is also equipped with interfaces for importing and visualizing real-time data from global oceanic modeling used for forcing the dynamic models of the straits.

## 11.6 Assessing the energy potential from tidal currents within the target sites in Indonesia

The adequacy of sites preselected by the Government of Indonesia for the generation of tidal current energy is investigated with the proposed procedure. Assessments are done for the 10 straits shown in Fig. 11.1. Three-dimensional models have been developed for all the straits following the guidelines presented before. Simulations covered 30 days. Horizontal axis turbines capable of harnessing bidirectional flows with cut-in velocities of 0.5 and 1 m $s^{-1}$ are considered. The criteria and threshold values listed before are adopted, leading to the identification of suitable areas. The extractable electric power from the energy hotspot is calculated, considering a series of water-to-wire efficiencies (EPRI, 2006). The total extractable power from the suitable domains is calculated based on a simplified variation of the "per-generator" method (Blunden et al., 2013). The wake effects of turbines have been ignored by setting the spacing to 2.5 times the rotor diameter laterally and 10 times the rotor diameter between the turbine rows to minimize the interferences (EMEC, 2009). The extracted energy is calculated from the flow fields in their natural (undisturbed) states, considering the applicable size, performance, and packing density of the devices within the energy hot spots.

Table 11.1 lists the results of resource characterization for the target sites shown in Fig. 11.1. The straits of Lamakera, Solor, and Lewotobi were proven unsuitable for tidal energy extraction due to the insufficient current velocities. The remaining straits were found suitable and might be able to deliver significant power. There is a considerable variation in the maximum kinetic power density among the straits. Values range from about 1.5 kW $m^{-2}$ in Sunda and Badung to up to about 15 kW $m^{-2}$ in Bali Strait. Despite the relatively low hydrokinetic power densities, Alas Strait has the most significant

**Table 11.1** Summary of resource mapping of the target sites in Indonesia.

| | | | Cut-in velocity = 1.0 m $s^{-1}$ | | Cut-in velocity = 0.5 m $s^{-1}$ | |
|---|---|---|---|---|---|---|
| **Strait** | **Maximum kinetic power density (kW $m^{-2}$)** | **Maximum extractable power per turbine (kW)** | **Suitable area ($km^2$)** | **Total extractable power (MW)** | **Suitable area ($km^2$)** | **Total extractable power (MW)** |
| 1.1 Sunda | 1.56 | 165 | 11 | 107 | 145 | 335 |
| 2.1 Bali | 14.75 | 1460 | 23 | 462 | 104 | 1045 |
| 2.2 Badung | 1.52 | 161 | 28 | 238 | 162 | 551 |
| 2.3 Lombok | 2.36 | 236 | 53 | 754 | 114 | 865 |
| 2.4 Alas | 3.07 | 396 | 104 | 1261 | 403 | 2258 |
| 3.1 Larantuka | 10.20 | 1250 | 3 | 199 | 6 | 299 |
| 3.4 Boling | 3.49 | 430 | 36 | 593 | 106 | 736 |

potential for energy production (ca. 1200 MW for cut-in velocities of 0.5 m $s^{-1}$). This is due to the size of the suitable area for the deployment of turbine arrays, which resulted considerably larger than in the other sites. According to the results, overall tidal power at the investigated sites might exceed 6000 MW. This corresponds to approximately 20% of Indonesia's current electric power demand. The results obtained here indicate that the potential of the country tidal energy is much higher than previously estimated by Alifdini et al. (2016). Nevertheless, more precise calculations taking into account the impacts of tidal turbines on flow conditions are required. Especially in the Larantuka Strait, where the geometric blockage effect is more significant, results might differ significantly.

## 11.7 Conclusion

A general procedure for resource characterization and estimation of energy potential from tidal currents has been successfully developed and applied to remote sites in Indonesia. The holistic procedure has been proved to be particularly appropriate for data-poor sites around the globe. Results obtained in this study point out the relevance of accounting for the main forcing phenomena for accurately predicting current velocities. Under some conditions, discrepancies of up to 100% resulted in the predicted current velocities by disregarding some of the main forcing mechanisms. It was found that the hydrodynamic impact of the tidal turbines must also be included in assessments to deliver realistic estimates of the power output. Significant reductions in the efficiency of turbines resulted in accounting for these effects. To advance the developments in the field, and increase the accuracy of estimates, on-site measurements of current velocities in the vicinity of arrays of turbines are required.

Results of the applications to the target sites identified by the Government of Indonesia led to the improvement of the information base regarding the potentials of tidal energy in Indonesia. Sunda, Bali, Badung, Lombok, Alas, Larantuka, and Boling straits in Indonesia were found to be suitable for the extraction of tidal power. Suitable areas and approximate estimates of the power output were provided. It was found that the overall power delivered by the potential straits under natural conditions might surpass 6000 MW. This is much higher than earlier estimates. To confirm predictions, detailed assessments taking into consideration the effect of the turbines on the flow patterns are under way. The variations in power output throughout the year observed in the study emphasize the need for long-term assessments.

Several issues have been identified for improving the proposed procedure. It is relevant to shift some of the problems currently handled within the EIA to the early stages of the procedure. In particular, maps addressing vulnerable habitats and species are needed and must be taken into account within SS. When made available, zoning plans to indicate areas prioritizing activities might positively improve the quality of resource characterization, which in turn provides better guidance for the development of tidal farms. Other aspects, such as the flow asymmetry and turbine misalignment, optimization of turbine locations, and power variability of the tidal current energy, should be addressed in future investigations. Extensions of the procedure to consider socioeconomic and environmental aspects at the feasibility stage of projects should be emphasized.

### Knowledge gaps and directions of future research

- There is a general agreement that pilot commercial-scale installations of tidal and wave power converters should be accompanied by research investigations evaluating performances as well as environmental impacts of the devices. So far, comprehensive life cycle assessments and cost–benefit analyses of converter arrays are missing. Studies concerning the integration of the different ORE resources, accounting for the fluctuation of power output, storage, and grid integration, are limited.
- Furthermore, socioeconomic impacts of ORE installations on issues such as well-being, quality of life, employment (job creation), household income, public budget, and GDP are not yet well understood (Uihlein and Magagna, 2016). Research and development efforts must be intensified to include the aforementioned aspects and develop a holistic approach for the assessment of ORE resources (Magagna and Uihlein, 2015; Magagna et al., 2016).
- Despite the advancements made in the numerical modeling field, there is a need for innovation to improve the quality and accessibility of the field surveys, especially in data-poor regions. Overcoming such problems might be possible through community efforts such as data platforms enabling the efficient sharing of measurements, or advancements in the remote sensing field to retrieve data from coastal areas with less noise and higher resolution (Xu et al., 2018). Such improvements might lead to development of models providing higher precision and therefore better assessment of ORE resources worldwide.

Implications/recommendations for policy and society

- The Indonesian government aims to increase the share of renewable energy to at least 23% by 2025 and 31% by 2050 (Boer et al., 2018, p. 45; UNEP, 2019, p. 80). ORE can help reaching this target, thus contributing to a cleaner and secure energy supply.
- According to the UN Convention on the Law of the Sea, territorial waters encompass the coastal waters up to 12 nautical miles from the coastline (Uihlein and Magagna, 2016). Setting up laws and regulating the use of coastal waters for the installation of ocean energy converters is vital alongside the intensification of efforts to promote ORE.
- Marine spatial planning (MSP) is essential to overcome problems with overlapping jurisdiction and hence to utilize the full potential of ORE resources. MSP also avoids multiple-user conflicts, improving the management of marine spatial claims toward a sustainable ecosystem-based management of marine areas (Maes, 2008).

# Acknowledgments

This study is funded by the German Ministry of Education and Research (BMBF) from March 2012 to February 2016 under grant number 03F0646A. We thank BMBF, the Project Management Jülich (PtJ), and the Leibniz Centre for Tropical Marine Research (ZMT) GmbH in Germany, as well as the State Ministry of Research and Technology (RISTEK), the Ministry of Marine Affairs and Fisheries (KKP), and the Agency for the Assessment and Application of Technology (BPPT) in Indonesia for coordination, administration, and financial support of the projects. For technical cooperation, we also thank Dr. Erwandi of the Indonesian Hydrodynamic Laboratory of BPPT-LHI, Indonesia, as well as the supporting administrative, scientific, and technical personal of the institutions involved.

# References

Ahmadian, R., Falconer, R.A., 2012. Assessment of array shape of tidal stream turbines on hydro-environmental impacts and power output. Renewable Energy 44, 318–327.

Alifdini, I., Widodo, A.B., Sugianto, D.N., Okta, Y., 2016. Identifikasi potensi energi Pasang Surut Menggunakan Alat Floating Dam di Perairan Kalimantan Barat, Indonesia identification of tidal energy potential using floating dam device in west Borneo waters (Indonesia).

Asif, M., Muneer, T., 2007. Energy supply, its demand and security issues for developed and emerging economies. Renewable and Sustainable Energy Reviews 11 (7), 1388–1413.

Backhaus, J.O., 1985. A three-dimensional model for the simulation of shelf sea dynamics. Deutsche Hydrografische Zeitschrift 38 (4), 165–187.

Blunden, L.S., Bahaj, A.S., Aziz, N.S., 2013. Tidal current power for Indonesia? An initial resource estimation for the Alas Strait. Renewable Energy 49, 137–142.

Boer, R., Dewi, R.G., Ardiansyah, M., Siagian, U.W., 2018. Indonesia, Second Biennial Update Report Under the United Nations Framework Convention on Climate Change. Directorate General of Climate Change, Ministry of Environment and Forestry, Indonesia, Jakarta, Indonesia.

Brown, A.J.G., Neill, S.P., Lewis, M.J., 2017. Tidal energy extraction in three-dimensional ocean models. Renewable Energy 114, 244–257.

Bryden, I.G., Grinsted, T., Melville, G.T., 2004. Assessing the potential of a simple tidal channel to deliver useful energy. Applied Ocean Research 26 (5), 198–204.

Burchard, H., Baumert, H., 1995. On the performance of a mixed-layer model based on the κ-ε turbulence closure. Journal of Geophysical Research: Oceans 100 (C5), 8523–8540.

Chen, Y., Lin, B., Lin, J., 2014. Modelling tidal current energy extraction in large area using a three-dimensional estuary model. Computers and Geosciences 72, 76–83.

Cornelissen, S.C., 2004. Numerical Modelling of Stratified Flows Comparison of the Sigma and Z Coordinate Systems (Master's Thesis). Delft University of Technology, Delft, the Netherlands.

Defne, Z., Haas, K.A., Fritz, H.M., 2011. GIS based multi-criteria assessment of tidal stream power potential: a case study for Georgia, USA. Renewable and Sustainable Energy Reviews 15 (5), 2310–2321.

Egbert, G.D., Erofeeva, S.Y., 2002. Efficient inverse modeling of barotropic ocean tides. Journal of Atmospheric and Oceanic Technology 19 (2), 183–204.

Electric Power Research Institute (EPRI), 2006. Methodology for Estimating Tidal Current Energy Resources and Power Production by Tidal in-Stream Energy Conversion (TISEC) Devices (USA).

Environmental Modeling Center, 2003. The GFS Atmospheric Model (USA).

Erwandi, A.K., Sasoko, P., Rina Wijanarko, B., Marta, E., Rahuna, D., 2011. Vertical axis marine current turbine development in Indonesian hydrodynamic laboratory-Surabaya for tidal power plant. In: International Conference and Exhibition on Sustainable Energy and Advanced Materials.

Gordon, A.L., 2005. Oceanography of the Indonesian seas and their Throughflow. Oceanography 18, 14–27.

Goward Brown, A.J., Lewis, M., Barton, B.I., Jeans, G., Spall, S.A., 2019. Investigation of the modulation of the tidal stream resource by ocean currents through a complex tidal channel. Journal of Marine Science and Engineering 7 (10), 341.

Haas, K.A., Fritz, H.M., French, S.P., Smith, B.T., Neary, V., 2011. Assessment of Energy Production Potential from Tidal Streams in the United States. Georgia Tech Research Corporation, Atlanta, GA (United States).

Hydraulics, D., 2006. Delft3D-FLOW User Manual (Delft, the Netherlands).

Kempener, R., Neumann, F., 2014. Tidal Energy Technology Brief. International Renewable Energy Agency (IRENA), pp. 1–34.

Lane, A., 1989. The Heat Balance of the North Sea.

Lesser, G.R., Roelvink, J.V., Van Kester, J.A.T.M., Stelling, G.S., 2004. Development and validation of a three-dimensional morphological model. Coastal Engineering 51 (8–9), 883–915.

Lewis, A., Estefen, S., Huckerby, J., Musial, W., Pontes, T., Torres-Martinez, J., 2011. Ocean Energy. IPCC Special Report on Renewable Energy Sources and Climate Change Mitigation.

Lewis, M., McNaughton, J., Márquez-Dominguez, C., Todeschini, G., Togneri, M., Masters, I., et al., 2019. Power variability of tidal-stream energy and implications for electricity supply. Energy 183, 1061–1074.

Lewis, M., Neill, S.P., Robins, P.E., Hashemi, M.R., 2015. Resource assessment for future generations of tidal-stream energy arrays. Energy 83, 403–415.

Maes, F., 2008. The international legal framework for marine spatial planning. Marine Policy 32 (5), 797–810.

Magagna, D., Monfardini, R., Uihlein, A., 2016. JRC Ocean Energy Status Report 2016 Edition. Publications Office of the European Union, Luxembourg.

Magagna, D., Uihlein, A., 2015. Ocean energy development in Europe: current status and future perspectives. International Journal of Marine Energy 11, 84–104.

Mayer, B., Damm, P.E., Pohlmann, T., Rizal, S., 2010. What is driving the ITF? An illumination of the Indonesian throughflow with a numerical nested model system. Dynamics of Atmospheres and Oceans 50 (2), 301–312.

Neill, S.P., Litt, E.J., Couch, S.J., Davies, A.G., 2009. The impact of tidal stream turbines on large-scale sediment dynamics. Renewable Energy 34, 2803–2812.

Orhan, K., Mayerle, R., 2017. Assessment of the tidal stream power potential and impacts of tidal current turbines in the Strait of Larantuka, Indonesia. Energy Procedia 125, 230–239.

Orhan, K., Mayerle, R., 2020. Potential hydrodynamic impacts and performances of commercial-scale turbine arrays in the strait of Larantuka, Indonesia. Journal of Marine Science and Engineering 8 (3), 223.

Orhan, K., Mayerle, R., Mayer, B., 2019. About the influence of density-induced flow on tidal stream power generation in the Sunda Strait, Indonesia. In: Proceedings of the 38th IAHR World Congress, Panama City, Panama, pp. 1–10.

Orhan, K., Mayerle, R., Narayanan, R., Pandoe, W.W., 2017. Investigation of the energy potential from tidal stream currents in Indonesia. Coastal Engineering Proceedings 1 (35), 10.

Orhan, K., Mayerle, R., Pandoe, W.W., 2015. Assessment of energy production potential from tidal stream currents in Indonesia. Energy Procedia 76, 7–16.

Piano, M., Neill, S.P., Lewis, M.J., Robins, P.E., Hashemi, M.R., Davies, A.G., et al., 2017. Tidal stream resource assessment uncertainty due to flow asymmetry and turbine yaw misalignment. Renewable Energy 114, 1363–1375.

Pohlmann, T., 2006. A meso-scale model of the central and southern North Sea: consequences of an improved resolution. Continental Shelf Research 26 (19), 2367–2385.

Postma, L., Stelling, G.S., Boon, J., 1998. 3-dimensional water quality and hydrodynamic modeling in Hong Kong, III. Stratification and water quality. Environmental Hydraulics 43–52.

Putri, M.R., 2005. Study of Ocean Climate Variability (1959–2002) in the Eastern Indian Ocean, Java Sea and Sunda Strait Using the HAMburg Shelf Ocean Model. Ph.D. thesis. University of Hamburg, Hamburg, Germany.

Roberts, A., Thomas, B., Sewell, P., Khan, Z., Balmain, S., Gillman, J., 2016. Current tidal power technologies and their suitability for applications in coastal and marine areas. Journal of Ocean Engineering and Marine Energy 2 (2), 227–245.

Sannino, G., Cavicchioli, C., 2013. Overcoming Research Challenges for Ocean Renewable Energy.

The European Marine Energy Center Ltd, 2009. Assessment of Tidal Energy Resource (UK).

Tozer, B., Sandwell, D.T., Smith, W.H.F., Olson, C., Beale, J.R., Wessel, P., 2019. Global bathymetry and topography at 15 arc sec: SRTM15+. Earth and Space Science 6 (10), 1847–1864.

Trust, C., 2011. Accelerating marine Energy: The Potential for Cost Reduction—Insights from the Carbon Trust marine Energy Accelerator. Carbon Thrust, London, UK.

Uihlein, A., Magagna, D., 2016. Wave and tidal current energy – a review of the current state of research beyond technology. Renewable and Sustainable Energy Reviews 58, 1070–1081 (Elsevier).

UNEP, 2019. Emissions gap Report 2019. United Nations Environment Programme (UNEP), Nairobi, Kenya.

Weeser, B., Jacobs, S., Kraft, P., Rufino, M.C., Breuer, L., 2019. Rainfall-Runoff modeling using crowd-sourced Water level data. Water Resources Research 55 (12), 10856–10871.

Xu, X.Y., Birol, F., Cazenave, A., 2018. Evaluation of coastal sea level offshore Hong Kong from Jason-2 altimetry. Remote Sensing 10 (2), 282.

Yang, Z., Wang, T., 2015. Modeling the effects of tidal energy extraction on estuarine hydrodynamics in a stratified estuary. Estuaries and Coasts 38 (1), 187–202.

12

# The governance of coastal and marine social–ecological systems: Indonesia and beyond

Marion Glaser[1,2], Luky Adrianto[3], Annette Breckwoldt[1], Nurliah Buhari[4], Rio Deswandi[1,5], Sebastian Ferse[1], Philipp Gorris[6], Sainab Husain Paragay[7], Bernhard Glaeser[8], Neil Mohammad[9], Kathleen Schwerdtner Máñez[1,10], Dewi Yanuarita[9]

[1]LEIBNIZ CENTRE FOR TROPICAL MARINE RESEARCH (ZMT), BREMEN, GERMANY; [2]INSTITUTE OF GEOGRAPHY, UNIVERSITY OF BREMEN, BREMEN, GERMANY; [3]FACULTY OF FISHERIES AND MARINE SCIENCES, IPB UNIVERSITY, BOGOR, INDONESIA; [4]MATARAM UNIVERSITY, WEST NUSA TENGGARA, INDONESIA; [5]CENTER FOR REGULATION POLICY AND GOVERNANCE, BOGOR, INDONESIA; [6]UNIVERSITY OF OSNABRÜCK, GERMANY; [7]ENLIGHTENING INDONESIA, MAKASSAR, SOUTH SULAWESI, INDONESIA; [8]SOCIETY FOR HUMAN ECOLOGY AND FREE UNIVERSITY OF BERLIN, GERMANY; [9]HASANUDDIN UNIVERSITY, MAKASSAR, INDONESIA; [10]PLACE NATURE CONSULTANCY, ASHAUSEN, GERMANY

## Abstract

*Human–nature interactions are at the root of sustainability problems worldwide. For the Indonesian coastal and marine realm, this chapter asks: How does governance affect, and how might it transform, human, and societal behavior in the coastal and marine realm toward greater ecological and social sustainability? Equipped with definitions of key social science terms (including governance, institutions, social–ecological systems), the reader is introduced to major changes, issues, and sectors in Indonesian coastal governance. Major policy recommendations at the end of subsections on the governance of (1) mangroves, (2) fisheries and aquaculture, (3) watersheds and land use, (4) marine and coastal conservation, (5) coastal livelihoods and social needs, and (6) pollution are presented. In line with the need for integrated planning, overlaps and trade-offs between these sectoral recommendations are identified and, within a theoretical framework of adaptive, evolutionary, and anticipatory governance, a set of generic suggestions for integrated inclusive governance approaches is proposed.*

## Abstrak

*Interaksi antara manusia dan alam adalah dasar dari masalah keberlanjutan di seluruh dunia. Untuk wilayah pesisir dan laut Indonesia, bab ini bertanya: bagaimana pengaruh tata kelola, dan bagaimana hal itu mengubah perilaku manusia dan masyarakat di wilayah pesisir dan kelautan menuju keberlanjutan ekologis dan sosial yang lebih baik? Dilengkapi dengan definisi istilah utama ilmu sosial (termasuk tata kelola, lembaga, sistem sosial-ekologis), pembaca diperkenalkan dengan*

Science for the Protection of Indonesian Coastal Ecosystems (SPICE). https://doi.org/10.1016/B978-0-12-815050-4.00008-0

*perubahan-perubahan, masalah dan sektor utama dalam tata kelola pesisir Indonesia. Di akhir subbagian, disajikan rekomendasi kebijakan utama tentang tata kelola dari 1. Hutan bakau, 2. Perikanan dan akuakultur, 3. Daerah aliran sungai dan penggunaan lahan, 4. Konservasi laut dan pesisir, 5. Mata pencaharian pesisir dan kebutuhan sosial, dan 6. Polusi. Sejalan dengan kebutuhan untuk perencanaan terpadu, tumpang tindih dan trade-off antara rekomendasi sektoral diidentifikasi, dan dalam kerangka teoritis tata kelola adaptif, serta evolusi dan antisipatif, serangkaian saran umum untuk pendekatan tata kelola inklusif terintegrasi ditampilkan dalam bab ini.*

## Chapter outline

**12.1 Introduction** ........ **409**
12.1.1 Some relevant definitions and understandings ........ 409
12.1.2 The governance of coastal and marine areas in Indonesia ........ 410
**12.2 Marine and coastal governance: sectors, issues, and options for intervention** ........ **411**
12.2.1 Coral reefs ........ 412
12.2.2 Mangroves ........ 413
12.2.3 Marine fisheries ........ 414
12.2.4 Watersheds and land use ........ 416
12.2.5 Marine and coastal conservation ........ 418
12.2.6 Coastal livelihoods and social needs ........ 422
12.2.7 Pollution ........ 423
**12.3 The need for integrated coastal and marine planning** ........ **425**
12.3.1 Participatory governance and comanagement ........ 426
12.3.2 Adaptive, evolutionary, and anticipatory governance ........ 427
12.3.3 Polycentric coastal and marine governance ........ 429
**12.4 Conclusions and outlook** ........ **430**
12.4.1 Critically examine the effects of polycentric governance ........ 430
12.4.2 Differentiate stakeholder perceptions and analyze links between perceptions and behaviour ........ 430
12.4.3 Work with social energy for sustainable human–nature relations ........ 431
12.4.4 Identify and use arenas of leverage against wickedly resilient system features ........ 432
12.4.5 Adopt explicit social equity objectives ........ 432
12.4.6 Employ innovative methods for linking knowledge types to generate synergies and innovation ........ 434
**12.5 Final remarks** ........ **434**
**Acknowledgments** ........ **435**
**References** ........ **436**

# 12.1 Introduction

Coasts and coastal oceans are hotspots for human populations, for biodiversity and for conflicts. Coastal and marine social–ecological systems (CM-SES) are thus more diverse in stakeholder groups, and in species than purely terrestrial or open ocean marine systems, and they supply a wide range of ecosystem services including food, energy, storm protection, and nutrient cycling (for a wider list, see Lotze and Glaser, 2008). In the Anthropocene, central issues in CM-SES, such as species depletion and resource degradation and associated problems for human well-being, such as ill-health and poverty, are caused by unsustainable human and societal behavior in multiple arenas (Crutzen, 2002; Glaser et al., 2012; Birkeland, 2015; Aswani et al., 2017). As a consequence, environmental governance that improves the resilience (Adger et al., 2005a,b; Brown, 2018) of vulnerable coastal and marine social-ecological systems (SES) is increasingly taking center stage, both in terms of a need for knowledge-building and as an applied action-oriented field.

This chapter discusses and connects some of the key social science-led findings from the second and third phases of the Science for the Protection of Indonesian Coastal Ecosystems (SPICE) project (2007–16). We provide a synthesis of over 20 relevant peer-reviewed social science and interdisciplinary articles produced during the SPICE project.[1] The first phase of SPICE (2004–06) involved social scientists only in an exploratory manner, with very little funding.

Our major question throughout the decade of social science research in the SPICE project was: How does governance affect, and how might it transform human and societal behavior in the coastal and marine realm toward greater ecological and social sustainability?

## 12.1.1 Some relevant definitions and understandings

This chapter appears in a book of mostly natural science contributions. To enable communication with readers across fields and disciplines, we start with working definitions of the key social science terms employed here.

> **Governance** goes beyond the realm of government (i.e., of public authority measures) to also include the market and civil society. It includes "all processes of governing,[2] whether undertaken by a government, market or network, whether over a family, tribe, formal or informal organization or territory and whether through laws, norms, power or language" (Bevir, 2013 cited in; Boyd et al., 2015, p. 153). **Environmental governance** relates to collectively binding decisions that affect human–nature relations.

[1]Since each of these papers reports its own set of methods, we have dispensed with a methods section here.

[2]In accordance with current usage, we understand "governing" as a synonym for "command, control, preside (over), rule" https://www.merriam-webster.com/thesaurus/govern.

**Management** has defined objectives and takes place within a governance framework. **Comanagement** refers to a range of possible distributions of management tasks between governments, ecosystem users, and other stakeholders (Adger et al., 2005a,b; Sen and Nielsen, 1996).

**Institutions** are humanly devised constraints that structure political, economic, and social interactions. Institutions include informal constraints (sanctions, taboos, customs, traditions, and codes of conduct) and formal rules (constitutions, laws, property rights) (North, 1991). Norms, rules, and regulations ranging from implicit values to formally encoded laws are thus all institutions.

**Organizations** consist of multiple people (e.g., firms, governments, nongovernment organizations [NGOs], armies, foundations, cooperatives) bound together by common objectives and relying on institutions for their functioning.

**SES** are integrated systems of people and nature "that could in theory be parsed in several different ways" (Cumming, 2011). SES have been defined in a dozen or more different ways, and each definition has its particular strengths and shortcomings (Cumming, 2011; Glaser et al., 2012; Ostrom, 2009a,b). Our SES working definition is problem-focused and multilevel. An SES is defined as "a*ll actors, institutions, organisms and ecosystem components associated with a particular biogeophysical system at multiple system levels, from the local community to the glo*bal" This framework is suitable for integrative governance-focused analyses (e.g., Glaser et al., 2010b). It envisages a set of collective and individual governance actors as interacting with nonhuman nature and focuses on social–ecological feedbacks between and across multiple system levels. These feedbacks generate and/or address governance and management issues (Glaser et al., 2010a,c; Glaser and Glaeser, 2014).

**CM-SES** have particularly diverse social and social–ecological feedbacks while, at the same time, displaying particularly segmented, complex, and often weak governance structures. CM-SES face problems of overfishing, multiple resource and space claims, ecosystem and resource degradation, and poverty in a context of rapid, often climate-related environmental change (Ferrol-Schulte et al., 2013; Glaser et al., 2010a,c; Glaser and Glaeser, 2014), with serious implications for coastal ecosystem users.

**System levels and scales.** To allow for concise argumentation in the analysis of complex SES, the terms scale and level need to be consistently distinguished and used. Scale relates to any dimension (e.g., spatial, temporal, institutional, analytical) used to measure and study a phenomenon. Levels are units of analysis located in different positions on a scale (Cash et al., 2006; Glaser and Glaeser, 2014; Schwerdtner Máñez et al., 2014).

### 12.1.2 The governance of coastal and marine areas in Indonesia

Indonesia with its about 17,000 islands is an archipelago and the largest country within the regional Coral Triangle (Ferse et al., 2012a). Even though most of the national territory is coastal by any definition of the term, Indonesian culture is, with few exceptions such as the Bajau people (Sopher, 1965; Lowe, 2013), mainly land-oriented (Geertz, 1960), and the country's policies have only fairly recently focused on the marine realm. The Ministry of Marine Affairs and Fisheries, a major player in coastal and marine governance, was, for instance, only established in 1999. At least 22 laws affect the coastal zone of Indonesia, and legal conflicts, inconsistencies, gaps and overlaps abound. For instance, "traditional fisheries" are defined by technique so that long-term artisanal fishers are not protected from new entrants to their fishery. There are multiple conflicts between sectoral laws and between national laws and regional regulation (Wever et al., 2012). Adrianto et al. (2009) discuss a number of fisheries management regimes including customary-based fisheries management such as *Panglima Laot* in Aceh as well as relatively recent Sea Farming Management around the Seribu Islands off Jakarta, with diverse policy decision processes and regulations.

In the course of the *reformasi* era of democratization and decentralization since 1998, central governmental powers over the coastal zone were decentralized to the provincial and district and city government levels, which were, respectively, allocated responsibilities in 12-mile and 4-mile zones from the shoreline out.[3]

The international governance-focused Coral Triangle Initiative (CTI) embeds Indonesia in global ocean governance efforts (Clifton, 2009), and the Government of Indonesia proclaimed one of the largest marine-protected areas (MPAs) on earth (the Savu Sea MPA) in 2009 at the Manado World Ocean conference.

Ocean and coastal governance in Indonesia is increasingly challenged by climate change, pollution, and the implications for vulnerable populations. National policy may address three aspects of vulnerability: exposure, sensitivity, and/or adaptive capacity. Ferrol-Schulte et al. (2015) find, however, that complexities and inconsistencies within Indonesian governmental structures, funding gaps, and poor coordination mean that national policies rarely benefit coastal communities and that policies to address coastal livelihood vulnerability are almost exclusively targeting adaptive capacity but do not aim to reduce exposure, arguably the first and predominant cause of vulnerability.

## 12.2 Marine and coastal governance: sectors, issues, and options for intervention

The governance context of CM-SES is thus particularly complex. This is exacerbated by the impacts of climate and other environmental changes. Diverse governance objectives have been formulated within this dynamic and complex context, including reducing resource overuse and degradation, poverty and inequity, pollution, improving conservation, and increasing natural resource-based production. Many of these governance objectives arise from strong sector interests, and there is congruence but also

[3]Following a revision of the regional autonomy law in 2014, all waters up to 12 miles from shore are under provincial authority.

competition and incompatibility between sectoral objectives and sector-specific recommendations in CM-SES governance. This section examines this for some of the major Indonesian coastal governance and policy arenas: coral reefs, mangroves, fisheries and aquaculture, watersheds and land use, marine and coastal conservation, coastal livelihoods and social needs, and pollution.

### 12.2.1 Coral reefs

Over 60% of the earth's coral reefs are under threat from local sources (Burke et al., 2012), and about one-fifth of global coral reefs are in Indonesia (Spalding et al., 2001). This gives Indonesia a planetary role in determining the future of coral reefs. At the same time, with the large majority of the Indonesian population living near the coast, anthropogenic pressure on Indonesian reefs is extremely high (Whittingham et al., 2003), with the result that more than 90% of the country's reefs are under local threat (Burke et al., 2012). Indonesian reef governance needs to address a number of issues at different scales. Impacts of climate change such as sea level rise or increasing storm surges are a real threat for coastal populations, and adaptation measures are of utmost importance here. Other more regional or local issues like overfishing threaten not only the reef ecosystem but also the livelihoods of reef-dependent populations, for instance, in the Spermonde Archipelago in South Sulawesi (Glaser et al., 2015, Chapter 5, this volume). The anthropogenic drivers of these problems act from multiple levels (local to global) of the reef-based SES.

A recent metaanalysis of over 1800 reef sites across the globe found a strong association between local population size and market accessibility and fish biomass on the reef (Cinner et al., 2016). Several important reef fisheries, such as those for sea cucumbers (Schwerdtner Máñez and Ferse, 2010) and live reef fish (Radjawali, 2011), target species with important ecological functions that underpin the health and resilience of the coral reef ecosystem. Under the influence of international markets and technological changes, several export-oriented fisheries have developed and collapsed in Indonesia. Over time, this has generated distinct sequential peaks and subsequent declines of fishing for different marine species associated with a growing danger that entire reef systems are being successively depleted (Ferse et al., 2014).

Policy recommendations for coral reef management that have arisen from our work in Indonesia are:

- to identify where undesirable resilient social–ecological cycles need to be undermined and interrupted (Glaser et al., 2018a,b,c);
- to protect core reef zones while also engaging stakeholders in the identification of key fishing grounds and designation of managed reef areas;
- to identify fishing gears and target species with particular impacts on the resilience of reefs and focus management efforts on these; and
- to enforce strict bans on destructive fishing methods.

## 12.2.2 Mangroves

At the interface between land and sea, mangroves protect against extreme events, such as the disastrous 2004 Tsunami and against erosion. Mangroves also provide important livelihood support to fishers and other coastal dwellers. Between 1980 and 2000, nearly 20% of global mangroves were lost (FAO, 2007), and this trend continues. This is crucial for Indonesia, as the country has 25% (2.93 million hectare) of the world's mangroves (Wells, 2006). Mangrove loss in Indonesia is mainly driven by shrimp culture, logging, and the expansion of palm oil plantations and coastal development. Remote village communities also use mangroves for local housing, but the impacts of this are not detectable in larger studies. Substantial mangrove loss occurs in the context of coastal city development (Ilman et al., 2016). In the SPICE study areas, destructive fishing techniques, logging, and farmers in search of cultivable land were main drivers of mangrove loss.

While the expansion of shrimp farming and oil plantations into mangrove areas has remained largely unaddressed over the past two decades (Ilman et al., 2016), mangrove protection efforts have emphasized replanting and restoration, and in this, mainly the biophysical complexity of the mangrove system. At the same time, socioeconomic and institutional issues were mostly neglected Rotich et al. (2016) with often negative outcomes. Reichel et al. (2009) find that a failure to take into account land tenure questions and local customary law in a major mangrove replanting program led to resource conflicts and the destruction of planted areas by local residents who, after initial enthusiasm for engaging in mangrove restoration, saw their resource access threatened by loss of tenure for areas of replanted mangrove, a rule which they considered illegitimate.

Mangroves are often subject to overlapping and competing interests of user groups and jurisdictional ambiguities between administrations (Glaser and Oliveira, 2004; Walters et al., 2008). The Segara Anakan lagoon on Java's south coast is a prime example illustrating how overlapping, contentious claims of local residents, various governmental agencies, and a state forest corporation, combined with lacking coordination, communication, and conflict resolution mechanisms render sustainable mangrove management ineffective (Heyde, 2016). Mangrove management is further challenged by the interplay between ecological levels where important ecosystem services are generated and often ill-matched institutional levels where management decisions that influence these services are made (for example, a management body at provincial level which is only responsible for a part of a mangrove ecosystem). Mangrove management is usually a governmental responsibility, often of a forest authority, which, within wider forest management tasks, often attributes marginal value and attention to these forests between land and sea.

Policy recommendations with regard to mangrove management that have arisen from our work (Reichel et al., 2009; Schwerdtner Máñez et al., 2014; Heyde, 2016; Heyde et al., 2017) are as follows:

- Protected zones for mangroves should be managed in ways more inclusive of local ecosystem user interests.

- Destructive fishing techniques in mangrove areas need to be prohibited.
- Additional income sources for the mangrove-dependent local population are needed.
- Active rights-based participation of local population in mangrove conservation planning and implementation is needed.
- Coordination between state agencies involved in mangrove management, and clarification of overlapping responsibilities is needed.
- Historically rooted resource claims and conflicts need to be resolved.
- Adequate mechanisms for the equitable sharing of benefits and costs in mangrove governance and management across system levels and between different stakeholders need to be developed.
- Policy mechanisms are needed, which ensure that local fishers are compensated for income losses associated with mangrove management rules.

### 12.2.3 Marine fisheries

Indonesia is a main fish producer in Southeast Asia and one of the main tuna fisheries on earth. About 95% of the country's multispecies fishery production comes from artisanal fisheries, and, in 2012, over 6 million people were directly or indirectly engaged in the fisheries sector (FAO, 2014). While Indonesia exports fish to all five continents, its fish is also an affordable protein source in remote locations of the country providing a significant contribution to national food security. In recent decades, growth in Indonesian fisheries has stagnated, and aquaculture production has increased. Fisheries, but even more so aquaculture, is becoming the focus of strong drives to increase commercial marine production and exports. Aquaculture practice has a large potential for the exacerbation of marine pollution.

The Indonesian Central Government's Ministry of Marine Affairs and Fisheries (MMAF) is responsible for the management of the fisheries sector of the country. MMAF issues fishing licenses for vessels larger than 30 GT, provincial governments for vessels of 10–30 GT, and regency governments for vessels of less than 10 GT. MMAF is supported by the Indonesian Navy and the Marine Police.

Central problems in the fishery sector are overfishing; low income and standard of living for fishers and fish farmers; lack of financial support in terms of credit schemes; weak practical fisheries management, particularly concerning monitoring, surveillance, and enforcement; and degradation of coral reefs and other marine environments that affect fisheries. Lack of alternatives combined with a cultural affinity to fishing causes school dropout for male children and adolescents from as young as 12 years of age (Glaser et al., 2015).

SPICE research on marine fisheries in the Spermonde Archipelago off the island of South Sulawesi illustrates much of the aforementioned. The region, once well known for its abundant fish stocks, has been used by Southeast Asia's famous marine fisher folk, the Bajau people, since at least the 16th century (Sopher, 1965). Dutch colonial records show Spermonde as an important source of valuable marine resources, such as sea cucumbers, determined for export. While fishing is also important for subsistence and for sale in local and regional markets, much of Spermonde's fish has long been sold in foreign markets. At least 22 distinctive fishing methods developed in the Archipelago, shaped by climate and ecosystem configurations as well as market and culture. Resource demands, technological innovations, knowledge, supply of equipment, policies, and formal as well as informal institutions including social networks all influence the development of the Spermonde fisheries (Ferse et al., 2012a).

Several export-oriented fisheries have developed in Spermonde that "fish the reefs empty" by producing distinct sequential peaks and declines of catch for different marine species (Ferse et al., 2014). Beside traditional sea cucumber fishing before the introduction of hookah diving in the 1970s, at least six distinct fisheries have been identified: (1) modern sea cucumber fishing, (2) live grouper, (3) ornamentals, (4) bamboo coral, (5) octopus, and (6) moray eel. When comparing these fisheries, it is evident that over time, exploitation waves become shorter. A recent exploitation wave was moray eel fishing, starting in Spermonde in 2012. Moray eels had not been previously exploited in the area as their flesh is locally considered non-palatable. Fishing then begun in response to a demand from China, where the fish is used in traditional medicine. Depletion took only a few months, caused by the low abundance and strong site affinity of moray eels. In 2013, fishers begun to fish outside the Archipelago, because Spermonde's reefs no longer yielded enough moray eel (Schwerdtner Máñez and Husain Paragay, 2013).

Indonesia's valuable marine fisheries resources are destined for foreign markets, and most fisheries are therefore driven by demands from beyond the local and national levels. For example, the rise of a wealthy consumer class in China drives Indonesian catches of live grouper (Fabinyi et al., 2012). Aquarium fashion trends are another important driver of Indonesian fisheries, manifest in Spermonde in the demand for ornamental species ranging from fish over invertebrates to corals (Schwerdtner Máñez and Ferse, 2010).

In Indonesia (Acciaoli, 2000; Ferse et al., 2012b; Pelras, 2000) and elsewhere (Ruddle, 2011), patrons provide fishers ("clients") in patron–client or *punggawa-sawi* relations) with loans, boats, fishing gear, and market access. In exchange, fishers are tied to "their" patron and obliged to accept lower product prices (see references in Ferrol-Schulte et al., 2014; Máñez and Pauwelussen (2016) for women fishers in patron–client relations).

The movement from one reef to another, also beyond the Archipelago, is a coping strategy developed by local fishers in response to stressors such as seasonality and overfishing of individual locations or species. In Spermonde, fishers' coping strategies also include borrowing from patrons, diversifying fishing methods, fishing migrations, and the self-organized, informal crafting of local institutions to regulate fishing. Some of these strategies lower the capacity of the system to adapt to future stressors and thus undermine the sustainability of the fishery and the livelihoods of those who depend on it (Ferse et al., 2014).

Fishing grounds in the Spermonde Archipelago are managed under three different institutional arrangements, each pertaining to different sections of the marine territory: open-access zones, island-restricted zones, and private-restricted zones. As part of a complex, adaptive system, institutional change in marine fisheries governance is subject to a range of forces, which need to be identified and managed (Deswandi et al., 2012). Suggested fisheries management strategies at different levels of the fishery system (Ferse et al., 2014; Glaser et al., 2015) cover three major themes:

1. Collaboration needs
   - Improved collaboration between the fisheries line agency and other ministries and line agencies is needed to ensure that fishery-related knowledge generation, public education, and sector planning including human livelihoods and new technologies are managed in an integrative manner.
2. Style and type of interventions
   - Additional livelihood options that are locally feasible and acceptable are needed for fishers and fish workers.
   - Aquaculture expansion, in particular, needs to be undertaken in equitable ways with special attention that small-scale, asset-poor coastal fishers obtain promising options for realizing benefits along the value chain. Ecolabeling should be explored for its ecological and social potentials.
   - Spatial and temporal flexibility needs to be built into fisheries management to include traditional user strategies that succeed in avoiding overexploitation.
   - Patron–client relationships need to be taken into account in the development of management strategies by, for example, providing social security and credit to fishers to reduce their dependence on patrons.
   - Measures are needed to prevent school dropout by boys who to enter fisheries at an early age.
3. Data on fisheries
   - Fishery data should be collected at fishing grounds and landing sites rather to replace assessment that are only based on exports.
   - An up-to-date database on fisheries in Spermonde and Indonesia, in particular at provincial and district levels to inform policy/intervention decisions, is needed.
   - Until this is realized, precautionary approaches could be considered in the management of data-poor fisheries.

### 12.2.4 Watersheds and land use

Land use changes and other human impacts in the upland areas of coastal watersheds affect coastal and marine systems in a variety of ways, including through their impact on sedimentation. Lukas (2015, 2017a,b) examines the dynamics and drivers of land use changes and watershed modifications and links them to a temporal record of sedimentation processes in the coastal Segara Anakan Lagoon on Java. Sedimentation of this lagoon has undermined local fishery incomes and has been regarded as a threat to the lagoon ecosystem. Lukas describes a complex, multidriver context within which forest and watershed management agencies and the state forest corporation in an alliance of interests with donors and development agencies have, over the past decades, maintained a simplified discourse that attributes the cause of lagoon sedimentation exclusively to the allegedly destructive, rainfed agricultural practices of increasing numbers of small-scale farmers in the uplands of the Segara Anakan Lagoon. This simplistic debate effectively obfuscates a number of other causes of lagoon sedimentation, including conflicts over state forest and plantation lands, the effects of which have been documented (Lukas, 2017a). The same predominant debate portrays the violent expansion of the state forest corporation into lands formerly pertaining to ousted local settlers and the repressive management of the state forests until the late 1990s as essential to watershed conservation and the control of sedimentation of the Segara Anakan Lagoon (Lukas, 2014, 2015). Lukas documents how the debate on watershed management and sedimentation in the Segara Anakan Lagoon and the whole of Java was so effectively shaped by the hegemonic views and economic interests of forest authorities and the state forest corporation that the effects of state forestry management practices, such as regular clear-cutting on steep slopes near streams and rivers, remained obfuscated and excluded from the debate. That decades of largely internationally funded work in tree planting and land terracing with small upland farmers did little to counteract lagoon sedimentation is thus not surprising. To support this line of argument, Lukas documents how some of the smallholders' lands are actually "pockets of good, soil-conserving practice" surrounded by contested and erosion-prone state forest areas, which significantly contribute to river sediment loads and lagoon sedimentation (Lukas, 2017b).

The complex, often not yet sufficiently understood linkages between upstream conditions and actions and downstream consequences in watersheds have remained an important theme in coastal and marine governance, and in recent years, "payments for environmental services" (PES) have been advocated as a policy instrument in this context. This is occurring within a supportive debate that develops despite a lack of evidence on PES effectiveness and despite criticisms from both the nature conservation and the political ecology fields (Heyde et al., 2012). In Indonesia, widespread insecure and contested land tenure, the small scale of existing schemes, and short time horizons are major obstacles to effective implementation of PES schemes (Heyde et al., 2012). Rather than promoting the implementation of PES, priority should be placed on resolving long-standing sociopolitical and institutional issues and resources conflicts, which are core issues of sustainable environmental governance (Heyde et al., 2017).

Policy recommendations on watershed governance that arise from SPICE work are as follows (Heyde et al., 2017; Lukas, 2015, 2017a,b):

- Fair resolution of historically rooted conflicts over forest and plantation lands and improved tenure security are needed for more sustainable watershed management.
- Attention needs to be paid to the effects of state forest management practices on erosion and sedimentation rates.

- Sediment reduction strategies need to take into account the entire range of sediment sources and their causes.
- PES have only limited potential and should not be prioritized as a means of watershed conservation in Java.

### 12.2.5 Marine and coastal conservation

MPAs are a major vehicle for marine conservation across the globe (Edgar et al., 2014; Gill et al., 2017; Roberts et al., 2017) and in Indonesia (Baitoninsingih, 2015). In a study on the then major national community-based marine-protected area (CB-MPA) program in the Spermonde island archipelago (the COREMAP program), Glaser et al. (2010c) find that, despite a sophisticated institutional structure at island level that is explicitly designed to facilitate local participation, local knowledge about the CB-MPA was low, and access to program resources limited to a minority, while the majority of interviewed island residents felt that they had no influence on decision-making in this conservation program. Moreover, most of the minority of islanders who knew of the CB-MPA program thought that it was a state/government undertaking with a few local elite members, in which the local community had no role and little commitment. Along with this critical evaluation of a major marine conservation approach in Indonesia, the authors also identify a number of effective marine management institutions (i.e., rules-in-use) at island level. These emergent and dynamic informal institutions reflect the local values, priorities, and rationales underlying fishing behavior in local waters (Deswandi et al., 2012). They were found to operate in parallel to, and often at odds with, legally established MPAs and their no-take areas. During our fieldwork, a local fisherman (about 30 years old, May 2009[4]) put it as follows: "Our rules were made before the CB-MPA was put into place and local people made these rules themselves."

Box 12.1 lists some informal, island-specific rules developed and applied by local people for the waters around different inhabited islands in the Spermonde Archipelago. These rules operated as effective, locally grounded, and, in some, but not all cases, conservation-supporting institutions. If adequately recognized and supported by government authority, the local social energy behind them may support more effective conservation. The following citations from our interviews[5] with fishers underline our argument:

> *"Rules are made by the people from this island. No bombs or cyanide are allowed in order to protect the island" (fisherman, about 50 years old, March 2009).*

> *"No one fishes around the corals, and outsiders are not allowed to come in. If outsiders come, they are chased off by villagers. Some outsiders asked for permission to bomb previously, but the villagers didn't allow it"(village head, May 2009).*

[4]Methods are described in Glaser et al. (2010c).

[5]The methods for this set of interviews are described in Glaser et al. (2010c).

**BOX 12.1 Rules-in-use in Spermonde waters for informally protected "Island exclusion zones" (IEZ) (SPICE results 2009–12, see Footnote 1)**

- Island 1: Only islanders may use traps and fish.
- Island 2: Island waters are open to those not using destructive methods (trawl, poison, bomb).
- Island 3: No fishing around the corals. Outsiders are "chased off" by islanders.
- Island 4: No trawls, blast or cyanide fishing. Outside fishers are moved on after 10 days.
- Island 5: Local reefs are protected by islanders. IEZ for long-line fishing by islanders only.

Since the informal marine management rules in Spermonde Archipelago evolved locally, they are in line with local rationales and well understood. They are adaptive, in that they have arisen, and change in response to technical, ecological, or demographic change. They relate to a region where the great majority of residents have no marine management traditions; their marine area/fisheries management rules evolved from current-day social–ecological and cultural constellations (Deswandi et al., 2012). Given these characteristics, locally emergent marine management rules have good potential as a link between scientific conservation frameworks and local rationalities and realities.

Marine area management rules differed between islands (see Box 12.1) in accordance with the main locally used fishing technology. Where several fishing techniques were used in one area, informal rules addressed the interferences and externalities between these fishing methods. For instance, hand-line fishermen on one island reported to have prohibited cyanide fishing in waters surrounding their island because divers chase off the fish and cyanide harms underwater resources. On one island, where fishers used the low-impact *bubu* (bamboo traps), outsiders were not allowed in local waters (which people could demarcate very clearly). This island received no formal support in enforcing their strict local no-entry rule to their local waters; they effectively "chased off" intruders on an entirely self-initiated and self-organized basis even when conflicts with other villages and gear damage were likely to ensue. Similar dynamics were reported in locations where locals employed hand lines for fishing.

In contrast to the formal official approach to MPA planning and implementation (the World Bank–funded and ADB-funded COREMAP program), the locally emergent management of marine territories around islands in Spermonde Archipelago promoted effective ecosystem user participation in marine management for three reasons: (1) island residents had a clear sense of ownership and responsibility for local rule implementation; they saw themselves to be implementing "their own rules"; (2) there were clear, locally known, and agreed boundaries for the managed marine territory; and (3) islanders had a local (sustainability or other) agenda for the marine territory in question. None of these characteristics were in evidence for the formal MPA.

A study by Gorris (2016) shows that while the lack of official authority to enforce locally emergent rules and the coordination problems between local communities and higher level state actors remain as key challenges,[6] fishers in the Spermonde area perceived their informal rules as effective. Gorris (2016) also shows that the motivation of community members to engage in the enforcement of their informal rules is driven by short-term economic considerations. For rules that are perceived to have a strong impact on individual fishing yields, the fear of potential short-term economic losses motivates local fishers to enforce, especially when fishers from other islands break the rules. Where rule-breaking is not perceived to decrease individual fishing yield, or where there are local benefits from rule breaking (as in bomb-fishing that also provides fish for "nonbombing" local fishers), the motivation of the community members to engage in rule enforcement ceases (Gorris, 2016).

The potentials of local informal marine management are higher still. Fig. 12.1 shows that adjacent marine territories that are informally but effectively claimed and managed by different local island communities as "their territory" (i.e., as island exclusion zones [IEZs] form into a larger continuous marine region. Such emergent forms of self-organized informal marine area protection can form the basis for new, "bottom-up" forms of more effective marine management across larger areas. That these findings are relevant more widely is documented, for instance, by Lowe (2013) who reports local

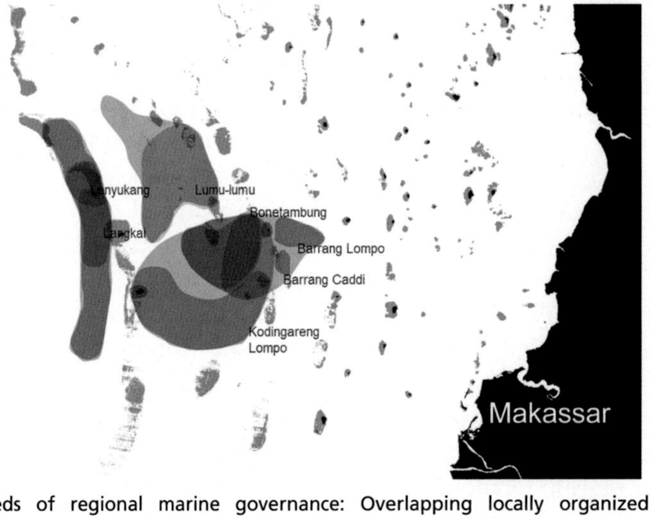


**FIGURE 12.1** Local seeds of regional marine governance: Overlapping locally organized informal marine management zones in the Spermonde Island Archipelago. *From Deswandi (2012).*

[6]This may in part be related to the changes in management authority brought about by the revision of the law on regional autonomy in 2014. This shifted management responsibility for the COREMAP MPAs, which were established at the district level, to the provincial level, which now has authority over all waters up to 12 nm from shore. According to researchers at UNHAS, the provincial level did not have the capacity to adequately manage these MPAs, further weakening their effectiveness (J. Jompa, pers. comm. April 2020).

demands in the Togean Islands for "villagers to have the right to defend their territories from outsiders who use harmful methods to meet short-term interests." Other cases of strong local engagement in marine management are reported from Eastern Indonesia (Resosudarmo and Jotzo, 2009) under traditional local management practices including *sasi* (temporal closures) in Maluku, *malimau pasie, malimau kapa and alek pasie* (traditional fishing ritual) in West Sumatra (Siry, 2011), *panglima laut* (traditional resource manager), and *awig awig* (customary prohibition of destructive techniques) in Bali (Muswar et al., 2019; Siry, 2011). Emerging, at least partly sustainability-supporting territoriality (not all self-organized local rules are sustainability-enhancing), is found even in areas without marine management traditions such as Spermonde Archipelago, and may support interlinked bottom-up regional networks for more sustainable marine management.

As part of political reforms since the late 1990s, authority over and within MPAs was decentralized, and community participation was emphasized. Baitoningsih (2015) examines the legal framework and actual practice of community involvement in MPA planning at local, regional, and national levels. She finds that while legally established procedures have supported a context-specific involvement of local marine user communities in consultation and information exchange with public authorities, current legal frameworks and policy implementation practice do not support a rights-based participation of ecosystem user communities in decision-making on MPA planning and establishment. Baitoningsih (2015) finds that local communities' priorities are marginalized in MPA practice since they lack formally established and implemented rights to participate in decisions and to assume MPA-related responsibilities. Despite being a major and much lauded conservation tool, MPAs thus often fail to sustain habitats, fisheries, and coastal livelihoods.

MPAs have been associated with socially divisive and inequitable effects in relation to, for instance, access to loans and other MPA-related benefits, with underfunded and inadequate methods for facilitating local participation, and with a participation concept, which falls short of assigning community rights in planning and decision-making. As a consequence, in design, procedures, and location, MPAs in Indonesia have often achieve little congruence with local priorities and rationales (Baitoningsih, 2009; 2015; Ferse et al., 2010; Glaser et al., 2010c).

Conservation-related policy recommendations that arise from our work are as follows:

- Even if the ambitious goal of covering 10% of marine territories with MPAs by 2020 (as per Conservation on Biological Diversity and SDG 14 objectives (Tittensor et al., 2014) were reached, 90% of marine territories remain unaccounted for. Marine and coastal planning needs to include conservation as an explicit objective for all marine territories.
- To mobilize local social energy for sustainable marine areas, a fuller, more rights-based concept and implementation of community participation in the planning and implementation of marine territories (including of MPAs) needs to replace

current practice, which confines community participation to merely receiving information from or, at best, offering views to government.

- On the basis of more meaningful local participation, clear procedures are needed to include appropriate locally emergent rules in management approaches and to coordinate among local communities and between communities and higher level government agencies.

### 12.2.6 Coastal livelihoods and social needs

Many Indonesians are highly dependent on marine resources so that their livelihoods are vulnerable to changes in the marine environment. National policy can mitigate this by addressing three facets of vulnerability: exposure, sensitivity, and adaptive capacity (Adger, 2006). Ferrol-Schulte et al. (2015) find that, for the period up to 2015, Indonesian coastal and marine policies address these three main dimensions of coastal livelihood vulnerability in a biased manner: Developing adaptive capacity was found to be a strong, near omnipresent policy focus; sensitivity was addressed to a lesser extent while exposure to livelihood risks, which may be regarded as the initial cause of social vulnerability, was entirely absent as a policy focus. To compound this clear neglect of human well-being concerns, the complexities and inconsistencies within the Indonesian governmental structures, funding gaps, and coordination failures often prevented that policies created at national level provided intended benefits to coastal communities.

Examining the options that arise for sustainable livelihood diversification from the marine aquarium trade for ornamental organisms, Ferse et al. (2012b) find that fishermen are organized in a tightly knit web of patron–client relationships, a key institution in Indonesian fisheries, which leaves little scope for them to initiate changes in livelihood strategies.

Acceptable livelihood alternatives for Indonesian coastal and, in particular, small island residents are needed to reduce the pressure on the marine environment and mitigate the multiple threats to marine resource–dependent livelihoods. With falling fisheries catches, mariculture is globally and in Indonesia seen as a major motor of economic growth and a crucial component of food security (Cabral et al., 2018). von Essen et al. (2013) find for two coastal villages in North Sulawesi that, despite interest in alternative livelihoods including sea cucumber cultivation, mariculture is still an unfamiliar activity, not yet sufficiently trusted to generate willingness to invest. Moreover, traditions and a sense of personal gratification from fishing cause fishers to stay with fishing despite falling returns so that farmers are more prepared to take up aquaculture than fishers. Thus, the authors find that sea cucumber mariculture is likely to remain an additional source of income rather than an actual alternative for those whose livelihoods currently depend on fisheries.

Adequate freshwater access is a basic human need and central component of sustainable livelihoods. The small coral islands of Spermonde, for instance, contain only thin freshwater lenses (often 10–20 cm). Surface soils are extremely thin, have a low water retention quality, and therefore hardly protect the underlying lenses (White and Falkland, 2010). These freshwater lenses are easily overexploited and especially susceptible to pollution by fuels, solid waste, and sewage disposal (Falkland and Brunel, 1993). While there was always a fragile relationship between water supply and water use, population growth combined with unregulated well drilling has increased water scarcity in many islands to such an extent that people at least partly depend on water imports from the mainland. Impacts of climate change such as sea level rise and overwash cause saltwater intrusion and thus are putting additional pressure on freshwater resources (Schwerdtner Máñez et al., 2012). The management of freshwater needs is an increasingly crucial component of coastal management.

Some of our policy recommendations with respect to coastal people's livelihoods are as follows:

- Practitioners and policy-makers engage in a more cohesive and balanced approach to address livelihood vulnerability in coastal management to reduce the causes of exposure, rather than only address the symptoms of vulnerability (Ferrol-Schulte et al., 2015).
- If livelihood alternatives are to reduce pressure on marine resources, fishers' acceptance of nonfishing income alternatives is examined and promoted, taking into account possible cultural values of fishing.
- To support sustainable marine resource use, management policies target both fishers and their patrons and regulate respective associated trade networks (Ferse et al., 2012b).
- Long-term support by people perceived as honest brokers for livelihood programs is ensured to avoid elite-capture, to ensure sustainable financing, and to make important knowledge accessible over the longer term.
- Island-specific solutions for freshwater management (including rainwater collection facilities) are codeveloped with village-level water committees and collaboratively constructed, supported, and maintained.

## 12.2.7 Pollution

Pollution is an increasingly pervasive and pressing issue in coastal and marine areas across the globe (e.g., Shahidul Islam and Kanaka, 2004; Vikas and Dwarakish, 2015), often related to the impact of megacities (Blackburn et al., 2014). In Indonesia, for instance, in the Siak River catchment area in Riau Province (Sumatra), poverty, health, and sustainable development are closely linked to the degraded water quality of the Siak River (Baum and Rixen, 2014), which reduces ecosystem services and thus the welfare and livelihoods of those who depend on river water for fisheries, washing, and even drinking. Contaminated water reduces local incomes, is linked to locally new health problems and job losses, and also undermines aspects of community culture for those who live in the Siak River basin. Women in riverside communities in particular perceived a close connection between river pollution and local people's health issues.

Around Jakarta Bay, in the Jakarta Metropolitan Area and the adjacent "Thousand Islands" (*Kepulauan Seribu*) region, about 30 million inhabitants are facing extreme pollution (Breckwoldt et al., 2016). At the same time, local coral reefs and marine resources provide crucial income and food, especially for the poorest sections of the coastal population. Baum et al. (2016) find that coastal residents on the islands and the mainland consider the species *Caesio cuning* (Redbelly yellowtail fusilier) as economically most valuable and *Rastrelliger kanagurta* (Indian mackerel) as degraded mostly by both pollution and overexploitation. Coastal residents in the Jakarta Metropolitan Area perceive pollution as the principal cause of marine resource degradation, ranking it as a more serious problem than overexploitation, tourism, natural disasters, destructive fishing methods, and poor enforcement (Baum et al., 2016). With often complete dependence on fishery resources without alternative livelihoods, coastal residents are vulnerable to declines in fishery resources as well as to their degradation through diverse harmful substances (e.g., Dsikowitzky et al., 2014). Alternative livelihoods that are not based on these marine and coastal areas are needed to reduce such vulnerability, complemented by infrastructure support such as soft loans as well as better access to markets and credit for poorer fishers (Baum et al., 2016). Government development projects conducted in a participatory manner with local communities would enable the joint definition and monitoring of sustainability indicators to identify and mitigate sources of vulnerability. An official emphasis on not only "treating the symptoms," i.e., helping people to adapt to and survive in their deteriorating environment, but more importantly on tackling the "root causes," i.e., exposure to anthropogenic stressors is crucial (Baum et al., 2016; Ferrol-Schulte et al., 2015).

Some policy recommendations (Suhren et al., 2014; Breckwoldt et al., 2016) on marine pollution that arise from our work are as follows:

- The sources of pollution and their accumulation in the food chain (Chapter 6, Schwarzbauer et al., this volume) need to be taken into account.
- Effective wastewater treatment still needs the highest priority—for quality of life in this megacity and to restore the marine coastal environment (e.g., UNESCO, 2000; Van der Wulp et al., 2016a,b).
- The threatened livelihoods of those who fish for and eat increasingly polluted and degraded marine products, and those who depend on polluted river water for daily needs need to be improved and secured.
- Participatory development of indicators for sustainable livelihoods needs to be codeveloped with vulnerable groups in local communities.
- The power constellations supporting the actions of major polluters need to be assessed and taken into account in policy development.
- Reliable information on contaminant accumulation in fishery products and on resultant dangers for both subsistence collectors and fishers and for market-reliant consumers will need to be generated and publicized.

## 12.3 The need for integrated coastal and marine planning

Over the course of the decade of research we report on here, in Indonesia, as in many other countries, centralized and sectoral and thus fragmented governance structures and processes interlaced with policies to foster the decentralization of authoritarian top-down, command-and-control state powers (Lukas, 2013; Wever et al., 2012). Current dynamics in Indonesia can be interpreted as an increase in polycentricism that renders governance systems more robust. Despite decentralization since the late 1990s, however, low connectivity between marine governance actors at the same system level persists: communication between different local system units (i.e., between small islands in the Spermonde Archipelago) and between actors and organizations at higher system levels (i.e., between *kabupatens* and between provinces) remains low while vertical (bottom-to-top, and top-to-bottom) governance interactions across system levels in the classical hierarchical pattern of governance interaction remain strong (Gorris, 2015). How the revision of the regional autonomy law in 2014, which was to address rent-seeking by authorities at the local/district level and which shifts authority over waters 0–4 nm from shore from the city/district level to the provincial level, affects management effectiveness, and the interactions between district and provincial levels remain to be seen. The case of the COREMAP CB-MPAs in Spermonde indicates that management effectiveness may have been negatively affected

The widespread lack of trust and effective communication between resource users on the ground, and the representatives of governmental agencies are also historically rooted; they are an important obstacle to sustainable natural resources management (Heyde, 2016; Lukas, 2015). The lack of trust and of effective coordination, communication, and conflict resolution mechanisms impede the resolution of resources conflicts and are therefore an important cause of environmental degradation (Heyde, 2016; Lukas, 2014, 2015).

Our selective review of some key coastal and marine sectors in Indonesia in Section 11.2 shows that while governance priorities and policy recommendations often point in similar directions, they can also be at odds with each other. For instance, while marine conservation and fisheries may conceivably evolve to codevelop an integrated strategy for sustainable production, in practice conflicts between production and conservation objectives and agencies prevail (Ferse et al., 2010). Collaboration between the education and conservation sectors on environmental education in coastal communities also appeared heavily obstructed in our fieldwork in South Sulawesi. Here, small underfunded NGOs worked alone in attempting to address the need for building and exchanging local knowledge on the dynamics of the coastal environment while both responsible ministries failed to engage (Glaser et al., 2015). Such target conflicts were exacerbated by competition within the public sector, between authorities and agencies about spheres of formal authority and influence, so that communications between public authorities and agencies appeared fraught with difficulty or absent and interministerial or interagency collaborations emerged only in exceptional "window of opportunity" periods (Wever et al., 2012; Heyde, 2016'[7]).

There is, however, the clear need for coastal governance and management planning and implementation to address the full range of key objectives as an interconnected whole system. This includes the integration of planning processes and structures across sectors and system levels, and at various system levels including at the planetary level, into a global process of defining and pursuing issues related to interconnected planetary boundaries (Galaz, 2014, p. 189) The global Sustainable Development Goals have made a start here, but integration across goals requires more work (Dodds et al., 2017; Griggs et al., 2017; Nilsson et al., 2018; Schmidt et al., 2017). Similarly, at the national level, governance planning in Indonesia and elsewhere needs to set the conditions for integrating coastal and marine planning processes by determining the character of the interlinkages and interactions between main policy arenas and objectives. Governance and management are affected by whether there is mutual reinforcement, neutrality, or competition between any two objectives, and the elaboration of trade-offs and synergies is an essential component of integrated planning.

An explicit effort at integrated systems thinking is required to move beyond single sector approaches. That system thinking can be fostered in "ordinary local populations" has been recognized (Krause and Welp, 2012), but the integration of locally rooted system understanding into decision-making remains a challenge for governance and management, not only in Indonesia. In line with increasingly diverse and interrelated challenges and growing complexity of human–nature interactions and feedbacks, environmental governance analysis has grown fast over recent decades. The widening field of theory and applied science offers potentials for the development of Indonesian coastal and marine governance and management, which we now explore briefly.

### 12.3.1 Participatory governance and comanagement

A near universal credo in environmental governance theory and debate is that local system users' participation that includes their knowledge and established rights in decision-making is needed to support long-term sustainable human–nature relations (e.g., Ferse et al., 2010; Glaeser, 2002, 2019). Participatory practice often falls short of such aspirations (e.g., Glaeser and Glaser, 2010, on the failed struggle of fishers to be included in debates relevant to their livelihoods at the World Ocean Conference in Manado in 2009). To achieve such inclusive governance, diverse knowledge systems need to be linked transparently, and well-structured open communications and interactions between stakeholders are needed. To achieve synergies between diverse, and differently situated, knowledge systems, these have to be understood, respected, and interlinked at eye level. Martens and Schnegg (2008) find that the mental constructs of local artisanal fisherfolk in the Spermonde region help them "to understand and predict

[7]SPICE project attempts to bring together different ministries to discuss research results of joint interest also met with only very limited success.

the ways in which species interact with each other and with human perturbations." The authors suggest that the complex "cognitive ecological networks" (i.e., the locally perceived relations between species) need to be discussed in terms of current natural science understandings of multitrophic relations within the same ecosystems.

A further emerging consensus is that options and tools for participatory governance and management need to be culturally sensitive. Based on fieldwork in South Sulawesi, Glaser et al. (2010c) suggest the concept of "nested participation" as suitable for cultural systems such as prevalent in Indonesia where most formal and informal institutions follow strongly hierarchical principles, which often hamper knowledge exchange between actors across different levels of the social hierarchy. The authors show how and where people are in various ways compelled to follow the opinions of "their seniors" in a hierarchical social system, and argue that the needed opportunities for formulating opinions and for naming conflicts can, and ought to be systematically generated in protected hierarchy-poor spaces in which some control for threats from powerful "senior" actors can be provided for knowledge holders in lower positions of the social hierarchy. Glaser et al. (2010c) show that communication pathways can then be established so that the "knowledge of the less privileged" can become part of knowledge-building and solution-seeking processes.

Comanagement as a concept means the sharing of rights and duties relating in pursuit a shared set of objectives. The comanagement concept gained ground as it became clear that an effective involvement of ecosystem users and other weaker stakeholders in environmental governance and management requires clear rights for ecosystem users as well as reliable public authority involvement. The authority to impose sanctions for major infringements is a particularly important public authority function in participatory governance and comanagement. On such a basis, the sectoral incompatibilities discussed in Section 11.2 can be handled in appropriately contextualized ways. Importantly for Indonesia, the island character of this archipelagic country offers special advantages for integrated marine and coastal governance in island-focused regimes (Glaser et al., 2018b).

## 12.3.2 Adaptive, evolutionary, and anticipatory governance

Adaptive governance theory, as developed in two early papers (Olsson et al., 2004a,b), indicates how local groups can learn and actively adapt to and shape change. Multiple studies have since shown that successful adaptation hinges on the capacity of local actors to self-organize for formulating, testing, and revising institutional innovations in response to change (Folke et al., 2005). Such flexibility is at the core of successful adaptive governance and management. The power of adaptive governance to generate context-appropriate solutions under conditions of change hinges upon the capacities of actors and institutions to respond to multiple shocks and stresses. Adaptive governance takes a stress or sudden shock as its analytical point of departure. This remains highly appropriate for the Indonesian coastal and marine environment, which is characterized by frequent major shocks such as earthquakes, tsunamis, and volcanic outbreaks, as well as by strong stress factors such as overfishing and pollution.

Evolutionary governance theory (EGT) has roots in adaptive governance. It focuses on how societies, markets, and public sector governance structures and processes coevolve subject to multiple forms of dependency (Beunen et al., 2015; van Assche et al., 2014). There are strong similarities to the dynamic approach of the older adaptive governance concept, but EGT adopts a more systems-oriented approach that focuses on how governance is shaped by the interactions and dependencies between institutions, actors, and visions over time. In a recent paper on the small island Gili Terawangan in Eastern Indonesia, Partelow and Nelson (2018) describe how early social networks among actors in the tourism sector facilitated the evolution of self-organized collective action and associated governance institutions, but how, at a later stage, these networks became unfit for governing an increasingly multiactor and complex field and thus evolved as governance instruments to meet new challenges. Partelow and Nelson (2018) find that strong social networks successfully initiated self-organized marine governance among the island's few early dive businesses in the 1990s and early 2000s, but that with a growing number of new dive businesses and hotels and more tourists, the initially successful social networks to support marine governance became less effective. The authors show how new governance institutions and collective action evolved from strong social networks and how, faced by economic growth, changing social—ecological conditions and the appearance of government (a previously locally absent actor), undermined initially effective social networks as the foundation for the island's governance success. The system-based logic of EGT helps in tracing the trajectory of governance evolution and the functions assumed by social networks and the linkages between them. Our field-based analyses of social networks in Spermonde (Glaser et al., 2018a,b,c; Gorris, 2015; Gorris et al., 2019) show that policy and practice need to clearly distinguish between networks that enhance sustainable human—nature relations and those that do not.

Anticipatory governance theory carries elements of both adaptive and evolutionary thinking. Clear differences exist though: While adaptation is about responses to change, "anticipation is about intentionality, action, agency, imagination, possibility, and about choice—but anticipatory governance is also about being doubtful, unsure, uncertain, fearful, and apprehensive" (Nutall, 2010). Under conditions of increasingly unpredictable, complex multilevel change that creates uncertain futures, "processes of foresight" (including scenario construction, future visioning, and similar techniques) are assuming an important role for anticipatory governance analysis. Anticipatory governance research can thus be described as the study of how foresight is organized and then linked to decision-making (Boyd et al., 2015). Anticipatory practice has been described as closely connected with "the evolution of steering mechanisms to shape an increasingly uncertain future" (Vervoort and Gupta, 2018). It is an important new field to enable governance practitioners to better respond to the complexities of Anthropocene challenges, which is gaining ground as new extreme global scenarios, including the global pandemic in 2020/2021, become reality. For the Indonesian coastal and marine realm, preparations for the disappearance of coral ecosystems driven by global warming, pollution, destructive, and overfishing is a possibility would need to be addressed by anticipatory governance practice.

Adaptive, evolutionary, and anticipatory governance share common roots and features. Boyd et al. (2015) suggest that anticipation is becoming a core mechanism for adaptation in SESs and that it is essential to unpack the theory of anticipation in regard to how it contributes to adaptiveness (op. cit. p. 157). The three schools of thought employ overlapping but complementary analytical lenses. All three can be usefully applied to the question social scientists in SPICE set out to investigate: *How does governance affect, and how might it transform, human and societal behavior in the coastal and marine realm toward greater ecological and social sustainability?*

All three aforementioed approaches to governance, however, are weak in addressing transformation objectives in which governance not only focuses on how to maintain existing system functions but also sets out to actively address the dysfunctionalities of the SES in question. For the Spermonde Archipelago, Glaser et al. (2018c) point out how wickedly resilient system features and processes need to weakened or dismantled to pave the way for sustainable human–nature relations in the marine realm.

### 12.3.3 Polycentric coastal and marine governance

The top-down "one serves all" country-wide approaches that predominated in Indonesia since colonial times through to the New Order Regime have proven ineffective (Glaser et al., 2015; Lukas, 2013). There is a wide consensus that particularly collective action problems are best addressed through concerted efforts that span multiple system scales and levels (Gerhardinger et al., 2018; Jones, 2014; Ostrom, 2009b). With the Indonesian reform period since the late 1990s (*reformasi*), the governance of the country's coastal and marine areas is growing more polycentric, with first evidence of governance through social networks (Gorris, 2015; Partelow and Nelson, 2018).

Environmental governance through efficient and effective social networks is commonly interpreted as sustainability-supporting in the increasingly complex world of multirooted anthropogenic social–ecological change. It is often assumed that if governance actors interact in well-connected networks, this will support more effective and sustainable outcomes (Ostrom, 1990). In his network-focused study of the marine governance approaches to address the degradation of coral reef systems in Brazil and Indonesia, Gorris (2015), however, finds both positive and negative outcomes of well-connected social networks: Polycentric multiactor governance systems are found to be robust at the regional level, but they also support illegal and destructive blast and poison fishing. Gorris et al. (2019) investigate distinct network configurations among actors in polycentric natural resources governance systems. This social network analysis study in the Spermonde Archipelago empirically compares interaction patterns of local islands with more successful governance outcomes to those in islands with less successful outcomes. Gorris and coauthors find more frequent interactions between local actors, a stronger integration of local nongovernmental actors in the polycentric governance network for islands with more successful marine governance outcomes, and a higher centralization of the local governance networks with rent-seeking actors in strong brokerage positions in islands where highly unsustainable governance outcomes prevailed. These results indicate the negative potentials of strong social network structures in governance: In our coastal study regions, strongly centralized local networks in which key actors linked closely to government actors at multiple levels helped to shield fish workers who engaged in highly unsustainable and illegal blast and poison fishing from legal prosecution. The patron–traders of such fishers colluded with local government officials to prevent or mitigate sanctions for "their" bomb and poison fishers, and villages with higher levels of in-group trust jointly enabled and entrenched destructive illegal fishing practices. Radjawali (2011) shows that the marine governance subnetworks for reef fisheries, marketing, and (informal) safeguarding against legal prosecution are closely connected and characterized by prevalent corruption and by indebtedness of producer–clients to patron–traders.

## 12.4 Conclusions and outlook

So as to assist marine and coastal governance decision-makers and to identify remaining key research gaps, we close by discussing some implications of our research results for management and policy development and then addressing current research gaps.

### 12.4.1 Critically examine the effects of polycentric governance

While polycentric governance with associated effective social networks may be a necessary condition for sustainability-enhancing governance outcomes, SPICE research indicates that it may also be associated with social–ecological dynamics that undermine sustainability. Whether the outcomes of polycentric governance enhance or undermine sustainability transformation thus needs to be carefully evaluated rather than assumed. Social networks in Indonesian reef fisheries have been found to support corruption and illegal marine resources while supplying marine products to distant luxury consumer markets with deleterious effects for local SES.

### 12.4.2 Differentiate stakeholder perceptions and analyze links between perceptions and behaviour

Perceptions are usefully defined as the subjective way people experience, think about, and understand someone or something (Beyerl et al., 2016). Perceptions of social and social–ecological processes, of the functioning and change of the natural environment, and of policies and other strategies to cope with such changes are crucial drivers of human behavior toward marine and coastal nature (Beyerl et al., 2016). How exactly their perceptions affect the behavior of marine and coastal governance actors toward nature, and their engagement in relevant institution building, is a relatively recent, as yet inadequately addressed, research question. Adding to the growing body of empirical evidence in the field of environmental behavior, Spranz (2017) suggests a set of normative, perception-based interventions for reducing plastic bag use in Indonesia. Walker-Springett et al. (2016) present nine case studies in which work on the perceptions of the societal importance of aquatic ecosystems has improved the governance of aquatic ecosystems. For community-based marine resource management, Beyerl et al. (2016) present important relevant perception types and call for trained local perception experts (PEs) to analyze the perceptions of involved stakeholders and to communicate them to decision-makers. If applied to different Indonesian marine governance and management regimes, this approach may throw new light on the drivers of dysfunctional "paper parks" (Baitoningsih, 2015) and projects.

### 12.4.3 Work with social energy for sustainable human–nature relations

Working against system dynamics is difficult and prone to fail. Diverse examples include dyke building against the intruding sea or behavioral advocacy against dominant values or incentives. While highly undesirable system features such as nepotism, corruption, social exploitation, or pollution might need to be directly addressed, system theory suggests that it is more effective to work with and not against system dynamics (Bertalanffy, 1949; Checkland, 1984). The pursuit of desirable social, biogeophysical, and social-ecological outcomes is thus likely to be more successful if strategies are adopted that rely on working with (rather than against) strategically selected prevalent system dynamics.

Social energy[8] is such a strategically important element of system dynamics. We here define social energy as system-immanent and self-organizing individual and collective efforts that support a society in organizing resources and meeting challenges and opportunities. Governance structures and processes may generate or undermine social energies that support sustainable human–nature relations. The stranglehold of local elites over local production systems, which has often been enforced with decentralization in Indonesia and elsewhere (Glaser et al., 2003; Glaser et al., 2010a), is a case in point. Wever et al. (2012) find, for both Indonesian and Brazilian coastal systems, that local ecosystem users' social energies and capacities exist and are likely to be essential for controlling the behavior toward nature of those who do not share local sustainability agendas. Multiple studies cite cases where local social energies for sustainability were undermined through alliances between powerful private and public sector actors leaving

[8]Social energy is here understood, in line with Hirschman A.O., (1983) The principle of conservation and mutation of social energy Grassroots Development Vol 7 No 2. & Hirschmann A.O. (1984) Getting Ahead Collectively: Grassroots Experiences in Latin America Pergamon Press) as selforganizing collective action by people to better their conditions; see also https://www.leibniz-zmt.de/en/research/research-projects/social-energy.html.

local ecosystems overexploited or polluted, and local ecosystem users marginalized (Glaser et al., 2003, 2015; Gorris, 2015; Lukas, 2013). Marine and coastal governance needs to create and safeguard the scope and flexibility to include the priorities in particular of those who directly interact with and depend on coastal and marine nature. Social energy for an intact marine environment may also be encountered or generated at other higher system levels and among nonlocal system actors. Spranz (2017), for instance, identifies individual consumer incentives to reduce the use of plastic bags that harm the marine environment, and initiatives for global marine governance such as the Ocean Knowledge to Action Network[9] rally organizations and civil society and market actors at the planetary level.

### 12.4.4 Identify and use arenas of leverage against wickedly resilient system features

Resilience usually enables the protection of a status quo. Most of the debate surrounding resilience assumes that resilient SES are always desirable since they foster fairness, inclusiveness, and diversity. The maintenance and building of resilience has thus become a central tenet of sustainability work (e.g., Olsson et al., 2014). SPICE research has, however, identified a number of resilient interlocking vicious cycles (e.g., overfishing, pests, and loss of biodiversity and reef structure driven by interlocking corruption, inequality, and poverty generation) as the main drivers of the overexploited and polluted state of coral reefs and of the impoverishment of fishing households. Such constellations are usefully described as "wickedly resilient," and the task of undermining and eliminating such undesirable forms of resilience is the "other side of the coin" of resilience management that is usually obscured (Glaser et al., 2018c). Adequate inclusive and rights-based governance processes and structures and well-designed strategic interventions are needed to weaken the feedbacks that trap SES in undesirable and entrenched "wickedly resilient" regimes (Cinner, 2011; Glaser et al., 2018c). That such "work against the system" is notoriously difficult is at least partly due to the accompanying need to engage with power. Developments following the transformation of the Indonesian "New Order", however, show that improvements can be achieved.

### 12.4.5 Adopt explicit social equity objectives

Equity in environmental governance is closely connected to the presence and quality of participation practice. Distributional equity can be analyzed through various lenses including socioeconomic group, gender, ethnic group, or occupational sector. The actual or perceived distribution of costs and benefits associated with governance structures, processes, and interventions affects implementation outcomes. Stakeholder perceptions of equity and (actual or perceived) equity outcomes centrally determine the prospects for the active participation of those most in need in governance and management.

[9]https://futureearth.org/wp-content/uploads/2017/03/leaflet_oceankan_developmentteam_final.pdf.

Gender equity levels are an outcome of (lack of) female participation in governance and management (Máñez and Pauwelussen, 2016). The social construction of male and female roles in coastal production affects what is considered a "male" and what a "female" activity, which real-life behaviors are obfuscated or ignored, who makes the rules and who gains and who loses, and to what extent. Socially constructed gender roles affect what men and women (can) do and are seen to do in coastal governance. For instance, women, although rare in Indonesian fisheries, glean and dive for giant clam among the seminomadic Bajau (*orang laut*) people. But this is an exception rather than the rule.

The more common exclusion of women can lead to their active subversion of governance and management structures and processes. Rohe et al. (2018) have reported for the South Pacific how women who are excluded from rule-making processes and institutions for marine spaces then undermine the implementation of these rules. Female fishers (gleaners) and, with this, their ecological knowledge are usually excluded from the consultation practices of Indonesian public agencies and ministries concerned with fisheries. Female roles in trade and financing marine activities along (pseudo-kin-based) social networks have also been noted in Indonesia (Manez and Pauwelussen, 2016), but there is no evidence of a female representation in formal decision-making relating to this sector. A gender-sensitive perspective on marine resource use that explicitly includes women's own voices on their knowledge and priorities is needed for more effective marine and coastal governance.

Environmental change processes, including climate change, have equity implications. Glaeser and Glaser (2010) show how the emerging "climate divide" further increases preexisting inequality and inequity in Indonesian society: the costs of climate change fall disproportionately on the poorest who also have the lowest capacity to respond and adapt, while benefits from carbon emissions trading schemes go elsewhere. The discursive and practical entanglement of watershed and mangrove conservation goals with the Indonesian state forest corporation's goals of wood production and of retaining territorial control in the Segara Anakan lagoon and its catchment illustrate a tension between these objectives of the Indonesian state on the one hand, and environmental sustainability and equity goals on the other. This has been shown to undermine both environmental sustainability and equity (Heyde, 2016; Lukas, 2015).

There is consensus that a strong focus on participation and inclusiveness is needed for sustainability transformations in governance. But there are pitfalls. A major one is the risk of increasing social inequity. It has been noted that the focus on the contributions of marginal actors prevalent in participatory approaches often shifts the burdens and costs of participation onto the most vulnerable actors and groups (Cooke and Kothari, 2003; Blythe et al., 2016) who, at the same time, gain few if any established rights associated to their participation. Where, as found for some explicitly participatory approaches in Indonesian coastal management (Glaser et al., 2015), a common agenda is not firmly established between the local stakeholders (whose participation is desired) and more powerful comanagement partners (typically government agencies), local populations come to interpret participation as a way for powerful actors to appropriate and instrumentalize their time and resources in pursuit of objectives that they had little or no part in establishing and thus often do not share.

### 12.4.6 Employ innovative methods for linking knowledge types to generate synergies and innovation

Methods for the analysis of SES and their governance remain scarce (Bundy et al., 2016; Glaeser, 2016, 2009; Glaser et al., 2018b); disciplinary research still prevails over approaches which are problem-focused and inclusive of multiple scientific and other forms of relevant knowledge. The diverse types of relevant knowledge of decision-makers at all levels of civil society and government, including the knowledge of the often impoverished and marginalized direct ecosystem users of tropical coasts and seas, need to be better included and understood. For the wider Oceania and Indo-Pacific region, Ross et al. (2019) find that community belief systems and practices are often poorly aligned with government even where comanagement schemes exist. Beyond the work on nested participation (Glaser et al., 2010c), inclusive deliberative practice in hierarchical social contexts needs to be further developed so as to identify how different forms of knowledge across system levels can be best included in coastal governance and management.

## 12.5 Final remarks

Indonesia is the largest archipelagic country on earth, with three time zones, encompassing more than 17,000 islands, 86,700 square kilometers of coral reefs, and 24,300 square kilometers of mangrove areas, supporting nearly 230 million people (Huffard et al., 2012). Even the Sulawesi Island waters to which much of the work here reported relates fall into three ecoregions, each with a diversity of associated human cultures:[10] To claim representativeness for any region-based set of studies such as the over-a-decade-long SPICE research we reported on here for the whole of Indonesia would thus be presumptuous. Rather, we regard our work as part of an important patchwork of interlocking studies that can inform decision-making in the marine and coastal realm. That our results relate to confirming cases from other regions of the global coasts and oceans, e.g., India (Kurien, 1992), Brazil (Glaser and Oliveira, 2004), and the South Pacific (Johannes et al., 2000, 2002) indicates that it places Indonesia into a global context, which is becoming increasingly relevant and indispensable for planetary sustainability.

[10]Namely (1) the Sulawesi Sea-Makassar Strait, (2) North-East Sulawesi and Tomini Bay, and (3) Banda Sea Ecoregions (Adhuri et al., 2019).

Knowledge gaps and directions of future research

- Where does effective polycentric governance enhance, and where does it undermine sustainability transformation?
- How do their perceptions affect the behavior of marine and coastal governance actors toward nature and their engagement in relevant institution building?
- How can perceptions be integrated in governance assessment?
- Which governance structures and processes generate and which undermine social energies for sustainability?
- What governance structures and processes and which strategic interventions are appropriate to weaken feedbacks that trap SES in undesirable and entrenched "wickedly resilient" regimes?
- How do perceptions of equity affect the prospects of system user participation in governance and management
- How are environmental sustainability and equity goals best combined in governance and management?
- How to prevent powerful actors from appropriating poor stakeholders' time and resources on forms of "participation" that support objectives that the latter had little part or interest in establishing?

Implications/recommendations for policy and society

- The engagement of key interested parties, in particular of ecosystem users in policy and rule development for the environments they live in and off is crucial for sustainable coastal governance and management.
- Such engagement needs to follow rights-based principles and include adequate mechanisms for the sharing of both benefits and compensating for costs associated with coastal governance and management decisions.
- Poorer ecosystem users require loans, social, and tenure security and additional livelihood options but also need to be enabled to secure their subsistence needs including good food and drinking water.
- The explicit inclusion of women's knowledge and priorities is key to effective marine and coastal governance.

## Acknowledgments

We would like to thank the inhabitants of Spermonde who kindly hosted us and agreed to participate in the various studies and surveys summarized here. We also thank the various students and colleagues from Indonesia, Germany, and other parts of the world, who contributed to our work with their theses, fieldwork, internships, with logistical support, and as guest researchers. A special thanks to Claudia Schultz for her ongoing support. The research reported here was made possible by the continuing financial and administrative support of the German Federal Ministry of Education and Research (Grant Nos. 03F0474A, 03F0641A, 03F0643A), the Indonesian Ministry for Research and Technology (RISTEK), the Indonesian Ministry for Maritime Affairs and Fisheries (KKP), and the German Academic Exchange Service (DAAD).

# References

Acciaoli, G., 2000. Kinship and Debt; the social organization of Bugis migration and fish marketing at lake Lindu, central Sulawesi. Bijdragen tot de Taal-, Landen Volkenkunde, Authority and enterprise among the peoples of South Sulawesi 156, 588–617.

Adger, N., 2006. Vulnerability. Global Environmental Change . 16 (3), 268–281.

Adger, W.N., Brown, K., Tompkins, E.L., 2005a. The political economy of cross-scale networks in resource co-management. Ecology and Society 10 (2), 9 [online] URL. http://www.ecologyandsociety.org/vol10/iss2/art9/.

Adger, W.N., Hughes, T.P., Carl, F., Carpenter, S.R., Rockström, J., 2005b. Social-ecological resilience to coastal disasters. Science 309, 1036–1039. https://doi.org/10.1126/science.1112122.

Adhuri, D.S., 2019. Socio-ecological diversities of the Sulawesi islands. Journal of Ocean and Culture 14. https://doi.org/10.33522/joc.2019.2.22. Available from: https://www.joac.org/archive/view_article_pubreader?pid=joc-2-0-22.

Adrianto, L., Arsyad Nawawi, M., Solihin, A., Irving Hartoto, D., Satria, A., 2009. Constructions of local fisheries management in Indonesia. International Collective in Support of Fisherworkers and the center for Coastal and marine Resources Studies. Bogor Agricultural University (IPB).

Aswani, S., Basurto, X., Ferse, S., Glaser, M., Campbell, L., Cinner, J.E., et al., 2017. Marine resource management and conservation in the Anthropocene. Environmental Conservation 1–11.

Baitoningsih, W., 2009. Community Participation in Designing Marine Protected Area in Spermonde Archipelago, South Sulawesi, Indonesia. Faculty for Biology and Chemistry, vol. 91. University of Bremen. Master of Science Course ISATEC, Bremen.

Baitoningsih, W., 2015. Let's Get Political: Community Participation in the MPA Establishment Process in Indonesia (Doctoral Thesis in Political Science). University of Bremen, Bremen, Germany.

Baum, A., Rixen, T., 2014. Dissolved inorganic nitrogen and phosphate in the human affected Blackwater river Siak, central Sumatra, Indonesia. Asian Journal of Water Environment and Pollution 11 (1), 13–24.

Baum, G., Kusumanti, I., Breckwoldt, A., Ferse, S.C.A., Glaser, M., Dwiyitno Adrianto, L., et al., 2016. Under pressure: investigating marine resource-based livelihoods in Jakarta Bay and the Thousand islands. Marine Pollution Bulletin 110, 778–789. https://doi.org/10.1016/j.marpolbul.2016.05.032.

Bertalanffy, L., 1949. General system theory. Biologia Generalis 1, 114–129.

Beunen, R., Van Assche, K., Duineveld, M., 2015. The search for evolutionary approaches to governance. In: Beunen, R., Van Assche, K., Duineveld, M. (Eds.), Evolutionary Governance Theory. Springer, Cham.

Bevir, M., 2013. Governance: A Very Short Introduction. Oxford University Press, Oxford.

Beyerl, K., Putz, O., Breckwoldt, A., 2016. The role of perceptions for community-based marine resource management. Frontiers in Marine Science 3, 238.

Birkeland, C. (Ed.), 2015. Coral Reefs in the Anthropocene. Springer Netherlands, Dortrecht, Netherlands.

Blackburn, S., Marques, C., de Sherbinin, A., Modesto, F., Ojima, R., Oliveau, S., et al., 2014. Mega-urbanisation on the coast. In: Pelling, M., Blackburn, S. (Eds.), Megacities and the Coast: Risk, Resilience and Transformation. Routledge, Oxon, US, pp. 1–21.

Blythe, J., Silver, J., Evans, L., Armitage, D., Bennett, N.J., Moore, M.-L., et al., 2016. Resilience, Development and Global Change. Routledge, Oxford, N:Y.

Boyd, E., Nykvist, B., Borgström, S., Stacewicz, I.A., 2015. Anticipatory governance for social-ecological resilience. Ambio 44 (S1), 149–161.

Breckwoldt, A., Dskikowitzky, L., Baum, G., Ferse, S.C.A., van der Wulp, S., Kusumanti, I., et al., 2016. A review of stressors, uses and management perspectives for the larger Jakarta Bay Area, Indonesia. Marine Pollution Bulletin 110 (2), 790–794.

Brown, K., 2018. The dark side of transformation: latent risks in contemporary sustainability discourse. Antipode 50 (5), 1206–1223.

Bundy, A., Chuenpagdee, R., Cooley, S.R., Defeo, O., Glaeser, B., Guillotreau, P., et al., 2016. A decision support tool for response to global change in marine systems: the IMBER-ADApT framework. Fish and Fisheries 17 (4), 1183–1193. https://doi.org/10.1111/faf.12110.

Burke, L., Reytar, K., Spalding, M., Perry, A. (Eds.), 2012. Reefs at Risk Revisited in the Coral triangle. World Resources Institute, Washington, DC, USA.

Cabral, R.B., Mayorga, J., Clemence, M., Lynham, J., Koeshendrajana, S., Muawanah, U., et al., 2018. Rapid and lasting gains from solving illegal fishing. Nature Ecology and Evolution 2, 650–658.

Cash, D.W., Adger, W.N., Berkes, F., Garden, P., Lebel, L., Olsson, P., et al., 2006. Scale and cross-scale dynamics: governance and information in a multilevel world. Ecology and Society 11 (2), 1–8.

Checkland, P.B., 1984. Systems thinking in management: the development of soft systems methodology and its implications for social science. In: Ulrich, H., Probst, G.J.B. (Eds.), Self-organization and Management of Social Systems - Insights, Promises, Doubts and Questions. Springer, Berlin, Heidelberg, New York, Tokio, pp. 94–104.

Cinner, J.E., 2011. Social-ecological traps in reef fisheries. Global Environmental Change 21, 835–839.

Cinner, J.E., Huchery, C., MacNeil, M.A., Graham, N.A.J., McClanahan, T.R., Maina, J., et al., 2016. Bright spots among the world's coral reefs. Nature 535, 416–419.

Clifton, J., 2009. Science, funding and participation: key issues for marine protected area networks and the coral Triangle initiative. Environmental Conservation 36, 91–96.

Cooke, B., Kothari, U., 2003. Participation: The New Tyranny, vol. 23. https://doi.org/10.1016/S0738-0593(02)00022-6. http://lst-iiep.iiep-unesco.org/cgi-bin/wwwi32.exe/[in=epidoc1.in]/?t2000=012096/(100).

Crutzen, P.J., 2002. Geology of mankind. Nature 2. https://doi.org/10.1038/415023a.

Cumming, G.S., 2011. Spatial Resilience in Social-Ecological Systems. Springer, New York.

Deswandi, R., 2012. Understanding institutional dynamics: The emergence, persistence, and change of institutions in fisheries in Spermonde archipelago, South Sulawesi, Indonesia. Ph.D. dissertation. Universität Bremen, Bremen, Germany, FB8.

Deswandi, R., Glaser, M., Ferse, S., 2012. What makes a social system resilient? In: Hornidge, A.K., Antweiler, C. (Eds.), Environmental Uncertainty and Local Knowledge. Southeast Asia as a Laboratory of Global Ecological Change. Transcript Publishers, Bielefeld, pp. 243–272.

Dodds, F., Donoghue, D., Roesch, J.L., 2017. Negotiating the Sustainable Development Goals - A Transformational Agenda for an Insecure World. Routledge.

Dsikowitzky, L., Dwiyitno Heruwati, E., Ariyani, F., Irianto, H.E., Schwarzbauer, J., 2014. Exceptionally high concentrations of the insect repellent N, N-diethyl-m-toluamide (DEET) in surface waters from Jakarta, Indonesia. Environmental Chemistry Letters 12 (3), 407–411.

Edgar, G., Stuart-Smith, R., Willis, T., Kininmonth, S., Baker, S., Banks, S., Thomson, R., 2014. Global conservation outcomes depend on marine protected areas with five key features. Nature 506, 216–220. https://doi.org/10.1038/nature13022.

Fabinyi, M., Pido, M., Harani, B., Caceres, J., Uyami-Bitara, A., De las Alas, A., et al., 2012. Luxury seafood consumption in China and the intensification of coastal livelihoods in Southeast Asia. The Live Reef Fish for Food Trade in Balabac, Philippines-Asia Pacific Viewpoint 53, 118–132.

Falkland, A.C., Brunel, J.P., 1993. Review of hydrology and water resources in humid tropical islands. In: Bonell, M., Hufschmidt, M.M., Gladwell, J.S. (Eds.), Hydrology and Water Management in the Humid Tropics. Cambridge University Press, Cambridge, pp. 135–166.

FAO, 2007. The World's Mangroves: A 1980–2005. A Thematic Study Prepared in the Framework of the Global forest Resources Assessment 2005 FAO Forestry Paper 153. Italy, Rome.

FAO, 2014. Fishery and Aquaculture Country Profiles: The Republic of Indonesia. http://www.fao.org/fishery/facp/IDN/en#CountrySector-Overview.

Ferrol-Schulte, D., 2014. Patron–client relationships, livelihoods and natural resource management in tropical coastal communities. Ocean and Coastal Management 100 (0), 63–73.

Ferrol-Schulte, D., Wolff, M., Ferse, S., Glaser, M., 2013. Sustainable livelihoods approach in tropical coastal and marine social–ecological systems: a review. Marine Policy 42, 253–258.

Ferrol-Schulte, D., Gorris, P., Baitoningsih, W., Adhuri, D.S., Ferse, S.C.A., 2015. Coastal livelihood vulnerability to marine resource degradation: a review of the Indonesian national coastal and marine policy framework. Marine Policy 52, 163–171.

Ferse, S., Máñez-Costa, M., Schwerdtner Máñez, K., Adhuri, D.S., Glaser, M., 2010. Allies, not aliens - increasing the role of local communities in MPA implementation. Environmental Conservation 37 (1), 23–34.

Ferse, S., Glaser, M., Schultz, C., Jompa, J., 2012a. Linking research to Indonesia's CTI action plan: the SPICE program. In: 12th International Coral Reef Symposium 9–13 July 2012, Cairns, Australia.

Ferse, S., Knittweis, L., Krause, G., Maddusila, A., Glaser, M., 2012b. Livelihoods of ornamental coral fishermen in South Sulawesi/Indonesia: implications for management. Coastal Management 40 (5), 525–555.

Ferse, S.C.A., Glaser, M., Neil, M., Schwerdtner Máñez, K., 2014. To cope or to sustain? Eroding long-term sustainability in an Indonesian coral reef fishery. Regional Environmental Change 14 (6), 2053–2065. https://doi.org/10.1007/s10113-012-0342-1.

Folke, C., Hahn, T., Olsson, P., Norberg, J., 2005. Annual Review of Environment and Resources 30, 441–473. https://doi.org/10.1146/annurev.energy.30.050504.144511. First published online as a Review in Advance on July 25, Volume publication date 21 November 2005.

Galaz, V., 2014. Global Environmental Governance, Technology and Politics: The Anthropocene gap. Edward Elgar Publishing, Cheltenham, UK.

Geertz, C., 1960. The Religion of Java. University of Chicago Press, p. 392p.

Gerhardinger, L.C., Gorris, P., Gonçalves, L.R., Herbst, D.F., Vila-Nova, D.A., de Carvalho, F.G., et al., 2018. Healing Brazil's blue Amazon: the role of knowledge networks in nurturing cross-scale transformations at the frontlines of ocean sustainability. Frontiers in Marine Science 4. https://doi.org/10.3389/fmars.2017.00395. Article 395.

Gill, D.A., Mascia, M.B., Ahmadia, G.N., Glew, L., Lester, S.E., Barnes, M., et al., 2017. Capacity shortfalls hinder the performance of marine protected areas globally. Nature 543, 665–669.

Glaeser, B., 2002. The changing human-nature relationships in the context of global environmental change. Volume 5: social and economic Dimensions of global environmental change. In: Encyclopaedia for Global Environmental Change. John Wiley, Chichester/UK, pp. 11–24.

Glaeser, B., 2016. From global sustainability research matrix to typology: a tool to analyze coastal and marine social-ecological systems. Regional Environmental Change 16 (2), 367–383. https://doi.org/10.1007/s10113-015-0817-y.

Glaeser, B., 2019. Human-nature relations in flux: two decades of research in coastal and ocean management. In: Wolanski, E., Day, J., Elliott, M., Ramachandran, R. (Eds.), Coasts and Estuaries – the Future. Elsevier, Cambridge, MA.

Glaeser, B., Glaser, M., 2010. Global change and coastal threats: the Indonesian case. Human Ecology Review 17 (2), 135–147.

Glaeser, B., Bruckmeier, K., Glaser, M., Krause, G., 2009. Social-ecological systems analysis in coastal and marine areas: a path toward integration of interdisciplinary knowledge. In: Lopes, P., Begossi, A. (Eds.), Current Trends in Human Ecology. Scholars Publishing, Cambridge, UK: Cambridge, pp. 183–203.

Glaser, M., Glaeser, B., 2014. Towards a framework for cross-scale and multi-level analysis of coastal and marine social-ecological systems dynamics. Regional Environmental Change 14 (6), 2039–2052. https://doi.org/10.1007/s10113-014-0637-5.

Glaser, M., Oliveira, R.S., 2004. Prospects for the co-management of mangrove ecosystems on the North Brazilian coast: whose rights, whose duties and whose priorities? Natural Resources Forum 28, 224–233.

Glaser, M., Berger, U., Macedo, R., 2003. Local vulnerability as an advantage: mangrove forest management in Pará state, North Brazil under conditions of illegality. Regional Environmental Change 3 (4), 162–172.

Glaser, M., Baitoningsih, W., Neil, M., Ferse, S., 2010a. Whose sustainability? Top-down participation and emergent rules in protected area management – local lessons for MPA design from Indonesia. Marine Policy 34, 1215–1225.

Glaser, M., Krause, G., Oliveira, R., Fontalvo-Herazo, M., 2010b. Mangroves and People: A Social-Ecological System. Mangrove Dynamics and Management in North Brazil. Springer, Berlin, pp. 307–351.

Glaser, M., Radjawali, I., Ferse, S., Glaeser, B., 2010c. Nested participation in hierarchical societies? Lessons for social-ecological research and management. International Journal of Sustainable Society 2 (4), 390–414.

Glaser, M., Krause, G., Halliday, A., Glaeser, B., 2012. Towards global sustainability analysis in the Anthropocene. In: Glaser, M., et al. (Eds.), Chapter 10, Human-Nature Interaction in the Anthropocene: Potentials of Social-Ecological Systems Analysis. Routledge, pp. 193–222.

Glaser, M., Breckwoldt, A., Deswandi, R., Radjawali, I., Baitoningsih, W., Ferse, S.C.A., 2015. Of exploited reefs and fishers – a holistic view on participatory coastal and marine management in an Indonesian archipelago. Ocean and Coastal Management 116, 193–213. https://doi.org/10.1016/j.ocecoaman.2015.07.022.

Glaser, M., Breckwoldt, A., Carruthers, T., Forbes, D., Costanzo, S., Kelsey, H., et al., 2018a. Towards a framework to support coastal change governance in small islands. Environmental Conservation 11.

Glaser, M., Breckwoldt, A., Gorris, P., Ferreira, B., 2018b. Analysing ecosystem user perceptions of the governance interactions surrounding a Brazilian near shore coral reef. Sustainability 10. Art. 1464.

Glaser, M., Plass-Johnson, J., Ferse, S., Neil, M., Yanuarita, S.D., Teichberg, M., et al., 2018c. Breaking resilience for a sustainable future: thoughts for governance and management in a coral reef archipelago in the anthropocene. Frontiers in Marine Science. https://doi.org/10.3389/fmars.2018.0034.

Gorris, P., 2015. Entangled? Linking Governance Systems for Regional-Scale Coral Reef Management: Analysis of Case Studies in Brazil and Indonesia (Doctoral Thesis). Jacobs University Bremen, Bremen.

Gorris, P., 2016. Deconstructing the reality of community-based management of marine resources in a small island context in Indonesia. Frontiers in Marine Science 3. https://doi.org/10.3389/fmars.2016.00120. Article 120.

Gorris, P., Glaser, M., Idrus, R., Yusuf, A., 2019. The role of social structure for governing natural resources in decentralized political systems: insights from governing a fishery in Indonesia. Public Administration 97 (3), 654–670. https://doi.org/10.1111/padm.12586.

Griggs, D., Nilsson, M., Stevance, A., McCollum, D. (Eds.), 2017. A Guide to SDG Interactions: From Science to Implementation. International Council for Science (ICSU), Paris. Accessed on August 22 2018. https://council.science/cms/2017/05/SDGs-Guide-to-Interactions.pdf.

Heyde, J., 2016. Environmental Governance and Resource Tenure in Times of Change: Experience from Indonesia. Ph.D. thesis. University of Bremen, Bremen, Germany.

Heyde, J., Lukas, M.C., Flitner, M., 2012. Payments for Environmental Services in Indonesia: A Review of Watershed-Related Schemes. Artec-Paper 186, Research center for Sustainability Studies (Artec). University of Bremen.

Heyde, J., Lukas, M.C., Flitner, M., 2017. Payments for environmental services: a new instrument to address long-standing problems? Policy paper. Artec-paper 213. Sustainability Research Center (Artec), University of Bremen.

Huffard, C.L., Erdmann, M.V., Gunawan, T.R.P. (Eds.), 2012. Geographic Priorities for marine Biodiversity Conservation in Indonesia. Ministry of Marine Affairs and Fisheries and Marine Protected Areas Governance Program, Jakarta-Indonesia, p. 105.

Ilman, M., Dargusch, P., Dart, P., Onrizal, O., 2016. A historical analysis of the drivers of loss and degradation of Indonesia's mangroves. Land Use Policy 54, 448–459. https://doi.org/10.1016/j.landusepol.2016.03.010.

Johannes, R.E., 2002. The renaissance of community-based marine resource management in Oceania. Annual Review of Ecology and Systematics 33, 317–340.

Johannes, R.E., Freeman, M.M.R., Hamilton, R.J., 2000. Ignore fishers' knowledge and miss the boat. Fish and Fisheries 1 (3), 257–273.

Jones, P.J.S., 2014. Governing marine protected areas: resilience through diversity. 1–240. https://doi.org/10.4324/9780203126295.

Krause, G., Welp, M., 2012. System thinking and social learning for sustainability. In: Glaser, M., Krause, G., Ratter, B., Welp, M., Halilday, A. (Eds.), Human-nature Interactions in the Anthropocene. Potentials of Social-Ecological Systems Analysis. Routledge, New York, pp. 13–33.

Kurien, J., 1992. Ruining the commons and responses of the commoners: coastal overfishing and fishermen's actions in Kerala state, India. In: Ghai, D., Vivian, J. (Eds.), Grassroots Environmental Action: Peoples Participation in Sustainable Development. Routledge, London, UK, pp. 221–258.

Lotze, H., Glaser, M., 2008. Ecosystem services of semi-enclosed marine systems. In: Urban, E.R., et al. (Eds.), Watersheds, Bays and Bounded Seas. The Science and Management of Semi-enclosed Marine Systems. Island Press, Washington DC, pp. 227–249.

Lowe, C., 2013. Wild profusion: biodiversity conservation in an Indonesian archipelago. 1–196, 9781400849703.

Lukas, M.C., 2013. Political transformation and watershed governance in Java: actors and interests. In: Muradian, R., Rival, L.M. (Eds.), Governing the Provision of Ecosystem Services. Springer, Dordrecht, New York, pp. 111–132.

Lukas, M.C., 2014. Eroding battlefields: land degradation in Java reconsidered. Geoforum 56 (0), 87–100. https://doi.org/10.1016/j.geoforum.2014.06.010.

Lukas, M.C., 2015. Reconstructing Contested Landscapes. Dynamics, Drivers and Political Framings of Land Use and Land Cover Change, Watershed Transformations and Coastal Sedimentation in Java, Indonesia (Doctoral Thesis). University of Bremen.

Lukas, M.C., 2017a. Widening the scope: linking coastal sedimentation with watershed dynamics in Java, Indonesia. Regional Environmental Change 17, 901–914. https://doi.org/10.1007/s10113-016-1058-4.

Lukas, M.C., 2017b. Konservasi daerah aliran sungai di Pulau Jawa, Indonesia: Terjebak dalam dikotomi kepemilikan lahan, artec-Paper 122. Research Center for Sustainability Studies (artec). University of Bremen.

Máñez, K.S., Pauwelussen, A., 2016. Fish is women's business too: looking at marine resource use through a gender lens. In: Schwerdtner Máñez, K., Poulsen, B. (Eds.), Perspectives on Oceans Past. Springer, Dordrecht. https://doi.org/10.1007/978-94-017-7496-3_11.

Martens, S., Schnegg, M., 2008. Kognitive ökologische netzwerke: Eine neue methode der kognitiven ethnologie. Ethnoscripts 10 (2), 43–54.

Muswar, H.S., Satria, A., Dharmawan, A., 2019. Political ecology analysis on marine ornamental fish eco-labelling in Les village, Bali, Indonesia. In: Human Ecology Conference, vol. 10, p. 13140/RG. University Putra Malaysia.

Nilsson, M., Chisholm, E., Griggs, D., Howden-Chapman, P., McCollum, D., Messerli, P., et al., 2018. Mapping interactions between the sustainable development goals: lessons learned and ways forward. Sustainability Science. https://doi.org/10.1007/s11625-018-0604-z.

North, D., 1991. Institutions. The Journal of Economic Perspectives 5 (1), 97–112.

Nutall, M., 2010. Anticipation, climate change and movement in Greenland. Les Inuit e le Changement Climatique/the Inuit and Climate Change 34, 21–37. https://doi.org/10.7202/045402ar.

Olsson, P., Folke, C., Berkes, F., 2004a. Adaptive co-management for building resilience in social-ecological systems. Environmental Management 34 (1), 75–90.

Olsson, P., Folke, C., Hahn, T., 2004b. Social-ecological transformation for ecosystem management: the development of adaptive co-management of a wetland landscape in southern Sweden. Ecology and Society 9 (4). Art. 2.

Olsson, P., Galaz, V., Boonstra, W.J., 2014. Sustainability transformations: a resilience perspective. Ecology and Society 19 (4), 11.

Ostrom, E., 1990. Governing the Commons. The Evolution of Institutions for Collective Action. Cambridge.

Ostrom, E., 2009a. A general framework for analyzing sustainability of social-ecological systems. Science 325 (5939), 419–422.

Ostrom, E., 2009b. A Polycentric Approach for Coping with Climate Change. Background Paper to the 2010 World Development Report. In: Policy Research Working Paper 5095. World Bank, Washington.

Partelow, S., Nelson, K., 2018. Social Networks, collective action and the evolution of governance for sustainable tourism on the Gili Islands, Indonesia. Marine Policy 12 (In press).

Pelras, C., 2000. Patron-client Ties among the Bugis and Makassarese of South Sulawesi. In: Bijdragen tot de Taal-, Land- en Volkenkunde, Authority and enterprise among the peoples of South Sulawesi, vol. 156, pp. 393–432.

Radjawali, I., 2011. Social networks and the live reef food fish trade: examining sustainability. Journal of Indonesian Social Sciences and Humanities 4, 65–100.

Reichel, C., Frömming, U., Glaser, M., 2009. Conflicts between stakeholder groups affecting the ecology and economy of the Segara Anakan region. Regional Environmental Change 9 (335). https://doi.org/10.1007/s10113-009-0085-9.

Resosudarmo, B.P., Jotzo, F., 2009. Development, resources and environment in Eastern Indonesia. In: Resosudarmo, B.P., Jotzo, F. (Eds.), Working with Nature against Poverty. Institute of Southeast Asian Studies, ISEAS Publishing, Singapore, pp. 1–18.

Roberts, C.M., O'Leary, B.C., McCauley, D.J., Cury, P.M., Duarte, C.M., Lubchenco, J., et al., 2017. Marine reserves can mitigate and promote adaptation to climate change. Proceedings of the National Academy of Sciences of the United States of America 114 (24), 6167–6175.

Rohe, J.R., Schlüter, A., Ferse, S.C.A., 2018. A gender lens on women's harvesting activities and interactions with local marine governance in a South Pacific fishing community. Maritime Studies 17 (2), 155–162.

Ross, Adhuri, H.D.S., Abdurrahim, A.Y., Phehan, A., 2019. Opportunities in community-government cooperation to maintain marine ecosystem services in the Asia-Pacific and Oceania. In: Ecosystem Services, vol. 38, pp. 1–9.

Rotich, B., Mwangi, E., Lawry, S., 2016. Where Land Meets the Sea: A Global Review of the Governance and Tenure Dimensions of Coastal Mangrove Forests. CIFOR; USAID Tenure and Global Climate Change Program, Bogor, Indonesia; Washington, DC.

Ruddle, K., 2011. "Informal" credit systems in fishing communities: issues and examples from Vietnam. Human Organization 70 (3), 224–232.

Schmidt, S., Neumann, B., Waweru, Y., Durussel, C., Unger, S., Visbeck, M., 2017. Sdg 14 – conserve and sustainable use the oceans, seas and marine resources for sustainable development. In: Griggs, D., Nilsson, M., Stevance, A., McCollum, D. (Eds.), A Guide to SDG Interactions: From Science to Implementation. International Council for Science (ICSU), Paris, pp. 174–218.

Schwerdtner Máñez, K., Ferse, S.C.A., 2010. The history of Makassan trepang fishing and trade. PLoS One 5 (6), 31134.

Schwerdtner Máñez, K., Husain Paragay, S., 2013. First evidence of targeted moray eel fishing in the Spermonde Archipelago, Sulawesi, Indonesia. Traffic Bulletin 25 (1).

Schwerdtner Máñez, K., Husain Paragay, S., Ferse, S.C.A., Máñez Costa, M., 2012. Water scarcityin the Spermonde archipelago, Sulawesi, Indonesia: past, present and future. Environmental Science and Policy 23, 74–84.

Schwerdtner Máñez, K., Krause, G., Ring, I., Glaser, M., 2014. The Gordian knot of mangrove conservation: disentangling the role of scale, services and benefits. Global Environmental Change 28 (0), 120–128.

Sen, S., Nielsen, R.J., 1996. Fisheries co-management: a comparative analysis. Marine Policy 20 (5), 405–418. https://EconPapers.repec.org/RePEc:eee:marpol:v:20:y:1996:i:5:p:405-418.

Shahidul Islam, M., Tanaka, M., 2004. Impacts of pollution on coastal and marine ecosystems including coastal and marine fisheries and approach for management: a review and synthesis. Marine Pollution Bulletin 48 (7), 624–649.

Siry, H.Y., 2011. In search of appropriate approaches to coastal zone management in Indonesia. Ocean and Coastal Management 54 (6), 469–477.

Sopher, D.E., 1965. The Sea Nomads: A Study Based on the Literature of the Maritime Boat People of Southeast Asia, Singapore.

Spalding, M.D., Ravilious, C., Green, E.P., 2001. World Atlas of Coral Reefs. University of California Press, Berkeley, CA, USA.

Spranz, R., 2017. Reducing Plastic Bag Use in Indonesia (PhD Thesis. Economics). Jacobs University, Bremen.

Suhren, E.J., Devita, F., Prüter, F., Tarumun, S., Glaeser, M., Glaeser, B., 2014. River health and community health: a collaborative action-oriented social-ecological analysis. Asian Journal of Water, Environment and Pollution 11 (1), 51–66.

Tittensor, D.P., Walpole, M., Hill, S.L., Boyce, D.G., Britten, G.L., Burgess, N.D., et al., 2014. A mid-term analysis of progress toward international biodiversity targets. Science 346 (6206), 241–244.

UNESCO, 2000. Reducing megacity impacts on the coastal environment – alternative livelihoods and waste management in Jakarta and the Seribu Islands. In: Coastal Region and Small Island Papers, vol. 6. UNESCO, Paris, p. 59pp.

van Assche, K., Beunen, R., Duineveld, M., 2014. Evolutionary Governance Theory. An Introduction. Springer, Heidelberg, New York, Dordrecht, London.

Van der Wulp, S.A., Dsikowitzky, L., Hesse, K.J., Schwarzbauer, J., 2016a. Master Plan Jakarta: The Giant Seawall and the Need for Structural Treatment of Municipal Waste Water.

Van der Wulp, S.A., Hesse, K.J., Ladwig, N., Damar, A., 2016b. Numerical simulations of River discharges, nutrient flux and nutrient dispersal in Jakarta Bay, Indonesia. Marine Pollution Bulletin 110, 675–685 (Special Issue Jakarta Bay Ecosystem).

Vervoort, J., Gupta, A., 2018. Anticipating climate futures in a 1.5°C era: the link between foresight and governance. Current Opinion in Environmental Sustainability 31, 104–111. https://doi.org/10.1016/j.cosust.2018.01.004.

Vikas, M., Dwarakish, G.S., 2015. Coastal pollution: a review. Aquatic Procedia 4, 381–388.

von Essen, L.M., Ferse, S., Glaser, M., Kunzmann, A., 2013. Attitudes and perceptions of villagers towards community-based mariculture in Minahasa, North Sulawesi, Indonesia. Ocean and Coastal Management 73, 101–112.

Walker-Springett, K., Jefferson, R., Böck, K., Breckwoldt, A., Comby, E., Cottet, M., et al., 2016. Ways forward for aquatic conservation: applications of environmental psychology to support management objectives. Journal of Environmental Management 166, 525–536.

Walters, B.B., Rönnbäck, P., Kovacsc, J.M., Crona, B., Hussain, S.A., Badolad, R., et al., 2008. Ethnobiology, socio-economics and management of mangrove forests: a review. Aquatic Botany 89, 220–236.

Wells, S., 2006. In the Front of the Line: Shoreline protection and Other Ecosystems Services from Mangroves and Coral Reefs. UNEP World Conversation Monitoring Centre, Cambridge.

Wever, L., Glaser, M., Gorris, P., Ferrol-Schulte, D., 2012. Decentralization and participation in integrated coastal management: policy lessons from Brazil and Indonesia. Ocean and Coastal Management 66, 63–72.

White, A.T., Falkland, T., 2010. Management of freshwater lenses on small Pacific islands. Hydrogeology Journal 18, 227–246.

Whittingham, E., Campbell, J., Townsley, P., 2003. Poverty and Reefs. DFID-IMM-IOC/UNESCO, Exeter, UK.

# Index

*Note*: 'Page numbers followed by "f" indicate figures, "t" indicate tables and "b" indicate boxes.'


**A**

Adaptive governance, 427–429
Advanced Very-High-Resolution Radiometer (AVHRR), 31–32
*Aegiceras corniculatum*, 266–267
Agriculture-dominated hinterland, 55f, 61
*Alpheus edamensis*, 212
*Alpheus macellarius*, 212
*Amphiprion percula*, 164
*Amphiprion polymnus*, 164
Anticipatory governance, 427–429
Aquatic ecosystems, 289
*Ascidia subterranea*, 212
Asian Development Bank (ADB), 56
Associated fish communities, 225–226
Atmospheric moisture distribution, 51
*Atrina vexillum*, 219
Available satellite data, 31–32
*Avicennia alba*, 266–267
Axiidean shrimps, 212

**B**

Bacterial communities, 151–152
Baroclinic mode, three-dimensional models in, 395
Barrang Lompo, 163–164
Benthic coral reef community dynamics, 150–151
Benthic/fish communities, 153, 154f
Berau Regency, 256
Biodiversity, 88
Biofilms, 151–152
Biogeochemistry, 258–260
Blackwater rivers, 35, 53–54
  dissolved organic carbon, 66–69, 67f
  dissolved oxygen, 66–69, 67f
  industrial plants, 70
  nutrients, sources and fate of, 69f, 70–71
  Siak River, 65–71, 66f
  Sumatra, 65–71
Brantas reservoirs, 69
Brantas River, 50–65
Burrowing shrimp, 211–215

**C**

Carbon accumulation, 91–93, 92t
Carbon accumulation rates, 265–266, 266f
Carbon sinks, seagrass beds as, 222–223
Carbon sources and storage, 260–266, 261f
Carbon stocks, 265–266, 265f
Carbon storage, 222
*Ceriops tagal*, 266–267
Circulation models accuracy assessment, 396
Climatological aspects, 38–40
Coastal and Disaster Risk Management for Extreme Events Impact Mitigation, 5–6
Coastal and marine social–ecological systems (CM-SES), 409–410
Coastal ecosystem health, 7–8, 48
Coastal governance, 410–411
  coastal livelihoods, 422–423
  coral reefs, 412
  Island exclusion zones (IEZ), 419b
  land use, 416–418
  Mangroves, 413–414
  marine and coastal conservation, 418–422
  marine fisheries, 414–416
  pollution, 423–424
  social needs, 422–423

Coastal governance (*Continued*)
watersheds, 416–418
Coastal ocean, $CO_2$ emissions from, 116–117, 117t
Coastal social–ecological systems, 9
Coastal upwelling regions, 19
$CO_2$ emissions
dissolved inorganic carbon (DIC), 101–102
ecosystem
disturbed peatlands, 123–127
net on-site ecosystem, 121
pristine peat swamps, 121–123, 122f
peatlands, 123
climate pledges/gaps, 126t, 127–131
$CO_2$ reduction potential, 125
cumulative emissions, climate response to, 123–125
dissolved inorganic carbon (DIC), 101–102
DOC emission factors, 99t
land losses, 126–127
off-site, 98–99
peat and forest fires, 97–98
peat soil oxidations, 98
Comanagement, 410, 426–427
Community composition, 60–61
Comprehensive life cycle assessments, 391
Coral reef social–ecological systems
bacterial communities, 151–152
benthic and fish communities, 153, 154f
benthic coral reef community dynamics, 150–151
biofilms, 151–152
biogeochemical processes, 149–150
coral physiology, 152
coral reef recruitment processes, 152
dissolved organic carbon (DOC), 149–150
disturbances consequences, 153–154, 155f–156f
functioning of, 148–154
*Galaxea fascicularis*, 152
genetic connectivity, coral triangle, 155–165, 158f, 159t
*Amphiprion percula*, 164
*Amphiprion polymnus*, 164
Barrang Lompo, 163–164
large-scale connectivity, 160f, 161–162
marine-protected area network design, connectivity data in, 164–165
Samalona, 163–164
self-recruitment, 163–164, 163t
small-scale connectivity, Spermonde Archipelago, 160f, 162–163
*Heliofungia actiniformis*, 150
Indonesia, 145–148
local anthropogenic threats, 146
operational taxonomic units (OTU), 151–152
*Porites lutea*, 152
Spermonde Archipelago, 145–148, 189t
benthic coral reef community dynamics, 150–151
fisheries on, 172–175
fishers, 168–169
fishing ground distribution, 176–178
fishing methods, 190f
food security, 169–171
gear choices of fishermen model, 175–176
management of, 171
marine feedbacks and dynamics, 167–168
reef-related livelihoods, 168–169
reef resources, perceptions of, 169–171
Spatially Explicit simulation model for Assisting the local MANagement of COral REefs (SEAMANCORE), 172–175
Spermonde's coral reef fisheries participatory assessment, 165–167
*Stylophora pistillata*, 152
*Stylophora subseriata*, 152
transparent exopolymer particles (TEPs), 149–150
water quality, 149–150

**D**

Data-poor sites, setting up circulation models, 393–397
Decision support system (DSS), 9, 375, 401

Deutscher Akademischer Austauschdienst, 5–6
Dissolved inorganic nitrogen (DIN), 61
Dissolved organic carbon (DOC), 104–106, 104t, 112–113, 149–150
  discharges, 113f, 114–115
  fate of, 114f, 115–116
  microbial organic carbon pump, 113
Dissolved organic matter (DOM), 32–33, 90

**E**

Earthquake, 51–52
Ecological carrying capacity (ECC), 373, 375, 382–384, 383f
Ecosystem
  $CO_2$ emissions
    disturbed peatlands, 123–127
    net on-site ecosystem, 121
    pristine peat swamps, 121–123, 122f
  degradation, 3
  Mangrove forests, 274–276, 276t
Ecosystem approach to aquaculture (EAA), 373
El Niño–Southern Oscillation (ENSO), 3
Energy estimation, tidal resources
  baroclinic mode, three-dimensional models in, 395
  circulation models accuracy assessment, 396
  comprehensive life cycle assessments, 391
  data-poor sites, setting up circulation models, 393–397
  decision support system (DSS), 401
  environmental impact assessment, 400
  Global Forecast System (GFS), 395
  gross domestic product (GDP), 391
  investigation sites, Indonesia, 392–393
  porous plates, 397
  site suitability (SS), 397–399, 398f
  technically extractable power, natural conditions, 399
  technology readiness level 8 (TRL 8), 390–391
  tidal currents, energy potential from, 401–402, 402t
  turbines, flow interactions with, 399–400
Environmental change, 7
Environmental governance, 409
Environmental impact assessment, 400
*Epinephelus fuscoguttatus*, 312–314, 313f
Estuaries and ocean, peatlands
  coastal ocean, $CO_2$ emissions from, 116–117, 117t
  dissolved organic carbon
    discharges, 113f, 114–115
    fate of, 114f, 115–116
    microbial organic carbon pump, 113
  emission factors, 119t, 120–123
  invisible carbon footprint, 118–119
  marine peat carbon budget, 119–120
  organic carbon burial, 117–118, 117t
Evapotranspiration, 95–96, 96f
Evolutionary governance, 427–429
Evolutionary governance theory (EGT), 428
Explicit social equity objectives, 432–434

**F**

Fish habitat utilization, 228–229
Flora and fauna, 266–268
Food and Agricultural Organization of the United Nations (FAO), 373
Food security, 169–171
Food web, 217–220
Freshwater, sources and sinks of, 28–31
  Malacca Strait region, 29
  NCEP/NCAR reanalysis, 29
  time series, 29
  verified global hydrological model WaterGAP, 29

**G**

*Galaxea fascicularis*, 152
Ganges/Brahmaputra, 66–68
Genetic connectivity, coral triangle, 155–165, 158f, 159t
  *Amphiprion percula*, 164
  *Amphiprion polymnus*, 164
  Barrang Lompo, 163–164
  large-scale connectivity, 160f, 161–162

Genetic connectivity, coral triangle (*Continued*)
  marine-protected area network design, connectivity data in, 164–165
  Samalona, 163–164
  self-recruitment, 163–164, 163t
  small-scale connectivity, Spermonde Archipelago, 160f, 162–163
German–Indonesian cooperation, 16
Glacial–interglacial and orbital timescales, 352–359, 353f–354f, 357f–358f
Global Forecast System (GFS), 395
*Glypturus armatus*, 213–214, 213f
Governance, 409
  programs, 71–74
Green turtles, 218
Gross domestic product (GDP), 391
The Gulf of Thailand, 28

**H**

Hamburg Shelf Ocean Model (HAMSOM), 16–17, 26–28
*Heliofungia actiniformis*, 150
Hexachlorobenzene (HCB), 324
High-frequency climate variations, 360–364
High suspended matter rivers, 50–65
Histosols, 86–87
Holocene, 360–364, 362f
Human interventions, 49–71
Human-seagrass interactions
  ecosystem services, 230
  fish/fisheries, 230–231
  high primary productivity, 230
  human-made infrastructure, 233
  invertebrates, 230–231
  seaweed farms, 231–233, 232f
Human–seagrass interactions
  current threats, 234–235, 234f
  ecological value, 230
Hydrological system, 52–53, 59–60, 290–296, 291f, 301f

**I**

Indian Ocean Dipole Mode (IOD), 348–349
Indonesian archipelago, 3
Indonesian–German workshop, 4
Indonesian peatlands
  carbon accumulation, 91–93, 92t
  cover changes, 93
  evapotranspiration, 95–96, 96f
  history of, 89–90
  hydrological cycle, 94–96, 94f
  land use, 93
  mature oil palm plantations, 95
  properties, 90–91
  soil organic matter, 89f
  vapor pressure deficit (VPD), 96
Indonesian Seas
  climate change, 14
  colonial period, 15
  freshwater, sources and sinks of, 28–31
    Malacca Strait region, 29
    NCEP/NCAR reanalysis, 29
    time series, 29
    verified global hydrological model WaterGAP, 29
  German–Indonesian cooperation, 16
  Hamburg Shelf Ocean Model (HAMSOM), 16–17
  Indonesian Throughflow (ITF), 15
  INSTANT program, 15
  marine circulation
    coastal upwelling regions, 19
    global context, 18
    Indonesian Throughflow (ITF), 18–19
    Makassar Strait throughflow, 18–19
    Ocean State University (OSU), 20–22
    regional circulation, 18–19
    single partial tidal amplitudes, 20
    tides, 20–22
    TOPEX/POSEIDON satellite altimeter data, 20–22
  Naga Report, 15
  oceanography of, 14
  remote sensing methods, 31–40
    Advanced Very-High-Resolution Radiometer (AVHRR), 31–32
    available satellite data, 31–32
    black-water rivers, 35
    climatological aspects, 38–40

dissolved organic matter (DOM), 32–33
Medium-Resolution Imaging Spectrometer (MERIS), 31–32
MERIS-derived products, 37
Moderate-resolution Imaging Spectroradiometer (MODIS), 31–32
MODIS-derived absorption of Gelbstoff, 38–39
Ocean and Land Color Instrument (OLCI), 31–32
ocean color, 32–35
Operational Land Imager (OLI), 32
phytoplankton distribution, 35–36
Rivers of SE-Sumatra transport, 34
satellite-based studies, 35–40
Sea-viewing Wide Field-of-view Sensor (SeaWiFS), 31–32
SE-Sumatra transport rivers, 34
spectral reflectance, 35
Suomi National Polar-orbiting Partnership spacecraft (Suomi NPP), 31–32
thermal infrared sensor (TIRS), 32
tidal and monsoon phases, coastal discharge, 36f, 37–38
Topex/Poseidon, 32
US Geological Survey (USGS), 32
Visible Infrared Imaging Radiometer Suite (VIIRS), 31–32
seasonal variability, 22–26
inter-tropical convergence zone (ITCZ), 22
long-term sea surface temperature, 26
seasonality of circulation, 22–25
sea surface salinity development, 26
sea surface temperature (SST), 23f
temperature and salinity, seasonality of, 25
water residence times, 26–28
The Gulf of Thailand, 28
HAMSOM, 26–28
South China Sea, 28
Vietnamese coast, 28
Indonesian Throughflow (ITF), 15, 18–19, 256, 348
Inorganic nitrogen fluxes, 61–64
INSTANT program, 15
Institutions, 410
Integrated coastal/marine planning, 425–430
International Bank for Reconstruction and Development (World Bank), 56
Inter-tropical convergence zone (ITCZ), 22
Invisible carbon footprint, 118–119
Island exclusion zones (IEZ), 419b

**J**

Jakarta Bay (JB)
aquatic ecosystems, 289
economic growth, 287f
fish parasites in, 317–321
biological indicators, 320–321
new applications, 320–321
species descriptions, 318–319
taxonomic treatments, 318–319
wildlife and maricultured fish, parasite biodiversity in, 319–320
zoonotic *Anisakis* spp., 319
hexachlorobenzene (HCB), 324
hydrological system, 290–296, 291f, 301f
implications, 330–333
maximum residue limit (MRL), 329
microbial diversity, 310–317
*Epinephelus fuscoguttatus*, 312–314, 313f
fish and shrimp, 316–317
metagenomic analysis, 312–314
*Penaeus monodon*, 314–316
Ministry of Marine Affairs and Fisheries (MMAF), 329
National Residue Monitoring Plan (NRMP), 329–330
nutrient dispersion, 290–296, 291f, 301f
organotins (OTs), 327–328
paralytic shellfish poisoning (PSP), 329
*Perna viridis*, 325–327
pollution
biota accumulation, 302–303
distribution, 297–298
emission sources, 298–300
flushing-out phenomenon, 301–302

Jakarta Bay (JB) (*Continued*)
hydrodynamic and dispersion models, 295b–296b
industrial emissions, 300–301, 301f
inorganic, 296–303
insect repellent N,N-diethyl-m-toluamide, 299–300
linear alkylbenzenes (LABs), 303
organic, 296–303
polycyclic aromatic hydrocarbons (PAHs), 303
quantity, 297–298
trace elements, source apportionment of, 298–299
trace hazardous elements, 297
types, 297–298
potential contaminants, 326t–327t
potential risk, 321–330
river loads determination, 293b–295b
saxitoxin (STX), 329
seafood consumption, 321–330
tolerable daily intake (TDI), 325
tributyltin (TBT), 327–328
water quality, 303–310
anthropogenic stressors, 305, 306f
biological responses, 303–310
coral reef organisms, 305–307
local *versus* regional stressors, 308–310, 309f
reef composition, 308
water-accumulated fraction (WAF), 305–306
water pollution, 303–305, 304f

**K**

*Kampung laut*, 268

**L**

Land–ocean continuum, 99–112
$CO_2$ emission from rivers, 109
dissolved inorganic carbon yields, 109–111
dissolved organic carbon (DOC), 104–106, 104t
dissolved organic carbon yields, 106–109, 107f
erosion, 111–112
leaching, 110f, 111–112
marine carbonate system, 101b–102b
priming, 112–120
SPICE study area, 96f, 102–103, 103f
Land reclamation, 273–274
Land runoff, buffers for, 221
Land uses, 93
Large-scale connectivity, 160f, 161–162
Local anthropogenic threats, 146
Long-term research program, 253–254
Long-term sea surface temperature, 65–71
Lower Brantas, dissolved oxygen regime of, 61–65, 62f
Low oxygen concentrations, 61
Low phosphate concentrations, 261–262
Lusi eruption, 60

**M**

Macrobenthic communities, 215–217
Madura Strait coastal waters, 69
Makassar Strait throughflow, 18–19
Malesia, 203–204
Mangrove Ecology and Sustainability, 253
Mangrove forests, 8
Berau Regency, 256
ecosystem services, 274–276, 276t
environmental change, Segara Anakan Lagoon (SAL), 269–274
fisheries, decline of, 269–270
land reclamation, 273–274
marine species, decline of, 269–270
new land, conflicts over, 273–274
sedimentation, 270–273, 272f
environmental settings, Segara Anakan Lagoon (SAL), 257–269
biogeochemistry, 258–260
carbon accumulation rates, 265–266, 266f
carbon sources and storage, 260–266, 261f
carbon stocks, 265–266, 265f
Flora and fauna, 266–268
natural resource use, 268–269
nitrogen supply, 264–265

physical setting, 257–258
pollution, 258–260
population, 268–269
water quality, 258–260
Indonesian Throughflow (ITF), 256
long-term research program, 253–254
management programs, 276–278
Mangrove Ecology and Sustainability, 253
threats to, 274–276
Togian Islands, 254, 254f, 256–257
Marine circulation
coastal upwelling regions, 19
global context, 18
Indonesian Throughflow (ITF), 18–19
Makassar Strait throughflow, 18–19
Ocean State University (OSU), 20–22
regional circulation, 18–19
single partial tidal amplitudes, 20
tides, 20–22
TOPEX/POSEIDON satellite altimeter data, 20–22
Marine finfish aquaculture
Bali, Indonesia, 376–377, 376f
decision support system (DSS), 375
dynamic model of, 377f, 379–380
ecological carrying capacity (ECC), 373
ecosystem approach to aquaculture (EAA), 373
Food and Agricultural Organization of the United Nations (FAO), 373
operation assessment, 380–384
ecological carrying capacity (ECC), 382–384, 383f
production carrying capacity (PCC), 381, 382f
site selection, 378f, 380, 381f
surveys and monitoring at, 378–379
sustainable management of
ecological carrying capacity (ECC), 375
production carrying capacity (PCC), 374–375
site selection (SS), 374
Marine peat carbon budget, 119–120
Marine Pollution Bulletin, 7
Marine-protected area network design, connectivity data in, 164–165
Marine social–ecological systems
coastal and marine social–ecological systems (CM-SES), 409–410
coastal governance, 410–411
coastal livelihoods, 422–423
coral reefs, 412
Island exclusion zones (IEZ), 419b
land use, 416–418
Mangroves, 413–414
marine and coastal conservation, 418–422
marine fisheries, 414–416
pollution, 423–424
social needs, 422–423
watersheds, 416–418
comanagement, 410
environmental governance, 409
governance, 409
institutions, 410
integrated coastal/marine planning, 425–430
adaptive governance, 427–429
anticipatory governance, 427–429
comanagement, 426–427
evolutionary governance, 427–429
explicit social equity objectives, 432–434
leverage against wickedly resilient system features, 432
linking knowledge, employ innovative methods for, 434
participatory governance, 426–427
polycentric coastal and marine governance, 429–430
polycentric governance, 430
social energy, 431–432
stakeholder perceptions, 430–431
sustainable human–nature relations, 431–432
organizations, 410
SES, 410
system levels and scales, 410
Marine species, decline of, 269–270
Maritime Continent, 2

Mature oil palm plantations, 95
Maximum residue limit (MRL), 329
Medium-Resolution Imaging Spectrometer (MERIS), 31–32
Megaherbivores loss, 211–215
MERIS-derived products, 37
MERMAID station, 63f–64f, 71
Microbial diversity, 310–317
- *Epinephelus fuscoguttatus*, 312–314, 313f
- fish and shrimp, 316–317
- metagenomic analysis, 312–314
- *Penaeus monodon*, 314–316

Microorganisms, 9
Millennial-scale climate variability, 359–360
Ministry of Marine Affairs and Fisheries (MMAF), 329
Ministry of Public Works, 56
Moderate-resolution Imaging Spectroradiometer (MODIS), 31–32
MODIS-derived absorption of Gelbstoff, 38–39
Monsoonal variability, 61
*Mytilus edulis*, 219

**N**

Naga Report, 15
National Residue Monitoring Plan (NRMP), 329–330
Natural control factors, 49–71
Natural hazards, 51–52
New land, conflicts over, 273–274
Nitrogen supply, 264–265
Non-Aligned Nations movement (NAM), 5–6
*Nypa fruticans*, 267

**O**

Ocean and Land Color Instrument (OLCI), 31–32
Ocean color, 32–35
Oceanic top reef habitats, 206–209
Ocean State University (OSU), 20–22
Ocean temperature variability, 363
Operational Land Imager (OLI), 32
Operational taxonomic units (OTU), 151–152
Organic carbon burial, 117–118, 117t
Organic pollutants, 9
Organotins (OTs), 327–328

**P**

*Paracleistostoma laciniatum*, 267
Paralytic shellfish poisoning (PSP), 329
Participatory governance, 426–427
Particulate organic carbon (POC), 64–66, 90
Particulate organic matter (POM), 90
Payment for environmental services (PES), 72
Peat carbon density (PCD), 90–91
Peatlands
- biodiversity, 88
- carbon losses, 97–99
- carbon storage, 88–89
- $CO_2$ emissions, 123
  - climate pledges/gaps, 126t, 127–131
  - $CO_2$ reduction potential, 125
  - cumulative emissions, climate response to, 123–125
  - dissolved inorganic carbon (DIC), 101–102
  - DOC emission factors, 99t
  - land losses, 126–127
  - off-site, 98–99
  - peat and forest fires, 97–98
  - peat soil oxidations, 98
- distribution, 88–89
- ecosystem $CO_2$ emissions
  - disturbed peatlands, 123–127
  - net on-site ecosystem, 121
  - pristine peat swamps, 121–123, 122f
- estuaries and ocean, 112–115
  - coastal ocean, $CO_2$ emissions from, 116–117, 117t
  - dissolved organic carbon, 112–113
  - emission factors, 119t, 120–123
  - invisible carbon footprint, 118–119
  - marine peat carbon budget, 119–120
  - organic carbon burial, 117–118, 117t

histosols, 86–87
Indonesian peatlands, 89–96
land–ocean continuum, 99–112
$CO_2$ emission from rivers, 109
dissolved inorganic carbon yields, 109–111
dissolved organic carbon (DOC), 104–109, 104t, 107f
erosion, 111–112
leaching, 110f, 111–112
marine carbonate system, 101b–102b
priming, 112–120
SPICE study area, 96f, 102–103, 103f
organic soils, 86–87
peat, 86–87
phenol oxidase, 87–88
socioeconomic implications, 127–129
REDD+, 129–131
SPICE field experiments, 131–132
Southeast (SE) Asia, 86
types, 87–88
vegetation, 88
*Perna viridis*, 325–327
Phenol oxidase, 87–88
Physical denudation, 53–54
Phytoplankton, 219
abundance, 60–61
biomass, 8
distribution, 35–36
productivity, 62t, 70
*Pinna bicolor*, 219
*Pinna muricata*, 219
Pollution
biota accumulation, 302–303
distribution, 297–298
emission sources, 298–300
flushing-out phenomenon, 301–302
hydrodynamic and dispersion models, 295b–296b
industrial emissions, 300–301, 301f
inorganic, 296–303
insect repellent N,N-diethyl-m-toluamide, 299–300
linear alkylbenzenes (LABs), 303
organic, 296–303
polycyclic aromatic hydrocarbons (PAHs), 303
quantity, 297–298
trace elements
hazardous elements, 297
source apportionment of, 298–299
types, 297–298
Polycentric governance, 430
Polycyclic aromatic hydrocarbons (PAHs), 260
*Porites lutea*, 152
Priority watersheds, 72
Production carrying capacity (PCC), 374–375, 381, 382f

**Q**

Quaternary environmental history, Indonesia
glacial–interglacial and orbital timescales, 352–359, 353f–354f, 357f–358f
high-frequency climate variations, 360–364
Holocene, 360–364, 362f
Indian Ocean Dipole Mode (IOD), 348–349
Indonesian Throughflow (ITF), 348
millennial-scale climate variability, 359–360
ocean temperature variability, 363
sea surface temperature (SST), 348
thermocline temperatures offshore, 356–359

**R**

REDD+, 129–131
Regional Environmental Change, 7
Remote sensing methods, 31–40
Advanced Very-High-Resolution Radiometer (AVHRR), 31–32
available satellite data, 31–32
black-water rivers, 35
climatological aspects, 38–40
dissolved organic matter (DOM), 32–33
Medium-Resolution Imaging Spectrometer (MERIS), 31–32

Remote sensing methods (*Continued*)
MERIS-derived products, 37
Moderate-resolution Imaging Spectro-radiometer (MODIS), 31–32
MODIS-derived absorption of Gelbstoff, 38–39
Ocean and Land Color Instrument (OLCI), 31–32
ocean color, 32–35
Operational Land Imager (OLI), 32
phytoplankton distribution, 35–36
Rivers of SE-Sumatra transport, 34
satellite-based studies, 35–40
Sea-viewing Wide Field-of-view Sensor (SeaWiFS), 31–32
SE-Sumatra transport rivers, 34
spectral reflectance, 35
Suomi National Polar-orbiting Partnership spacecraft (Suomi NPP), 31–32
thermal infrared sensor (TIRS), 32
tidal and monsoon phases, coastal discharge, 36f, 37–38
Topex/Poseidon, 32
US Geological Survey (USGS), 32
Visible Infrared Imaging Radiometer Suite (VIIRS), 31–32
*Rhizophora apiculata*, 266–268
River fluxes, human interventions
agriculture-dominated hinterland, 55f, 61
Asian Development Bank (ADB), 56
blackwater rivers, 53–54
dissolved organic carbon, 66–69, 67f
dissolved oxygen, 66–69, 67f
industrial plants, 70
nutrients, sources and fate of, 69f, 70–71
Siak River, 65–71, 66f
Sumatra, 65–71
Brantas River, 50–65
coastal ecosystem health, 47
community composition, 60–61
composition, 53–57
dissolved inorganic nitrogen (DIN), 61
environmental changes
atmospheric moisture distribution, 51
earthquake, 51–52
hydrology, 52–53
natural hazards, 51–52
erosion, 53–54
estuarine and coastal ecosystem services, 47
extreme events, 49–71
fate of nutrients variations, 53–57
Ganges/Brahmaputra, 66–68
governance programs, 71–74
high suspended matter rivers, 50–65
human interventions, 49–71
hydrological regime, 59–60
inorganic nitrogen fluxes, 61–64
International Bank for Reconstruction and Development (World Bank), 56
largest reservoir, 60
lower Brantas, dissolved oxygen regime of, 61–65, 62f
low oxygen concentrations, 61
Lusi eruption, 60
Madura Strait coastal waters, 69
management programs, 71–74
MERMAID station, 63f–64f, 71
Ministry of Public Works, 56
monsoonal variability, 61
natural control factors, 49–71
particulate organic carbon (POC), 64–66
payment for environmental services (PES), 72
physical denudation, 53–54
phytoplankton
abundance, 60–61
productivity, 62t, 70
priority watersheds, 72
seasonal variations, 61
Segara Anakan Lagoon, 61
sources variations, 53–57
total suspended matter (TSM), 58f, 64–66
tropical coastal zones, 50–51
weathering, 53–54
Rivers of SE-Sumatra transport, 34
*Rostronia stylirostris*, 212

## S

Samalona, 163–164
Satellite-based studies, 35–40
Saxitoxin (STX), 329
Science for the Protection of Indonesian Coastal Ecosystems (SPICE)

Action Plan, 5
Coastal and Disaster Risk Management for Extreme Events Impact Mitigation, 5–6
Coastal Ecosystem Health, 7–8
Coastal Ecosystems: Hazards Management and Rehabilitation, 5–6
coastal social–ecological systems, 9
decision support system (DSS), 9
Deutscher Akademischer Austauschdienst, 5–6
ecosystem degradation, 3
education, 5–8
El Niño–Southern Oscillation (ENSO), 3
Environmental change, 7
Indonesian archipelago, 3
Indonesian–German workshop, 4
Mangrove forests, 8
Marine Pollution Bulletin, 7
Maritime Continent, 2
microorganisms, 9
Non-Aligned Nations movement (NAM), 5–6
organic pollutants, 9
phytoplankton biomass, 8
Regional Environmental Change, 7
research, 5–8
research and education program, 4–5
research clusters, thematic foci and geographical locations of, 6t
social–ecological systems, 5–6
Southeast Asia Center for Ocean Research and Monitoring (SEACORM), 6–7
south Sulawesi, Spermonde Archipelago in, 8
synthesis of, 8–9
thermohaline circulation (THC), 3
twirling cities, 2
Seafood consumption, 321–330
Seagrass canopy, 226–228
Seagrass ecosystems
*Alpheus edamensis*, 212
*Alpheus macellarius*, 212
*Ascidia subterranea*, 212
axiidean shrimps, 212
burrowing shrimp, 211–215
carbon sinks, seagrass beds as, 222–223
carbon storage, 222
food web, 217–220
*Glypturus armatus*, 213–214, 213f
green turtles, 218
human–seagrass interactions
current threats, 234–235, 234f
ecological value, 230
ecosystem services, 230
fish/fisheries, 230–231
high primary productivity, 230
human-made infrastructure, 233
invertebrates, 230–231
seaweed farms, 231–233, 232f
land runoff, buffers for, 221
macrobenthic communities, 215–217
megaherbivores loss, 211–215
*Mytilus edulis*, 219
phytoplankton, 219
policy/management implications, 238b
research, 235b
*Rostronia stylirostris*, 212
seagrass meadows, trophic transfers from, 224–225, 224f
spatial distribution, 216
*Syringodium isoetifolium*, 218
trophic pyramid, 217–220
tropical seagrass beds, fish species
associated fish communities, 225–226
fish habitat utilization, 228–229
seagrass canopy, 226–228
tropical Southeast Asian seagrass meadows
coral triangle, 203
few fossil records, 204
high biodiversity, tropical Indo-West Pacific, 203–204
Malesia, 203–204
Spermonde Archipelago. *See* Spermonde Archipelago; seagrass ecosystems
water filters, 221
Seagrass meadows, trophic transfers from, 224–225, 224f
Seasonality of circulation, 22–25
Seasonal variability, 22–26, 61

Seasonal variability (*Continued*)
inter-tropical convergence zone (ITCZ), 22
long-term sea surface temperature, 26
seasonality of circulation, 22–25
sea surface salinity development, 26
sea surface temperature (SST), 23f
temperature and salinity, seasonality of, 25
Sea surface salinity development, 26
Sea surface temperature (SST), 23f, 348
Sea-viewing Wide Field-of-view Sensor (SeaWiFS), 31–32
Sedimentation, 270–273, 272f
Segara Anakan Lagoon (SAL), 61, 253–256, 255f
environmental changes, 269–274
fisheries, decline of, 269–270
land reclamation, 273–274
marine species, decline of, 269–270
new land, conflicts over, 273–274
sedimentation, 270–273, 272f
environmental settings, 257–269
biogeochemistry, 258–260
carbon accumulation rates, 265–266, 266f
carbon sources and storage, 260–266, 261f
carbon stocks, 265–266, 265f
Flora and fauna, 266–268
natural resource use, 268–269
nitrogen supply, 264–265
physical setting, 257–258
pollution, 258–260
population, 268–269
water quality, 258–260
Self-recruitment, 163–164, 163t
SE-Sumatra transport rivers, 34
Single partial tidal amplitudes, 20
Site selection (SS), 374, 378f, 380, 381f
Site suitability (SS), 397–399, 398f
Small-scale connectivity, Spermonde Archipelago, 160f, 162–163
Social–ecological systems, 5–6
Social energy, 431–432
Soil organic matter, 89f
*Sonneratia caseolaris*, 266–267
Sources variations, 53–57
South China Sea, 28
Southeast Asia Center for Ocean Research and Monitoring (SEACORM), 6–7
South Sulawesi, Spermonde Archipelago in, 8
Spatially Explicit simulation model for Assisting the local MANagement of COral REefs (SEAMANCORE), 172–175
Spectral reflectance, 35
Spermonde Archipelago, 145–148, 189t
benthic coral reef community dynamics, 150–151
fisheries on, 172–175
fishers, 168–169
fishing ground distribution, 176–178
fishing methods, 190f
food security, 169–171
gear choices of fishermen model, 175–176
management of, 171
marine feedbacks and dynamics, 167–168
reef-related livelihoods, 168–169
reef resources, perceptions of, 169–171
seagrass ecosystems, 204–206, 205f
area estimation, 206–209
oceanic top reef habitats, 206–209
tropical seagrass bed systems, 209–211
small-scale connectivity, 160f, 162–163
Spatially Explicit simulation model for Assisting the local MANagement of COral REefs (SEAMANCORE), 172–175
Spermonde's coral reef fisheries participatory assessment, 165–167
Stakeholder perceptions, 430–431
*Stylophora pistillata*, 152
*Stylophora subseriata*, 152
Suomi National Polar-orbiting Partnership spacecraft (Suomi NPP), 31–32

Sustainable human–nature relations, 431–432
*Syringodium isoetifolium*, 218

T

Taxonomic treatments, 318–319
Technically extractable power, natural conditions, 399
Technology readiness level 8 (TRL 8), 390–391
Temperature and salinity, seasonality of, 25
Thermal infrared sensor (TIRS), 32
Thermocline temperatures offshore, 356–359
Thermohaline circulation (THC), 3
Tidal and monsoon phases, coastal discharge, 36f, 37–38
Tidal currents, energy potential from, 401–402, 402t
Tides, 20–22
Tolerable daily intake (TDI), 325
Topex/Poseidon, 32
TOPEX/POSEIDON satellite altimeter data, 20–22
Total suspended matter (TSM), 58f, 64–66
Transparent exopolymer particles (TEPs), 149–150
Tributyltin (TBT), 327–328
Trophic pyramid, 217–220
Tropical seagrass beds, fish species
  associated fish communities, 225–226
  fish habitat utilization, 228–229
  seagrass canopy, 226–228
Tropical Southeast Asian seagrass meadows
  coral triangle, 203
  few fossil records, 204
  high biodiversity, tropical Indo-West Pacific, 203–204
  Malesia, 203–204
  Spermonde Archipelago. *See* Spermonde Archipelago; seagrass ecosystems
Turbines, flow interactions with, 399–400
Twirling cities, 2

U

US Geological Survey (USGS), 32

V

Vapor pressure deficit (VPD), 96
Vegetation, 88
Vietnamese coast, 28
Visible Infrared Imaging Radiometer Suite (VIIRS), 31–32

W

Water-accumulated fraction (WAF), 305–306
Water filters, 221
Water quality, 303–310
  anthropogenic stressors, 305, 306f
  biological responses, 303–310
  coral reef organisms, 305–307
  local *versus* regional stressors, 308–310, 309f
  pollution, 303–305, 304f
  reef composition, 308
Water residence times, 26–28
  The Gulf of Thailand, 28
  HAMSOM, 26–28
  South China Sea, 28
  Vietnamese coast, 28
Wildlife/maricultured fish, parasite biodiversity in, 319–320

Z

Zoonotic *Anisakis* spp., 319

Printed in the United States
by Baker & Taylor Publisher Services